T0338261

EMOTION RECOGNITION

EMOTION RECOGNITION

A Pattern Analysis Approach

Edited by

AMIT KONAR
Artificial Intelligence Laboratory
Department of Electronics and Telecommunication Engineering
Jadavpur University
Kolkata, India

ARUNA CHAKRABORTY
Department of Computer Science & Engineering
St. Thomas' College of Engineering & Technology
Kolkata, India

Copyright © 2015 by John Wiley & Sons, Inc. All rights reserved.

Published by John Wiley & Sons, Inc., Hoboken, New Jersey.
Published simultaneously in Canada.

No part of this publication may be reproduced, stored in a retrieval system, or transmitted in any form or by any means, electronic, mechanical, photocopying, recording, scanning, or otherwise, except as permitted under Section 107 or 108 of the 1976 United States Copyright Act, without either the prior written permission of the Publisher, or authorization through payment of the appropriate per-copy fee to the Copyright Clearance Center, Inc., 222 Rosewood Drive, Danvers, MA 01923, (978) 750-8400, fax (978) 646-8600, or on the web at www.copyright.com. Requests to the Publisher for permission should be addressed to the Permissions Department, John Wiley & Sons, Inc., 111 River Street, Hoboken, NJ 07030, (201) 748-6011, fax (201) 748-6008.

Limit of Liability/Disclaimer of Warranty: While the publisher and author have used their best efforts in preparing this book, they make no representations or warranties with respect to the accuracy or completeness of the contents of this book and specifically disclaim any implied warranties of merchantability or fitness for a particular purpose. No warranty may be created or extended by sales representatives or written sales materials. The advice and strategies contained herein may not be suitable for your situation. You should consult with a professional where appropriate. Neither the publisher nor author shall be liable for any loss of profit or any other commercial damages, including but not limited to special, incidental, consequential, or other damages.

For general information on our other products and services please contact our Customer Care Department with the U.S. at 877-762-2974, outside the U.S. at 317-572-3993 or fax 317-572-4002.

Wiley also publishes its books in a variety of electronic formats. Some content that appears in print, however, may not be available in electronic format.

Library of Congress Cataloging-in-Publication Data:
Konar, Amit.
Emotion recognition : a pattern analysis approach / Amit Konar, Aruna Chakraborty.
pages cm
Includes index.
ISBN 978-1-118-13066-7 (hardback)
1. Human-computer interaction. 2. Artificial intelligence. 3. Emotions–Computer simulation. 4. Pattern recognition systems. 5. Context-aware computing. I. Chakraborty, Aruna, 1977- II. Title.
QA76.9.H85K655 2014
004.01'9–dc23

2014024314

Printed in the United States of America.

10 9 8 7 6 5 4 3 2 1

To our parents

CONTENTS

PREFACE

Emotion represents a psychological state of the human mind. Researchers from different domains have diverse opinions about the developmental process of emotion. Philosophers believe that emotion originates as a result of substantial (positive or negative) changes in our personal situations or environment. Biologists, however, consider our nervous and hormonal systems responsible for the development of emotions. Current research on brain imaging reveals that the cortex and the subcortical region in the frontal brain are responsible for the arousal of emotion. Although there are conflicts in the developmental process of emotion, experimental psychologists reveal that a change in our external or cognitive states carried by neuronal signals triggers our hormonal glands, which in turn excites specific modules in the human brain to develop a feeling of emotion.

The arousal of emotion is usually accompanied with manifestation in our external appearance, such as changes in facial expression, voice, gesture, posture, and other physiological conditions. Recognition of emotion from its external manifestation often leads to inaccurate inferences particularly for two reasons. First, the manifestation may not truly correspond to the arousal of the specific emotion. Second, measurements of external manifestation require instruments of high precision and accuracy. The first problem is unsolvable in case the subjects over which experiments are undertaken suppress their emotion, or pretend to exhibit false emotion. Presuming that the subjects are conducive to the recognition process, we only pay attention to the second problem, which can be solved by advanced instrumentation.

This single volume on *Emotion Recognition: A Pattern Analysis Approach* provides through and insightful research methodologies on different modalities of emotion recognition, including facial expression, voice, and biopotential signals. It is primarily meant for graduate students and young researchers, who like to initiate their doctoral/MS research in this new discipline. The book is equally useful to professionals engaged in the design/development of intelligent systems for applications in psychotherapy and human–computer interactive systems. It is an edited volume written by several experts with specialized knowledge in the diverse domains of emotion recognition. Naturally, the book contains a thorough and in-depth coverage on all theories and experiments on emotion recognition in a highly comprehensive manner.

The recognition process involves extraction of features from the external manifestation of emotion on facial images, voice, and biopotential signals. All the features extracted are not equally useful for emotion recognition. Thus, the next step to feature extraction is to reduce the dimension of features by feature reduction techniques. The last step of emotion recognition is to employ a classifier or clustering method to classify the measured signals into one specific emotion class. Several techniques of computational intelligence and machine learning can be used here for recognition of emotion from its feature space.

The book includes 20 contributory chapters. Each chapter starts with an abstract, followed by introduction, methodology, experiments and results, conclusions, and references. A biography and photograph of individual contributors are given at the end of each chapter to inspire and motivate young researchers to start his/her research career in this new discipline of knowledge through interaction with these researchers.

Chapter 1 serves as a prerequisite for the rest of the book. It examines emotion recognition as a pattern recognition problem and reviews the commonly used techniques of feature extraction, feature selection, and classification of emotions by different modalities, including facial expressions, voice, gesture, and posture. It also reviews the commonly used techniques for general pattern recognition, feature selection, and classification. Lastly, it compares the different techniques used in recognition of single and multimodal emotions.

In Chapter 2, Tong and Ji propose a systematic approach to model the dynamic properties of facial actions, including both temporal development of each action unit and dynamic dependencies among them in a spontaneous facial display. In particular, they employ a Dynamic Bayesian Network to explicitly model the dynamic and semantic relationship among the action units. The dynamic nature of the facial action is characterized by directed temporal links among the action units. They consider representing the semantic relationships by directed static links among action units. They employ domain knowledge and training data to automatically construct the Dynamic Bayesian Network (DBN) model. In this model, action units are recognized by generating probabilistic inference over time. Experiments with real images reveal that explicit modeling of the dynamic dependencies among action units demonstrates that the proposed method outperforms the existing techniques for action unit recognition for spontaneous facial displays.

Chapter 3 by Saha *et al.* provides a new forum for cross-cultural studies for facial expressions. A psychological study of facial expression for different cultural groups reveals that facial information representing a specific expression varies across cultures. Here, the authors demonstrate that the occurrence of action units proposed by Ekman possesses inter-culture variations. The rule base is generated for classification of six basic expressions using a decision tree. Experiments reveal that the performance of the classifier improves when the rule base becomes culture specific.

In Chapter 4, Chang and Huang propose a novel approach to design a subject-dependent facial expression recognition system. Facial expressions representing a particular emotion vary widely across the people. Naturally designing a general strategy to correctly recognize the emotion of people still remains an unsolved problem. Chang and Huang employ Radial Basis Function (RBF) Neural Network to classify seven emotions including neutral, happy, angry, surprised, sad, scared, and disgusted. Experimental results given to substantiate the classification methodology indicate that the proposed system can accurately identify emotions from facial expressions.

Zia Uddin and Kim in Chapter 5 present a new method to recognize facial expressions from time sequential facial images. They consider employing enhanced Independent Component Analysis to extract independent component features and use Fisher Linear Discriminant Analysis (FLDA) for classification of emotions.

Yang and Wang in Chapter 6 propose a new technique for feature selection in facial expression recognition problem using rough set theory. They consider designing a self-learning attribute reduction algorithm using rough sets and domain-oriented data mining theory. It is indicated that rough set methods outperform genetic algorithm in connection with feature selection problem. It is also found that geometrical features concerning mouth are found to have the highest importance in emotion recognition.

Halder *et al.* in Chapter 7 propose a novel scheme for facial expression recognition using type 2 fuzzy sets. Both interval and general type 2 fuzzy sets (IT2FS and GT2FS) are used independently to model fuzzy face spaces for different emotions. The most important research findings in their research include automated evaluation of secondary membership functions from the ensemble of primary membership functions obtained from different sources. The evaluated secondary memberships are used subsequently to transform a GT2FS into an equivalent IT2FS. The reasoning mechanism for classification used in IT2FS is extended by transforming GT2FS by IT2FS. Experiments undertaken reveal that GT2FS-based recognition outperforms the IT2FS with respect to classification accuracy at the cost of additional computational complexity.

Chapter 8 by Zheng *et al.* provides a survey on the recent advances on emotion recognition by non-frontal 3D and multi-view 2D facial image analysis. Feature extraction is the most pertinent issue in non-frontal facial image analysis for emotion recognition. Zheng *et al.*employ geometric features, appearance-based features, including scale invariant feature transform (SIFT) and local binary pattern feature (LBP) and Gabor wavelet features. LBP feature attempts to capture local image information and also proves its excellence in the fields of facial emotion descriptions. SIFT features are invariant to image translation, scaling, rotation and also partially invariant to illumination changes. SIFT features have earned popularity for their robustness in local geometric distortions. Gabor wavelet features are also proved to be effective for face and facial expression recognition system. The Gabor filter usually employs a Kernel function constructed by taking a product of Gaussian envelop with a harmonic oscillation function. The rest of the chapter provides a thorough discussion on 3D non-frontal face databases, including BU-3DFE and its dynamic version BU-4DFE along with Multi-PIE and Bosphorous databases. The chapter ends with a discussion on major issues to be considered for future researchers.

Cen *et al.* in Chapter 9 propose a method for speech emotion recognition by employing maximum a posteriori based fusion technique. The proposed method is capable of effectively combining the strengths of several classification techniques for recognition of emotional states in speech signals. To examine the effectiveness of the proposed method they consider probabilistic neural net, k-Nearest Neighbor (k-NN), and Universal Background Model-Gaussian Mixture Model (UBM-GMM) as base classifiers in their numerical experiments. Experiments undertaken also reveal that the composite classifier results in higher classification accuracy than those obtained by the individual base classifiers. It is also shown that the proposed method works well with a small training dataset.

Weninger *et al.* in Chapter 10 provides an overview of recent developments in naturalistic emotion recognition based on acoustic and linguistic cues. They consider a number of use cases, where emotion recognition helps in improving quality of service and quality of life. The other important reviews introduced in the chapter include description of existing corpora of emotional speech data and the underlying theory of emotion modeling. The chapter also introduces a novel approach for implementation of automatic emotion recognition system. Main focus is given on the challenges for real-life applications highlighting the importance of non-prototypicality, lack of solid ground truth and data sparsity, requirement of real-time and incremental processing, and robustness with respect to acoustic conditions. The chapter ends with a discussion on novel strategies to augment training data by semi-unsupervised learning.

Chapter 11 by Petrantonakis *et al.* provides a scheme for EEG-based emotion recognition. The association of EEG signals to the activation of the brain in connection with elicitation of emotion and changes in facial expression is an interesting field of study under the framework of affective computing. This chapter attempts to handle the above problem by judiciously employing sophisticated signal processing tools drawn from the field of Multidimensional Directed Information analysis. The chapter addresses two important issues in the field of EEG-based emotion recognition. The first issue is concerned with quantitatively estimating the degree of emotion elicitation in subjects under suitable stimulus. The latter issue of primary focus is to study the overall enhancement in performance of the EEG-based emotion recognition system by introducing new feature vectors. This deals with designing novel feature extraction algorithms, structured in the time–frequency domain.

Chapter 12 by Murugappan provides an interesting study on emotion classification using multiple physiological signals. Here, the author considers electromyogram (EMG) and electrocardiogram (ECG) signals to classify four different emotions including happiness, fear, disgust, and neutral. Physiological signals are collected from 20 subjects in the age group of 21–26 using six electrode signals. Discrete wavelet transform has been used to decompose physiological signals into different frequency bands. Four different frequency bands in 4–16 Hz, 16–31 Hz, 31–63 Hz, and 4–63 Hz are used to study emotion classifications. Similarly three different frequency bands called very low frequency component (0.015–0.04 Hz), low frequency component (0.04–0.14 Hz), and high frequency component (0.14–0.5Hz) are used to recognize emotions using ECG signals. Butterworth band-pass filters are employed to eliminate the effect of noise and other interferences from the EEG signals. Four different wavelet functions including db5, db7, Sym8, and Coif5 have been used to determine the average spectral power from the preprocessed signals. A k-Nearest Neighbor and linear discriminant classifier are used to map the statistical features into emotions. Experiments undertaken reveal that EMG signals provide better classification accuracy than the ECG counterpart.

Lin *et al.* in Chapter 13 provides an overview of state-of-the-art EEG-based emotion recognition techniques. They also examine neuro-physiological EEG dynamics associated with affective responses. Previous studies achieved high classification accuracy by considering all available channels and frequency bands. This chapter, however, aims at resolving EEG feature selection and electrode reduction issues by the generation of subject-independent feature/electrode set extraction techniques. Moreover, the present study addresses several practical issues, and potential challenges for affective brain–computer interfaces as well. The authors conclude that a user-friendly EEG cap with a small number of electrodes can efficiently detect affective states, and therefore significantly promote practical affective brain–computer interface applications in daily life.

Gunes *et al.* in Chapter 14 provides representative answers to how body expression analysis helps affect recognition by considering the following three case studies. The first one is concerned with data acquisition and annotation of the first publicly available databases of affective face and body displays. Second, they propose a representative approach for affective state recognition from face and body display by detecting the space–time interest points in video. Third, they provide a representative approach for explicit detection of the temporal phases and affective states from bodily expressions. The chapter concludes with a discussion on how the authors could advance the state of the art in the field.

Wagner *et al.* in Chapter 15 propose the design and development of a robust multimodal ensemble-based system for emotion recognition. The proposed emotion recognition system includes: emotion modeling, data segmentation and annotation, feature extraction and selection, and lastly classification and multimodal fusion. Special emphasis is given on a commonly ignored problem of temporarily missing data in one or more observed modalities. Both offline and online systems to recognize emotion of subjects have been developed. In offline design, the above issue has been avoided by excluding those parts of the corpus where one or more channels are corrupted or not suitable for evaluation. In online design, the above problem has been solved in the multimodal fusion step. The application of different annotation schemes is performed on the well-known CALLAS Expressivity Corpus that features both facial and vocal modalities. Lastly, the authors present an application, called the affective virtual listener: Alfred, which they had developed using their online recognition principles.

Datcu *et al.* in Chapter 16 provide a scheme for semantic audiovisual data fusion for automatic emotion recognition. They employ Hidden Markov Models (HMMs) to learn and describe the temporal dynamic of emotional clues in visual and acoustic channels. The novelty of the approach lies in dynamic modeling of emotions using

HMMs, which maps LBPs and Mel-frequency cepstral coefficients (MFCCs) as audio features. They also propose a new method for visual feature selection based on multiclass Adaboost classifier. The latter part of the chapter deals with discussions on the possible ways to continue the research on multimodal emotion recognition.

Chetty *et al.* in Chapter 17 propose a novel scheme for multilevel fusion for audiovisual emotion recognition. The primary focus of this chapter is to detect subtle and micro-expressions in facial images, and next to utilize multiple channels of information for automatic emotion recognition. A simple approach to quantify the intensity of emotion has been adopted. Facial deformation features and marker-based facial features are considered to quantify emotion. Experiments indicate that shape transformation features offer better quantification of facial expression in comparison to marker-based features. The performance of the proposed system is examined with the rich expression data available in the DaFEx database. The chapter further examines two commonly used methods of fusing multiple channels of information, known as feature fusion and score fusion, and proposes a new multilevel fusion approach for enhancing the performance for both person-independent and person-dependent emotion recognition. Possible extension of the work, as expected by the authors, might include an exploration of the temporal progression with respect to facial motions and contours of pitch/energy as important attributes for emotion recognition.

Most of the traditional emotion recognition systems rely on static emotion information acquired through a single channel. However, our daily experience of natural emotion recognition reveals that people capture both dynamic and static information in facial expression, voice, and gestures and also consider all the modalities to recognize emotion of a subject.

Hupont *et al.* in Chapter 18 demonstrate the scope of dynamic affect recognition through multiple modalities. They provide an algebra to fuse the different sources of affective information through mathematical formulation and obtain a 2D dynamic emotional path representing the users' emotional evolution as the final output. They also examine the scope of the proposed affect sensing scheme in real human–computer interaction systems.

Wu *et al.* in Chapter 19 propose an interesting approach to automatically recognize human emotional states from audiovisual bimodal signals. They first introduce the current data fusion strategies among audiovisual signals for bimodal emotion recognition. Later they propose a novel state-based alignment strategy, employed in a Semi-Coupled Hidden Markov Model (SC-HMM) to align the temporal relation of states between audiovisual strings. Because of this alignment strategy SC-HMM is capable of alleviating the problem of data sparseness and achieves better statistical dependency between states of audiovisual HMMs in most real-world scenarios. The acquired audiovisual signals from seven subjects with four emotional states are used for performance evaluation of the overall system. The subjects are asked to utter 30 types of sentences twice to generate emotional speech and facial expression for each emotion. Subject-independent expressions are conducted to demonstrate that the proposed SC-HMM outperforms other fusion-based bimodal emotion recognition techniques.

Katsis *et al.* in Chapter 20 examine the scope of emotion recognition in automotive fields. It is apparent that the perception and judgment of the drivers are impaired, when they are overwhelmed by anger or stress. In fact, high stress influences adversely drivers' reaction in critical conditions, and thus is primarily responsible for accidents. Further, the driving event sometimes alters driver's emotions, which subsequently may affect drivers' performance. Chapter 20 introduces recent advances in emotion recognition, focusing on elicitation of emotion by the driving task and also external influences that affect drivers' performance. The state-of-the-art research on stimulus (information) used for emotion recognition and the methods used for emotion recognition are discussed. Lastly, two exemplar systems are presented in the context of emotion recognition of drivers.

AMIT KONAR AND ARUNA CHAKRABORTY

ACKNOWLEDGMENTS

The editors gratefully acknowledge the contributions of the chapter authors for spending their valuable time and energy in preparing the manuscript. They also acknowledge the support they received from Jadvapur University (JU), Kolkata, India, and St. Thomas' College of Engineering and Technology (STCET), Kolkata, India, during the preparatory phase of the book.The editors wish to express their deep gratitude to Prof. Abhijit Chakrabarti, the Vice Chancellor of JU and Prof. Sivaji Bandyopadhyay, Dean of the Faculty of Engineering and Technology, JU, Dr. Sailesh Mukhopadhyay, the Founding Principal of STCET, Prof. Swapna Sen, Ex-Principal, STCET, Prof. Gautam Banerjea, Director (Administration), and Dr. Subir Chowdhury, Director (Academics) of STCET for providing them the necessary support to complete the book in its present form.

The editors would like to thank Prof. Iti Saha Mishra, HOD, ETCE Department, JU, andf Prof. Subarna Bhattacharya, HOD, CSE Department, STCET, and Prof. Amit Kr. Siromoni, HOD, IT Department, STCET, who always stood by the editors to complete the book. They would also like to thank their Ph.D. student Ms. Anisha Halder for her active support of the book. The book contains several facial images for illustration. These images are collected from standard image databases. The editors are grateful to the image database managers/owners for giving permission to print the images in the book free of cost/service charges. Lastly, the editors acknowledge the support they received from their family members to successfully complete the book in its present form.

AMIT KONAR

Artificial Intelligence Laboratory
Department of Electronics and Telecommunication Engineering
Jadavpur University
Kolkata, India

ARUNA CHAKRABORTY

Department of Computer Science & Engineering
St. Thomas' College of Engineering & Technology
Kolkata, India

CONTRIBUTORS

Amit Konar, Aruna Chakraborty and Anisha Halder Electronics and Tele-Communication Engineering Department, Jadavpur University and Department of Computer Science and Engineering, St. Thomas' College of Engineering and Technology Kolkata, India.

Yan Tong and Qiang Ji Department of Computer Science and Engineering, University of South Carolina, Columbia, South Carolina and Department of Electrical, Computer, and Systems Engineering Rensselaer Polytechnic Institute, Troy, New York.

Chandrani Saha, Washef Ahmed, Soma Mitra, Debasis Mazumdar and Sushmita Mitra Centre for Development of Advanced Computing (CDAC), Kolkata, and Indian Statistical Institute, Kolkata, India.

Chuan-Yu Chang and Yan-Chiang Huang Department of Computer Science and Information Engineering, National Yunlin University of Science and Technology, Douliou, Yunlin, Taiwan.

Md. Zia Uddin and Tae-Seong Kim Inha University, Republic of Korea and Kyung Hee University, Republic of Korea.

Yong Yang and Guoyin Wang Institute of Computer Science and Technology, Chongqing University of Posts and Telecommunications, Chongqing.

Anisha Halder, Amit Konar, Aruna Chakraborty and Atulya K. Nagar Electronics and Tele-Communication Engineering Department, Jadavpur University, Kolkata, Department of Computer Science and Engineering, St. Thomas' College of Engineering and Technology Kolkata, India and Department of Mathematics and Computer Science, Liverpool Hope University, Liverpool, UK.

Wenming Zheng, Hao Tang and Thomas S. Huang Key Laboratory of Child Development and Learning Science, Ministry of Education, Southeast University, Nanjing 210096, P.R. China, HP Labs, 1501 Page Mill Road, Palo Alto, CA, USA, and Beckman Institute, University of Illinois at Urbana Champaign, 405 North Mathews Avenue, Urbana, IL, USA.

Ling Cen, Zhu Liang Yu and Wee Ser Institute for Infocomm Research (I2R), A*STAR, Singapore, South China University of Technology, China and Nanyang Technological University, Singapore.

Felix Weninger, Martin Wollmer and Bjorn Schuller Institute for Human Machine Communication, Technische Universitat Munchen, Germany.

Panagiotis C. Petrantonakis and Leontios J. Hadjileontiadis Department of Electrical and Computer Engineering, Aristotle University of Thessaloniki, Thessaloniki, Greece.

M.Murugappan School of Mechatronic Engineering, Universiti Malaysia Perlis (UniMAP), Campus Ulu Pauh, 02600, Arau, Perlis, Malaysia.

Yuan-Pin Lin, Tzyy-Ping Jung, Yijun Wang and Julie Onton Brain Research Center, National Chiao Tung University, Hsinchu, Taiwan and Swartz Center for Computational Neuroscience, University of California San Diego, CA, USA.

Hatice Gunes, Caifeng Shan, Shizhi Chen and YingLi Tian School of Computer Science and Electronic Engineering, Queen Mary University of London, U.K., Philips Research, High Tech Campus, Eindhoven, The Netherlands and Department of Electrical Engineering, The City College of New York, USA.

Johannes Wagner, Florian Lingenfelser and Elisabeth Andre Department of Human Centered Multimedia, University of Augsburg, Germany.

Dragos Datcu and Leon J.M. Rothkrantz Faculty of Military Sciences, Netherlands Defence Academy, Den Helder, The Netherlands, Delft University of Technology, Delft, The Netherlands.

Girija Chetty, Michael Wagner and Roland Goecke Faculty of Information Sciences and Engineering, University of Canberra, Australia.

Isabelle Hupont, Sergio Ballano, Eva Cerezo and Sandra Baldassarri Aragon Institute of Technology, Zaragoza, Spain and University of Zaragoza, Zaragoza, Spain.

Chung-Hsien Wu, Jen-Chun Lin and Wen-Li Wei Department of Computer Science and Information Engineering, National Cheng Kung University, Tainan, Taiwan, R.O.C.

Christos D. Katsis, George Rigas, Yorgos Goletsis and Dimitrios I. Fotiadis
Dept. of Applications of Information Technology in Administration and Economy, Technological Educational Institute of Ionian Islands, Lefkada, Unit of Medical Technology and Intelligent Information Systems, Dept. of Materials Science and Engineering, University of Ioannina, Ioannina,and Dept. of Economics, University of Ioannina, Ioannina, Greece.

1

INTRODUCTION TO EMOTION RECOGNITION

AMIT KONAR AND ANISHA HALDER

Artificial Intelligence Laboratory, Department of Electronics and Telecommunication Engineering, Jadavpur University, Kolkata, India

ARUNA CHAKRABORTY

Department of Computer Science & Engineering, St. Thomas' College of Engineering & Technology, Kolkata, India

A pattern represents a characteristic set of attributes of an object by which it can be distinguished from other objects. Pattern recognition aims at recognizing an object by its characteristic attributes. This chapter examines emotion recognition in the settings of pattern recognition problems. It begins with an overview of the well-known pattern recognition techniques, and gradually demonstrates the scope of their applications in emotion recognition with special emphasis on feature extraction, feature reduction, and classification. Main emphasis is given to feature selection by single and multiple modalities and classification by neural, fuzzy, and statistical pattern recognition techniques. The chapter also provides an overview of stimulus generation for arousal of emotion. Lastly, the chapter outlines the methods of performance analysis and validation issues in the context of emotion recognition.

1.1 BASICS OF PATTERN RECOGNITION

A pattern is a representative signature of an object by which we can recognize it easily. Pattern recognition refers to mapping of a set of patterns into one of several object classes. Occasionally, a pattern is represented by a vector containing the features of

Emotion Recognition: A Pattern Analysis Approach, First Edition. Edited by Amit Konar and Aruna Chakraborty.
© 2015 John Wiley & Sons, Inc. Published 2015 by John Wiley & Sons, Inc.

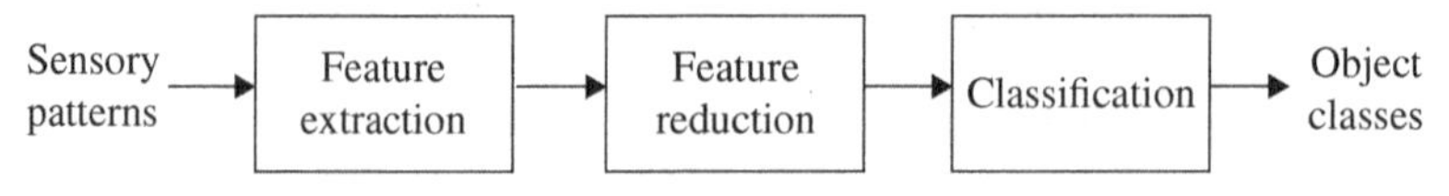

FIGURE 1.1 Basic steps of pattern recognition.

an object. Thus, in general, the pattern recognition process can be described by three fundamental steps, namely, feature extraction, feature selection, and classification. Figure 1.1 provides a general scheme for pattern recognition. The feature extraction process involves using one or more sensors to measure the representative features of an object. The feature selection module selects more fundamental features from a list of features. The classification module classifies the selected features into one of several object classes.

The pattern recognition problem can be broadly divided into two main heads: (i) supervised classification (or discrimination), and (ii) unsupervised clustering. In supervised classification, usually a set of training instances (or data points) comprising a set of measurements about each object along with its class is given. These data points with their class labels are used as exemplars in the classifier design. Given a data point with unknown class, the classifier once trained with the exemplary instances is able to determine the class label of the given data point. The classifier thus automatically maps an unknown data point to one of several classes using the background knowledge about the exemplary instances.

Beginners to the subject often are confronted with the question: how does the classifier automatically determine the class label of an unknown data point, which is not present in the exemplary instances. This is due to the inherent generalization characteristics of the supervised classifier.

In unsupervised classification, the class labels of the data points are not known. The learning system partitions the whole set of data points into (preferably) nonoverlapping subsets based on some measure of similarity of the data points under each subset. Each subset is called a class/cluster. Because of its inherent characteristics of grouping data points into clusters, unsupervised classification is also called clustering.

Both statistical decision theory and machine learning have been employed in the literature to design pattern recognition algorithms [1, 2]. Bayes' theorem is the building stone of statistical classification algorithms. On the other hand there exists a vast literature on supervised and unsupervised learning [3], algorithms, which capture the inherent structural similarity [4] of the data points for application in pattern recognition problems.

1.2 EMOTION DETECTION AS A PATTERN RECOGNITION PROBLEM

Emotion represents the psychological state of the human mind and thought processes. Apparently, the process of arousal of emotion has a good resemblance with its manifestation as facial, vocal, and bodily gestures. This phenomenon has attracted researchers to determine the emotion of a subject from its manifestation. Although the one-to-one correspondence from manifestation of emotion to a particular emotional state is yet to be proved, researchers presume the existence of such mapping to recognize the emotion of a subject from its manifestation.

Given the manifestation of an emotion, the task of recognizing the emotion, thus, is a pattern recognition problem. For example, facial expression–based emotion recognition requires extraction of a set of facial features from the facial expression of a given subject. Recognition of emotion here refers to classification of facial features into one of several emotion classes. Usually, a supervised classifier pretrained with emotional features as input and emotion class as output is used to determine the class of an unknown emotional manifestation.

Apparently, the emotional state of the human mind is expressed in different modes including facial, voice, gesture, posture, and biopotential signals. When a single mode of manifestation is used to recognize emotion, we call it a unimodal approach. Sometimes all modes are not sufficiently expressed. Naturally, recognition from a less expressed mode invites the scope of misclassification. This problem can be avoided by attempting to recognize an emotion from several modalities. Such a process is often referred to as multimodal emotion recognition.

1.3 FEATURE EXTRACTION

Feature extraction is one of the fundamental steps in emotion recognition. Features are obtained in different ways. On occasion features are preprocessed sensory readings. Preprocessing is required to filter noise from measurements. Sensory readings during the period of emotion arousal sometimes have a wide variance. Statistical estimates of the temporal readings, such as mean, variance, skewness, kurtosis, and the like, are usually taken to reduce the effect of temporal variations on measurements. Further, instead of directly using time/spatial domain measurements, frequency domain transforms are also used to extract frequency domain features. For example, frequency domain information is generally used for EEG (electroencephalogram) and voice signals. Frequency domain parameters are time invariant and less susceptible to noise. This attracted researchers to use frequency domain features instead of time domain.

Frequency domain features have one fundamental limitation in that they are unable to tag time with frequency components. Tagging the time with frequency contents of a signal is important, particularly for a certain class of signals, often labeled as nonstationary signals. EEG, for instance, is a nonstationary signal, the frequency contents of which change over time because of asynchronous firing of the neurons. Wavelet transform coefficients of an EEG signal represent time–frequency correlations and thus deserve to be one of the fundamental features for nonstationary EEG signals. We now briefly outline the features used in different modalities of emotion recognition.

1.3.1 Facial Expression–Based Features

The most common modality of emotion recognition is by facial expression analysis. Traditionally there exist two major classes of techniques for face/facial expression representation and relevant feature extraction. The first one is called geometrical features. They rely on parameters of distinctive facial features such as eyes, mouth, and nose. On the other hand appearance-based approach considers a face as an array of intensity values suitably preprocessed. This array is then compared with a face template using a suitable matrix.

1.3.1.1 Geometric Model–Based Feature Extraction Deformable templates have been used for locating facial features. For example, Kass *et al.* [5] suggested the use of active contour models—snakes—for tracking lips in image sequence. They initialized snake on the lips in the facial image and showed that it is able to track lip movement accurately. It however fails, if there exists occlusion or other structure in the image.

Yuille *et al.* [6] employed deformable templates based on simple geometrical shapes for locating the eye and mouth. Yuille's model incorporated shape constraints, but there is no proof that the form of a given model is sufficiently general to capture the deformable geometric shapes.

Researchers are taking a keen interest to represent geometric relations between facial information to extract facial features. In Reference 7, Craw, Tock, and Bennett, considered positional constraints in facial expressions to extract necessary features for emotion recognition.

Brunelli and Poggio considered a number of high dimensional [8] measurements or location of a number of key points in a single image or an image sequence for facial image interpretation.

Kirby and Sirovich [9] took attempts to decompose facial image into a weighted sum of basis images or eigen faces using Karhunen and Loeve expansion. They considered 50 expansion coefficients and were able to reconstruct an approximation of the facial expansion using these parameters.

1.3.1.2 Appearance-Based Approach to Feature Extraction Appearance-based approach involves preprocessing followed by a compact coding through statistical redundancy reduction. The preprocessing in most cases is required to align the geometry in face image, for instance, by having the two eyes and nose tip at fixed positions through affine texture warping [10]. Optical flow or Gabor wavelets are used to capture facial appearance motion and robust registration, respectively, for successful recognition.

Pixel-based appearance is often represented by a compact coding. Usually statistical reduction principle is used to represent this coding. The unsupervised learning techniques used for compact coding include Principal Component Analysis (PCA), Independent Component Analysis (ICA), Kernel-PCA (KPCA), local feature analysis, and probability density estimation. Supervised learning techniques including Linear Discriminant Analysis (LDA) and Kernel Discriminant Analysis (KDA) are also used for compact coding representation.

The main drawback of PCA-based compact coding is that it retains some unwanted variations. It is also incapable of extracting local features that offer robustness against changes in local region or occlusions. ICA produces basis vectors that are more spatially local than those of PCA. Thus, ICA is sensitive to occlusion and pose variations. ICA retains higher order statistics and maximizes the degree of statistical independence of features [11, 12].

Recently Scholkopf *et al.* [13] extended the conventional PCA to KPCA, which is able to extract nonlinear features [14]. However, like PCA, KPCA captures the overall variance of 11 patterns and is not necessarily optimal for discrimination.

Statistical supervised learning such as LDA attempts to find the basis vectors maximizing the interclass distance and minimizing the intraclass distance. Similarly, KDA determines the most significant nonlinear basis vector to maximize the interclass distance while minimizing the intraclass distance. Among the other interesting works, the following need special mentioning.

Cohn *et al.* [15] proposed a facial action recognition technique that employs discriminant function analysis of individual facial regions, including eyebrows, eyes, and mouth. They used two discriminant functions for three facial actions of the eyebrow region, two discriminant functions for three facial actions of the eye region, and five discriminant functions for nine facial actions of the nose and mouth region. The classification accuracy for the eyebrow, eye and nose, mouth regions are 92, 88, and 88%, respectively.

In Reference 16 Essa and Pentland proposed a novel control-theoretic method to extract the spatiotemporal motion–energy representation of facial motion in an observed expression. They generated the spatiotemporal templates for six different facial expressions, considering two facial actions, including smile and raised eyebrows for two subjects. Templates are formed by averaging the patterns of motion generated by two subjects exhibiting a certain expression. The Euclidean norm of the difference between the motion–energy template and the observed motion energy is defined as a metric of similarity of the motion energies. A recognition accuracy of 98% is achieved while experimenting with 52 frontal-view image sequences of eight people having distinct expressions.

Kimura and Yachida in Reference 17 modeled facial images by a potential net and attempted to fit the net to each frame of a facial image sequence. The deformed version of the potential net is used to match the expressionless face, typically the first frame of the sequence. The variation in the nodes of the deformed net is used for subsequent processing. In their own experiments, the authors in Reference 17 considered a six image sequence of emotional expressions experienced by a subject with gradual variation in the strength of expressions from expressionless (relaxed) to a maximum expression. The experiment was repeated for three emotions: anger, happiness, and surprise. PCA has been employed here to classify three emotions using standard eigen space analysis.

In Reference 18 Lucey *et al.* detected pain from the movement of facial muscles into a series of action units (AUs), based on the Facial Action Coding System (FACS) [19]. For this novel task, they considered three types of Active Appearance Model (AAM) features: (i)similarity-normalized shape features (SPTS), (ii) similarity-normalized appearance features (SAPP), and (iii) canonical-normalized appearance features (CAPP). AAM features are used here to track the face and to extract visual features, based on facial expressions using the FACS. They obtained classification accuracy of 75.1, 76.9, and 80.9% using SAPP, SPTS, and CAPP, respectively, using Support Vector Machine (SVM) classifier first and then improving the performance (Fusion of Scores) by linear logistical regression (LLR).

Tian *et al.* in Reference 20 proposed a new method for recognizing AUs for facial expression analysis. They used both permanent and transient features for their work. Movement of eyebrow, cheek, eyes, and mouth are considered permanent features. On the other hand, deeping of facial furrows is considered a transient features. They used different feature extraction algorithms for different features. For the lips, they used lip-tracking algorithm. For eyes, eyebrows, and cheeks, they considered Lucas–Kanade algorithm, and for the transient features they employed Canny edge detector algorithm. Two neural network (NN) based classifiers are considered to recognize the changes in AUs: one for six upper-face AUs and the other for ten lower-face AUs of the FACS. A percentage accuracy of 95.4% is obtained for upper-face AUs and 95.6% for lower-face AUs.

In Reference 21 Kim and Bien designed a personalized classifier from facial expressions using soft computing techniques. They used degree of mouth openness (f_1), degree of eye openness (f_2), the vertical distance between the eyebrows and the eyes (f_3), degree of nasolabial root wrinkles (NLR) (f_4), and degree of nasolabial furrows (NLF) (f_5) in their classifier design. These features are extracted from facial expressions by different techniques. For example, f_1 and f_2 are extracted by a human visual system based approach. f_1 is measured by combining global features (the height ratio and the area ratio between the whole face and the mouth region) and a local feature (Gabor–Gaussian feature). For f_2 they used the "dip" feature in log-polar mapped image and the Gabor-filter coefficients. For f_3, f_4, and f_5, which are transient components, they used the Euclidean distance (f_3) and the Gabor-filtered coefficients (f_4 and f_5). Image features are extracted from four sets of facial expression data to show effectiveness of the proposed method, which confirms considerable enhancement of the whole performance by using Fuzzy Neural Nets (FNN) based classifier.

Huang *et al.* in Reference 22 proposed a novel approach to recognize facial expression using skin wrinkles. They considered many features like eyes, mouth, eyebrows, nostrils, nasolabial folds, eye pouches, dimples, forehead, and chin furrows for their research and used Deformable Template Model (DTM) and Active Wavelet Network (AWN) for extracting those features. Classification accuracy obtained by using Principle Component Analysis and Neural Network is around 70%.

In Reference 23, Kobayashi and Hara recognized basic facial expressions by using 60 facial characteristic points (FCP) from three components of the face (eyebrows, eyes, and mouth). These features are extracted by manual calculation and emotions are classified by neural network.

In Reference 24 a real-time automated system was modeled by Anderson and McOwan, for recognition of human facial expressions. Here, the muscle movements of the human face are considered as features after tracking the face. A modification of spatial ratio template tracker algorithm is used here for tracking the face first and later to determine the motion of the face by optical flow algorithm. A percentage accuracy of 81.82% has been obtained by using SVM as classifier.

Otsuka and Ohya [25] considered a matching of temporal sequence of the 15D feature vector to the models of the six basic facial expressions by using a specialized

Hidden Markov Model (HMM). The proposed HMM comprises five states, namely, relaxed (S1, S5), contracted (S2), apex (S3), and relaxing (S4). The recognition of a single image sequence here is realized by considering transition from the final state to the initial state. Further, the recognition of multiple sequences is accomplished by considering transitions from a given final state to initial states of all feasible categories. The state-transition probability and output probability of each state are obtained from sampled data by employing Baum–Welch algorithm. The k-means clustering algorithm here has been used to estimate the initial probabilities. The method was tested on the same subjects for whom data was captured. Consequently, the feasibility of the proposed technique for an unknown subject is questionable. Although the proposed method was labeled as good, no justification was given in favor of its goodness. Besides the above, the works of Ekman [27–38], Pantic [39–46], [48–51], Cohn [52–57], Konar [58–69] and some others [70–73] deserve special mention.

1.3.2 Voice Features

Voice features used for emotion recognition include prosodic and spectral features. Prosodic features are derived from pitch, intensity, and first formant frequency profiles as well as voice quality measures. Spectral features include Mel-frequency cepstral coefficients (MFCC), linear prediction cepstral coefficients (LPC), log frequency power coefficients (LFPC), perceptual linear prediction (PLP) coefficients. We now briefly provide an overview of the voice features below.

1. Pitch represents the perceived fundamental frequency of a sound. Fundamental frequency is defined as the frequency at which the vocal cords vibrate during speech.
2. A formant is a peak in a frequency spectrum that results from the resonant frequencies of any acoustical system. For human voice, formants are recognized as the resonance frequencies of the vocal tracts. Formant regions are not directly related to the fundamental frequency and may remain more or less constant as the fundamental changes. If the fundamental is low in the formant range, the quality of the sound is rich, but if the fundamental is above the formant regions, the sound is thin. The first three formants: F_1, F_2, and F_3 are more often used to disambiguate the speech.
3. Power spectral density describes the distribution of power of a speech signal with frequency and also shows the strength (signal energy is strong or weak at different frequency) of the signal as a function of frequency. The energy or power (average energy per frame) in a formant comes from the sound source (vibration of the vocal folds, frequency of the vocal tract, movement of lips and jaw). The energy in the speech signal $x(n)$ is computed as

$$E_x = \sum_{n=0}^{N-1} x^2(n) \tag{1.1}$$

The power of the signal $x(n)$ is the average energy per frame:

$$P_x = \frac{1}{N} \sum_{n=0}^{N-1} x^2(n) \tag{1.2}$$

where N is total no. of samples in a frame.

4. "Jitter is defined as perturbations of the glottal source signal that occurs during vowel phonation and affect the glottal pitch period" [75]. Let $u[n]$ be a pitch period sequence. Then we define absolute jitter by

$$\frac{1}{N+1} \sum_{n=0}^{N-1} |u(n+1) - u(n)|$$

5. "Shimmer is defined as perturbations of the glottal source signal that occur during vowel phonation and affect the glottal energy" [75]. Let $u[n]$ be a peak amplitude sequence of N samples. Then absolute shimmer is given by

$$\frac{1}{N+1} \sum_{n=0}^{N-1} |u(n+1) - u(n)|$$

6. MFCC [76] is a widely used term in speech and speaker recognition. However, the definition of MFCC requires defining two important parameters: "Mel scale" and Mel-frequency spectrum. The Mel scale is defined as

$$f_{mel} = 1125 ln(1 + f/700)$$

where f is the actual frequency in Hz. Mel-frequency cepstrum (MFC) is one form of representation of the short-term power spectrum of sound, based on a linear cosine transformation of a log-power spectrum on a nonlinear "mel" scale of frequency.

Mel-frequency cepstral coefficients are coefficients that collectively make up an MFC. They are derived from a type of cepstral representation of the audio clip (a nonlinear "spectrum-of-a-spectrum"). The difference between the cepstrum and the Mel-frequency cepstrum is that in the MFC, the frequency bands are equally spaced on the Mel scale, which approximates the human auditory system's response more closely than the linearly spaced frequency bands used in the normal cepstrum.

7. A speech sample can be modeled as a linear combination of its past samples. A unique set of predictor coefficients is determined by minimizing the sum of the squared differences between the actual speech samples and the linearly predicted ones. These predictor coefficients are referred to as linear prediction-based cepstral coefficient (LPCC) [77].

Busso *et al.* in Reference 78, presented a novel approach for emotion detection from emotionally salient aspects of the fundamental frequency in the speech signal. They selected pitch contour (mean, standard deviation, maximum, and minimum range of sentence- and voice-level features of pitch) as features for their experiment. Pitches obtained from emotional and neutral speech are compared first by symmetric KLD (Kullback–Leibler Distance). Then pitch features are quantified by comparing nested Logistic Regression Models. They used GMM (Gaussian Mixture Model) and LDC (Linear Discriminant Classifier) for classification process and obtained accuracy over 77%.

In Reference 79, Lee *et al.* detected emotions in spoken dialogues. Features, they used in the paper are pitch, formant frequencies, energy, and timing features like speech duration rate, ratio of duration of voiced and unvoiced region, and duration of the longest voiced speech. In this paper, irrelevant features are eliminated from the base feature set by forward selection (FS) method, and then a feature set is calculated by PCA. This novel approach improved emotion classification by 40.7% for males and 36.4% for females using LDC and k-NN (k-Nearest Neighborhood classifier) for emotion classification.

Wu *et al.* [80] proposed a new method for emotion recognition of affective speech based on multiple classifiers using acoustic–prosodic information and semantic labels. Among the acoustic–prosodic features, they selected pitch, intensity, formants and formant bandwidth, jitter-related features, shimmer-related features, harmonicity-related features, and MFCC. They derived Semantic Labels from HowNet (Chinese Knowledge Base) to extract EAR (Emotion Association Rules) from the recognized word sequence of the affective speech. They used multiple classifiers like GMM, SVM, MLP, (Multilayer Perceptron), MDT (Meta Decision Tree), and Maximum Entropy Model (MaxEnt) and got an overall accuracy of 85.79%.

Kim *et al.* [81] have developed an improved emotion recognition scheme with a novel speaker-independent feature. They employed orthogonal–linear discriminant analysis (OLDA) for extracting speech features, that is, ratio of a spectral flatness measure (SFM) to a spectral center (RSS), pitch, energy, and MFCC. They used GMM as a classifier for emotion recognition. An average recognition rate of 57.2% ($\pm$5.7%) at a 90% confidence interval can be obtained by their experiment. Among the other research works on speech, the work of Mower [82–86], Narayanan [78–94], Wu [95–98], Schuller [99–116], and some others [117–133] deserve special mention.

1.3.3 EEG Features Used for Emotion Recognition

Electroencephalogram is an interesting modality for emotion recognition. Under a hostile environment, people sometimes attempt to conceal the manifestation of their emotional states in facial expression and voice. EEG, on the other hand, gives a more realistic modality of emotion recognition, particularly, due to its temporal changes during arousal of emotion, and thus, concealment of the emotion in EEG is not feasible.

Usually the frontal lobe of the human brain is responsible for cognitive and emotion processing. There exists an internationally accepted 10-20 system for electrode placement on the scalp. Such placement of electrodes ensures that most of the

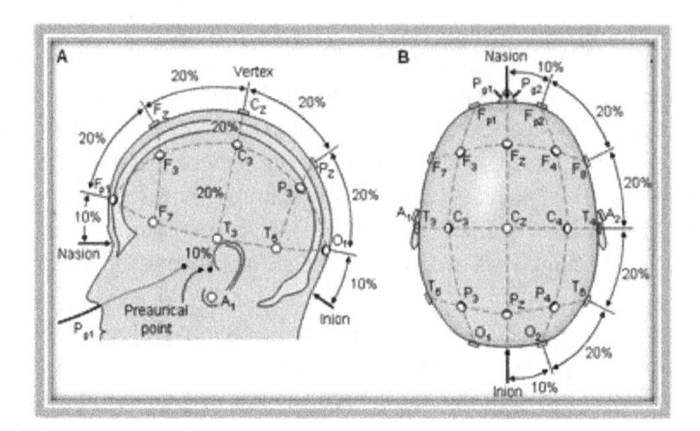

FIGURE 1.2 The international 10-20 electrode placement system.

brain functions, such as motor activation, emotion processing, reasoning, etc., can be retrieved correctly from the EEG signals obtained from these channels. In the 10-20 system (shown in Figure 1.2) of electrode placement, the channels F_3, F_4, F_{p_1}, and F_{p_2} are commonly used for emotion recognition.

Both time- and frequency-domain parameters of EEG are used as features for the emotion classification problems. Among the time-domain features, adaptive autoregressive (AAR) and Hzorth parameters, and among the frequency-domain features power-spectral density is most popular. EEG being a nonstationary signal, its frequency contents change widely over time. Time–frequency correlated features thus carry essential information of the EEG signal. Wavelet transform coefficients are important examples of time–frequency correlated features. In our study [134, 135], we considered wavelet coefficients, power spectral density, and also AAR parameters [136] for feature extraction. Typically the length of such feature vectors is excessively high, and thus, a feature reduction technique is employed to reduce the length of vectors without losing essential features.

Petrantonakis *et al.* [137] proposed a novel approach to recognize emotion from brain signals using a novel filtering procedure, namely, hybrid adaptive filtering (HAF) and higher order crossings (HOC) analysis. HAF was introduced for an efficient extraction of the emotion-related EEG characteristics, developed by applying Genetic Algorithms (GA) to the Empirical Mode Decomposition (EMD) based representation of EEG signals. HOC analysis was employed for feature extraction from the HAF-filtered signals. They introduced a user-independent EEG-based emotion recognition system for the classification of six typical emotions, including happiness, surprise, anger, fear, disgust, and sadness. The EEG signals were acquired from F_{p_1}, F_{p_2}, F_3, and F_4 positions, according to the 10-20 system from 16 healthy subjects using three EEG channels through a series of facial-expression image projections, as a Mirror Neuron System based emotion elicitation process. For an extensive evaluation of the classification performance of the HAF–HOC scheme, Quadratic Discriminant Analysis (QDA), k-Nearest Neighbor (k-NN), Mahalanobis Distance (MD), and Support Vector Machines (SVMs) were adopted. For the individual-channel case, the best results were obtained by the QDA (77.66% mean classification rate), whereas for the combined-channel case, the best results were obtained using SVM (85.17% mean classification rate).

In Reference 138, Petrantonakis *et al.* proposed a novel method for evaluating the emotion elicitation procedures in an EEG-based emotion recognition setup. By employing the frontal brain asymmetry theory, an index, namely Asymmetry index (AsI), is introduced, in order to evaluate this asymmetry. This is accomplished by a multidimensional directed information analysis between different EEG sites from the two opposite brain hemispheres. The proposed approach was applied to three-channel (F_{p_1}, F_{p_2}, and F_3/F_4 10/20 sites) EEG recordings drawn from 16 healthy right-handed subjects. For the evaluation of the efficiency of the AsI, an extensive classification process was conducted using two feature-vector extraction techniques and an SVM classifier for six different classification scenarios in the valence/arousal space. This resulted in classification results up to 62.58% for the user-independent case and 94.40% for the user-dependent one, confirming the efficacy of AsI as an index for the emotion elicitation evaluation.

Yuan-Pin Lin *et al.* [139] developed a new idea to recognize emotion from EEG signals while listening to music. In this study, EEG data were collected through a 32-channel EEG module, arranged according to the international 10-20 system. Sixteen excerpts from Oscar's film soundtracks were selected as stimuli, according to the consensus reported from hundreds of subjects. EEG signals were acquired from 30 channels. The features selected include power spectrum density (PSD) of all the 30 channels. SVM was successfully employed to classify four emotional states (joy, anger, sadness, and pleasure) using the measured PSDs. The best result for classification accuracy obtained is found to have mean 82.29% with a variance of 3.06% using a 10 times of 10-fold cross-validation scheme across 26 subjects. A few more interesting works on EEG-based emotion recognition that need special mention include References 140–153.

1.3.4 Gesture- and Posture-Based Emotional Features

Gestures are expressive and meaningful motions, involving hands, face, head, shoulders, and/or the complete human body. Gesture recognition has a wide range of applications, such as sign language for communication among the disabled, lie detection, monitoring emotional states or stress levels of subjects, and navigating and/or manipulating in virtual environments.

Recognition of emotion from gestures is challenging as there is no generic notion to represent a subject's emotional states by her gestures. Further, the gestural pattern has a wider variation depending on the subject's geographical origin, culture, and the power and intensity of his or her expressions.

Gestures can be static, considering a single pose or dynamic with a prestroke, stroke, and poststroke phases [154]. Automatic recognition of continuous gestures requires temporal segmentation. The start and end points of a continuous gesture are often useful to segregate it from the rest. Segmentation of a gesture sometimes is difficult as the preceding and the following gestures often are similar.

The most common gestural pattern, often used in emotion recognition, is the hand movements. Glowinski *et al.* [155] proposed an interesting technique for hand (and head) gesture analysis for emotion recognition. They considered a "bounding" triangle formed by the centroids of the head and hands, and determined several parameters of the 3D triangle to extract the features of individual hand gesture representative of emotions. A set of triangles obtained from a motion cue is analyzed to extract a large feature vector, the dimension of which is reduced later by PCA. A classification technique is used to classify emotion from a reduced 4D data space. Methodologies of feature reduction and classification will be discussed in a subsequent section later.

Camurri *et al.* [156] classified expressive gestures from the human full body movement during the performance of the subject in a dance. They identified motion cues and measured overall duration, contraction index, quantity of motion, and motion fluency. On the basis of these motion cues, they designed an automated classifier to classify four emotions (anger, fear, grief, and joy).

Castellano *et al.* [157] employed hand gestures for emotion recognition. They considered five different expressive motion cues, such as quantity of motion and contraction index of the body(degree of contraction and expansion of the body), velocity, acceleration, and fluidity (uniformity of motion) using Expressive Gesture Processing Library [158], and determined emotion from the cues by direct classification of time series.

In Reference 154, the authors considered both static and dynamic gesture recognition. While static gestures can be recognized by template matching, dynamic gesture is represented as a collection of time-staggered states, and thus can be modeled with HMM, sequential state machines (SSM), and discrete time differential neural nets (DTDNN). Preprocessing of the dynamic gestures includes tracking of important points or regions of interest in the image frames describing temporal states of the gesture. The tracking is often performed using particle filtering and level sets. Although particle filters and level sets have commonality in the usage of tracking in images, the principles used therein have significant difference. Usually a particle filter tracks a geometric-shaped region, circle or rectangle, on the image. So, in a nonrigid video, if the points enclosed in a region (of an image frame) have different directions of velocity, all the points of the reference frame cannot be tracked in subsequent frames. However, in a level set, the points enclosed in a region (of a frame) can be tracked by a nonlinear boundary, which may change its shape over the subsequent frames to keep track of all the points in the region of interest. Both the methods referred to above can track the hand and head gestures. For other important works on emotion recognition using gesture features, readers may consult the references aside 159–169.

1.3.5 Multimodal Features

Multimodality refers to analysis of different manifestations of emotion, including facial expression, voice, brain signals, body gesture, and physiological reactions. A few well-known multimodal schemes for emotion recognition are outlined below.

1.3.5.1 ***AudioVisual*** Zheng *et al.* [170] proposed an interesting approach to audio-visual emotion recognition. They considered 12 predefined motions of facial features, called Motion Units (MU) and 20 prosodic features, including pitch, RMS energy, formants F1–F4 and their bandwidths, and all of their corresponding derivatives. They achieved a person-independent classification accuracy of 72.42% using multi-stream HMM.

Mower *et al.* [171] designed a scheme for audiovisual emotion recognition. They considered both distance features between selective facial points and prosodic/spectral features of speech to recognize the emotion of an unknown subject. The distance features are selected within and between two of the following four regions: cheek, mouth, forehead and eyebrow. For example, the metrics considered refer to the distance from top of the check to the eyebrow, lower part of the cheek to mouth/nose/chin, relative distance features between pairs of selected points on cheek, and average positional features. Similarly they considered mouth features, such as mouth opening/closing, lips puckering, and distance of lip corner and top from the nose. The prosodic features extracted from speech include pitch and energy, and the spectral features employed include MFCC. They employed Emotion-Profile Support Vector Machines (EP-SVM) to obtain classification accuracy of 68.2%.

Busso *et al.* in Reference 172, proposed a new method to recognize emotion using facial expressions, speech, and multimodal information. They considered two approaches indicating fusion of two modalities at decision and feature levels. They employed 102 markers on the given facial image to determine the motion and alignment of the marked data points during utterance of 258 sentences expressing the emotions. They also considered prosodic features, such as pitch and intensity and reduced the feature dimensions by PCA. They classified the emotions by considering both individual visual and voice features, and obtained a classification accuracy of 85% for face data, 70.9% for voice data, and 89.1% for bimodal (face plus voice together) information.

1.3.5.2 ***Facial Expression–Body Gesture*** In a recent paper, Gunes and Picardi 173 consider automatic temporal segment detection and affect recognition from facial and bodily manifestation of emotional arousal. The main emphasis of the paper lies in the following thematic study. First, they demonstrate through experiments that affective faces and bodily gestures need not be strictly synchronous, although apparently they seem to occur jointly. Second, they observed that explicit detection of the temporal phases improves the accuracy of affect recognition. Third, experimental results obtained by them reveal that multimodal information including facial expression and body gesture together perform a better recognition of affect than only facial or body gestures. Last, they noticed that synchronous feature-level fusion achieves better performance than decision-level fusion.

1.3.5.3 ***Facial Expression–Voice–Body Gesture*** Nicolaou, Gunes, and Pantic, in Reference 174, have taken 20 facial feature points as facial features, MFCC, energy, RMS energy, pitch as speech features, and 5 shoulder points as body gesture feature for Continuous Prediction of Spontaneous Affect from Multiple Cues and Modalities in Valence-Arousal Space. They introduced Bidirectional Long Short-Term Memory Neural Networks (BLSTM-NN) and Support Vector Regression (SVR) classifier for emotion classification and concluded that BLSTM-NN gives better performance than SVR.

Castellano *et al.* in Reference 175 introduced a novel approach to emotion recognition using multiple modalities, including face, body gesture, and speech. They selected 19 facial feature points as facial features, MFCC, pitch values, and lengths of voiced segments as speech features, and 80 motion features for each gesture as body gesture features. They trained and tested a model with a Bayesian classifier, using a multimodal corpus with 8 emotions and 10 subjects. To fuse facial expressions, gestures, and speech information, two different approaches were implemented: feature-level fusion, where a single classifier with features of the three modalities is used; and decision-level fusion, where a separate classifier is used for each modality and the outputs are combined a posteriori. Lastly, they concluded that the fusion performed at the feature level provided better results than the one performed at the decision level.

1.3.5.4 EEG–Facial Expression Chakraborty *et al.* [176] correlated stimulated emotion extracted from EEG and facial expression using facial features, including eye-opening, mouth-opening, and eyebrow constriction and EEG features, including frequency domain, time domain, and spatiotemporal features. They considered frequency domain features, such as peak power and average powers of $\alpha, \beta, \gamma, \theta$, and δ bands, time domain features, including 16 Kalman filters coefficients, and spatiotemporal features including 132 wavelet coefficients. They employed a feed-forward neural network to train it with a set of experimental instances using the well-known Back-propagation algorithm. The resulting network on convergence is capable of classifying instances to a level of 95.2% classification accuracy.

1.3.5.5 Physiology In Reference 177, Picard *et al.* proposed that for developing a machine's ability to recognize human affective state, machines are expected to possess emotional intelligence. They performed an experiment considering four physiological signals: electromyogram (EMG), blood pressure volume, Hall effect, and respiration rate, taken from four sensors. One additional physiological signal, heart rate (H) here has been calculated as a function of the inter-beat intervals of the blood volume pressure. In their analysis, they used a combination of sequential floating forward search (SFFS) and Fisher projection (FP), called SFFS-FP, for selecting and transforming the features. A classification accuracy of 81% was obtained by using maximum a posteriori (MAP) classifier for SFFS-FP.

1.3.5.6 Facial Expression–Voice–Physiology Soleymani *et al.* [178] introduced a multimodal database for affect recognition and implicit tagging. They chose 27 subjects and recorded their videos of facial and bodily responses while watching 20 emotional videos. Features used for their experiment include distance metrics of the eye, eyebrow, and mouth as facial features, audio and vocal expressions, eye gaze, pupil size, electrocardiograph (ECG), galvanic skin response (GSR), respiration amplitude, and skin temperature as physiological features. The main contribution of this research lies in the development of a large database of recorded modalities with high qualitative synchronization between them making it valuable to the ongoing development and benchmarking of emotion-related algorithms. The resulting database would provide support to a wide range of research on emotional intelligence, including data fusion, synchronization studies of modality, and many others.

1.3.5.7 EEG–Physiological Signals Takahashi, in Reference 179, undertook an interesting research on emotion recognition from multimodal features including EEG, Pulse, and Skin Conductance. They collected psychological data of 12 subjects. The experimental setup contains a set of three sensors and two personal computers; one PC being used to present stimulus to a subject, while the other is used to acquire biopotential signals stated above. They used SVM classifier for emotion recognition from biopotential signals and acquired a classification accuracy of 41.7% for five emotions including joy, anger, sadness, fear, and relaxation. Among the other works on emotion recognition References 180–188 deserve special mention.

1.4 FEATURE REDUCTION TECHNIQUES

EEG and voice features usually have a high dimensionality and many of the experimentally obtained features are not independent. The speed of classifiers is often detrimental to the dimension of features. Feature reduction algorithms are required to reduce the dimensionality of features. Both linear and nonlinear feature reduction techniques are employed for emotion recognition.

Linear reduction techniques employ the characteristics of real symmetric matrices to extract independent features. In other words, the eigen vectors of real symmetric matrices are orthogonal (independent) to each other. Further, the larger eigen values of a system carry more information than the others. So, the eigen vectors corresponding to the large eigen values are used to reduce data dimensionality of a given linear system.

Linear reduction principles have gained popularity for their simplicity in use. However, on occasion researchers prefer nonlinear reduction techniques to their linear counterparts to improve precision and reliability of the classifier. Among the nonlinear feature reduction techniques, the most popular is rough set–based feature reduction. In this section, we briefly outline a few well-known linear and nonlinear feature reduction techniques.

1.4.1 Principal Component Analysis

Principal Component Analysis is one of the most popular linear feature reduction techniques. PCA represents N measurements for M subjects as a $M \times N$ matrix A, and computes $A^{T}A$ to obtain a real symmetric matrix B of dimension $N \times N$. Now the N eigen values of B are evaluated, and the results are sorted in descending order.

It is known that the larger eigen values have higher contribution in representing system characteristics, and thus to reduce features, we take k no. of eigen vectors corresponding to the first k eigen values of the list. The eigen vectors are arranged in columns, and the matrix is called Eigen Vector (EV) matrix of $(N \times k)$ dimension. Now, for each measurement vector a_i, taken from the ith row of the matrix A, we take a projection of a_i on the eigen space by multiplying a_i by EV matrix, and thus obtain $a_i^/$, where $a_i^/$ has dimension $(1 \times k)$. This is repeated for all $i = 1$ to N, and thus the feature vectors a_i are mapped to k-dimensional vectors, where $k \ll N$.

PCA is good for feature reduction of linear systems. If PCA is used for feature reduction of nonlinear systems, where the functional relationship between any two features is nonlinear, PCA sometimes loses important information. Researchers use PCA for its high efficiency and accuracy. Some important work using PCA include References 17, 79, 87, and 189–193.

1.4.2 Independent Component Analysis

Independent Component Analysis is a good choice to separate the sources from mixed signals. Particularly, in EEG and ECG the time series data x_t at time t is a nonlinear function $f(.)$ of previous time samples, that is, $x_t = f(x_0, x_1, \dots, x_{t-1})$. EEG signals taken from the forehead of a subject are often contaminated with eye-blinking signals, called electrooculogram (EOG). Further, the signal obtained at an electrode located on the scalp/forehead is due to the contribution of a number of signal sources at the neighborhood of the electrode. Elimination of EOG from EEG data and identification of the source signals can be performed together by ICA. One fundamental (although logical) restriction of using ICA lies in the inequality: $C >= S$, where C represents the number of EEG channels, and S stands for the number of independent signal sources. ICA has been widely used in the literature [194–198] to recognize emotion from facial expressions.

1.4.3 Evolutionary Approach to Nonlinear Feature Reduction

Evolutionary algorithms are population-based meta-heuristic optimization algorithms, which rest on the Darwinian principle of the survival of the fittest. The primary aim of this class of algorithms is to determine near optimal solutions, if not global, from a set of trial solutions through an evolutionary process determined by a set of operators like crossover, mutation, and selection. The most popular member of this class is Genetic Algorithm [199–201], devised by Prof. Holland approximately a half century ago. Among other members Differential Evolution (DE) [202, 203] is most popular for its structural and coding simplicity and exceptional performance in optimization problems.

Given a set of feature vectors (also called data points) and class labels for each vector, to implement the evolutionary feature reduction, we first use a supervised learning–based classifier to classify the data points into a fixed number of classes c. Next we reduce the dimension of data points by dropping one feature randomly at a time, and again classify the data points into c number of classes. If the resulting classes do not differ significantly with the previous classes, then the dropped feature has no major significance. The GA or DE is used to randomly select k number of features at a time, and classify the data points into c classes, and test whether the classes generated have significant difference with the classes obtained from the original dataset after classification. Since k is randomly selected in $[1, n]$, where n is the total number of features in the original dataset, we at the end of the search process expect to find a suitable value of k, for which the classes would be similar with the classes of the original dataset. Thus high dimensional features are reduced to k-dimension for $k \ll n$.

1.5 EMOTION CLASSIFICATION

This section provides principles of several approaches to emotion classification by pattern recognition techniques.

1.5.1 Neural Classifier

Neural networks have widely been used in emotion classification by facial expressions and voice. Both supervised and unsupervised neural architectures are employed in emotion classifiers. The supervised neural networks require a set of training instances. During the training process, the network encodes the connection weights in a manner, such that for all the input components of the training instances, the network can reproduce the output components of the corresponding training instances correctly as listed in the training instances. After the encoding is completed, the trained network can be used for testing. In the testing phase, an unknown input instance is submitted to the network, and the network generates the output instance using the encoded weights. In case of emotion recognition, the output of the neural net usually represents emotion classes, whereas input of the neural net represents a set of features extracted from facial expression/voice/gesture of the subjects. Naturally, a neural network pretrained with emotional features as the input and emotion classes as the output would be able to classify a specific emotional expression into one of several emotion classes.

1.5.1.1 Back-Propagation Algorithm Among the well-known neural topologies, the Back-propagation is most common. Weight adaptation in the Back-propagation neural net is performed by Newtonian Gradient/Steepest descent learning principle. Let W_{i_j} be the connection weight between neuron N_i and neuron N_j, and E be the error function representing the root mean square error between desired output and computed output for each input training instance. Then the weight adaptation policy is formally given by

$$\delta W_{i_j} = -\eta \delta E / \delta W_{i_j}$$

where ΔW_{i_j} denotes change in weight and η is the learning rate. In Back-propagation algorithm, the weight adaptation for each layer is derived using the above equation.

Computation of ΔW_{i_j} in the output layer of a multilayered feed-forward neural network is straightforward as the error function E involves the weight W_{i_j}. However, computation of ΔW_{i_j} in the intermediate and input layers is not easy as the error function E does not involve W_{i_j}, and a chain formula of known partial derivatives is used to compute ΔW_{i_j}.

Once ΔW_{i_j} computation is over, we add it to W_{i_j} to obtain its new value. The process of layerwise computation of weights always starts at the output layer, and continues up to the input layer, and this is usually referred to as one pass. Several passes are required for convergence of weights toward steady-state values. After convergence of the weights, the trained network can be used for the application phase. During this phase, the network is excited with a new instance.

One fundamental limitation of the back-propagation algorithm is trapping at local optima on the error (energy) surface. Several methods have been proposed to address the issue. The most common is adding momentum to the weight adaptation dynamics. This helps the dynamics to continue movement even after coming in close vicinity of any local optima. Once the dynamics pass the local optima, their speeds are increased so that the motion is continued until the global optimum is identified. Among the enormous work on emotion recognition using Back-propagation neural network, References 204–208 need special mention.

1.5.1.2 Radial Basis Function Based Neural Net Radial Basis Function (RBF) neurons employ a specialized basis function to map an input pattern $\vec{X}$ to two soft levels 0 and 1. A pattern classifier with k classes usually has k basis functions, designed to map an unknown input vector to one of k classes. When a pattern falls in a class, its RBF function yields a value close to one. On the other hand when an input pattern does not fall in a class, the function returns a small value close to zero. Among the popular RBF functions, the Gaussian function is most common for its wide applications in science and engineering. Let X_c be the center of an RBF function. Then for any input vector X_c, we define the RBF function Y as given below:

$$Y = exp - \|(X - X_c)\|$$

where $\|.\|$ denotes an Euclidean norm.

A typical RBF neural net consists of two layers, the first layer being the RBF layer, and the last layer being realized by a perceptron neuron, the weights of which are determined by the perceptron learning algorithm. When an unknown input instance is supplied, the response of the first layered RBF neurons become close to zero for most of the neurons and close to one for one or fewer neurons. The weights generated by perceptron learning algorithm are later used to map an intermediate pattern into pattern class. A few research works employing Radial Basis Function Based Neural Network as a classifier for emotion recognition, include References 209–211.

1.5.1.3 Self-Organizing Feature Map Neural Net In a Self-Organizing Feature Map (SOFM) Neural Net, we need to map input patterns onto a 2D array of neurons based on the similarity of inputs with the patterns stored in individual neurons. The patterns stored by neurons have the same dimension as that of input patterns. These patterns are called weights of the respective neurons. A given input pattern is mapped onto a neuron with the shortest Euclidean distance. A neighborhood around the selected neuron is considered, and the weights of all the neurons in the neighborhood are adapted by the following equation:

$$W_{i_j}(t+1) \leftarrow W_{i_j}(t) + \eta(X_k - W_{i_j}(t))$$

where $W_{i_j}(t)$ is the weight of neuron (i, j) in the neighborhood of the selected neuron in the 2D array at time t, X_k is the kth input vector, and η is the learning rate. After the weights are adapted, a new input vector is mapped onto the array by the distance criterion, and the process of neighborhood selection around the neuron and weight adaptation of the neurons in the neighborhood is repeated for all the inputs. The whole process of mapping input vectors onto a 2D array and weight adaptation of neurons is aimed at topological clustering of neurons, so that similar input vectors are mapped at close vicinity on the 2D array.

During the recognition phase, we need to retrieve one or more fields of a given vector, presuming that the remaining fields of the vector are known. Generally, the unknown vector has the same dimension to that of other input vectors used for weight adaptation. The unknown vector first is mapped onto a neuron in the 2D array based on the measure of minimum Euclidean distance between the input vector and all weight vectors, considering only the known fields of the unknown vector during Euclidean distance evaluation. The neuron having the best match, that is, having the smallest Euclidean distance between its weight vector and the unknown input vector is identified. The unknown fields of the input vector are retrieved from the corresponding fields of the weight vector of the selected neuron. SOFM can be used for emotion recognition from face [212], speech [213], EEG [214], as well as from gesture [215], for its high efficiency and accuracy as a classifier.

1.5.1.4 Support Vector Machine Classifiers Support Vector Machines have been successfully used for both linear and nonlinear classification. A linear SVM separates a set of data points into two classes with class labels +1 and −1. Let $X = [x_1 x_2 ... x_n]^{\mathrm{T}}$ be any point to be mapped into +1, −1 by a linear function $f(X, W, b)$, where $W = [w_1 w_2 ... w_n]$ is a weight vector and b is a bias term. Usually, $f(X, W, b) = \mathrm{Sign}(WX + b) = \mathrm{Sign}(\sum_i w_i x_i + b)$. Figure 1.3 illustrates classification of 2D data points. In 2D, the straight line that segregates the two pattern classes is usually called a hyperplane. Further, the data points that are situated at the margins of the two boundaries of the linear classifier are called support vectors. Figure 1.3 describes a support vector for a linear SVM.

Let us now select two points $X+$ and $X-$ as two support vectors such that for $X = X^+$, $WX^+ + b = +1$. Similarly, for $X = X^-$, $WX^- + b = -1$. Now, the separation between the two support vectors lying in the class +1 and class −1, called marginal width is given by

$$M = (WX^+ + b) - (WX^- + b)/\|W\| = 2/\|W\|$$

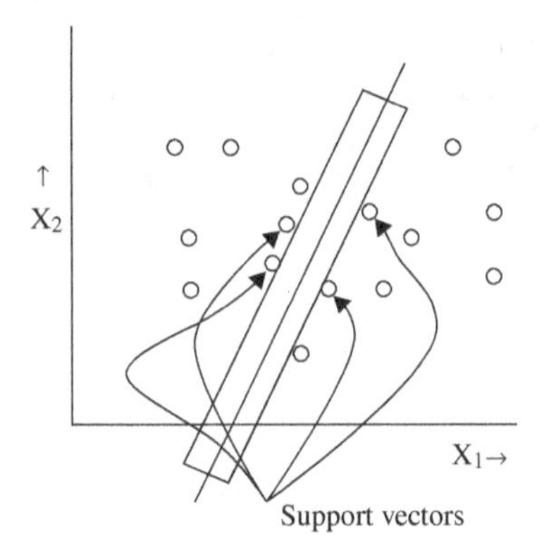

FIGURE 1.3 Defining support vector for a linear SVM system.

The main objective in a linear SVM is to maximize M, that is, to minimize $||W||$, which is same as minimizing $1/2W^{\mathrm{T}}W$. Thus, the linear SVM can be mathematically described by

$$\begin{aligned} &\text{Minimize } \phi(W) = 1/2W^{\mathrm{T}}W \\ &\text{subject to } y_i(WX_i + b) \geq 1 \text{ for all } i \end{aligned} \tag{1.3}$$

where y_i is either 1 or -1 depending on the class that X_i belongs to.

Here, the objective is to solve W and b that satisfies the above equation. The solution to the optimization problem is not given here due to space limitation. However, inquisitive readers can have it in the standard literature. Linear SVM has a wide range of applications in supervised classifiers. It is currently one of the best popular algorithms for pattern classification. Many researchers choose SVM [18,24,26,216–219] as a classifier in emotion classification for its higher accuracy.

1.5.1.5 Learning Vector Quantization In scalar quantization, the random occurrence of a variable x in a given range $[x_{\min}, x_{\max}]$ is quantized into few fixed levels. For example, suppose there are m quantization levels, uniformly spaced in $[x_{\min}, x_{\max}]$. So, the quantization step height is determined as $q = (x_{\max} - x_{\min})/m$. The kth quantization level has a value $x_k = x_{\min} + h.(k - 1)$. An analog signal x having a value greater than kth quantization level, but less than $(k + 1)$th quantization level is quantized to the kth level. This particular feature of the quantization process is called truncation. Sometimes, we use roundoff characteristics of the quantizer. In case of roundoff, an analog signal having a value less than $x_k + q/2$ but greater than x_k would be quantized to x_k. But if the analog signal $x > x_k + q/2$, but less than x_{k+1}, it will be quantized to x_{k+1}.

In vector quantization, vectors V of dimension n are quantized to fixed vectors V_i of the same dimension. If all components of V are close enough with respect to corresponding components of vector V_i, then V would be quantized to V_i. Usually, such quantization is widely used in data compression. In learning vector quantization

(LVQ), we use a two-layered neural net. The first layer is used for data reception, while the second is a competitive layer, where one of several neurons only fires, and the weights are reinforced using the input. Let X_i be the ith input vector, whose components are mapped to the neurons at the input layer of the neural net. Let us assume that there are p neurons in the second layer. Now let the weight vector of the neurons in the second layer be $W_1, W_2, \ldots, W_p$. Suppose

$$\|W_k - X_i\| <= \|W_j - X_i\| \text{ for all } j$$

Then we would adapt W_k by the following update rule:

$$W_k \leftarrow W_k + \eta(X_i - W_k)$$

where η is the learning rate between (0, 1). The weights are thus reinforced by all the inputs. After the learning with all the input instances are over, we identify the unknown components of a given test instance vector by identifying the trained weight vector, with which the input instance has the smallest Euclidean distance, considering only the known components. The unknown fields of the selected weight vector are then used for subsequent applications. Researchers used LVQ as a classifier for emotion recognition from facial expression [220,221] or speech [222].

1.5.2 Fuzzy Classifiers

Measurements obtained from facial expression, voice, and gesture/posture for emotion recognition often are found to be contaminated with various forms of uncertainty. For example, repeated measurements of facial, vocal, and bodily gestures of a subject experiencing the same emotion have wider variance. This is often referred to as an intrapersonal level of uncertainty. Further, when the measurements are taken from different subjects experiencing same/similar emotion, the variance in measurements is found to have a large value, causing interpersonal level of uncertainty. Classical type 1 fuzzy logic considers a single membership function to represent the uncertainty involved over a given measurement space, but fails to model the true spirit of intra- and interpersonal level of uncertainty [223]. Type 2 fuzzy sets provide an opportunity to represent both inter- and intrapersonal variations in uncertainty, and thus has immense scope in fuzzy classifiers, capable of correctly classifying emotions from measurements suffering from uncertainty.

In classical fuzzy rule–based classifiers, fuzzy rules are employed to map fuzzy encoded measurements into emotion classes with different degrees of certainty. Thus individual emotions support a given set of emotional features to different degrees, and naturally the class with the highest support is considered as the winner. Type 2 fuzzy rules on the other hand map a set of imprecise fuzzy encoded measurements obtained from different sources into emotion classes. The class offering maximum support to the measurement space is considered as the winning class.

An alternative approach to fuzzy pattern recognition is to cluster a set of features based on their similarity. Since measurements are noisy, data points lying on the boundary of two classes can be categorized to both the classes with certain degrees. Fuzzy C-means clustering algorithm has widely been used as a basic tool of pattern clustering to overcome clustering of data points suffering from noisy measurements or incomplete specification of data dimensions. A few important research works employing fuzzy logic in classifiers include [60, 224, 225].

1.5.3 Hidden Markov Model Based Classifiers

Let X, Y, and Z be three random variables where they may take up any value from $x_1, x_2, \ldots, x_n$, $y_1, y_2, \ldots, y_m$, and $z_1, z_2, \ldots, z_r$, respectively. Suppose Y depends on X and Z depends on Y. So, we can represent the dependence relationship among X, Y, and Z by a directed graph, where X is a predecessor of Y and Y is a predecessor of Z. Now suppose, by experiments we have the conditional probabilities: $P(Y/X)$ and $P(Z/Y)$ for any X, Y, and Z. Now in a Markov process we consider the state-transition probabilities in one level only, that is, we consider $P(Z/Y,X) = P(Z/Y)$.

In an HMM, the probability of occurrence of a class is determined by a sequence of state transitions. For example, suppose if $X = x_2$ then $Y = y_3$, and if $Y = y_3$ then $Z = z_4$. Suppose, the sequence of state transitions from $X = x_2$ through $Y = y_3$ and $Z = z_4$ denotes class 1 of a pattern recognition problem. This state transition probability is $P(Z = z_4/Y = y_3) \cdot P(Y = y_3/X = x_2)$. Now, if $P(X = x_2)$ is the probability of occurrence of $X = x_2$, then $P(Z = z_4)$ following the sequence $X = x_2$ through $Y = y_3$ is given by $P(X = x_2) \cdot P(Y = y_3/X = x_2) \cdot P(Z = z_4/Y = y_3)$. Thus for all known sequences passing through X, Y, and Z, we know the probability of the pattern class.

Now, for an unknown sequence, we compute the state transition probability of the sequence, and check whether the probability of the sequence matches enough with the probability of the standard nearest sequence. If so, we consider the unknown sequence to lie in the class of the nearest matched sequence. HMM is vastly used in research for recognition of emotion from face [25], speech [226–228], and sometimes from both face and speech [229–231].

1.5.4 k-Nearest Neighbor Algorithm

k-Nearest Neighbor algorithm is a simple method to determine the class of an unknown pattern. It presumes that the given data points are pre-classified and the label of each class is known. Now, for an unknown pattern, represented by a data point X, the distance between X and all nearest neighbors Y_j of X are identified, and the count of Y_j lying in different classes is determined. The class having the largest number of nearest neighbors of X is declared as the class of the unknown data point X.

The distance metric selection is an important problem in the k-NN algorithm. Usually, Euclidean distance is used in most of the literature [232]. However, when the data points have a large dimension or the components are not scaled properly, the distance between any two data points sometimes is too large, and consequently the results of classification is not free from errors. The given set of data points therefore should be normalized by scaling d_{i_j} by $d_{i_j}/\mathrm{Max}_j(d_{i_j})$, where d_{i_j} is the jth component of the ith data point d_i.

k-NN algorithm works fine for lower dimensional data, typically when dimension is less than 5. It has popularity in the pattern recognition community particularly for its simplicity. The complexity of the algorithm grows with an increase in the dimension of the data points. To avoid this increase in computational complexity, feature reduction algorithm is first employed to identify the independent components of the data points by using a feature reduction algorithm, and k-NN is applied on the data classes developed with reduced dimensionality. k-NN has been used in many research works as a classifier to recognize emotion [79, 90, 233].

1.5.5 Naïve Bayes Classifier

Suppose, there are n objects, each having a set of m features $f_1, f_2, \ldots, f_m$ and the objects are classified into k classes: $c_1, c_2, \ldots, c_k$ based on the similarity of their features/attributes. Now, by Bayes' conditional probability theory, we have the conditional probability

$$P(c_j \mid f_1, f_2, \ldots, f_m) = P(f_1, f_2, \ldots, f_m \mid c_j)P(c_j)/P(f_1, f_2, \ldots, f_m)$$

where the probabilities have their standard meaning. Now, for any j, the expression on the right-hand side contains a constant denominator: $P(f_1, f_2, \ldots, f_m)$. So, we instead of computing $P(f_1, f_2, \ldots, f_m)$, simply represent it by a scalar constant $1/A$. Therefore, the last result may be reconstructed as

$$P(c_j \mid f_1, f_2, \ldots, f_m) = A.P(f_1, f_2, \ldots, f_m \mid c_j)P(c_j) \tag{1.4}$$

Now, $P(f_1, f_2, \ldots, f_m \mid c_j)$ denotes the probability of the joint occurrence of $f_1, f_2, \ldots, f_m$, given the object class c_j. However, computation of $P(f_1, f_2, \ldots, f_m \mid c_j)$ is not simple. One simple assumption is to consider the features independent, that is,

$$P(f_1, f_2, \ldots, f_m \mid c_j) = P(f_1/c_i).P(f_2/c_i)...P(f_m/c_i) = \prod_{i=1}^{m} P(f_i/c_j) \tag{1.5}$$

Combining (1.3) and (1.4) we obtain,

$$P(c_j \mid f_1, f_2, \ldots, f_m) = A.P(f_1, f_2, \ldots, f_m \mid c_j)P(c_j) = A.\prod_{i=1}^{m} P(f_i/c_j).P(c_i)$$

Now, from experimental instances, if we can prepare a table $P(f_i/c_j)$ for all i, j, we can determine $P(c_j \mid f_1, f_2, \ldots, f_m)$ for all j, and identify the class k, for which

$$P(c_k \mid f_1, f_2, \ldots, f_m) >= P(c_j \mid f_1, f_2, \ldots, f_m) \text{ for all } j$$

The class k is the desired class for the given combination of features. This is how a Naïve Bayes classifier works. The current literature employing Naïve Bayes Classifier to recognize emotion includes References 234–236.

1.6 MULTIMODAL EMOTION RECOGNITION

Emotion of a subject is recognized from his or her facial expression, voice, gesture, and posture. The term modality refers to modes of recognition. Early research in emotion recognition stressed the importance of single modality. Single modality, however, cannot always provide sufficient information for emotion recognition. Current research on emotion recognition thus is targeted to multimodal information extraction to improve the reliability of the recognition process.

One common metric to represent the performance of the recognition process is to construct a confusion matrix, where each grid (i,j) of this matrix represents the percentage of emotional expression correctly classified in class j, when the actual emotion is i. Experiments undertaken on different modality of the same subjects reveal that the confusion matrices constructed for each modality have significant differences, indicating that the individual modality of classification is not highly reliable.

Multimodal fusion is to integrate all single modalities into a combined single representation. There are basically two types of fusion techniques that have been used in most of the literature to improve reliability in emotion recognition from multimodal information. They are (1) feature-level fusion and (2) decision-level fusion. Feature-level (early) fusion classifies emotion of subjects from their individual modalities in the first step, and then attempts to fuse the results of unimodal classification with an aim to reduce uncertainty in classification. One simple method of fusion is to check whether the individual classifiers for multiple modalities yield the same class for a subject. If so, the decision about the emotion class of the subject is certain. However, if the classifiers produce conflicting class labels for a subject, then the class label obtained from majority of the classifiers is honored. Decision-level (late) fusion feeds all measurements obtained from sensors employed for different modalities to a pretrained classifier, which in turn determines the class label of the unknown measurement.

In Reference 173, the authors applied both feature-level and decision-level fusion on face and bodily gestures and concluded that feature-level fusion achieves better performance than decision-level fusion. On the other hand, in Reference 172, the authors observed that the overall performance of the feature-level bimodal classifier and decision-level bimodal classifier is similar. But by analyzing with confusion matrix, they also observed that the recognition rate for each emotion type was totally different. In the decision-level bimodal classifier, the recognition rate of each emotion is increased (except happiness, which is decreased by 2%) compared to the facial expression classifier. Recognition rate of emotion anger and neutral are significantly increased in the feature-level bimodal classifier whereas recognition rate of happiness is decreased by 9% by using the same. Therefore, they concluded that "the best approach to fuse the modalities will depend on the application."

1.7 STIMULUS GENERATION FOR EMOTION AROUSAL

Elicitation of emotion usually is performed by presenting selective stimuli to the subjects. Selection of the modality of emotion elicitation and identification of the right stimuli for a given modality are equally important. Among the well-known modalities of emotion elicitation, visual, audio, audiovisual, music are most popular. Traces of using ultrasonic signals for emotion elicitation are also found in the literature [237]. Proximity signals, such as touch of a mother to stop crying of babies, have also been used in the literature for emotion control.

Considering a particular modality of emotion elicitation, we can identify some fundamental basis for selection of stimulus. In Reference 60, the authors sought the opinion of several people to determine the candidate audiovisual stimulus, which seems to be most effective for elicitation of a particular emotion. To briefly outline their scheme, suppose we have n audiovisual stimuli presented before a group of m subjects. Each subject is asked to provide his or her feeling about possible percentage of elicitation of five emotions by the stimuli. After the score sheet containing the opinions of m subjects for five emotions excited by n stimuli is prepared, the sum of percentage support obtained from m subjects for each stimulus favoring individual emotion class is determined. The emotion class that receives highest support for each stimulus is selected for elicitation of that emotion.

Selection of stimuli and their modalities are important issues for arousal of emotion. The modalities include stimulation by audio, visual, audiovisual, consolation message, music, and dance. Among these modalities, significant research has been undertaken on audio [238], visual [239, 240], audiovisual [241–243] and music [244–248] based stimuli for emotion arousal. In Reference 244 the authors studied the effect of certain features in music as a qualitative measure of emotion arousal. They noticed that pitch range, tone quality, rhythmic regularity, and tempo are some of the important features capable of originating and regulating emotion. Some well-known music pieces are often used for generating particular emotion in subjects. Baumgartner *et al.* [240] selected some classical orchestral pieces like Gustav Holst: *Mars—The Bringer of War from the Planets*, Samuel Barber: *Adagio for Strings*, and Beethoven: *Symphony No. 6 (3rd mvt.)* for arousal of three basic emotions: fear, happiness, and sadness. The tones of Holst evoke fear, music of Barber evokes sadness, and the one by Beethoven evokes happiness.

Visual mode of excitation includes voiceless movie, particularly action movies describing thematic issues of a particular real-life context. The modality of video presentation, context, and climax keep the subject attracted toward the stimulus, and the context of the movie excites a specific faculty of the mind, arousing an emotion. For example, a set of consecutive pathetic scenes causes an arousal of sadness, while a scene of amusements causes relaxation/happiness. In Reference 240, the authors have shown 48 still pictures containing humans or human faces to 70 subjects for arousal of three basic emotions: fear, happiness, and sadness. They have selected pictures of a man attacking a woman with a knife or a man pointing a pistol to the viewer for arousing fear. Pictures of a man holding his smiling baby, laughing children playing on the beach, or athletes in a victory pose are used for making subjects happy. For arousing sadness, picture of a crying little boy standing in front of a destroyed house or a couple standing at a gravestone are shown.

Audiovisual modality of excitation has been found to very effective for control of emotional state-transitions. The onset time (i.e., the duration of stimulation after which the brain responds) and offset time (i.e., the minimum time brain remains excited after removal of the stimulus) [31, 249] are important issues in audiovisual stimulation. These times are not always constant for a given subject. They are sensitive to both context and climax.

1.8 VALIDATION TECHNIQUES

After a system design is complete, we go for validation of the system performance using test samples. In case performance of the system is found satisfactory, the system design is certified for field applications. Otherwise, we should identify the free system variables/parameters, which are not direct inputs or outputs, for tuning. Machine learning and/or evolutionary algorithms are often used to tune the parameters of physical systems in order to improve the system performance with known test samples. In Reference 60, the authors have employed evolutionary algorithm to adapt the parameters of fuzzy membership functions to improve the classification accuracy of the fuzzy rule–based classifier. The classifier in that paper has been tested with facial features of subjects with known emotion classes.

A good system validation requires identifying right test instances taken from all classes of emotion with wider variance in feature space even for the same class. If the test instances are misclassified, we need to adapt system parameters. Selection of non-input/output parameters having a wider range of variation with a power of controlling the performance of the classification algorithm is also an important issue. Lastly, the evolutionary/learning algorithm to be incorporated should have the ability to tune system parameters with a motive of improving the performance index of the classifier. Usually system validation is done offline. So, no restriction is imposed on the convergence time of the validation algorithm. However, a fast algorithm is always preferred without allowing degradation in system performance.

In Reference 177, authors used leave-one-out cross-validation method for selecting the correct classifier between SFFS, FP, and a combination of SFFS and FP, called SFFS-FP. In each case, the feature vector per emotion to be classified was excluded from the dataset before the SFFS, FP, or SFFS-FP was run. The best set of features or best transform (determined from the training set) was then applied to the test data point to determine classification accuracy. They observed that SFFS-FP improves the performance and provides the highest recognition rate on eight classes of emotion.

Busso *et al.* [78], compares the performance of their proposed approach with the conventional classifier when there is a mismatch between the training and testing conditions. For this purpose, they separate the emotional databases into two language groups: English and German. One of these groups is used for training, and the other one for testing. The system is then trained and tested with different databases recorded in three different languages (English, German, and Spanish) to validate the robustness of the proposed approach. It has been seen that the performance of the conventional classifier without the neutral models decreases up to 17.9%. So they concluded that "using neutral models for emotional recognition is hypothesized to be more robust and, therefore, to generalize better than using a direct emotion classification approach."

1.8.1 Performance Metrics for Emotion Classification

Performance metrics are used to measure the performance of an emotion classifier. The simplest metric for performance measure is classification accuracy, which indicates the percentage of success in emotion classification from a given set of measurements of the subjects during the period of their emotion arousal. Classification accuracy cannot provide the subjective details about how many instances of emotional expressions falling under an emotion class are misclassified in all other emotion classes. For example, if out of 60 facial expressions, 30 appear in anger, 20 in fear, and 10 in anxiety classes, then we understand that the set of features is either incomplete or the emotional expressions are not sufficiently distinguishable. Confusion matrix offers the details of percentage misclassification of an emotion into other classes in percentage or absolute values, and so it is a popularly used metric to represent classification accuracy in emotion recognition problem.

Statistical tests are undertaken to compare the relative performance of emotion classifiers. Two types of tests are generally performed. The first type compares the performance of two classifiers, while the latter type determines ranks of individual classifiers based on some metric, usually classification accuracy. The merit of using these statistical tests is to ensure robustness in performance analysis. The tests require some measure of individual performance of a classifier tested with different databases, and ultimately determine the relative performance of the classifiers as obtained from all the databases together. Among the well-known statistical tests, McNemar's test and Friedman test [250, 251] need special mention. McNemar's test considers two classifier algorithms at a time, and identifies the relatively better between them based on a specific measure determined by the count of misclassified items by the individual algorithms. Friedman test considers all the classifier algorithms together and ranks them best on the individual performance of the algorithms.

1.9 SUMMARY

Pattern recognition at present is an established discipline of knowledge. A general pattern recognition problem comprises three or more steps: feature extraction, feature reduction, and classification. Emotion recognition falls under the basic category of pattern recognition problems. Thus, the basic steps of pattern recognition are equally applicable here. However, the main challenge in emotion recognition lies in extraction of essential features, and selecting a suitable classifier. The chapter provides a thorough review of the existing techniques of feature extraction and classification, commonly used in emotion recognition.

A review of the existing techniques of emotion recognition from a pattern classification perspective is given in the chapter. Special emphasis is given to modality-based recognition as for individual modalities the feature extraction and classification techniques have significant differences. An outline to different validation techniques for emotion classification is given here. In addition, the statistical techniques used to compare the relative performance in emotion classification are narrated at the end of the chapter.

On occasion single modality cannot correctly yield good accuracy. Multimodality becomes an important necessity under the context. The chapter provides an extensive discussion on multimodal emotion recognition. Stimulus generation for elicitation of emotion is an open problem, and very few works in this regard are available in the literature. The chapter provides a thorough overview on stimulus generation. Lastly, the chapter cites an exhaustive list of references [1–264] to emotion research, including both fundamental research and applications.

REFERENCES

[1] J. P. Marques de Sá, *Pattern Recognition: Concepts, Methods and Applications*, Springer, 2001.

[2] S. Theodoridis and K. Koutroumbas, *Pattern Recognition*, Academic Press, 2008.

[3] T. Mitchell, *Machine Learning*, McGraw Hill, 1997.

[4] J. C. Bezdek and S. K. Pal, *Fuzzy Models for Pattern Recognition*, IEEE Press, New York, 1992.

[5] M. Kass, A. Witkin, and D. Terzopoulus, "Snakes: active contour models," *Int. J. Comput. Vis.*, 1(4): 321–331, 1988.

[6] A. Yuille, D. Cohen, and P. Halliman, "Extraction from faces using deformable templates," *Int. J. Comput. Vis.*, 8: 104–109, 1992.

[7] I. Craw, D. Tock, and A. Bennett, "Finding face features," in Proceedings of the Second European Conference on Computer Vision, Santa Margherita Ligure, Italy, May 19–22, 1992, pp. 92–96.

[8] R. Brunelli and T. Poggio, "Face recognition: features versus templates," *IEEE Transactions on Pattern Analysis and Machine Intelligence*, 15(10): 1042–1052, 1993.

[9] M. Kirby and L. Sirovich, "Application of the Karhunen–Loeve procedure for the characterization of human faces," *IEEE Transactions on Pattern Analysis and Machine Intelligence*, 12(1): 103–108, 1990.

[10] B. Moghaddam and A. Pentland, "Beyond Euclidean Eigenspaces: Bayesian Matching for Visual Recognition," in *Face Recognition: From Theories to Applications*, Springer, Berlin, 1998.

[11] B. Draper, K. Baek, M. S. Bartlett, and R. Beveridge, "Recognizing faces with PCA and ICA," *Comput. Vis. Image Underst.* (Special Issue on Face Recognition), 91: 115–137, 2003.

[12] B. Moghaddam, "Principal manifolds and probabilistic subspaces for visual recognition," *IEEE Transactions on Pattern Analysis and Machine Intelligence*, 24(6): 780–788, 2002.

[13] B. Scholkopf, A. Smola, and K. M. Uller, "Nonlinear component analysis as a kernel eigenvalue problem," *Neural Comput.* 10(5): 1299–1319, 1998.

[14] M.-H. Yang, "Face recognition using kernel methods," in *Advances in Neural Information Processing Systems* (eds S. Becker, T. G. Ditterich, and Z. Ghahramani), 2002, pp. 215–220.

[15] J. F. Cohn, A. J. Zlochower, J. J. Lien, and T. Kanade, "Feature-point tracking by optical flow discriminates subtle differences in facial expression," in Proceedings of the Third International Conference on Automatic Face and Gesture Recognition, Nara, Japan, April 14–16, 1998, pp. 396–401.

[16] I. Essa and A. Pentland, "Coding, analysis interpretation, recognition of facial expressions," *IEEE Transactions on Pattern Analysis and Machine Intelligence*, 19(7): 757–763, 1997.

[17] S. Kimura and M. Yachida, "Facial expression recognition and its degree estimation," in Proceedings of IEEE Computer Society Conference on Computer Vision and Pattern Recognition, San Juan, Puerto Rico, June 17–19, 1997, pp. 295–300.

[18] P. Lucey, J. F. Cohn, I. Matthews, S. Lucey, S. Sridharan, J. Howlett, and K. M. Prkachin, "Automatically detecting pain in video through facial action units," *IEEE Transactions on Systems, Man and Cybernetics, Part B (Cybernetics)*, 41(3): 664–674, 2011.

[19] P. Ekman, "The argument and evidence about universals in facial expressions of emotion," in *Handbook of Social Psychophysiology* (eds H. Wagner and A. Manstead), John Wiley & Sons, 1989, pp. 143–164.

[20] Y. Tian, T. Kanade, and J. Cohn, "Recognizing action units for facial expression analysis," *IEEE Transactions on Pattern Analysis and Machine Intelligence*, 23(2): 97–115, 2001.

[21] D. J. Kim and Z. Bien, "Design of 'personalized' classifier using soft computing techniques for 'personalized' facial expression recognition," *IEEE Transactions on Fuzzy Systems*, 16(4): 874–885, 2008.

[22] Y. Huang, Y. Li, and N. Fan, "Robust symbolic dual-view facial expression recognition with skin wrinkles: local versus global approach," *IEEE Transactions on Multimedia*, 12(6): 536–543, 2010.

[23] H. Kobayashi and F. Hara, "The recognition of basic facial expressions by neural network," *Trans. Soc. Instrum. Contr. Eng.*, 29(1): 112–118, 1993.

[24] K. Anderson and P. W. McOwan, "A real-time automated system for the recognition of human facial expressions," *IEEE Transactions on Systems, Man and Cybernetics, Part B (Cybernetics)*, 36(1): 96–105, 2006.

[25] T. Otsuka and J. Ohya, "Spotting segments displaying facial expression from image sequences using HMM," in Proceedings of the Third IEEE International Conference on Automatic Face and Gesture Recognition, Nara, Japan, April 14–16, 1998, pp. 442–447.

[26] N. Cristianini and J. Shawe-Taylor, *An Introduction to Support Vector Machines and Other Kernel-Based Learning Methods*, Cambridge University Press, 2000.

[27] P. Ekman, J. Hager, C. H. Methvin, and W. Irwin, *Ekman–Hager Facial Action Exemplars*, Human Interaction Laboratory, University of California, San Francisco, CA.

[28] P. Ekman and W. F. Friesen, "The repertoire of nonverbal behavioral categories—origins, usage, and coding," *Semiotica*, 1: 49–98, 1969.

[29] P. Ekman, "Universals and cultural differences in facial expressions of emotion," in *Nebraska Symposium on Motivation* (ed J. Cole), University of Nebraska Press, 1972, pp. 207–283.

[30] P. Ekman and W. Friesen, *Facial Action Coding System*, Consulting Psychologists Press, Palo Alto, CA, 1978.

[31] P. Ekman and E. L. Rosenberg, *What the Face Reveals: Basic and Applied Studies of Spontaneous Expression Using the Facial Action Coding System*, Oxford University Press, Oxford, 1997.

[32] P. Ekman, *Face of Man: Universal Expression in a New Guinea Village*, Garland, New York, 1982.

[33] J. Cohn and P. Ekman, "Measuring facial action by manual coding, facial EMG, and automatic facial image analysis," in *Handbook of Nonverbal Behavior Research Methods in the Affective Sciences* (eds J. A. Harrigan, R. Rosenthal, and K. Scherer), Oxford University Press, New York, 2005, pp. 9–64.

[34] P. Ekman, T. S. Huang, T. J. Sejnowski, and J. C. Hager (eds), *Final Report to NSF of the Planning Workshop on Facial Expression Understanding*, Human Interaction Laboratory, University of California, San Francisco, 1993.

[35] M. S. Bartlett, J. C. Hager, P. Ekman, and T. J. Sejnowski, "Measuring facial expressions by computer image analysis," *Psychophysiology*, 36: 253–263, 1999.

[36] G. Donato, M. S. Bartlett, J. C. Hager, P. Ekman, and T. J. Sejnowski, "Classifying facial actions," *IEEE Transactions on Pattern Analysis and Machine Intelligence*, 21: 974–989, 1999.

[37] P. Ekman, W. V. Friesen, and J. C. Hager, *Facial Action Coding System*, The Human Face, Salt Lake City, UT, 2002.

[38] P. Ekman, "Darwin, deception, and facial expression," *Ann. New York Acad. Sci.*, 1000: 105–221, 2003.

[39] M. Pantic and L. J. M. Rothkrantz, "Automatic analysis of facial expressions: the state of the art," *IEEE Transactions on Pattern Analysis and Machine Intelligence*, 22(12): 1424–1445, 2000.

[40] M. Pantic and L. J. M. Rothkrantz, "Toward an affect-sensitive multimodal human–computer interaction," *Proceedings of the IEEE*, 91(9): 1370–1390, 2003.

[41] M. Pantic and L. Rothkrantz, "Facial action recognition for facial expression analysis from static face images," *IEEE Transactions on Systems, Man and Cybernetics, Part B (Cybernetics)*, 34(3): 1449–1461, 2004.

[42] M. Pantic and I. Patras, "Dynamics of facial expressions—recognition of facial actions and their temporal segments from face profile image sequences," *IEEE Transactions on Systems, Man and Cybernetics, Part B (Cybernetics)*, 36(2): 433–449, 2006.

[43] M. F. Valstar and M. Pantic, "Fully automatic recognition of the temporal phases of facial actions," *IEEE Transactions on Systems, Man and Cybernetics, Part B (Cybernetics)*, 42(1): 28–43, 2012.

[44] M. F. Valstar, M. Pantic, Z. Ambadar, and J. F. Cohn, "Spontaneous vs. posed facial behavior: automatic analysis of brow actions," in Proceedings of the ACM International Conference on Multimodal Interfaces, Banff, Canada, November 2–4, 2006, pp. 162–170.

[45] M. F. Valstar, M. Pantic, and H. Gunes, "A multimodal approach to automatic recognition of posed vs. spontaneous smiles," in Proceedings of the ACM International Conference on Multimodal Interfaces, Nagoya, Japan, November 12–15, 2007, pp. 38–45.

[46] M. Pantic and M. Bartlett, "Machine analysis of facial expressions," in *Face Recognition* (eds K. Delac and M. Grgic), I-Tech Education and Publishing, Vienna, Austria, 2007, pp. 377–416.

[47] M. K. Pitt and N. Shephard, "Filtering via simulation: auxiliary particle filters," *J. Am. Stat. Assoc.*, 94(446): 590–599, 1999.

[48] M. Pantic, M. F. Valstar, R. Rademaker, and L. Maat, "Web-based database for facial expression analysis," in Proceedings of the IEEE International Conference on Multimedia and Expo, Amsterdam, The Netherlands, July 6, 2005, pp. 317–321.

[49] M. F. Valstar and M. Pantic, "Induced disgust, happiness and surprise: an addition to the MMI facial expression database," in Proceedings of the International Conference on

Language Resources and Evaluation, Workshop on EMOTION, Valletta, Malta, May 17–23, 2010, pp. 317–321.

[50] I. Patras and M. Pantic, "Tracking deformable motion," in Proceedings of IEEE International Conference on Systems, Man and Cybernetics, Waikoloa, Hawaii, October 10–12, 2005, pp. 1066–1071.

[51] I. Patras and M. Pantic, "Particle filtering with factorized likelihoods for tracking facial features," in Proceedings of IEEE International Conference on Automatic Face and Gesture Recognition, Seoul, South Korea, May 17–19, 2004, pp. 97–102.

[52] Y. L. Tian, T. Kanade, and J. F. Cohn, *Handbook of Face Recognition*, Springer, New York, 2005.

[53] J. Lien, T. Kanade, J. Cohn, and C. Li, "Subtly different facial expression recognition and expression intensity estimation," in Proceedings of IEEE Conference on Computer Vision and Pattern Recognition, Santa Barbara, CA, June 23–25, 1998, pp. 853–859.

[54] T. Kanade, J. Cohn, and Y. Tian, "Comprehensive database for facial expression analysis," in Proceedings of IEEE International Conference on Automatic Face and Gesture Recognition, Grenoble, France, March 26–30, 2000, pp. 46–53.

[55] J. F. Cohn and K. L. Schmidt, "The timing of facial motion in posed and spontaneous smiles," *J. Wavelets, Multi-Resolution Inf. Process.*, 2(2): 121–132, 2004.

[56] J. F. Cohn, J. F. Reed, Z. Ambadar, J. Xiao, and T. Moriyama, "Automatic analysis and recognition of brow actions in spontaneous facial behavior," in Proceedings of the IEEE International Conference on Systems, Man, and Cybernetics, Vol. 1, 2004, pp. 610–616.

[57] J. F. Cohn, "Foundations of human computing: facial expression and emotion," in Proceedings of the ACM International Conference on Multimodal Interfaces, Vol. 1, 2006, pp. 610–616.

[58] A. Chakraborty and A. Konar, *Emotional Intelligence: A Cybernetic Approach*, Springer, 2009.

[59] A. Chakraborty, A. Konar, and A. Chatterjee, "Controlling human moods by fuzzy logic," *Int. J. Math Sci.*, 6(3–4): 401–416, 2007.

[60] A. Chakraborty, A. Konar, U. Chakraborty, and A. Chatterjee, "Emotion recognition from facial expressions and its control using fuzzy logic," *IEEE Transactions on Systems, Man and Cybernetics, Part A (Systems and Humans)*, 39(4): 726–743, 2009.

[61] A. Halder, A. Konar, R. Mandal, P. Bhowmik, A. Chakraborty, and N. R. Pal, "General and interval type-2 fuzzy face-space approach to emotion recognition," in IEEE Transactions on Systems, Man and Cybernetics, Part A (Systems and Humans), 2011.

[62] A. Halder, R. Mandal, A. Konar, and L. C. Jain, "Fuzzy based hierarchical algorithm for template matching in emotional facial images," *J. Intell. Fuzzy Syst.*, 2012.

[63] B. Biswas, A. Konar, and A. K. Mukherjee, "Image matching with fuzzy moment descriptors," *Eng. Appl. Artif. Intell.*, 14: 43–49, 2001.

[64] A. Halder, P. Rakshit, S. Chakraborty, A. Konar, E. Kim, and A. K. Nagar, "Reducing uncertainty in interval type-2 fuzzy sets for qualitative improvement in emotion recognition from facial expressions," in Proceedings of the IEEE International Conference on Fuzzy Systems, Brisbane, Australia, June 10–15, 2012.

[65] A. Halder, R. Mandal, A. Chakraborty, A. Konar, and R. Janarthanan, "Application of general type-2 fuzzy set in emotion recognition from facial expression," in Joint International Conference on Fuzzy and Neural Network (FANCCO), Vishakhapatnam, India, 2011.

[66] A. Halder, A. Jati, G. Singh, A. Konar, A. Chakraborty, and R. Janarthanan, "Facial action point based emotion recognition by principal component analysis," in SocPros, Indian Institute of Technology, Roorkey, India, 2011.

[67] A. Konar, A. Chakraborty, A. Halder, R. Mandal, and R. Janarthanan, "Interval type-2 fuzzy model for emotion recognition from facial expression," in PerMIn, CDAC, Kolkata, India, 2012.

[68] A. Chakraborty, P. Bhowmik, S. Das, A. Halder, A. Konar, and A. K. Nagar, "Correlation between stimulated emotion extracted from EEG and its manifestation on facial expression," in IEEE Systems, Man and Cybernetics Conference, San Antonio, TX, 2009.

[69] S. Das, A. Halder, P. Bhowmik, A. Chakraborty, A. Konar, and A. K. Nagar, "Voice and facial expression based classification of emotion using linear support vector machine," in International Conference on Developments in eSystems Engineering (DESE), Abu Dhabi, UAE, 2009.

[70] T. Cootes, G. Edwards, and C. Taylor, "Active appearance models," *IEEE Transactions on Pattern Analysis and Machine Intelligence*, 23(6): 681–685, 2001.

[71] I. Kotsia and I. Pitas, "Facial expression recognition in image sequences using geometric deformation features and support vector machines," *IEEE Transactions on Image Processing*, 16(1): 172–187, 2007.

[72] A. Jain and D. Zongker, "Feature selection: evaluation, application, and small sample performance," *IEEE Transactions on Pattern Analysis and Machine Intelligence*, 19(2): 153–158, 1997.

[73] P. Yang, Q. Liu, and D. N. Metaxas, "Boosting encoded dynamic features for facial expression recognition," *Pattern Recogn. Lett.*, 30(2): 132–139, 2009.

[74] J. Whitehill and C. Omlin, "Haar features for FACS AU recognition," in Proceedings of the Seventh IEEE International Conference on Automatic Face and Gesture Recognition, Southampton, UK, April 10–12, 2006, pp. 97–101.

[75] Y. Stylianou, "Modeling of speech signal for analysis purposes or mathematical modeling of jitter and shimmer", in eTERFACE, 2008.

[76] S. H. Chen and Y. R. Luo, "Speaker verification using MFCC and support vector machine," in International Multi Conference of Engineers and Computer Scientists (IMECS), Vol. 1, Hong Kong, 2009.

[77] K. R. Aida-Zade, C. Ardil, and S. S. Rustamov, "Investigation of combined use of MFCC and LPC features in speech recognition systems," *Int. J. Comput. Inf. Syst. Control Eng.*, 2(7): 6–12, 2008.

[78] C. Busso, S. Lee, and S. Narayanan, "Analysis of emotionally salient aspects of fundamental frequency for emotion detection," *IEEE Transactions on Speech and Audio Processing*, 17(4): 582–596, 2009.

[79] C.-M. Lee and S. S. Narayanan, "Toward detecting emotions in spoken dialogs," *IEEE Transactions on Speech and Audio Processing*, 13(2): 293–303, 2005.

[80] C. H. Wu and W. B. Liang, "Emotion recognition of affective speech based on multiple classifiers using acoustic–prosodic information and semantic labels," *IEEE Transactions on Affective Computing*, 2(1): 10–21, 2011.

[81] E. H. Kim, K. H. Hyun, S. H. Kim, and Y. K. Kwak, "Improved emotion recognition with a novel speaker-independent feature," *IEEE/ASME Transactions on Mechatronics*, 14(3): 317–325, 2009.

[82] E. Mower, M. J. Mataric, and S. Narayanan, "A framework for automatic human emotion classification using emotion profiles," *IEEE Transactions on Speech and Audio Processing*, 19(5): 1057–1070, 2011.

[83] E. Mower, S. Lee, M. J. Mataric, and S. Narayanan, "Human perception of synthetic character emotions in the presence of conflicting and congruent vocal and facial expressions," in Proceedings of the IEEE International Conference on Acoustics, Speech, and Signal Processing (ICASSP), Las Vegas, NV, March 30–April 4, 2008, pp. 2201–2204.

[84] E. Mower, S. Lee, M. J. Mataric, and S. Narayanan, "Joint-processing of audio–visual signals in human perception of conflicting synthetic character emotions," in Proceedings of the IEEE International Conference on Multimedia and Expo (ICME), Hannover, Germany, June 23–26, 2008, pp. 961–964.

[85] E. Mower, M. J. Mataric, and S. Narayanan, "Selection of emotionally salient audio–visual features for modeling human evaluations of synthetic character emotion displays," in Proceedings of IEEE International Symposium on Multimedia (ISM), Berkeley, CA, December 15–17, 2008.

[86] E. Mower, M. J. Mataric, and S. Narayanan, "Human perception of audio–visual synthetic character emotion expression in the presence of ambiguous and conflicting information," *IEEE Transactions on Multimedia*, 11(5): 843–855, 2009.

[87] C. Busso and S. S. Narayanan, "Interrelation between speech and facial gestures in emotional utterances: a single subject study," in IEEE Transactions on Speech and Audio Processing, Vol. 15, No. 8, 2007, pp. 1–16.

[88] S. Ananthakrishnan and S. Narayanan, "Automatic prosody labeling using acoustic, lexical, and syntactic evidence," *IEEE Transactions on Speech and Audio Language Processing*, 16(1): 216–228, 2008.

[89] C. Busso, S. Lee, and S. Narayanan, "Using neutral speech models for emotional speech analysis," in Proceedings of Interspeech'07 Eurospeech, Antwerp, Belgium, 2007, pp. 2225–2228.

[90] C. Busso, Z. Deng, S. Yildirim, M. Bulut, C. M. Lee, A. Kazemzadeh, S. Lee, U. Neumann, and S. Narayanan, "Analysis of emotion recognition using facial expressions speech and multimodal information," in International Conference on Multimodal Interfaces, State College, PA, 2004, pp. 205–211.

[91] C. Busso and S. Narayanan, "Interplay between linguistic and affective goals in facial expression during emotional utterances," in Proceedings of the Seventh International Seminar Speech Production (ISSP'06), Ubatuba-SP, Brazil, 2006, pp. 549–556.

[92] C. Busso and S. Narayanan, "Joint analysis of the emotional fingerprint in the face and speech: a single subject study," in Proceedings of the IEEE International Workshop on Multimedia Signal Process (MMSP 2007), Chania, Crete, Greece, 2007, pp. 43–47.

[93] C. Busso and S. Narayanan, "Recording audio–visual emotional databases from actors: a closer look," in Proceedings of the Second International Workshop Emotion: Corpora for Research on Emotion and Affect, International Conference Language Resources and Evaluation (LREC 2008), Marrakech, Morocco, 2008, pp. 17–22.

[94] M. Grimm, K. Kroschel, E. Mower, and S. Narayanan, "Primitives-based evaluation and estimation of emotions in speech," *Speech Commun.*, 49: 787–800, 2007.

[95] C. H. Wu and G. L. Yan, "Speech act modeling and verification of spontaneous speech with disfluency in a spoken dialogue system," *IEEE Transactions on Speech and Audio Processing*, 13(3): 330–344, 2005.

[96] J. C. Lin, C. H. Wu, and W. L. Wei, "Error weighted semi-coupled hidden Markov model for audio–visual emotion recognition," *IEEE Transactions on Multimedia*, 14(1): 142–156, 2012.

[97] C. H. Wu, Z. J. Chuang, and Y. C. Lin, "Emotion recognition from text using semantic label and separable mixture model," *ACM Transactions on Asian Language Information Processing*, 5(2): 165–182, 2006.

[98] C.-H. Wu and Z.-J. Chuang, "Emotion recognition from speech using IG-based feature compensation," *Int. J. Comput. Linguist. Chin. Lang. Process.*, 12(1): 65–78, 2007.

[99] B. Schuller, "Recognizing affect from linguistic information in 3D continuous space," *IEEE Transactions on Affective Computing*, 2(4): 192–205, 2011.

[100] B. Schuller, B. Vlasenko, F. Eyben, M. Wollmer, A. Stuhlsatz, A. Wendemuth, and G. Rigoll, "Cross-corpus acoustic emotion recognition: variances and strategies," *IEEE Transactions on Affective Computing*, 1(2): 119–131, 2010.

[101] B. Schuller, G. Rigoll, and M. Lang, "Speech emotion recognition combining acoustic features and linguistic information in a hybrid support vector machine-belief network architecture," in Proceedings of the IEEE International Conference on Acoustics, Speech, and Signal Processing, 2004, pp. 17–21.

[102] B. Schuller, C. Hage, D. Schuller, and G. Rigoll, "Mister D.J., Cheer Me Up!': musical and textual features for automatic mood classification," *J. New Music Res.*, 39(1): 13–34, 2010.

[103] M. Schröder, R. Cowie, D. Heylen, M. Pantic, C. Pelachaud, and B. Schuller, "Towards responsive sensitive artificial listeners," in Proceedings of the Fourth International Workshop on Human-Computer Conversation, 2008.

[104] B. Schuller, S. Steidl, and A. Batliner, "The INTERSPEECH 2009 emotion challenge," in Proceedings of the International Speech Communication Association, 2009, pp. 312–315.

[105] B. Schuller, R. Müller, M. Lang, and G. Rigoll, "Speaker independent emotion recognition by early fusion of acoustic and linguistic features within ensembles," in Proceedings of the International Conference on Speech Communication Association, 2005, pp. 805–808.

[106] A. Batliner, B. Schuller, S. Schaeffler, and S. Steidl, "Mothers, adults, children, pets-towards the acoustics of intimacy," in Proceedings of the IEEE International Conference on Acoustics, Speech, and Signal Processing, 2008, pp. 4497–4500.

[107] B. Schuller, "Speaker, noise, and acoustic space adaptation for emotion recognition in the automotive environment," in 2008 ITG Conference on Voice Communication (SprachKommunikation), 2008, pp. 1–4.

[108] B. Schuller, G. Rigoll, S. Can, and H. Feussner, "Emotion sensitive speech control for human–robot interaction in minimal invasive surgery," in Proceedings of the 17th IEEE International Symposium on Robot and Human Interactive Communication, 2008, pp. 453–458.

[109] B. Schuller, B. Vlasenko, D. Arsic, G. Rigoll, and A. Wendemuth, "Combining speech recognition and acoustic word emotion models for robust text-independent emotion recognition," in Proceedings of the International Conference on Multimedia and Expo, 2008.

[110] B. Vlasenko, B. Schuller, A. Wendemuth, and G. Rigoll, "On the influence of phonetic content variation for acoustic emotion recognition," in Proceedings of the Fourth IEEE

Tutorial and Research Workshop on Perception and Interactive Technologies for Speech-Based Systems, 2008.

[111] B. Schuller, B. Vlasenko, R. Minguez, G. Rigoll, and A. Wendemuth, "Comparing one and two-stage acoustic modeling in the recognition of emotion in speech," in Proceedings of the IEEE Workshop Automatic Speech Recognition and Understanding, 2007, pp. 596–600.

[112] B. Vlasenko, B. Schuller, A. Wendemuth, and G. Rigoll, "Combining frame and turn-level information for robust recognition of emotions within wpeech," in Proceedings of the INTERSPEECH, 2007, pp. 2249–2252.

[113] B. Vlasenko, B. Schuller, A. Wendemuth, and G. Rigoll, "Frame vs. turn-level: emotion recognition from speech considering static and dynamic processing," in Proceedings of the International Conference on Affective Computing and Intelligent Interaction (ed. A. Paiva), 2007, pp. 139–147.

[114] A. Batliner, S. Steidl, B. Schuller, D. Seppi, K. Laskowski, T. Vogt, L. Devillers, L. Vidrascu, N. Amir, L. Kessous, and V. Aharonson, "Combining efforts for improving automatic classification of emotional user states," in Proceedings of the First International Conference on Language Technologies, 2006, pp. 240–245.

[115] M. Wöllmer, F. Eyben, S. Reiter, B. Schuller, C. Cox, E. Douglas-Cowie, and R. Cowie, "Abandoning emotion classes—towards continuous emotion recognition with modelling of long-range dependencies," in Proceedings of the Ninth Annual Conference of the International Speech Communication Association, 2008, pp. 597–600.

[116] G. Rigoll, R. Müller, and B. Schuller, "Speech emotion recognition exploiting acoustic and linguistic information sources," in Proceedings of the 10th International Conference on Speech and Computer, Vol. 1, 2005, pp. 61–67.

[117] E. H. Kim, K. H. Hyun, S. H. Kim and Y. K. Kwak, "Improved emotion recognition with a novel speaker-independent feature," *IEEE/ASME Transactions on Mechatronics*, 14(3): 317–325, 2009.

[118] L. Siegel, "A procedure for using pattern classification techniques to obtain a voiced/unvoiced classifier," *IEEE Transactions on Acoustics, Speech and Signal Processing*, ASSP 27: 83–88, 1979.

[119] S. Ntalampiras and N. Fakotakis, "Modeling the temporal evolution of acoustic parameters for speech emotion recognition," *IEEE Transactions on Affective Computing*, 3(1): 116–125, 2012.

[120] T. Tolonen and M. Karjalainen, "A computationally efficient multipitch analysis model," *IEEE Transactions on Speech and Audio Processing*, 8(6): 708–716, 2000.

[121] D. Ververidis, C. Kotropoulos, and I. Pitas, "Automatic emotional speech classification," in Proceedings of the 29th International Conference on Acoustics, Speech, and Signal Processing (ICASSP' 04), 2004, pp. 593–596.

[122] D. Ververidis and C. Kotropoulos, "Emotional speech recognition: resources, features, and methods," *Speech Commun.*, 48(9): 1162–1181, 2006.

[123] L. R. Rabiner, "A tutorial on hidden Markov models and selected applications in speech recognition," *Proceedings of the IEEE*, 77(2): 257–286, 1989.

[124] R. Cowie and R. Cornelius, "Describing the emotional states that are expressed in speech," *Speech Commun.*, 40: 5–23, 2003.

[125] E. N. Luengo, I. Hernáez, and J. Sánchez, "Automatic emotion recognition using prosodic parameters," in Proceedings of the INTERSPEECH, 2005, pp. 493–496.

[126] S. G. Kooladugi, N. Kumar, and K. S. Rao, "Speech emotion recognition using segmental level prosodic analysis," in Proceedings of the International Conference on Devices and Communications, 2011, pp. 1–5.

[127] Y. Zhou, Y. Sun, J. Zhang, and Y. Yan, "Speech emotion recognition using both spectral and prosodic features," in International Conference on Information Engineering and Computer Science (ICIECS), Wuhan, December 19–20, 2009.

[128] B. Vlasenko and A. Wendemuth, "Tuning hidden Markov model for speech emotion recognition," in Proceedings of the DAGA, March 2007.

[129] D. Datcu and L. J. Rothkrantz, "The recognition of emotions from speech using gentle-boost classifier. A comparison approach," in Proceedings of the International Conference on Computer Systems and Technologies, Vol. 1, 2006, pp. 1–6.

[130] D. Ververidis and C. Kotropoulos, "Automatic speech classification to five emotional states based on gender information," in Proceedings of the EUSIPCO, 2004, pp. 341–344.

[131] H. Meng, J. Pittermann, A. Pittermann, and W. Minker, "Combined speech-emotion recognition for spoken human-computer interfaces," in Proceedings of the IEEE International Conference on Signal Processing and Communications, 2007.

[132] D. Morrison, R. Wang, and L. C. De Silva, "Ensemble methods for spoken emotion recognition in call-centres," *Speech Commun.*, 49(2): 98–112, 2007.

[133] O. W. Kwon, K. Chan, J. Hao, and T. W. Lee, "Emotion recognition by speech signals," in Proceedings of the 8th European Conference on Speech Communication and Technology (EUROSPEECH), 2003.

[134] A. Khasnobish, "Emotion recognition from stimulated EEG signals", M.E. Thesis, May 2012.

[135] G. Singh, A. Jati, A. Khasnobish, S. Bhattacharyya, A. Konar, D. N. Tibarewala, and R. Janarthanan, "Negative emotion recognition from stimulated EEG signals", in Third International Conference on Computing Communication & Networking Technologies (ICCCNT), July 26–28, 2012.

[136] S. Bhattacharyya, A. Sengupta, T. Chakraborty, D. Banerjee, A. Khasnobish, A. Konar, D. N. Tibarewala, and R. Janarthanan, "EEG controlled remote robotic system from motor imagery classification", in Third International Conference on Computing Communication & Networking Technologies (ICCCNT), July 26–28, 2012.

[137] P. C. Petrantonakis and L. J. Hadjileontiadis, "Emotion recognition from brain signals using hybrid adaptive filtering and higher order crossings analysis," *IEEE Transactions on Affective Computing*, 1(2): 81–97, 2010.

[138] P. C. Petrantonakis and L. J. Hadjileontiadis, "A novel emotion elicitation index using frontal brain asymmetry for enhanced EEG-based emotion recognition," *IEEE Transactions on Information Technology in Biomedicine*, 15(5): 737–746, 2011.

[139] Y. P, Lin, C. H. Wang, T. P. Jung, T. L. Wu, S. K. Jeng, J. R. Duann, and J.-H. Chen, "EEG-based emotion recognition in music listening," *IEEE Transactions on Biomedical Engineering*, 57(7): 1798–1806, 2010.

[140] J. J. B. Allen, J. A. Coan, and M. Nazarian, "Issues and assumptions on the road from raw signals to metrics of frontal EEG asymmetry in emotion," *Biol. Psychol.*, 67(1/2): 183–218, 2004.

[141] L. A. Schmidt and L. J. Trainor, "Frontal brain electrical activity (EEG) distinguishes valence and intensity of musical emotions," *Cognit. Emot.*, 15(4): 487–500, 2001.

[142] W. Heller, "Neuropsychological mechanisms of individual differences in emotion, personality and arousal," *Neuropsychology*, 7: 476–489, 1993.

[143] Y. W. Chen and C. J. Lin, "Combining SVMs with various feature selection strategies," in *Feature Extraction, Foundations and Applications*, Springer, New York, 2006.

[144] R. Palaniappan, "Utilizing gamma band to improve mental task based brain-computer interface design," *IEEE Transactions on Neural Systems and Rehabilitation Engineering*, 14(3): 299–303, 2006.

[145] E. Altenmuller, K. Schurmann, V. K. Lim, and D. Parlitz, "Hits to the left, flops to the right: different emotions during listening to music are reflected in cortical lateralisation patterns," *Neuropsychologia*, 40(13): 2242–2256, 2002.

[146] A. Choppin, "EEG-based human interface for disabled individuals: emotion expression with neural networks," Master's Thesis, Tokyo Institute of Technology, 2000.

[147] K. Takahashi, "Remarks on emotion recognition from bio-potential signals," in Proceedings of the Second International Conference on Autonomous Robots and Agents, 2004, pp. 186–191.

[148] G. Chanel, J. Kronegg, D. Grandjean, and T. Pun, "Emotion assessment: arousal evaluation using EEG and peripheral physiological signals," Technical report, University of Geneva, 2005.

[149] D. O. Bos. EEG-based emotion recognition: the influence of visual and auditory stimuli, 2006, http://hmi.ewi.utwente.nl/verslagen/capita-selecta/CS-Oude_Bos-Danny.pdf. Last accessed on Sept. 18, 2014.

[150] Z. Khalili and M. Moradi, "Emotion detection using brain and peripheral signals," in Proceedings of the Biomedical Engineering Conference, 2008, pp. 1–4.

[151] R. Horlings, D. Datcu, and L. J. M. Rothkrantz, "Emotion recognition using brain activity," in Proceedings of the International Conference on Computer Systems and Technologies, 2008, pp. 1–6.

[152] M. Murugappan, M. Rizon, R. Nagarajan, S. Yaacob, I. Zunaidin and D. Hazry, "Lifting scheme for human emotion recognition using EEG," in Proceedings of the International Symposium on Information Technology, 2008, pp. 1–7.

[153] K. Schaaff and T. Schultz, "Towards an EEEG-based emotion recognizer for humanoid robots," in Proceedings of the 18th IEEE International Symposium on Robot and Human Interactive Communication, 2009, pp. 792–796.

[154] S. Mitra and T. Acharya, "Gesture recognition: a survey," *IEEE Transactions on Systems, Man and Cybernetics*, 37(3): 311–324, 2007.

[155] D. Glowinski, N. Dael, A. Camurri, G. Volpe, M. Mortillaro, and K. Scherer, "Toward a minimal representation of affective gestures," *IEEE Transactions on Affective Computing*, 2(2): 106–118, 2011.

[156] A. Camurri, I. Lagerlöf, and G. Volpe, "Recognizing emotion from dance movement: comparison of spectator recognition and automated techniques," *Int. J. Hum. Comput. Stud.*, 59(1–2): 213–225, 2003.

[157] G. Castellano, S. D. Villalba, and A. Camurri, "Recognising human emotions from body movement and gesture dynamics," in *Affective Computing and Intelligent Interaction*, Volume 4738: Lecture Notes in Computer Science (eds A. Paiva, R. Prada, and R. W. Picard), Springer, Berlin, 2007, pp. 71–82.

[158] A. Camurri, B. Mazzarino, and G. Volpe, "Analysis of expressive gesture: the eyesweb expressive gesture processing library," in *Gesture-Based Communication in*

Human–Computer Interaction, Volume 2915: Lecture Notes in Computer Science (eds A. Camurri and G. Volpe), Springer, Heidelberg, 2004, pp. 460–467.

[159] B. Hartmann, M. Mancini, S. Buisine, and C. Pelachaud, "Design and evaluation of expressive gesture synthesis for embodied conversational agents," in Proceedings of the Third International Joint Conference on Autonomous Agents Multi-Agent Systems, 2005, pp. 1095–1096.

[160] A. Ramamoorthy, N. Vaswani, S. Chaudhury, and S. Banerjee, "Recognition of dynamic hand gestures," *Pattern Recogn.*, 36: 2069–2081, 2003.

[161] W. T. Freeman and M. Roth, "Orientation histograms for hand gesture recognition," in Proceedings of the IEEE International Workshop on Automatic Face and Gesture Recognition, Zurich, Switzerland, June 1995, pp. 296–301.

[162] H. S. Yoon, J. Soh, Y. J. Bae, and H. S. Yang, "Hand gesture recognition using combined features of location, angle and velocity," *Pattern Recogn.*, 34: 1491–1501, 2001.

[163] J. Daugman, "Face and gesture recognition: an overview," *IEEE Transactions on Pattern Analysis and Machine Intelligence*, 19(7): 675–676, 1997.

[164] M. De Meijer, "The contribution of general features of body movement to the attribution of emotions," *J. Nonverbal Behav.*, 13: 247–268, 1989.

[165] H. G. Wallbott, "Bodily expression of emotion," *Eur. J. Soc. Psychol.*, 28: 879–896, 1998.

[166] N. Bianchi-Berthouze and A. Kleinsmith, "A categorical approach to affective gesture recognition," *Connect. Sci.*, 15: 259–269, 2003.

[167] H. Gunes and M. Piccardi, "Bi-modal emotion recognition from expressive face and body gestures," *J. Netw. Comput. Appl.*, 30(4): 1334–1345, 2007.

[168] H. Gunes and M. Piccardi, "Creating and annotating affect data bases from face and body display: a contemporary survey," in Proceedings of the IEEE International Conference on Systems, Man, and Cybernetics, 2006, pp. 2426–2433.

[169] H. Gunes and M. Piccardi, "Observer annotation of affective display and evaluation of expressivity: face vs. face-and-body," in Proceedings of the HCSNet Workshop on Use Vision Human–Computer Interaction, 2006, pp. 35–42.

[170] Z. Zeng, J. Tu, M. Liu, T. S. Huang, B. Pianfetti, D. Roth, and S. Levinson, "Audio-visual affect recognition," *IEEE Transactions on Multimedia*, 9(2): 424–428, 2007.

[171] E. Mower, M. J. Matarić, and S. Narayanan, "A framework for automatic human emotion classification using emotion profiles," *IEEE Transactions on Speech and Audio Processing*, 19(5): 1057–1070, 2011.

[172] C. Busso, Z. Deng, S. Yildirim, M. Bulut, C. M. Lee, A. Kazemzadeh, S. Lee, U. Neumann, and S. Narayanan, "Analysis of emotion recognition using facial expressions, speech and multimodal information," in *International Conference on Multimodal Interfaces*, State College, PA, October 2004.

[173] H. Gunes and M. Piccardi, "Automatic temporal segment detection and affect recognition from face and body display," *IEEE Transactions on Systems, Man and Cybernetics, Part B (Cybernetics)*, 39(1): 64–84, 2009.

[174] M. A. Nicolaou, H. Gunes, and M. Pantic, "Continuous prediction of spontaneous affect from multiple cues and modalities in valence-arousal space," *IEEE Transactions on Affective Computing*, 2(2): 92–105, 2011.

[175] G. Castellano, L. Kessous, and G. Caridakis, "Emotion recognition through multiple modalities: face, body gesture, speech," in *Affect and Emotion in Human–Computer*

Interaction, Volume 4868: Lecture Notes in Computer Science (eds C. Peter and R. Beale), Springer-Verlag, Berlin, 2008, pp. 92–103.

[176] A. Chakraborty, P. Bhowmik, S. Das, A. Halder, A. Konar, and A. K. Nagar, "Correlation between stimulated emotion extracted from EEG and its manifestation on facial expression," in Proceedings of the IEEE International Conference on Systems, Man, and Cybernetics, San Antonio, TX, October 2009.

[177] R. W. Picard, E. Vyzas, and J. Healey, "Toward machine emotional intelligence: analysis of affective physiological state," *IEEE Transactions on Pattern Analysis and Machine Intelligence*, 23(10): 1175–1191, 2001.

[178] M. Soleymani, J. Lichtenauer, T. Pun, and M. Pantic, "A multimodal database for affect recognition and implicit tagging," *IEEE Transactions on Affective Computing*, 3(1): 2012.

[179] K. Takahashi, "Remarks on emotion recognition from bio-potential signals," in 2nd International Conference on Autonomous Robots and Agents, Palmerston North, New Zealand, December 2004.

[180] J. Russell and J. Fernandez-Dols, *The Psychology of Facial Expression*, Cambridge University Press, New York, 1997.

[181] M. Cohen, *Perspectives on the Face*, Oxford University Press, Oxford, 2006.

[182] P. A. Devijver and J. Kittler, *Pattern Recognition: A Statistical Approach*, Prentice-Hall, 1982.

[183] R. S. Lazarus and B. N. Lazarus, *Passion and Reason: Making Sense of Our Emotions*, Oxford University Press, 1996.

[184] S. Li and A. Jain, *Handbook of Face Recognition*, Springer, New York, 2005.

[185] J. Pittermann, A. Pittermann, and W. Minker, *Handling Emotions in Human–Computer Dialogues*, Springer, 2008.

[186] A. Young, *Face and Mind*, Oxford University Press, Oxford, 1998.

[187] V. Vapnik, *Statistical Learning Theory*, John Wiley & Sons, New York, 1998.

[188] H. Gunes and M. Piccardi, "A bimodal face and body gesture database for automatic analysis of human nonverbal affective behavior," in Proceedings of the 18th International Conference on Pattern Recognition (ICPR'06), 2006.

[189] T. Sucontphunt, X. Yuan, Q. Li, and Z. Deng, "A novel visualization system for expressive facial motion data exploration," in IEEE Conference on Visualization Symposium, PacificVIS '08, 2008, pp. 103–109.

[190] Y. H. Yang, Y. C. Lin, Y. F. Su, and H. H. Chen, "A regression approach to music emotion recognition," *IEEE Transactions on Speech and Audio Processing*, 16(2): 448–457, 2008.

[191] G. J. Edwards, T. F. Cootes, and C. J. Taylor, "Face recognition using active appearance models," in Proceedings of the European Conference on Computer Vision, Vol. 2, 1998, pp. 581–695.

[192] C. L. Huang and Y. M. Huang, "Facial expression recognition using model-based feature extraction and action parameters classification," *J Vis. Commun. Image Represent.*, 8(3): 278–290, 1997.

[193] M. J. Lyons, J. Budynek, and S. Akamatsu, "Automatic classification of single facial images," *IEEE Transactions on Pattern Analysis and Machine Intelligence*, 21(12): 1357–1362, 1999.

[194] W. Sato, T. Kochiyama, S. Yoshikawa, and M. Matsumura, "Emotional expression boosts early visual processing of the face: ERP recording and its decomposition by independent component analysis," *Cogn. Neurosci. Neuropsychol.*, 12(4): 709–714, 2001.

[195] M. S. Bartlett and T. J. Sejnowski, "Independent components of face images: a representation for face recognition," in Proceedings of the 4th Annual Joint Symposium on Neural Computation, 1997.

[196] C. F. Chuang and F. Y. Shih, "Recognizing facial action units using independent component analysis and support vector machine," *Pattern Recogn.*, 39(9): 1795–1798, 2006.

[197] M. S. Bartlett and T. J. Sejnowski, "Viewpoint invariant face recognition using independent component analysis and attractor networks," *Neural Inf. Process. Syst. Nat. Synth.*, 9: 817–823, 1997.

[198] M. S. Bartlett, J. R. Movellan, and T. J. Sejnowski, "Face recognition by independent component analysis," *IEEE Transactions on Neural Networks*, 13(6): 1450–1464, 2002.

[199] J. H. Holland, "Outline for a logical theory of adaptive systems," *J. ACM*, 9(3): 297–314, 1962.

[200] J. H. Holland, "Genetic algorithms and the optimal allocation of trials," *SIAM J. Comput.*, 2(2): 88–105, 1973.

[201] J. H. Holland, "Erratum: genetic algorithms and the optimal allocation of trials," *SIAM J. Comput.*, 3(4): 326, 1974.

[202] R. A. Fisher, *The Genetical Theory of Natural Selection*, Clarendon Press, Oxford, 1930.

[203] R. Storn, "System design by constraint adaptation and differential evolution," *IEEE Transactions on Evolutionary Computation*, 3(1): 22–34, 1999.

[204] H. Kobayashi and F. Hara, "Facial interaction between animated 3D face robot and human beings," in Proceedings of the International Conference on Systems, Man, and Cybernetics, 1997, pp. 3732–3737.

[205] C. Padgett and G.W. Cottrell, "Representing face images forn emotion classification," in Proceedings of the Conference on Advances in Neural Information Processing Systems, 1996, pp. 894–900.

[206] Z. Zhang, M. Lyons, M. Schuster, and S. Akamatsu, "Comparison between geometry-based and gabor wavelets-based facial expression recognition using multi-layer perceptron," in Proceedings of the International Conference on Automatic Face and Gesture Recognition, 1998, pp. 454–459.

[207] H. Kobayashi and F. Hara, "Recognition of six basic facial expressions and their strength by neural network," in IEEE International Workshop on Robot and Human Communication, 1992, pp. 381–386.

[208] J. Zhao and G. Kearney, "Classifying facial emotions by neural networks with fuzzy inputs," in Proceedings of the Conference on Neural Information Processing, Vol. 1, 1996, pp. 454–457.

[209] M. Rosenblum, Y. Yacoob, and L. Davis, "Human expression recognition from motion using a radial basis function network architecture," *IEEE Transactions on Neural Networks*, 7(5): 1121–1138, 1996.

[210] L. W. Chew, K. P. Seng, L. M. Ang, V. Ramakonar, and A. Gnanasegaran, "Audio-emotion recognition system using parallel classifiers and audio feature analyzer," in Third International Conference on Computational Intelligence, Modelling and Simulation (CIMSiM), 2011, pp. 210–215.

[211] S. Zhang, X. Zhao, and B. Lei, "Spoken emotion recognition using radial basis function neural network," *Adv. Comput. Sci., Environ., Ecoinf., Edu.*, 214: 437–442, 2011.

[212] B. Takacs and H. Wechsler, "Locating facial features using SOFM," in Proceedings of the 12th IAPR International Conference on Pattern Recognition, Vol. 2, Conference B: Computer Vision and Image Processing, 1994.

[213] J. He and H. Leich, "Speech trajectory recognition in SOFM by using Bayes theorem," in International Symposium on Speech, Image Processing and Neural Networks, 1994, pp. 109–112.

[214] R. Khosrowabadi, H. C. Quek, A. Wahab, and K. K. Ang, "EEG-based emotion recognition using self-organizing map for boundary detection," in 20th International Conference on Pattern Recognition (ICPR), 2010, pp. 4242–4245.

[215] G. Caridakis, K. Karpouzis, C. Pateritsas, A. Drosopoulos, A. Stafylopatis, and S. Kollias, "Hand trajectory based gesture recognition using self-organizing feature maps and markov models," in IEEE International Conference on Multimedia and Expo, 2008, pp. 1105–1008.

[216] T. Tabatabaei, S. Krishnan, and A. Guergachi, "Emotion recognition using novel speech signal features," in Proceedings of the IEEE International Symposium on Circuits and Systems, 2007, pp. 345–348.

[217] C. Ye, J. Liu, C. Chen, M. Song, and J. Bu, "Speech emotion classification on a riemannian manifold," in Advances in Multimedia Information Processing - PCM, 2008, pp. 61–69.

[218] O. W. Kwon, K. Chan, J. Hao, and T. Lee, "Emotion recognition by speech signals," in Proceedings of the 8th European Conference on Speech Communication and Technology, 2003, pp. 125–128.

[219] B. Schuller, R. Mller, M. Lang, and G. Rigoll, "Speaker independent emotion recognition by early fusion of acoustic and linguistic features within ensembles," in Proceedings of the Interspeech, 2005, pp. 805–809.

[220] S. Bashyal and G. K. Venayagamoorthy, "Recognition of facial expressions using Gabor wavelets and learning vector quantization," *Eng. Appl. Artif. Intell.*, 1–9, 2007.

[221] C. Y. Chang, J. S. Tsai, C. J. Wang, and P. C. Chung, "Emotion recognition with consideration of facial expression and physiological signals," in IEEE Symposium on Computational Intelligence in Bioinformatics and Computational Biology, 2009, pp. 278–283.

[222] J. Nicholson, K. Takahashi, and R. Nakatsu, "Emotion recognition in speech using neural networks," in Proceedings of the 6th International Conference on Neural Information Processing (ICONIP '99), Vol. 2, 1999, pp. 495–501.

[223] J. M. Mendel and D. Wu, *Perceptual Computing*, IEEE Press and Wiley Publications, 2010.

[224] N. Amir and S. Ron, "Towards an automatic classification of emotions in speech," in Proceedings of the ICSLP, 1998, pp. 555–558.

[225] M. Grimm, E. Mower, K. Kroschel, and S. Narayanan, "Combining categorical and primitives-based emotion recognition" in Proceedings of the 14th European Signal Processing Conference, 2006.

[226] B. S. Kang, C. H. Han, S. T. Lee, D. H. Youn, and C. Lee, "Speaker dependent emotion recognition using speech signals," in Proceedings of the ICSLP, 2000, pp. 383–386.

[227] T. L. Nwe, F. S. Wei, and L. C. De Silva, "Speech-based emotion classification," in Proceedings of the TENCON, Vol. 1, 2001, pp. 297–301.

[228] T. S. Polzin, "Detecting verbal and non-verbal cues in the communications of emotions," Ph.D. dissertation, School of Computer Science, Carnegie Mellon University, Pittsburgh, PA, 2000.

[229] L. C. De Silva and P. C. Ng, "Bimodal emotion recognition," in Proceedings of the IEEE International Conference on Automatic Face and Gesture Recognition, 2000, pp. 332–335.

[230] Y. Yoshitomi, S. Kim, T. Kawano, and T. Kitazoe, "Effect of sensor fusion for recognition of emotional states using voice, face image and thermal image of face," in Proceedings of the Ninth IEEE International Workshop on Robot and Human Interactive Communication, 2000, pp. 178–183.

[231] C. Lee, S. Yildirim, M. Bulut, A. Kazemzadeh, C. Busso, Z. Deng, S. Lee, and S. Narayanan, "Emotion recognition based on phoneme classes" in Proceedings of the Interspeech, 2004, pp. 205–211.

[232] X. Wu, V. Kumar, J. R. Quinlan, J. Ghosh, Q. Yang, H. Motoda, G. J. McLachlan, A. Ng, B. Liu, P. S. Yu, Z. H. Zhou, M. Steinbach, D. J. Hand, and D. Steinberg, "Top 10 algorithms in data mining," *Knowl. Inf. Syst.* 14(1): 1–37, 2008.

[233] S. Scherer, F. Schwenker, and G. Palm, "Classifier fusion for emotion recognition from speech," in Proceedings of the International Conference on Intelligent Environments, 2007, pp. 152–155.

[234] M. H. Yang, D. Roth, and N. Ahuja, "A SNoW-based face detector," in Proceedings of the Neural Information Processing Systems, Vol. 12, 2000, pp. 855–861.

[235] F. Dellaert, T. Polzin, and A. Waibel, "Recognizing emotion in speech," in Proceedings of the International Conference on Spoken Language Processing, 1996, pp. 1970–1973.

[236] N. Sebe, M. S. Lew, I. Cohen, A. Garg, and T. S. Huang, "Emotion recognition using a Cauchy Naive Bayes classifier," in 16th International Conference on Pattern Recognition, Vol. 1, 2002, pp. 17–20.

[237] O. Lowery, Technical Report of Silent Sounds. Inc., 2004.

[238] R. Huber, A. Batliner, J. Buckow, E. Noth, V. Warnke, and H. Niemann, "Recognition of emotion in a realistic dialogue scenario," in Proceedings of the International Conference on Spoken Language Processing, 2000, pp. 665–668.

[239] J. D. Mayer, M. Dipaolo, and P. Salove, "Perceiving affective content in ambiguous visual stimuli: a component of emotional intelligence," *J Pers. Assess.*, 772–781, 1990.

[240] T. Baumgartner, M. Esslen, and L. Jäncke, "From emotion perception to emotion experience: emotions evoked by pictures and classical music," *Int. J. Psychophysiol.*, 34–43, 2005.

[241] A. Chakraborty and A. Konar, "Stability and chaos in cognitive reasoning," in *Reasoning in Intelligent Systems* (ed. J. Nakamatshu), World Scientific, 2010.

[242] M. Ghosh, A. Chakraborty, A. Konar, and A. Nagar, "Detection of chaos and limit cylces in the emotional dynamics from the facial expression of subjects stimulated with audio-visual movies," in *Computational Intelligence, Control and Computer Vision in Robotics and Automation* (ed. B. Subudhi), Narosa Publisher.

[243] I. Kanluan, M. Grimm, and K. Kroschel, "Audio-visual emotion recognition using an emotion space concept," in 16th European Signal Processing Conference (EUSIPCO 2008), Lausanne, Switzerland, August 25–29, 2008.

[244] T. Fritz, S. Jentschke, N. Gosselin, D. Sammler, I. Peretz, R. Turner, A. D. Friederici, and S. Koelsch, "Universal recognition of three basic emotions in music," *Curr. Biol.*, 19(7): 573–576, 2009.

[245] P. Eckerdal and B. Merker "'Music' and the 'action song' in infant development: an interpretation," in *Communicative Musicality. Exploring the Basis of Human Companionship* (eds S. Malloch and C. Trevarthen), Oxford University Press, Oxford, 2009, pp. 241–262.

[246] S. Koelsch, T. Fritz, D. Y. von Cramon, K. Muller, and A. D. Friederici, "Investigating emotion with music: an fMRI study." *Hum. Brain Mapp.*, 27: 239–250, 2006.

[247] D. Sammler, M. Grigutsch, T. Fritz, and S. Koelsch, "Music and emotion: electrophysiological correlates of the processing of pleasant and unpleasant music," *Psychophysiology*, 44: 293–304, 2007.

[248] P. N. Juslin, "Communicating emotion in music performance: a review and a theoretical framework," in *Music and Emotion: Theory and Research* (eds P. N. Juslin and J. A. Sloboda), Oxford University Press, New York, 2001, pp. 309–337.

[249] V. Bettadapura, "Face expression recognition and analysis: the state of the art", Technical Report, arXiv:1203.6722, April 2012.

[250] J. Demsar, "Statistical comparisons of classifiers over multiple data sets," *J. Mach. Learn. Res.*, 7: 1–30, 2006.

[251] T. G. Dietterich, "Approximate statistical tests for comparing supervised classification learning algorithms," *Neural Comput.*, 10(7): 1895–1923, 1998.

[252] B. Abbound, F. Davoine, and M. Dang, "Facial expression recognition and synthesis based on an appearance model," *Signal Processing: Image Communication*, 19(8): 723–740, 2004.

[253] M. Wang, Y. Iwai, and M. Yachida, "Expression recognition from time-sequential facial images by use of expression change model," in Proceedings of the International Conference on Automatic Face and Gesture Recognition, 1998, pp. 324–329.

[254] R. Mandal, A. Halder, P. Bhowmik, A. Chakraborty, and A. Konar, "Uncertainty management in type-2 fuzzy face-space for emotion recognition," in Proceedings of the IEEE International Conference on Fuzzy Systems, Taiwan, China, 2011.

[255] R. Mandal, A. Halder, A. Konar, and A. K. Nagar "A fuzzy condition sensitive hierarchical algorithm for template matching in emotionally expressive facial images", in IPCV'11, Las Vegas, Nevada, July 18–21, 2011.

[256] M. Yang, D. Kriegman, and N. Ahuja, "Detecting faces in images: a survey," *IEEE Transactions on Pattern Analysis and Machine Intelligence*, 24(1): 34–58, 2002.

[257] A. Samal and P. A. Iyengar, "Automatic recognition and analysis of human faces and facial expressions: a survey," *Pattern Recogn.*, 25(1): 65–77, 1992.

[258] A. Konar, *Artificial Intelligence and Soft Computing: Behavioral and Cognitive Modeling of the Human Brain*, CRC Press LLC, January 2000, 788 pages, ISBN 0-8493-1385-6.

[259] A. Konar, *Computational Intelligence: Principles, Techniques and Applications*, Springer, Heidelberg, March 2005, 708 pages, ISBN 3-540-20898-4.

[260] A. Konar and L. C. Jain, *Cognitive Systems Engineering: A Distributed Computational Intelligence Approach*, Springer, London, July 2005, 350 pages, ISBN 10: 1-85233-975-6.

[261] A. Bhattacharya, A. Konar, and A. K. Mandal, *Parallel and Distributed Logic Programming: Towards the Design of a Framework for the Next Generation Database Machines*, Springer, 2006, 291 pages, ISBN 10: 3-540-33458-0.

[262] R. W. Picard, *Affective Computing*, MIT Media Laboratory Perceptual Computing Section, Cambridge, MA, Technical Report 321, November 1995.

[263] A. Halder, A. Chakraborty, A. Konar, and A. K. Nagar, "Computing with words model for emotion recognition by facial expression analysis using interval type-2 fuzzy sets," in FUZZ-IEEE, Hyderabad, India, 2013.

[264] M. J. Hasegawa, S. Levinson, and T. Zhang, "Children's emotion recognition in an intelligent tutoring scenario," in Proceedings of the Interspeech, 2004.

AUTHOR BIOGRAPHIES

Amit Konar received the B.E. degree from Bengal Engineering and Science University, Howrah, India, in 1983, and the M.E., M.Phil., and Ph.D. (Engineering) degrees from Jadavpur University, Kolkata, India, in 1985, 1988, and 1994, respectively.

In 2006, he was a visiting professor with the University of Missouri, St. Louis. He is currently a professor with the Department of Electronics and Telecommunication Engineering (ETCE), Jadavpur University, where he is the founding coordinator of the M.Tech. program in Intelligent Automation and Robotics. He has supervised fifteen Ph.D. theses. He has over 200 publications in international journals and conference proceedings.

He is the author of eight books, including two popular texts *Artificial Intelligence and Soft Computing* (CRC Press, 2000) and *Computational Intelligence: Principles, Techniques and Applications* (Springer, 2005). He serves as the Associate Editor of *IEEE Transactions on Systems, Man and Cybernetics, Part A* and *IEEE Transactions on Fuzzy Systems*. His research areas include the study of computational intelligence algorithms and their applications to the domains of electrical engineering and computer science. Specifically, he has worked on fuzzy sets and logic,' neurocomputing, evolutionary algorithms, Dempster–Shafer theory, and Kalman filtering, and has applied the principles of computational intelligence in image understanding, VLSI design, mobile robotics, pattern recognition, brain–computer interfacing, and computational biology. He was the recipient of All India Council for Technical Education (AICTE) accredited 1997–2000 Career Award for Young Teachers for his significant contribution in teaching and research.

Anisha Halder received the B.Tech. degree in Electronics and Communication Engineering from the Haldia Institute of Technology, Midnapore, India, and the M.E. degree in Control Engineering from Electronics and Telecommunication Engineering Department, Jadavpur University, Kolkata, India, in 2007 and 2009 respectively. She is currently pursuing her Ph.D. degree under the guidance of Prof. Amit Konar and Dr. Aruna Chakraborty at Jadavpur University.

Her principal research interests include artificial intelligence, pattern recognition, cognitive science, and human–computer interaction. She is the author of over 25 papers published in top international journals and conference proceedings.

Aruna Chakraborty received the M.A. degree in Cognitive Science and the Ph.D. degree in Emotional Intelligence and Human–Computer Interactions from Jadavpur University, Kolkata, India, in 2000 and 2005, respectively. She is currently an associate professor in the Department of Computer Science & Engineering, St. Thomas' College of Engineering & Technology, Kolkata. She is also a visiting faculty of Jadavpur University, where she offers graduate-level courses on intelligent automation and robotics, and cognitive science.

She has coauthored a book with Prof. A. Konar titled *Emotional Intelligence: A Cybernetic Approach* (Springer, Heidelberg, 2009). She serves as an editor to the *International Journal of Artificial Intelligence and Soft Computing*, Inderscience, UK. Her current research interests include artificial intelligence, emotion modeling, and their applications in next-generation human–machine interactive systems. She is a nature lover, and loves music and painting.

2

EXPLOITING DYNAMIC DEPENDENCIES AMONG ACTION UNITS FOR SPONTANEOUS FACIAL ACTION RECOGNITION

Yan Tong

Department of Computer Science and Engineering, University of South Carolina, Columbia, SC, USA

Qiang Ji

Department of Electrical, Computer, and Systems Engineering, Rensselaer Polytechnic Institute, Troy, NY, USA

An automated system for real-time facial action unit (AU) recognition has many applications. Most of the existing algorithms are limited to either recognizing each AU statically or only considering the temporal evolution of each AU, while ignoring the dynamic relationships among AUs. However, recent psychological research in human behavior analysis shows that the dynamic property of facial actions is an important factor to interpreting naturalistic human behavior.

In this work, we propose to systematically model the dynamic properties of facial actions including not only the temporal development of each AU, but also the dynamic dependencies among AUs in a spontaneous facial display. Specifically, a Dynamic Bayesian Network (DBN) is employed to explicitly model the dynamic and semantic relationships among AUs, where the dynamic nature of facial action is characterized by directed temporal links among AUs; and the semantic relationships are represented by directed static links among AUs. The DBN model is automatically constructed

Emotion Recognition: A Pattern Analysis Approach, First Edition. Edited by Amit Konar and Aruna Chakraborty.
© 2015 John Wiley & Sons, Inc. Published 2015 by John Wiley & Sons, Inc.

using both domain knowledge and the training data. Given the dynamic model, AUs are recognized through probabilistic inference over time. Experiments with real images demonstrate that by explicitly modeling the dynamic dependencies among AUs, the proposed method improves AU recognition over the existing methods, especially for spontaneous facial displays.

2.1 INTRODUCTION

The Facial Action Coding System (FACS) [1] has been widely used for human behavior analysis. Based on FACS, facial behavior could be described by 44 AUs, which are the smallest visually detectable facial mucular movements that produce facial appearance changes [2]. Recent psychological research in human behavior analysis shows that the dynamic characteristic of facial action (the activating order and duration of the AUs) is an important factor in interpreting naturalistic human behavior [3]. The facial expression dynamics include the temporal evolution (i.e., the duration and temporal pattern) of each AU and the dynamic relationships among various AUs (i.e., the occurrence order of multiple AUs). Since facial activity is a dynamic event, the activation and movement of facial muscles can be expressed as a function of time. Hence, the temporal evolution (the timing and the duration) of an AU can be described by four temporal states [4]: (1) the onset state (the intensity of the facial action grows stronger, and the facial appearance changes caused by the facial action become more significant); (2) the apex state (there are no more facial appearance changes because of the facial action); (3) the offset state (the facial action is relaxing and returns to its neutral state); and (4) the neutral state (there are no facial appearance changes because of the facial action).

Nishio *et al.* [5] have shown that when the mouth moves prior to the eye, a smile expression is mostly interpreted as a smile of enjoyment. On the contrary, when the eyes move prior to the mouth, it is mostly interpreted as a dampened smile. More recently, Schmidt and Cohn [6, 7] found that spontaneous smile usually has a relatively slower and smoother onset, and that the intensity of lip corner motion is a strong linear function of time in contrast to the posed smile. Furthermore, their study [7] also shows that there are dynamic relationships among AUs in the spontaneous smiles: AU12 (lip corner raiser) appears either simultaneously or closely followed by other AUs such as AU6 (cheek raiser), AU15 (lip corner depressor), and AU17 (chin raiser) within 1 second. Valstar *et al.* [8] find that the dynamic properties of AUs such as the activating speed, magnitude, and the occurrence orders of AUs are critical to distinguishing the spontaneous eyebrow motion from posed eyebrow motion. Therefore, it is desirable to exploit the dynamic dependencies among AUs for spontaneous facial action analysis. Furthermore, there are semantic relationships among AUs such as the co-occurrence and mutually exclusive relationships as described in Reference 9. It is these spatiotemporal relationships among AUs that produce a meaningful facial display.

However, most of the existing AU recognition approaches either recognize each AU statically or only consider its temporal development individually while ignoring the dynamic dependencies among AUs. The recent work by Tong *et al.* [9] models the semantic and static relationships among AUs from posed facial expressions, whereas only two dynamic relationships among AUs are manually and heuristically specified. In contrast, this work focuses exclusively on modeling the dynamic dependencies among AUs from spontaneous facial expressions. Specifically, we propose to model two types of dynamic dependencies: self-evolution and pairwise dynamic dependencies among AUs. Such dynamic relationships could be well modeled by a DBN, which characterizes the dynamic nature of facial action by directed temporal links among AUs across consecutive frames. Furthermore, we propose to use a machine learning method to systematically learn the DBN model based on both training data and the prior domain knowledge. The experiment results demonstrate that the proposed system with the modeling of dynamic AU relationships yields significant improvements over the method that only models semantic AU relationships for both posed and spontaneous facial expressions.

2.2 RELATED WORK

In general, AU recognition can either be spatial or spatiotemporal. A detailed discussion on AU recognition can be found in References 2, 10, 11. Spatial methods recognize AUs using the features extracted from a single frame. Spatiotemporal approaches try to capture the temporal properties of facial actions for recognition including recognition using dynamic features [12–14], the HMM-based methods [15–18], temporal rule-based method [4], and the DBN-based methods [9, 19, 20].

The dynamic features are explicitly employed in References 12–14 to characterize the temporal properties of facial actions. Specifically, Valstar *et al.* [12] use the motion history images; Yang *et al.* [13] develop dynamic Harr features; and Zhao and Pietikäinen [14] propose a local binary pattern combining local information from appearance and motion. However, these methods utilize the whole video sequence to compute the dynamic features and can only work well on the image sequence containing a single facial expression. As a result, they are not applicable under real-world circumstances. Although these approaches analyze the temporal properties of facial features, they ignore the temporal evolution of AU itself as well as the dynamic dependencies among AUs.

Hidden Markov Models (HMMs) have been used to recognize AUs [16] or facial expressions [15, 17, 18]. Each AU or facial expression (AU combination) is modeled by an HMM. The classification is performed by choosing the AU or facial expression, which maximizes the likelihood of the corresponding HMM, given the extracted facial features. There are several limitations for these HMM-based methods. First, since the number of AUs and their combinations is large, it is practically difficult to build an HMM for each AU or AU combination. Second, they often require an AU to evolve in a complete cycle before it can be recognized, which is often not the case for spontaneous expression. Finally, existing HMM-based methods only model the dynamic evolution of a single AU, whereas they lack the ability to model the dynamic relationships among AUs.

Pantic and Patras [4] recognize AUs based on a neutral-expressive-neutral sequential facial expression model with a set of deterministic temporal rules. An AU is recognized as "presence" if its full temporal pattern (onset-apex-offset) is observed continuously. However, this method is susceptible to inaccurate facial motion extraction where the temporal states like onset, apex, and offset are not detected correctly. Furthermore, they only consider the temporal evolution of each AU individually, whereas the dynamic dependencies among AUs are ignored.

In addition, DBNs have been increasingly used for computer vision problems such as activity recognition and facial expression recognition. Among them, Zhang and Ji [20] exploit a DBN to classify six basic facial expressions with a dynamic fusion strategy; Gu and Ji [19] use a similar idea for facial event classification such as fatigue. However, AUs are not recognized explicitly in their models. Moreover, they neither consider the dynamics of each AU nor the dynamic relationships among AUs. Tong *et al.* [9] employ a DBN to model the relationships among AUs mainly in the same time slice. In contrast, this work focuses on automatically and systematically modeling and learning the dynamic dependencies among AUs. In addition, Reference 9 is limited to recognizing AUs from posed facial expressions, whereas this work focuses on recognizing AUs in a spontaneous facial display.

In summary, the dynamic modeling of AUs in the existing work is limited and often *ad hoc*. In contrast, the proposed system employs advanced learning techniques to learn both the structure and parameters of the DBN from the training data. Therefore, the dynamic evolution of each AU and the dynamic relationships among different AUs can be systematically and stochastically modeled in a spontaneous facial display. As far as we know, this is an area that has not been attempted by anybody so far.

2.3 MODELING THE SEMANTIC AND DYNAMIC RELATIONSHIPS AMONG AUs WITH A DBN

In a spontaneous facial behavior, besides the spatial relationships among AUs as described in Reference 9, there are also dynamic dependencies among AUs since multiple AUs often proceed in sequence to represent different naturalistic facial expressions. Generally speaking, there are two types of temporal causalities among AUs. First, each AU is a dynamic event that develops over time. For each AU, its temporal evolution usually consists of a complete temporal segment including onset, apex, and offset states continuously and sequentially lasting from a quarter of a second (e.g., a blink) to several minutes (e.g., a jaw clench) [4]. Second, the relationships among AUs also undergo a temporal evolution, that is, an AU will be contracted following the activating of another AU. For example, in a spontaneous smile, usually AU12 (lip corner puller) is first contracted to express a slight emotion, then with the increasing of emotion intensity, AU6 (cheek raiser) would be contracted in an average of 0.4 seconds after the contracting of AU12 [7], and after both the actions reach their apexes simultaneously, AU6 would be relaxed; finally, AU12 would be gradually released, and both of them return to the neutral state. Furthermore, due to the variability among individuals and different contexts, the dynamic relationships

TABLE 2.1 A list of AUs and their interpretations

AU1	AU2	AU4	AU5
Inner brow raiser	Outer brow raiser	Brow Lowerer	Upper lid raiser
AU6	AU7	AU9	AU12
Cheek raiser	Lid tighten	Nose wrinkle	Lip corner puller
AU15	AU17	AU23	AU24
Lip corner depressor	Chin raiser	Lip tighten	Lip presser
AU25	AU27		
Lips part	Mouth stretch		

Source: Adapted from Reference 20.

among AUs are stochastic. Therefore, such dynamic dependencies among AUs can be well modeled by a DBN.

For subsequent discussions, Table 2.1 summarizes a list of commonly occurring AUs and their interpretations, although the proposed system is not restricted to recognizing these AUs. Details about AUs and their definitions may be found in Reference 1.

2.3.1 A DBN for Modeling Dynamic Dependencies among AUs

In this work, we propose to use a DBN to model and learn the dynamic dependencies among AUs. A DBN is a directed graphical model, which models the temporal evolution of a set of random variables $\mathbf{X}$ over time [21]. It is defined as $B = (G, \Theta)$, where G is the model structure, and Θ represents the set of model parameters, that is, the conditional probability tables (CPTs) for all nodes. Let $\mathbf{X}^t$ represent a set of random variables at a discrete time slice t. There are two assumptions in the DBN model: first, the system is first-order Markovian, that is, $P(\mathbf{X}^{t+1}|\mathbf{X}^0, \dots, \mathbf{X}^t) = P(\mathbf{X}^{t+1}|\mathbf{X}^t)$; second, the transition probability $P(\mathbf{X}^{t+1}|\mathbf{X}^t)$ is same for all t. Therefore, a DBN B can be also defined by a pair $(B_0, B_{\rightarrow})$, where $B_0 = (G_0, \Theta_0)$ as shown in Figure 2.1a captures the static distribution over all variables $\mathbf{X}^0$, for example, the

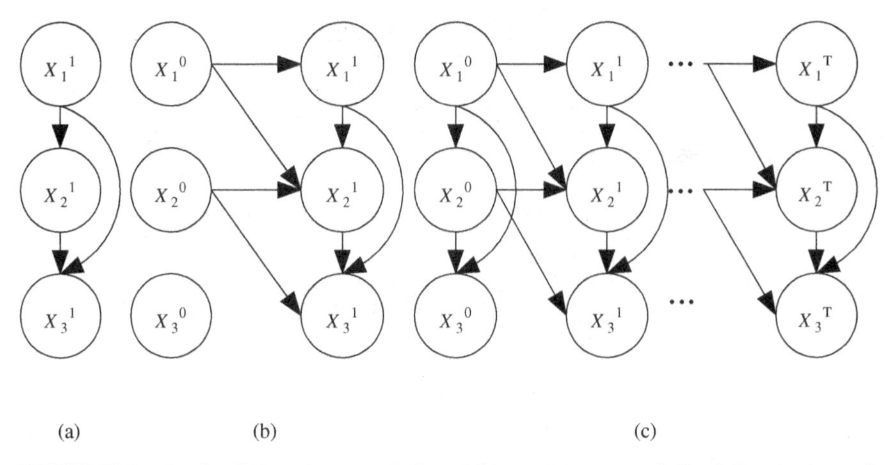

FIGURE 2.1 A pair of (a) static network B_0 and (b) transition network $B_{\rightarrow}$ define the dynamic dependencies for three random variables X_1, X_2, X_3. (c) The corresponding "unrolled" DBN for $T + 1$ time slices.

static relationships among AUs, and the transition network $B_{\rightarrow} = (G_{\rightarrow}, \Theta_{\rightarrow})$ as shown in Figure 2.1b specifies the transition probability $P(\mathbf{X}^{t+1}|\mathbf{X}^t)$ for all t in a finite time slices T, for example, the dynamic dependencies among AUs.

Given a DBN model, the joint probability over all variables $\mathbf{X}^0, \ldots, \mathbf{X}^T$ can be factorized by "unrolling" the DBN into an extended static BN as shown in Figure 2.1c, whose joint probability is computed as

$$P(\mathbf{x}^0, \ldots, \mathbf{x}^T) = P_{B_0}(\mathbf{x}^0) \prod_{t=0}^{T-1} P_{B_{\rightarrow}}(\mathbf{x}^{t+1}|\mathbf{x}^t) \tag{2.1}$$

where $\mathbf{x}^t$ represents the sets of values taken by $\mathbf{X}^t$; $P_{B_0}(\mathbf{x}^0)$ captures the joint probability of all variables in the static BN B_0; and $P_{B_{\rightarrow}}(\mathbf{x}^{t+1}|\mathbf{x}^t)$ represents the transition probability and can be decomposed as follows based on the conditional independence encoded in the DBN:

$$P_{B_{\rightarrow}}(\mathbf{x}^{t+1}|\mathbf{x}^t) = \prod_{i=1}^{N} P_{B_{\rightarrow}}(x_i^{t+1}|pa(X_i^{t+1})) \tag{2.2}$$

where $pa(X_i^{t+1})$ represents the parent configuration of variable X_i^{t+1} in the transition network $B_{\rightarrow}$; and N is the number of nodes in a time slice.

In this research, we propose to employ a DBN to model the directed temporal relationships among AUs. Specifically, we intend to learn and model two types of temporal relationships for AUs in two adjacent time slices. In the first type, an arc linking each AU_i node at time $t - 1$ to that at time t depicts how a single AU_i develops over time. In the second type, an arc from AU_i node at time $t - 1$ to AU_j at time t

depicts the dynamic dependencies among AUs such that AU_i at the previous time step affects $AU_j (j \neq i)$ at the current time step. Since the multiple relationships are hard to capture, and they vary significantly from people to people, we intend to capture the strong pairwise dynamic dependencies among AUs that are true for most people while ignoring the weak relationships to avoid using incorrect relationships for inference.

Although we define four temporal states (neutral, onset, apex, and offset) for depicting the dynamic evolution of each AU, spontaneous facial expression does not always follow the temporal pattern of neutral-onset-apex-offset-neural [4]. On the contrary, it may have multiple apexes in a single facial expression, and the pattern neutral-expression1-····-expression*N*-neutral is more often in a spontaneous facial activity. For example, in a conversation, AU25 (lips part) would be activated frequently without returning to the neutral state. Moreover, the temporal pattern will vary due to the individual differences and different contexts. In this work, each AU is represented by a binary value [0, 1] representing its absence/presence status; therefore, the AU recognition does not depend on the order and timing of the temporal states. An AU is in the "presence" status, when it is activated and is at one of the three temporal states (onset, apex, and offset); whereas it is in the "absence" status, when it is at the neutral state.

Furthermore, we should notice that the temporal relationships are critically dependent on the temporal resolution of the image sequences, that is, the frame rate of the recorded videos. The learned temporal relationships may not be valid on the image sequences collected under a different frame rate. Furthermore, as we mentioned above, although some AUs would be present in sequence, the duration between the occurrences of the two AUs varies due to different facial expressions and individual differences. Therefore, we consider the temporal relationships between two fixed-size time durations instead of the temporal relationships between two successive frames. This way, if AU_i appears within the tth time duration, it is counted as "presence" at t, that is, $AU_i^t = 1$. For example, the image sequences in Cohn–Kanade database [22] are recorded with 12 frames per second (fps), whereas those in ISL database [23] are collected with 30 fps. Therefore, if we set a time duration with $1/6$ seconds[1], each time duration consists of two frames for Cohn–Kanade database, but five frames for ISL database. Hereafter, t represents the tth time duration, whereas $t + 1$ represents the next time duration. And we use "slice" instead of "duration" for a generalized representation in the later presentation.

In addition, since the duration of an AU varies from facial expression to facial expression and is also subject to the individual difference, and also since each AU is defined by two states "presence/absence," we employ a two-slice DBN ($T = 2$) to model the temporal evolution of the facial activity in two adjacent time slices in this work. On the other hand, if each AU is defined by multiple states such as the four temporal states, the AU may be better recognized through a multislice DBN.

[1]The time duration is set to $1/6$ seconds in the current work because it can deal with the different temporal resolutions in two databases and the time duration is enough to show the transition for a single AU.

TABLE 2.2 The pairwise dynamic relationships among AUs are derived from database [22], where each entry $a_{i,j}$ represents $P(AU_j^t{=}1|AU_j^{t-1}{=}0, AU_i^{t-1}{=}1)$

	AU1,t-1	AU2,t-1	AU4,t-1	AU5,t-1	AU6,t-1	AU7,t-1	AU9,t-1	AU12,t-1	AU15,t-1	AU17,t-1	AU23,t-1	AU24,t-1	AU25,t-1	AU27,t-1
AU1,t	0	0	0.02	0.16	0	0.01	0.01	0.01	0.02	0.02	0	0.01	0.02	0.04
AU2,t	0	0	0.01	0.1	0	0.01	0.01	0	0.01	0.01	0	0	0.02	0.04
AU4,t	0.02	0.02	0	0.02	0	0.03	0.08	0.01	0.03	0.04	0.04	0.04	0.01	0.01
AU5,t	0.09	0.15	0.01	0	0	0.01	0	0	0.01	0.01	0.01	0	0.03	0.07
AU6,t	0	0	0.02	0	0	0.03	0.04	0.1	0.01	0.01	0.01	0.01	0.02	0
AU7,t	0.02	0.01	0.06	0.02	0.01	0	0.23	0.01	0.02	0.03	0.04	0.04	0.01	0
AU9,t	0	0	0.02	0	0	0.03	0	0	0.01	0.01	0.01	0.01	0	0
AU12,t	0.01	0.01	0.01	0	0.01	0.01	0.01	0	0.01	0.01	0.02	0.01	0.01	0
AU15,t	0.01	0.01	0.02	0.01	0	0.01	0.01	0	0	0.03	0.03	0.04	0	0
AU17,t	0.01	0.01	0.07	0.01	0.01	0.06	0.12	0	0.06	0	0.13	0.07	0.01	0
AU23,t	0	0	0.02	0	0	0.02	0.01	0	0	0.01	0	0.01	0.01	0
AU24,t	0	0	0.02	0	0	0.02	0	0	0.02	0.02	0.03	0	0	0
AU25,t	0.12	0.14	0.02	0.12	0.07	0.01	0.01	0.09	0	0.01	0.01	0.01	0	0
AU27,t	0.09	0.13	0	0.12	0	0	0	0	0	0	0	0	0.04	0

2.3.2 Constructing the Initial DBN

Since the "presence/absence" of an AU at time t depends on both its own state and the states of other AUs in the previous time slice, $P(AU_j^t|AU_j^{t-1}, AU_i^{t-1})$ is used to capture the dynamic relationships of two AUs (AU_i and AU_j) $(i \neq j)$ and the dynamic evolution of a single AU AU_j $(i = j)$. In this work, each AU is represented by a binary value [0, 1] representing its "presence/absence" status for model simplicity. For initialization, the two types of dynamic relationships among AUs are partly learned from Table 2.2, which is constructed from two posed facial expression databases: namely, Cohn–Kanade facial expression database [22] and ISL facial expression database [23]. On the other hand, in order to recognize AUs in spontaneous facial expression, we should also refine the dynamic relationships among AUs in the spontaneous facial expression using other database containing natural facial expressions such as [28].

For example, each entry $a_{i,j}$ in Table 2.2 represents the probability $P(AU_j^t = 1|AU_j^{t-1} = 0, AU_i^{t-1} = 1)$ computed as below:

$$P\big(AU_j^t = 1|AU_j^{t-1} = 0, AU_i^{t-1} = 1\big) = \frac{N_{AU_j^t + \neg AU_j^{t-1} + AU_i^{t-1}}}{N_{\neg AU_j^{t-1} + AU_i^{t-1}}}, \tag{2.3}$$

where $N_{\neg AU_j^{t-1} + AU_i^{t-1}}$ is the total number of the events that AU_i is present and AU_j is absent in the $(t-1)$th slice, regardless of the presence of other AUs, and $N_{AU_j^t + \neg AU_j^{t-1} + AU_i^{t-1}}$ is the total number of the events that AU_i is present and AU_j is absent in the $(t-1)$th slice, and AU_j is present in the tth slice in the databases. This probability implies that AU_j is activated following AU_i. The other probabilities $P(AU_j^t = 1|AU_j^{t-1} = 0, AU_i^{t-1} = 0)$, $P(AU_j^t = 1|AU_j^{t-1} = 1, AU_i^{t-1} = 0)$, and $P(AU_j^t = 1|AU_j^{t-1} = 1, AU_i^{t-1} = 1)$ are computed similarly.

If the probability $P(AU_j^t = 1|AU_j^{t-1} = 0, AU_i^{t-1} = 1)$ is higher than a predefined threshold T_{up} or the probability $P(AU_j^t = 1|AU_j^{t-1} = 1, AU_i^{t-1} = 0)$ is lower than a predefined threshold T_{bottom}, we assume that there is strong dynamic dependency between AU_i and AU_j, which can be modeled with an inter-slice link from AU_i^{t-1} to AU_j^t in the DBN. For example, the link from AU_{12}^{t-1} to AU_6^t represents the dynamic dependency between AU6 (cheek raiser) and AU12 (lip corner puller), since AU6 is present mostly after AU12 is activated. This dynamic dependency can be also derived from the psychological studies [7]. The link from AU_2^{t-1} to AU_5^t means that AU5 (lid raiser) is activated mostly after AU2 (outer brow raiser) is activated. This way, an initial transition network is manually constructed as in Figure 2.2a.

2.3.3 Learning DBN Model

Given a set of observed data $D = \{D_1, \dots, D_M\}$, we could refine the DBN model with a structure learning algorithm, that is, finding a DBN structure G that best fits the observed data. For learning a DBN model, the training data D should be divided into S sequences, each of which contains certain facial expressions with length M_s, such that $\sum_s M_s = M$, where M is the total number of training images. As mentioned above, a DBN consists of two parts (B_0 and $B_{\rightarrow}$), therefore, we should learn both of them from the training data. To evaluate the fitness of the network, we need to define a scoring function. The score for a DBN model can be defined as

$$Score(B) = logP(B, D) = logP(B) + logP(D|B) \tag{2.4}$$

where $logP(B)$ is the log prior probability of the network structure; and $logP(D|B)$ is the log likelihood, which can be computed approximately based on the Bayesian Information Criteria (BIC) [24] as follows:

$$logP(D|B) \approx logP(D|G, \widehat{\Theta_G}) - \frac{logM}{2} Dim_G \tag{2.5}$$

where the first term evaluates how well the network B fits the data D; the second term is a penalty relating to the complexity of the network; $\widehat{\Theta_G}$ is the set of parameters of G that maximizes the likelihood of the data; M is the number of training data; and Dim_G is the number of parameters.

Instead of giving an equal prior probability $P(B)$ to all possible structures, we assign a higher probability to the initial structure B_{init}, which is manually constructed. Let $Pa_B(X_i)$ be a set of parent nodes of X_i in B. Assume δ_i is the number of nodes in the symmetric difference of $Pa_B(X_i)$ and $Pa_{B_{init}}(X_i)$. Then the prior probability of any other network ($B \neq B_{init}$) is computed as $P(B) = c\kappa^{\delta}$ [25], where c is a normalization constant, $0 < \kappa \leq 1$ is a predefined constant factor, $\delta = \Sigma_{i=1}^{N} \delta_i$, and N is the number of nodes in the network.

Given the scoring function, the structure learning is performed by searching the network with the highest score among the possible structures. Furthermore, the score of a DBN can be decomposed into two parts as follows:

$$Score(B) = Score_{B_0} + Score_{B_{\rightarrow}} \tag{2.6}$$

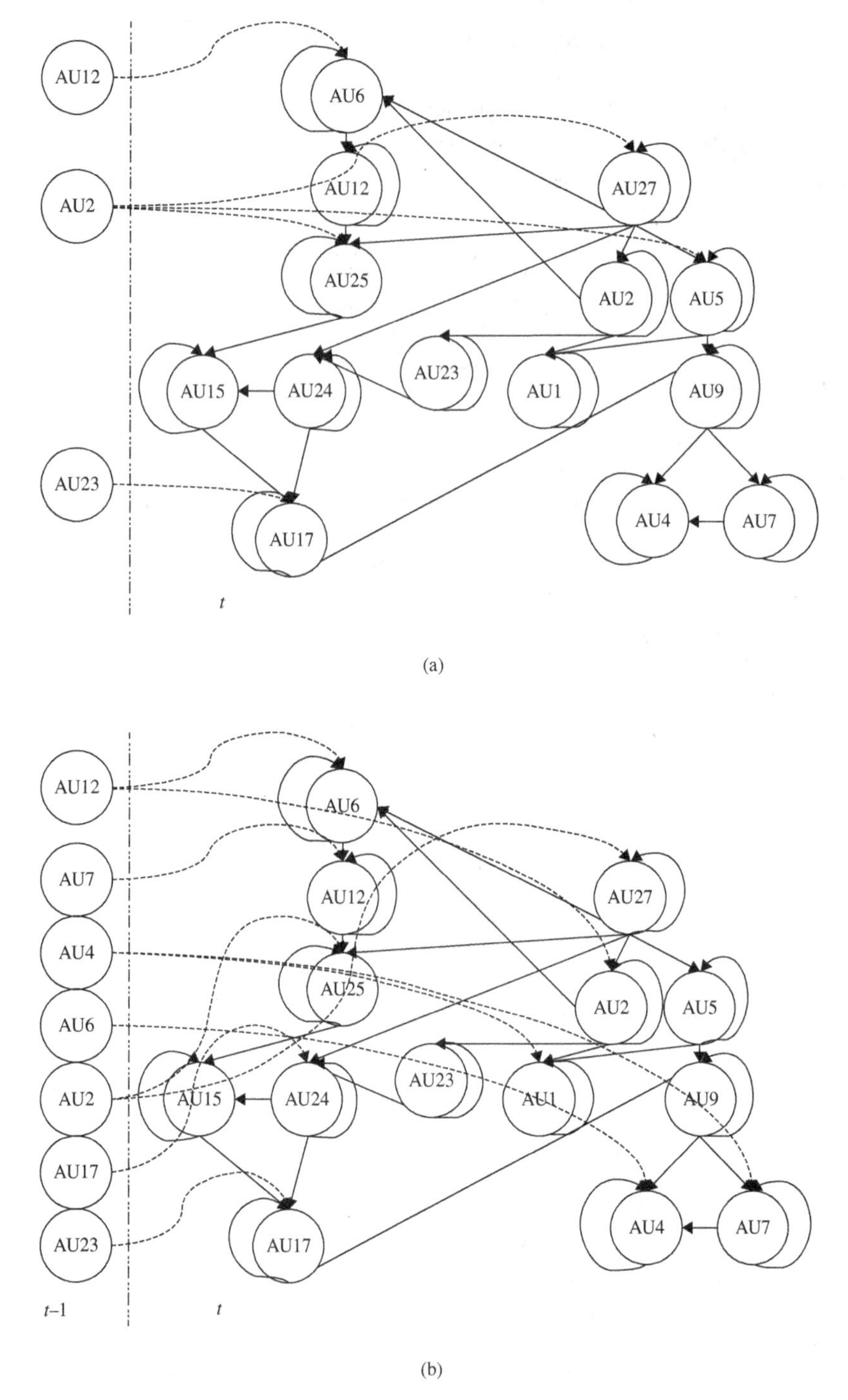

FIGURE 2.2 (a) The initial transition network and (b) the learned transition network by the proposed algorithm for AU modeling. The self-arrow at each AU node indicates the temporal evolution of a single AU from the previous time slice to the current time slice. The dashed line with arrow from AU_i at time $t-1$ to AU_j ($j \neq i$) at time t indicates the dynamic dependency between different AUs.

where $Score_{B_0}$ represents the score of the static network; while $Score_{B_\rightarrow}$ represents that of the transition network. Therefore, we can learn the structure of B_0 and the structure of $B_\rightarrow$ separately.

2.3.3.1 *Learning the Static Network* Instead of recognizing AUs from a pre-segmented image sequence, we intend to recognize AUs under more realistic circumstances, where the image sequences do not necessarily start from the neutral expression. Therefore, B_0 models the static relationships among AUs within a time slice. Hence, we learn B_0 by using all the data $D_1, \ldots, D_M$, instead of using only the first slices from the image sequences. Following Equation 2.4, the score for the static network is defined as follows:

$$Score_{B_0} = logP(B_0) + \sum_{i,j,k} N^0_{ijk} log\hat{\theta}^0_{ijk} - \frac{logM}{2} Dim_{B_0} \tag{2.7}$$

where θ^0_{ijk} represents the parameters for the static network B_0, that is, $\theta^0_{ijk} = P(X^0_i = k|pa_j(X^0_i))$ for node X^0_i at kth state given its jth parent configuration in B_0. And N^0_{ijk} is defined as $N^0_{ijk} = \sum_M I(X^0_i = k, pa_j(X^0_i))$, where $I(.)$ is an indicator function such that $I(.) = 1$ if the argument is true, otherwise $I(.) = 0$.

Having defined a score as Equation 2.7, we need to identify a static network structure with the highest score by a hill-climbing searching algorithm as discussed below. First, an initial network structure $B_{0\text{init}}$ is manually constructed by combining the data analysis from the database [22] and the domain knowledge from FACS rules [1]. Then starting from $B^0_0 = B_{0\text{init}}$, $Score_{B_0}$ is computed as Equation 2.7 for each nearest neighbor of B^0_0, which is generated from B^0_0 by adding, deleting, or reversing a single arc, subject to the acyclicity constraint and the upper bound of parent nodes. In this way, the structure that has the maximum score among all of the nearest neighbors is selected as the static network B_0.

2.3.3.2 *Learning the Transition Network* Learning the transition network is more complicated than learning the static network. The transition network $B_\rightarrow$ consists of two types of links: inter-slice links and intra-slice links. Inter-slice links are the dynamic links connecting the temporal variables of two successive time slices; in contrast, intra-slice links connect the variables within a time slice, which are same as the static network structure. The score of the transition network is defined as below:

$$Score_{B_\rightarrow} = logP(B_\rightarrow) + \sum_{i,j,k} N^\rightarrow_{ijk} log\hat{\theta}^\rightarrow_{ijk} - \frac{log(M-S)}{2} Dim_{B_\rightarrow} \tag{2.8}$$

where $M - S$ is the total number of pairwise transitions between two successive slices in the training data; $\theta^\rightarrow_{ijk}$ represents the parameters for the transition network $B_\rightarrow$, that is, $\theta^\rightarrow_{ijk} = P(X^t_i = k|pa_j(X^t_i))$ for node X^t_i at kth state given its jth parent configuration in

$B_{\rightarrow}$; and $N_{ijk}^{\rightarrow}$ accounts for the number of the instances of transitions in the sequences such that $N_{ijk}^{\rightarrow} = \sum_s \sum_t I(X_i^t = k, pa_j(X_i^t))$, where $I(.)$ is an indicator function.

Given the definition of a score for the transition network $B_{\rightarrow}$ as Equation 2.8, we need to identify a transition network structure with the highest score by a searching algorithm. There are some coherent constraints on $B_{\rightarrow}$. First, the variables $\mathbf{X}^0$ in $B_{\rightarrow}$ as shown in Figure 2.1b do not have parents. Second, the inter-slice links exist only from the previous slice to current slice. Finally, based on the stationary assumption, both the inter- and intra-slice links should be repeated for all $t \in [1, T]$. Furthermore, since we intend to capture the strong pairwise dynamic dependencies between AUs that are true for most people while ignoring the weak relationships, an additional constraint is imposed that each node X_i^{t+1} has at most two parents from the previous time slice t. In order to learn the transition network, we can only learn the inter-slice links, while fixing the intra-slice links, which are same as the static network structure. Specifically, learning the transition network is performed as follows.

Step 1: Starting from $B_{\rightarrow}^0 = B_{\rightarrow \text{init}}$, which is is manually constructed as Figure 2.2a, compute the score of each nearest neighbor of $B_{\rightarrow}^0$, which is generated from $B_{\rightarrow}^0$ by adding or deleting an inter-slice link, subject to the acyclicity constraint, the constraints mentioned above, and the limitation on the upper bound of the number of parent nodes. Furthermore, adding an inter-slice link from AU_i^{t-1} to AU_j^t is not allowed, if there is an intra-slice link from AU_i^t to AU_j^t.

Step 2: Update $B_{\rightarrow}^0$ with the network structure that has the maximum score among all of the nearest neighbors, and go back to previous step until no neighbors have a higher score than the current structure.

The DBN parameters θ_{ijk}^0 and $\theta_{ijk}^{\rightarrow}$ are learned simultaneously with learning the model structure.

Figure 2.2b shows the learned transition network by the proposed learning algorithm. It is shown that the learned structure can better reflect the dynamic relationships among AUs in the training data. For example, the dynamic link from AU_4^{t-1} to AU_7^t means that the eyelids intend to be narrowed by activating AU7 (lid tightener) with the increasing intensity of AU4 (brow lowerer). And the dynamic link from AU_{17}^{t-1} to AU_{24}^t means that before the lips are pressed together, the chin boss most likely is already moved upward by activating AU17 (chin raiser).

Figure 2.3 gives the complete DBN model including the measurement nodes, which are represented with shaded circles. The measurement nodes provide evidence obtained through some AU recognition techniques. The AU nodes are hidden nodes, whose states need to be inferred from the DBN given the evidence. The link between the measurement node and the AU node models the measurement uncertainty with the specific AU recognition technique, that is, its recognition accuracy. Specifically, we first perform face and eyes detection in live video automatically. Given the knowledge of eye centers, the face region is normalized and convolved pixel by pixel by a set of multi-scale and multi-orientation Gabor filters. Then we extract the measurement for each AU through a general purpose learning mechanism based on Gabor feature representation and AdaBoost classifiers similar to Reference 26. Therefore,

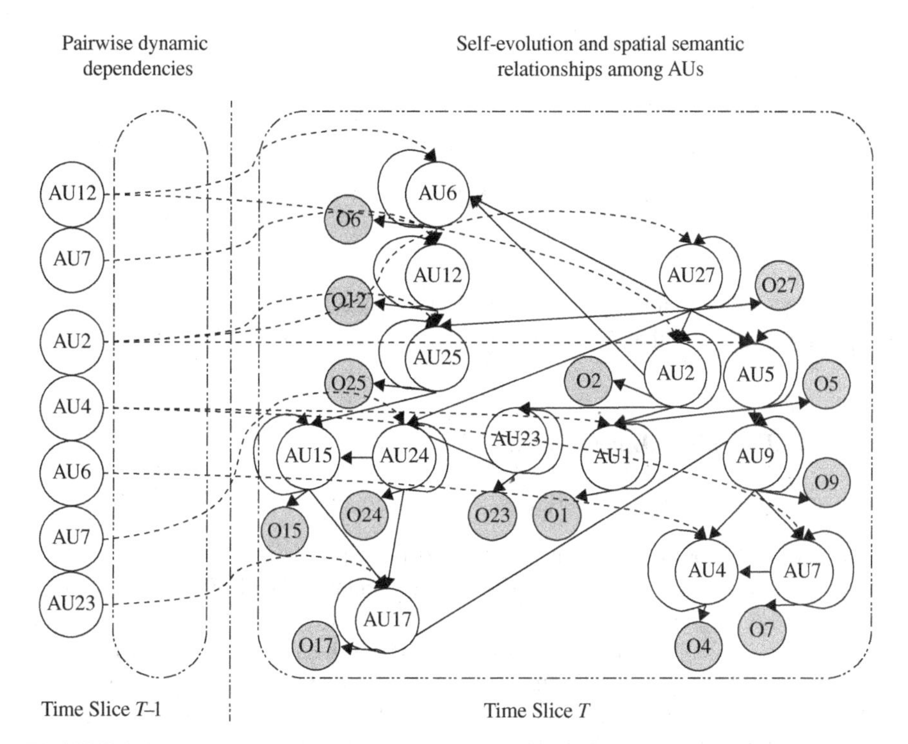

FIGURE 2.3 The complete DBN model for AU modeling. The self-arrow at each AU node indicates the temporal evolution of a single AU from the previous time slice to the current time slice. The dashed line with arrow from AU_i at time $t - 1$ to AU_j ($j \neq i$) at time t indicates the dynamic dependency between different AUs. And the shaded circle indicates the measurement O_i for each AU_i.

such a DBN is capable of accounting for uncertainties in the AU recognition process, representing semantic and dynamic relationships among AUs, and providing principled inference solutions.

2.3.4 AU Recognition Through DBN Inference

Once the AU measurements are obtained, the states of AU nodes can be inferred from the DBN by finding the most probable explanation (MPE) of the evidence at each time slice as follows:

$$\mathbf{AU}^{t*} = \underset{\mathbf{AU}^t}{\operatorname{argmax}}\, p(\mathbf{AU}^t | \mathbf{O}^{1:t}) \tag{2.9}$$

where $\mathbf{AU}^t$ indicate all AU nodes at time slice t, and $\mathbf{O}^t = \{O_1^t, O_2^t, \ldots, O_N^t\}$ are the measurements for the N target AUs at tth time slice. Thus, the probability of $\mathbf{AU}^t$, given the available evidence $\mathbf{O}^{1:t}$, can be computed by performing the dynamic BN inference (filtering) [27], that is, estimating the posterior probability $P(\mathbf{AU}^t|\mathbf{O}^{1:t})$ using Bayes' rule:

$$\begin{aligned} P(\mathbf{AU}^t|\mathbf{O}^{1:t}) &\propto P(\mathbf{O}^t|\mathbf{AU}^t)P(\mathbf{AU}^t|\mathbf{O}^{1:t-1}) \\ &= P(\mathbf{O}^t|\mathbf{AU}^t)\sum_{\mathbf{AU}^{t-1}} P(\mathbf{AU}^t|\mathbf{AU}^{t-1})P(\mathbf{AU}^{t-1}|\mathbf{O}^{1:t-1}) \end{aligned} \tag{2.10}$$

In this way, we obtain the true states of the AUs given the AU measurements through probabilistic inference over time.

2.4 EXPERIMENTAL RESULTS

2.4.1 Facial Action Unit Databases

In this work, the proposed system is trained on FACS-labeled images from two databases. The first database is the Cohn and Kanade's DFAT-504 database [22], which consists of posed facial expressions from more than 100 subjects covering different races, ages, and genders. In order to extract the temporal relationships, the Cohn–Kanade database is coded into AU labels frame by frame in our work. The experimental validation on this database would be used to evaluate the proposed system by comparing with other state-of-the-art AU recognition techniques.

In the second set of experimental validation, we extend our work to recognize AUs from natural facial expressions. Although the publicly available facial expression databases provide a large number of posed facial expression data, the resource of spontaneous databases is limited. Therefore, we create a new spontaneous facial expression database, where the images are collected by Multiple Aspects of Discourse (MAD) research lab at the University of Memphis [28] and coded into AU labels frame by frame in our work. This database consists of image sequences recorded in true color with a frame rate of 9 fps. In the image sequences, the subjects are providing technical support to some user via headphones. Hence, there are various natural facial expressions other than the six basic facial expressions.

From the MAD database, we find that the spontaneous facial expressions are different from the posed facial expressions in the Cohn–Kanade database in several ways.

- Most of the spontaneous facial expressions are activated without significant facial appearance changes, that is, the amplitudes of the spontaneous facial expressions are smaller than those of the posed facial expressions.
- The spontaneous facial expression often has a slower onset phase and a slower offset phase compared to the posed facial expression.
- The spontaneous facial expression may have multiple apexes, and the expression does not always follow a neutral-expression-neutral temporal pattern [4]. On

the contrary, a neutral-expression 1 to neutral-expression N temporal pattern is more often visible, especially during conversation.

- In the database, there are large out-of-plane head movements overlapping with the facial expression changes: most of the images contain quarter-right or half-right profile faces. In addition, faces are partly occluded in some images. These make it more challenging for AU recognition.

These differences also make it difficult to recognize AUs individually and statically. Therefore, explicitly employing the semantic and dynamic dependencies among AUs will help to recognize AUs in spontaneous facial expression.

2.4.2 Evaluation on Cohn and Kanade Database

We first evaluate our system on the Cohn–Kanade database [22] for recognizing 14 target AUs as shown in Figure 2.3 to demonstrate the system performance on the standard database. The database is divided into eight sections, each of which contains images from different subjects. Each time, we use one section for testing and the remaining sections for training, so that the training and testing set are mutually exclusive. The average recognition performance, that is, the true skill score (the difference between the positive rate and the false positive rate), is computed on all sections.

Figure 2.4 shows the performance for generalization to novel subject in Cohn–Kanade database of using the AdaBoost classifiers alone, using the semantic AU

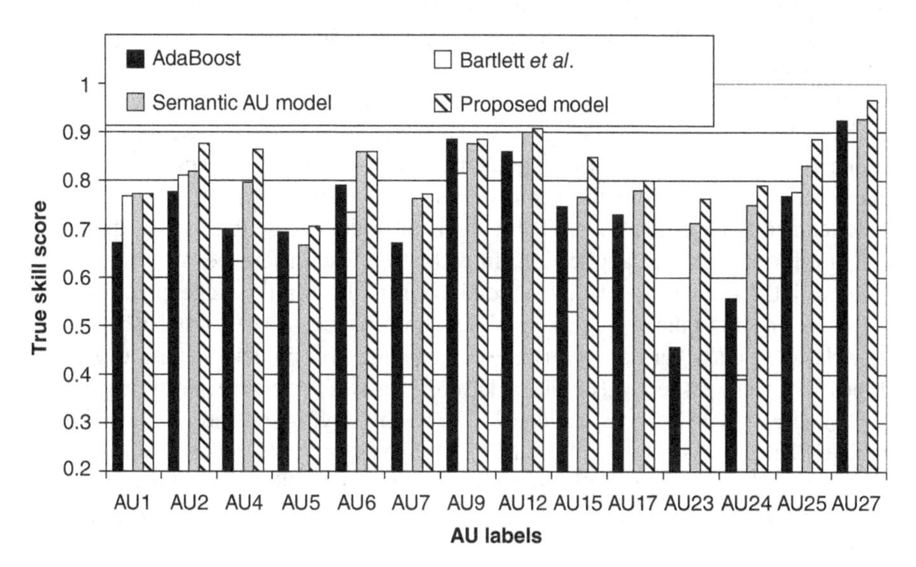

FIGURE 2.4 Comparison of AU recognition results on the novel subjects in Cohn–Kanade database using the AdaBoost classifier, the results of Bartlett *et al.* [29], the semantic AU model [9], and proposed method based on the true skill score.

model [9], using the proposed model, and Bartlett *et al.*'s method [29], respectively. The AdaBoost classifiers achieve an average positive recognition rate (PRR) 80.6% and false positive rate (FPR) 7.84% for the 14 target AUs. With the proposed DBN model, the average PRR further increases to 88.6% and FPR decreases to 5.03%.

As shown in Figure 2.4, for the AUs that are difficult to be recognized, the improvements are impressive, which exactly demonstrates the benefit of using the DBN. For example, since AU23 and AU24 are contracted by the same set of facial muscles, the probability of the two AUs copresence is high. Furthermore, AU24 is activated mostly after the chin boss is moved upward, that is, activating AU17 (chin raiser). By employing such semantic and dynamic relationships in DBN, the PRR of AU23 increases from 54% to 80.2% with the FPR decreasing from 8.7% to 3.7%; and the PRR of AU24 increases from 63.4% to 84.2% with the FPR decreasing from 7.9% to 5.4%. Similarly, by employing the co-absence relationship between AU25 (lips apart) and AU15 (lip corner depressor) as well as the dynamic relationship between AU2 (outer brow raiser) and AU25, the FPR of AU25 reduces from 13.3% to 4.6% and the PRR increases from 89.8% to 93.6%.

2.4.3 Evaluation on Spontaneous Facial Expression Database

Instead of recognizing AUs from the posed facial expressions, it is more important to recognize AUs from natural facial expressions. Therefore, the proposed system is trained and tested on a spontaneous facial expression database (MAD database [28]). Currently, the database contains a total of 4093 FACS-labeled images from five subjects (four female and one male, three African American and two European). In this work, we intend to recognize 12 target AUs as shown in Figure 2.5, which frequently occur in the image sequences. We use 2856 images for training, while 1237 images for testing.

Figure 2.5 shows the learned DBN model from the spontaneous facial expression database. We can find that although most of the semantic and dynamic relationships among AUs are true for both the spontaneous and posed facial expressions, some relationships learned from the spontaneous facial expressions are different from those learned from posed ones. That demonstrates the necessity to refine the DBN structure for a particular application. In addition, these differences can be used to distinguish the spontaneous facial expressions from the posed facial expressions.

Figure 2.6 shows the average recognition performance on MAD database of using the AdaBoost classifiers alone, using the semantic AU model [9], and using the proposed model, respectively. The AdaBoost classifiers achieve an average PRR 72.6% and FPR 8.85% for the 12 target AUs. By employing semantic AU model, the average PRR increases to 76.3% and FPR decreases to 7.02%. With the use of the proposed model, the average PRR increases to 82.9% and FPR decreases to 6.7%. This demonstrates the effectiveness and importance of modeling dynamic dependencies among AUs.

As we mentioned previously, AU recognition from spontaneous facial expressions is much more difficult than recognition from posed facial expressions especially due to the low intensities of AUs and the multiple facial expressions that appear

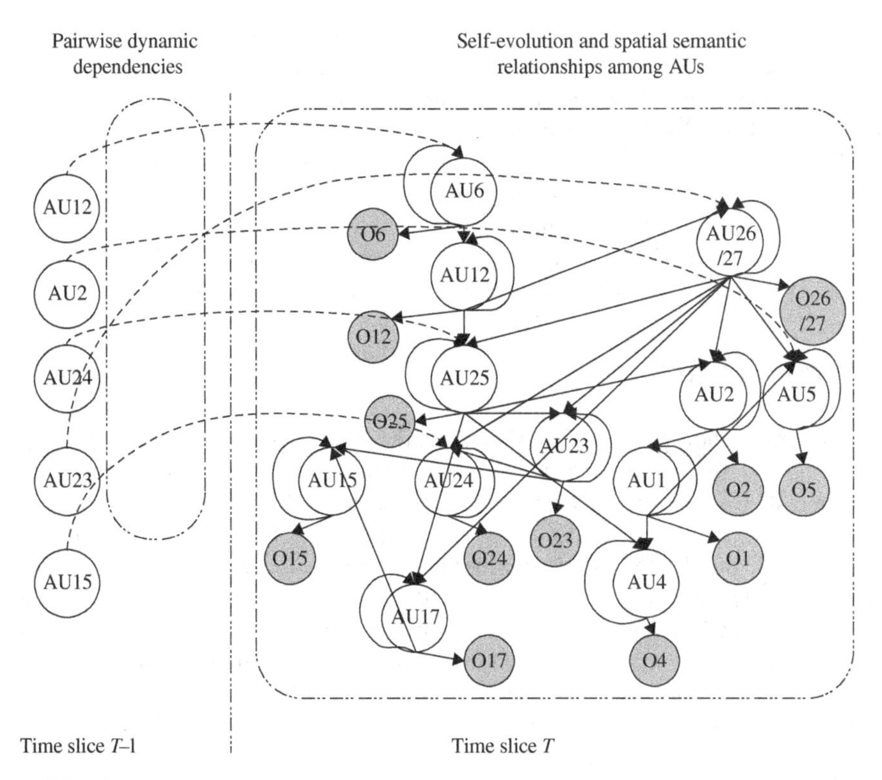

FIGURE 2.5 The complete DBN model for AU modeling in spontaneous facial expression database.

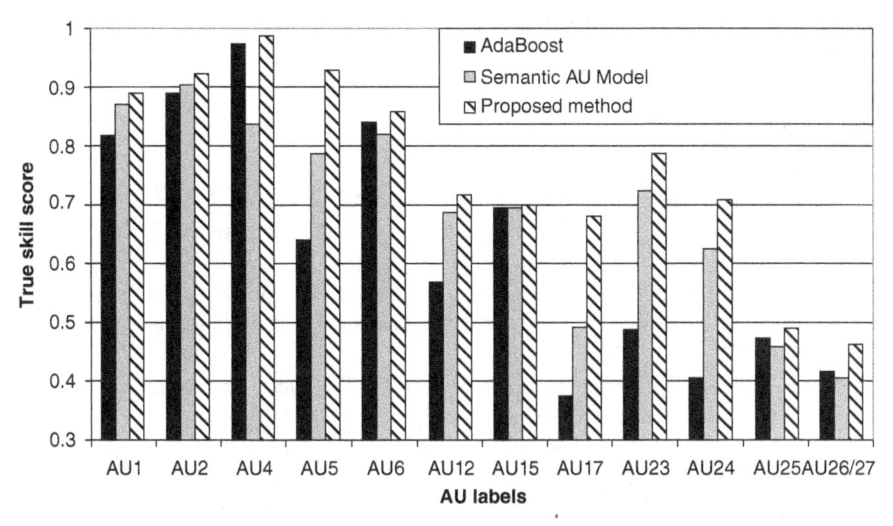

FIGURE 2.6 Comparison of AU recognition results in MAD database using the AdaBoost classifier, the semantic AU model, and the proposed model based on the true skill score.

sequentially. By employing the dynamic relationships between AUs, the recognition performance of some difficult AUs improves significantly. For example, AU25 (lips part) will appear mostly after AU24 (lip presser) is activated in the speech-related context. And the lip will be first tightened by activating AU23 (lip tightener), and then dropped down by contracting AU26 (jaw drop). Therefore, by employing dynamic relationships in DBN, the PRR of AU23 increases from 54.3% to 83.3% with an FPR decreasing from 5.6% to 4.3%; and the PRR of AU24 increases from 48.7% to 78.8% with an FPR decreasing from 8.2% to 7.88%. And the PRR of AU17 (chin raiser) increases from 45.5% to 72.7% with an FPR decreasing from 7.94% to 4.83%.

2.5 CONCLUSION

The recent psychological research shows that the dynamic dependencies among AUs are important for interpreting naturalistic human behavior. However, most of the existing AU recognition approaches often ignore the dynamic relationships of AUs. In this work, we systematically model the strong and common temporal causality between pair of AUs by a DBN. Furthermore, advanced learning algorithm is proposed to systematically learn the DBN model based on both training data and the prior subjective knowledge. The experiment results demonstrate that the proposed system with the learned dynamic AU relationships yields significant improvements over the method that only focus on modeling semantic AU relationships in recognizing AUs from both posed and spontaneous facial expressions. The future work will focus on two aspects. Theoretically, we plan to extend the work to recognizing the full temporal pattern of the AU including neutral, onset, apex, and offset states, which would be more challenging for the DBN modeling procedure and learning approach. For applications, we would like to employ the learned spatiotemporal relationships among AUs to distinguish fake facial expressions from genuine and natural facial expressions.

REFERENCES

[1] P. Ekman, W. V. Friesen, and J. C. Hager, *Facial Action Coding System: The Manual*, Research Nexus Division, Network Information Research Corporation, Salt Lake City, UT, 2002.

[2] M. Pantic and M. S. Bartlett, "Machine analysis of facial expressions," in *Face Recognition* (eds K. Delac and M. Grgic), I-Tech Education and Publishing, Vienna, Austria, 2007, pp. 377–416.

[3] J. Russell and J. Fernandez-Dols, *The Psychology of Facial Expression*, Cambridge University Press, New York, 1997.

[4] M. Pantic and I. Patras, "Dynamics of facial expression: recognition of facial actions and their temporal segments from face profile image sequences," *IEEE Transactions on Systems, Man, and Cybernetics – Part B: Cybernetics*, 36(2): 433–449, April 2006.

[5] S. Nishio, K. Koyama, and T. Nakamura, "Temporal differences in eye and mouth movements classifying facial expressions of smiles," in Proceedings of the Third IEEE

International Conference on Automatic Face and Gesture Recognition, April 1998, pp. 206–211.

[6] J. F. Cohn, L. I. Reed, Z. Ambadar, J. Xiao, and T. Moriyama, "Automatic analysis and recognition of brow actions and head motion in spontaneous facial behavior," in Proceedings of the IEEE International Conference on Systems, Man, and Cybernetics, vol. 1, 2004, pp. 610–616.

[7] K. Schmidt and J. Cohn, "Dynamics of facial expression: normative characteristics and individual differences," in IEEE International Conference on Multimedia and Expo, 2001, pp. 728–731.

[8] M. F. Valstar, M. Pantic, Z. Ambadar, and J. F. Cohn, "Spontaneous vs. posed facial behavior: automatic analysis of brow actions," in Proceedings of the Eighth International Conference on Multimodal Interfaces, 2006, pp. 162–170.

[9] Y. Tong, W. Liao, and Q. Ji, "Facial action unit recognition by exploiting their dynamic and semantic relationships," IEEE Transactions on Pattern Analysis and Machine Intelligence, 29(10): 1683–1699, October 2007.

[10] B. Fasel and J. Luettin, "Automatic facial expression analysis: a survey," *Pattern Recognition*, 36: 259–275, 2003.

[11] M. Pantic and L. J. M. Rothkrantz, "Automatic analysis of facial expressions: the state of the art," *IEEE Transactions on Pattern Analysis and Machine Intelligence*, 22(12): 1424–1445, 2000.

[12] M. Valstar, M. Pantic, and I. Patras, "Motion history for facial action detection in video," in Proceedings of the IEEE International Conference on Systems, Man, and Cybernetics, vol. 1, 2004, pp. 635–640.

[13] P. Yang, Q. Liu, and D. N. Metaxas, "Boosting coded dynamic features for facial action units and facial expression recognition," in Proceedings of the IEEE International Conference on Computer Vision and Pattern Recognition, June 2007, pp. 1–6.

[14] G. Zhao and M. Pietiäinen, "Dynamic texture recognition using local binary patterns with an application to facial expressions," *IEEE Transactions on Pattern Recognition and Machine Intelligence*, 29(6): 915–928, June 2007.

[15] I. Cohen, N. Sebe, A. Garg, L. Chen, and T. S. Huang, "Facial expression recognition from video sequences: temporal and static modeling," *Comput. Vis. Image Underst.* (Special Issue on Face Recognition), 91(1–2): 160–187, 2003.

[16] J. J. Lien, T. Kanade, J. F. Cohn, and C. Li, "Detection, tracking, and classification of action units in facial expression," *J. Robot. Auton. Syst.*, 31: 131–146, 2000.

[17] N. Oliver, F. Berard, and A. Pentland, "LAFTER: lips and face tracking," in Proceedings of the IEEE International Conference on Computer Vision and Pattern Recognition, 1997.

[18] T. Otsuka and J. Ohya, "Recognizing multiple persons' facial expressions using HMM based on automatic extraction of significant frames from image sequences," in Proceedings of the IEEE International Conference on Image Processing, vol. 2, 1997, pp. 546–549.

[19] H. Gu and Q. Ji, "Facial event classification with task oriented dynamic bayesian network," in Proceedings of the IEEE International Conference on Computer Vision and Pattern Recognition, vol. 2, 2004, pp. 870–875.

[20] Y. Zhang and Q. Ji, "Active and dynamic information fusion for facial expression understanding from image sequences," *IEEE Transactions on Pattern Analysis and Machine Intelligence*, 27(5): 699–714, May 2005.

[21] T. Dean and K. Kanazawa, "Probabilistic temporal reasoning," in AAAI-88, 1988, pp. 524–528.

[22] T. Kanade, J. F. Cohn, and Y. Tian, "Comprehensive database for facial expression analysis," in Proceedings of the Fourth IEEE International Conference on Automatic Face and Gesture Recognition, 2000, pp. 46–53.

[23] Intelligent Systems Lab (ISL) database, "RPI ISL facial expression databases," http://www.ecse.rpi.edu/homepages/cvrl/database/database.html. Last accessed on Sept. 20, 2014.

[24] G. Schwarz, "Estimating the dimension of a model," *Ann. Stat.*, 6: 461–464, 1978.

[25] D. Heckerman, D. Geiger, and D. M. Chickering. "Learning bayesian networks: the combination of knowledge and statistical data," *Mach. Learn.*, 20(3): 197–243, 1995.

[26] M. S. Bartlett, G. C. Littlewort, M. G. Frank, C. Lainscsek, I. R. Fasel, and J. R. Movellan, "Automatic recognition of facial actions in spontaneous expressions," *J. Multimed.*, 1(6): 22–35, 2006.

[27] K. B. Korb and A. E. Nicholson, *Bayesian Artificial Intelligence*, Chapman & Hall/CRC, London, UK, 2004. Last accessed on Sept. 20, 2014.

[28] Multiple Aspects of Discourse research lab. http://madresearchlab.org/. Last accessed on Sept. 20, 2014.

[29] M. S. Bartlett, G. C. Littlewort, J. R. Movellan, and M. G. Frank, "Fully automated facial action coding," 2007. http://mplab.ucsd.edu/grants/project1/research/Fully-Auto-FACS-Coding.html. Last accessed on Sept. 20, 2014.

AUTHOR BIOGRAPHIES

Yan Tong received her Ph.D. degree in Electrical Engineering from Rensselaer Polytechnic Institute, Troy, New York. She is currently an assistant professor in the Department of Computer Science and Engineering, University of South Carolina. Before joining USC, she had been working in the Visualization and Computer Vision Lab of GE Global Research Center since January 2008.

Dr. Tong's research interests focus on computer vision and pattern recognition, especially on human–computer interaction including but not limited to affective computing, statistical shape modeling and their applications in face image interpretation and medical image analysis, and probabilistic graphical models and their applications in computer vision including modeling, reasoning, information fusion, and learning under uncertainty. She has authored/coauthored 20 peer-reviewed publications and 4 book chapters.

Qiang Ji received his Ph.D. degree in Electrical Engineering from the University of Washington. He is currently a professor with the Department of Electrical, Computer, and Systems Engineering, Rensselaer Polytechnic Institute (RPI). He recently served as a program director at the National Science Foundation (NSF), where he managed NSF's computer vision and machine learning programs. He has also held teaching and research positions with the Beckman Institute at the University of Illinois at Urbana-Champaign, the Robotics Institute at Carnegie Mellon University, the Department of Computer Science at the University of Nevada at Reno, and the U.S. Air Force Research Laboratory. Prof. Ji currently serves as the director of the Intelligent Systems Laboratory (ISL) at RPI.

Prof. Ji's research interests are in computer vision, probabilistic graphical models, information fusion, and their applications in various fields. He has published over 150 papers in peer-reviewed journals and conferences. His research has been supported by major governmental agencies including NSF, NIH, DARPA, ONR, ARO, and AFOSR as well as by major companies including Honda and Boeing. Prof. Ji is an editor on several related IEEE and international journals and he has served as chair, technical area chair, and program committee member in numerous international conferences/workshops.

3

FACIAL EXPRESSIONS: A CROSS-CULTURAL STUDY

CHANDRANI SAHA, WASHEF AHMED, SOMA MITRA, AND DEBASIS MAZUMDAR

Centre for Development of Advanced Computing (C-DAC), Kolkata, India

SUSHMITA MITRA

Indian Statistical Institute, Kolkata, India

The current chapter will present a cross-cultural study on facial expressions. Different cross-cultural psychological studies reveal that facial expressions of a person's emotions vary across cultures. In this chapter we have computationally shown that the occurrence of Action Units (proposed by Ekman) also possess inter-culture variations. We have generated a rule base for the classification of six basic facial expressions using a decision tree and shown that the performance of the rule-based classifier improves when the rule base becomes culture specific.

3.1 INTRODUCTION

Facial expression analysis has been an active research topic for psychologists and behavioral scientists since the work of Darwin [1]. In pattern recognition, Automatic Facial Expression Analysis (AFEA) has now become a vibrant area of research. A facial expression is a human behavior that reflects the facial changes in response to a person's internal emotional states, intentions, or social communications. Appearance or dynamical change in the muscular arrangement of the face in response to the

Emotion Recognition: A Pattern Analysis Approach, First Edition. Edited by Amit Konar and Aruna Chakraborty.
© 2015 John Wiley & Sons, Inc. Published 2015 by John Wiley & Sons, Inc.

internal emotional state of the mind are designated by different emotional facial expressions. It has been established from the study of behavioral scientists [2] that the verbal part (i.e., spoken words) of a message contributes only 7% to the effect of the message as a whole, and the vocal part (e.g., voice intonation) contributes 38%, while the facial expression contributes nearly 55% to the effect of spoken message. This establishes the importance of facial expression analysis in the field of psychology and behavioral science. AFEA receives attention in the context of widespread application areas like human–computer interface (HCI), cross-lingual communication, etc. Research in the area of AFEA was brought into focus through the pioneering work of Bruce [3], Takeuchi and Nagao [4], and Hara and Kobayashi [5]. The principal task of AFEA is to analyze the basic emotional facial expressions based on visual cues available in an image either still or video. The literature in the area of AFEA is very rich [6,7].

Most of the algorithms mentioned in the literature depend on the theory of basic emotions proposed by Ekman [8]. They conducted various experiments on human judgment on still photographs of posed facial behavior and concluded that six discrete basic emotional expressions can be recognized universally, namely, happiness, surprise, disgust, fear, anger, and sadness. They introduced action unit (AU) based description of facial expressions [9]. To date, pattern recognition researchers mostly rely on Ekman's AU–based theory for AFEA concerning classification of six discrete basic emotional facial expressions. In spite of a rich literature on AU-based facial expression analysis, we feel, there exists little information in literature addressing the effect of different ethnic cultures on the occurrence of a set of action units expressing certain facial expressions. In Reference 10, experiments were conducted by Ekman to explore whether there are universal facial expressions of emotion irrespective of culture. In Reference 11, Tsai *et al.* in their study explored the variability of emotional expressions across different cultures. Both of the works are description based, and as a result, are somewhat subjective.

This chapter introduces a computational approach to detect the cross-cultural variations in occurrence of AUs amongst different basic expressions. The regions of prominent muscle movement on face have been extracted using optical flow and appropriate mapping has been established to convert these into AUs proposed by Ekman. Rule bases have been generated using decision tree for classification of facial expressions for three different cultures. The dependence of the performance of the rule base on cultures has been analyzed. It has been established that the performance of the rule-based classifier improves when the rule base becomes culture-specific.

The chapter organization is as follows: Section 3.2 describes the extraction of facial regions bearing prominent muscle movement for different emotional facial expressions. These facial regions are then mapped with Ekman's action units. Section 3.3 explores the cultural variability in the occurrence of different AUs in six basic facial expressions. Section 3.4 describes the generation of rule base using decision tree for facial expression classification, the performance of which is evaluated in the context of cultural variability. The conclusion is given in Section 3.5.

3.2 EXTRACTION OF FACIAL REGIONS AND EKMAN'S ACTION UNITS

Emotional facial expressions are a manifestation of different muscle movements on the face. The Facial Action Coding System (FACS) defined by Ekman [12], is the best known and the most frequently used framework developed for human observers to describe facial activities. FACS uses visually observable facial muscle actions, termed as action units, to describe the facial expressions. It is a comprehensive, anatomy-based system for describing six basic facial expressions. Cardinality of the set of basic AUs used by Ekman *et al.* to describe facial actions is 44. Ordinality of the AUs are as per the list prepared by Ekman *et al.* Out of these 44 AUs, researchers [13] adapted 18 as the most significant in describing the six basic emotional expressions with optimum distinguishability (as shown in Figure 3.1). FACS coding procedures allow for linguistic description and coding of facial expressions in terms of events. An event is the AU-based description of each facial expression, which may consist of a single AU or many AUs in combination [15]. Classic psychological studies, like the EMFACS (emotional FACS) [12], suggest the possibility of mapping AUs onto the basic emotion categories using a finite number of rules. Table 3.1 summarizes the rule of association between action units and six basic emotional expressions [13].

In this section we describe the methodology to extract AUs from image sequences of basic expressions. The image sequences are obtained by splitting videos into successive frames. The frame in the video sequence that contains the first instance of facial deformation is termed as the ***first significant frame***. The next frame showing further deformation is called the ***second significant frame***. Gradient-based optical flow, developed by Horn and Schunck [16], is used for extracting the motion vectors of different pixels in a facial image. Thirteen facial regions, like forehead, eye, lip,

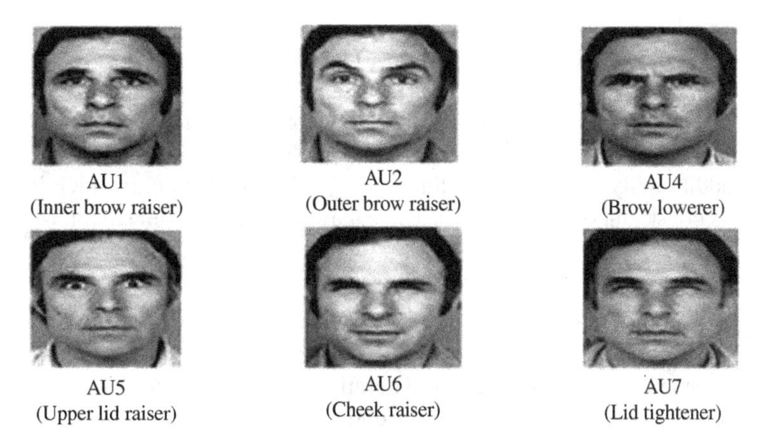

FIGURE 3.1 Set of most significant action units to describe six basic expressions. Adapted with permission from FACS manual [14]. © Joseph C. Hager (2002).

TABLE 3.1 Rule of association between AUs and basic expressions

Facial expressions	Action units
Happiness	6, 12, 16, 25, 26
Surprise	1, 2, 5, 26, 27
Disgust	9, 10, 17, 25, 26
Anger	2, 4, 7, 16, 17, 23, 24, 25, 26
Fear	1, 4, 5, 7, 20, 25, 26
Sadness	1, 4, 7, 15, 17, 25, 26

cheek, have been identified from the magnitude of the optical flow vectors. Those regions bear maximum deformation during the display of facial expressions. The feature vectors are calculated within these regions, and are used by decision tree to classify each basic expression. The six generated two-class decision trees are analyzed to identify the AUs occurring in each basic facial expression, when compared to the neutral face.

3.2.1 Computation of Optical Flow Vector Representing Muscle Movement

Optical flow is the apparent motion of brightness patterns in an image sequence. A differential method or gradient-based method developed to compute the components of optical flow (u, v) from a pair of images was proposed by Horn and Schunck in their seminal paper [16]. The algorithm assumes that as points move, their brightness $E(x, y)$ at a point (x, y) does not change significantly within a small time interval (here it is inter-frame delay). This imposes a brightness constraint, which generates an ill-posed problem for obtaining two components of optical flow vector. The linear equation obtained from brightness constraint is given by

$$uE_x + vE_y + E_t = 0 \tag{3.1}$$

Here, E_x, E_y, and E_t are the derivatives of $E(x, y)$ along x, y and time (t), respectively. The components of optical flow vector are denoted by (u, v), respectively. Horn and Schunck additionally assumed smoothness of the flow vector field, which imposes smoothness constraint on the problem. The equation obtained from smoothness constraint is given by

$$\nabla^2 u + \nabla^2 v = 0 \tag{3.2}$$

These two constraint equations are solved iteratively to obtain a numerical estimate of optical flow velocity components. Here the optical flow vectors quantify the deformation resulting from muscle movements during the display of emotional expressions. Figure 3.2 depicts the optical flow vectors on a face of Indian origin, displaying the happiness expression.

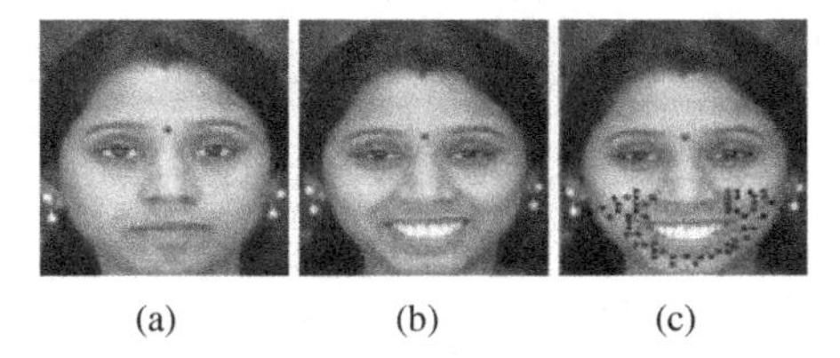

FIGURE 3.2 Happiness expression of Indian subject: (a) neutral face, (b) face with happiness expression, (c) generated optical flow.

3.2.2 Computation of Region of Interest

Emotional facial expressions cause deformation of different regions of the face for a particular facial expression. The deformations are anatomically related to movement of a specific facial muscle group [12]. This results in movement in the surrounding tissues, which in turn results in a change in frame intensity in a video sequence capturing a facial expression display. Here, we manually choose, from the input expression video sequences, the frame with neutral face and that with the apex of expression. The facial regions are then detected using a Haar wavelet–based face detector [17]. The gradient-based optical flow is then computed globally on the face region between the neutral face and the face with the apex of the expression. An experiment has been conducted with six basic emotional facial expressions on 50 Indian subjects. Upto 20% of the maximum magnitude of the flow vector has been considered, for each expression, to identify the maximally deformed face region. Any flow vector less than this threshold is relegated as noise. The resulting optical flow vector for the six basic expressions is shown in Figure 3.3, to highlight their impact with respect to the neutral face.

Magnitude of optical flow for Indian faces is found to be high in 13 major regions, which display maximum deformations for all six basic expressions. These regions

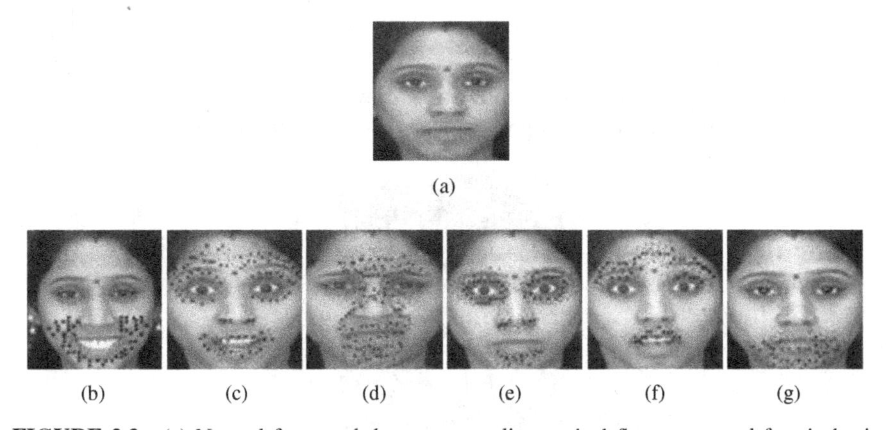

FIGURE 3.3 (a) Neutral face, and the corresponding optical flow generated for six basic emotional facial expressions: (b) happiness, (c) surprise, (d) disgust, (e) anger, (f) fear, and (g) sadness.

TABLE 3.2 List of ROI and abbreviation of computed feature vectors

S. no.	Regions	Feature vector (Mean, standard deviation)
0	Forehead	A_0,A_1
1	Right eyebrow	A_2,A_3
2	Left eyebrow	A_4,A_5
3	Right eye	A_6,A_7
4	Left eye	A_8,A_9
5	Middle of two inner brows	A_{10},A_{11}
6	Nasion (root of nose)	A_{12},A_{13}
7	Right cheek	A_{14},A_{15}
8	Left cheek	A_{16},A_{17}
9	Right nasalobial region	A_{18},A_{19}
10	Left nasalobial region	A_{20},A_{21}
11	Mouth	A_{22},A_{23}
12	Chin	A_{24},A_{25}

are listed in the second column of Table 3.2. Figure 3.4 shows the local windows on an Indian face, created around the regions of interest (ROI) containing optical flow of high magnitude. The numbers marked on the windows over the face correspond to the first column of Table 3.2. The advantage of such local computation of optical flow within the windowed regions arise from the more compact and reliable mapping to AU.

3.2.3 Computation of Feature Vectors Within ROI

The direction of orientation of the windows with global horizontal axis is taken as the window axis. Projection P_{ij} of optical flow vectors $\vec{w}_{ij}(\vec{X})$ is taken on the long symmetry axis of the ith window for the jth pixel as

$$P_{ij} = \vec{w}_{ij}(\vec{X}).\vec{n_i} \tag{3.3}$$

where $\vec{w}_{ij}(\vec{X}) = (u_{ij}(x,y), v_{ij}(x,y))$ is the optical flow vector in jth pixel of ith window computed from two successive significant frames. Here the frames carrying large

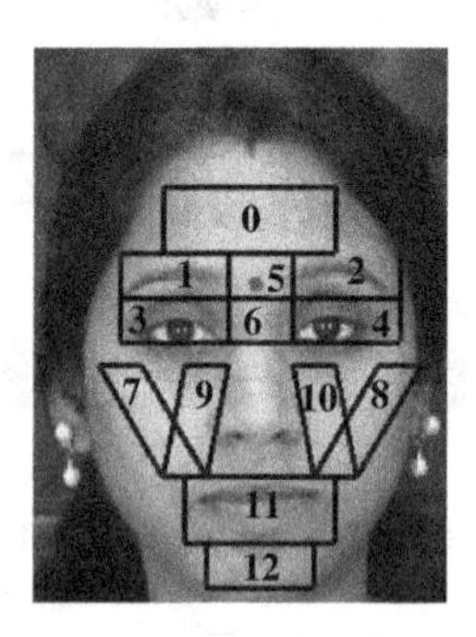

FIGURE 3.4 Windowed regions in an Indian face bearing maximum deformation during basic facial expression display.

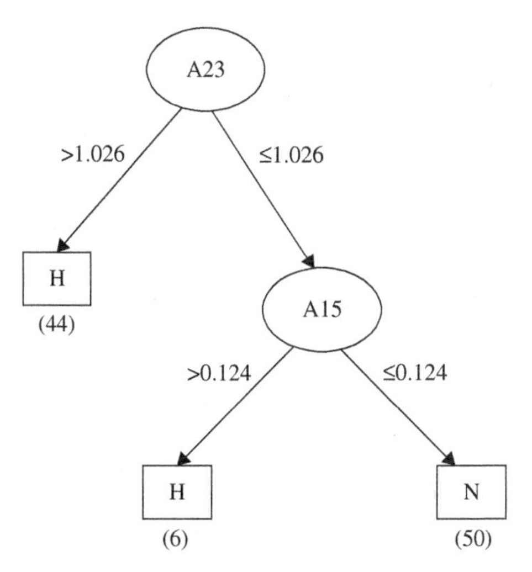

FIGURE 3.5 Decision tree for classifying happiness from the corresponding neutral faces for Indian subjects.

deformation due to expression are termed as significant frames, and $\vec{n_i}$ is the unit vector along the axis of the ith window ($i = 0,\ldots,12$). We use the mean and standard deviation of the flow projections, within the 13 windows, to compute the feature vectors. So there is 26D feature vector denoted by $A_0, A_1,\ldots,A_{25}$, and listed in third column of Table 3.2.

3.2.4 Facial Deformation and Ekman's Action Units

We use decision trees to map facial deformation with Ekman's action units. Decision tree [18] is a popular classification method that results in a flowchart-like tree structure, where each node denotes a test on an attribute value and each branch represents an outcome of the test. The tree leaves represent the classes. In the current work, the discriminatory power of each attribute is evaluated using Shanon's entropy. The decision tree program C5.0 developed by J. Ross Quinlan[1] has been used here. The 26 dimensional feature vectors $A_0, A_1,\ldots,A_{25}$, mentioned in third column of Table 3.2, are used as attributes to the decision tree. The discrimination between the neutral and each basic expression is made using a projection of optical flow along the window axis as feature vector. The features prioritized by decision tree are mapped with Ekman's AUs for that specific expression.

Figure 3.5 shows the decision tree for classifying happiness of 50 Indian subjects with respect to the corresponding neutral faces. The maximum information carrying discriminatory attributes A_{23} and A_{15} are found to belong to the mouth and right cheek regions (Table 3.2) implying that the AUs for happiness expression lie in the mouth and cheek regions. The histogram of direction of optical flow in mouth and cheek regions is illustrated in Figure 3.6. The histogram of mouth region shows three

[1] http://www.rulequest.com/index.html

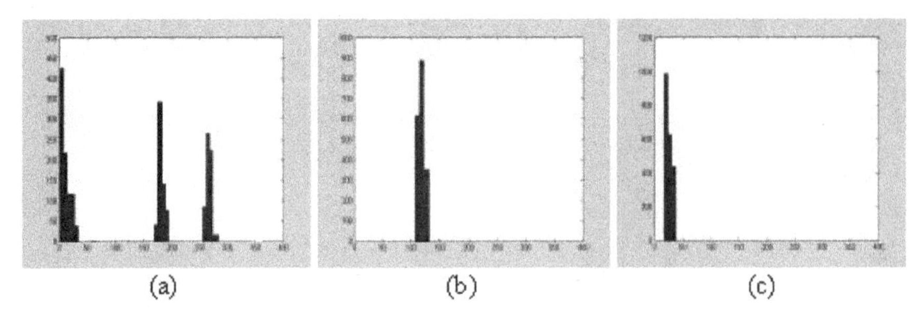

FIGURE 3.6 Optical flow direction histogram of: (a) mouth, (b) right cheek, and (c) left cheek.

distinct peaks around 0°–25°, 170°–190°, and 250°–290°. The first and second peak arise due to horizontal movement of the lip, and hence, signifies the occurrence of AU "lip corner puller" (AU 12) of Figure 3.1. The last peak arises due to downward movement of mouth, and is therefore associated with AU 25 and AU 26 (lips part and jaw drop, respectively). One prominent peak around 120°–150° appears in right cheek region and another around 30°–60° appears in left cheek region. These correspond to the occurrence of AU 6, that is, cheek raiser of Figure 3.1. Table 3.3 presents the complete list of AUs involved in happiness expression, obtained by observing the decision tree and optical flow directions.

The same procedure has been repeated for the other basic expressions of Indian subjects, followed by the extraction of the associated AUs. The second column of Table 3.4 presents the AUs associated with the six basic expressions for Indian subjects.

It is observed from Tables 3.4 and 3.1 that there is a variation in the rule of associations between the basic expressions and Ekman's AUs. This variation has been studied in detail for three different cultures, in the next section, over the six basic facial expressions.

3.3 CULTURAL VARIATION IN OCCURRENCE OF DIFFERENT AUS

Culture receives different descriptions by different researchers in the field of sociology, psychology, and behavioural sciences. Z. Ruttkay [19] interpreted culture as a

TABLE 3.3 Action units corresponding to deformed regions for "happiness" expression in case of Indian subjects

Facial region deformed due to action	Direction of deformation	Mapped action units for happiness
Right cheek	120°–150°	AU 6—cheek raiser
Left cheek	30°–60°	
Mouth	0°–25°	
	170°–190°	AU 12—lip corner puller
	250°–290°	AU 25, AU 26—lips part, jaw drop

TABLE 3.4 Rule of association between AUs and basic expressions for different cultures

Facial expressions	AU (Indian)	AU (Japanese)	AU (Euro-American)
Happiness	**6, 12, 25, 26**	**6, 12, 25, 26**	**6, 12, 25, 26**
Surprise	**1, 2, 5, 25, 26, 27**	**1, 2, 5, 25, 26, 27**	**1, 2, 5,** 6, **25, 26, 27**
Disgust	**4, 9, 10,** 17, **20,** 25	**4, 9, 10, 20,** 24	**4, 9, 10, 20**
Anger	2, **4,** 5, 7, 17, **23, 24,** 25	4, 7, 9, 10, 20, **23, 24,** 25	**4,** 5, 9, 10, 17, 20, **23, 24**
Fear	**1, 2, 4,** 5, **25, 26, 27**	**1, 2, 4,** 10, 16, 20, 24, **25, 26, 27**	**1, 2, 4,** 5, 10, 15, 20, 24, **25, 26, 27**
Sadness	1, **4, 7, 15, 17,** 25	**4, 7, 15, 17,** 25	**4, 7, 15, 17**

set of characteristics which form a "common denominator" among groups of people, including both mental and communicative characteristics, as values in life and multimodal language usage. Culture is manifested in all levels of processes guiding social interaction, including emotional facial expressions. In study of cross-cultural psychology, an important hypothesis is experimentally verified by Tsai *et al.* [11] regarding whether Euro-Americans (EA) having their origin in Scandanavian countries (EA-S) show difference in emotional facial expressions compared to Euro-Americans from Ireland (EA-I). The test result revealed that EA-S are less expressive than EA-I while reliving various emotions, especially happiness and love. This suggests that, EAs continue to be influenced by their cultural heritages. EA-S people emphasize on emotional control whereas EA-I people put more value on emotional expressions. This study supports the role of cultural difference while displaying emotional expressions on the face.

In AU-based automatic facial expression analysis, the important question is whether the rule base proposed by Zhang (Table 3.1) is universal or there exists any cross-cultural variation that introduces uncertainty during the placement of class boundaries by any classifier. To study the cross-cultural variation, experiment was conducted with the experimental data-set consisting of facial expressions of three distinct cultural origin, namely,

1. Indian (C-DAC, Kolkata database)[2];
2. Japanese (NICT, Japan, database,[3] and JAFFE database[4]);
3. Euro-American (Cohn–Kanade database)[5].

C-DAC, Kolkata database contains video sequences of six basic facial expressions of 50 subjects—27 males and 23 females in the age group 25–35. The NICT, Japan, database contains video sequences of six basic facial expressions of 50 subjects. There

[2]C-DAC, Kolkata, India, Facial Expression Video database, chandrani.saha@cdac.in
[3]NICT, Japan, Facial Expression Video database, yamazaki@nict.go.jp
[4]Japanese Female Facial Expression database, http://www.kasrl.org/jaffe_download.html
[5]Cohn–Kanade AU-Coded Facial Expression database, http://www.consortium.ri.cmu.edu/ckagree/

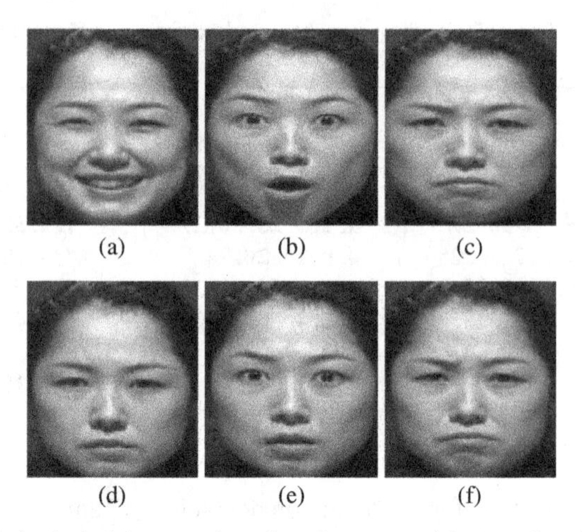

FIGURE 3.7 Six basic facial expressions for a Japanese subject: (a) happiness, (b) surprise, (c) disgust, (d) anger, (e) fear, and (f) sadness. Extracted with permission from NICT database.

are 24 males and 26 females of the age group 25–40. In both the databases, the videos are grabbed under constrained illumination by high definition video camera (Sony HVR-V1P). The JAFFE database contains 213 still images of 7 facial expressions (6 basic facial expressions + 1 neutral) posed by 10 Japanese female models of the age group 30–35. Each image has been rated on 6 emotion adjectives by 60 Japanese subjects. Six basic expressions posed by 50 subjects of the age group 25–35 have been chosen from NICT and JAFFE database, for our study. Figure 3.7 depicts sample Japanese faces with the basic facial expressions.

The Cohn–Kanade database provides facial behavior recorded in 210 adults [21, 22], between the ages of 18 and 50 years. They were 69% female, 31% male, 81% Euro-American, 13% Afro-American, and 6% other groups. The images were grabbed by a Panasonic camera and video recorder under constrained illumination. From 182 subjects, 1917 digitized sequences were generated. Six basic expressions posed by 50 subjects of Euro-American culture of the age group 25–35 are chosen from the Cohn–Kanade database for the experiment. Figure 3.8 presents the sample Euro-American faces with basic expressions.

The AUs occurring in these six basic expressions of 50 subjects belonging to each of the above-mentioned cultures is extracted using the methodology described in Section 3.2.4. The rules of association between AUs and basic expressions have been derived by conglomerating the study result over all 50 subjects and is tabulated in Table 3.4. The cross-cultural variability is evident from the results, as indicated by the numerals not highlighted in columns 2–4 of Table 3.4. It is observed that although AU 6 (cheek raiser) occurs in surprise expression of Euro-Americans, it is absent in the cases of Indian and Japanese. Again, AU 17 (chin raiser), AU 24 (lip pressor), and AU 25 (lips part) are absent in disgust expression for Euro-Americans, while these are present among either Indian or Japanese subjects.

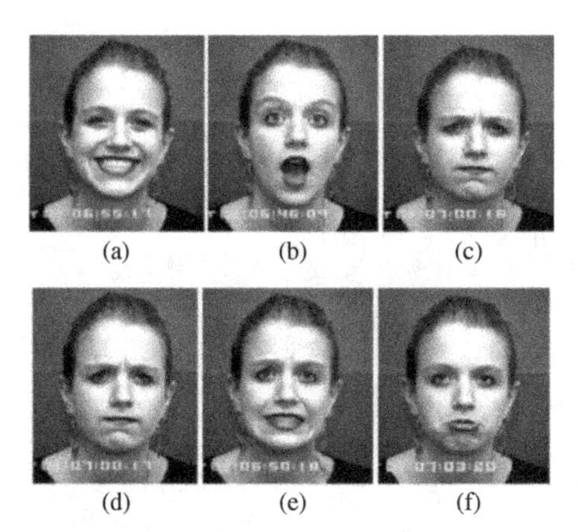

FIGURE 3.8 Six basic facial expressions for a Euro-American subject: (a) happiness, (b) surprise, (c) disgust, (d) anger, (e) fear, and (f) sadness. Extracted with permission from Cohn–Kanade database [21,22].

3.4 CLASSIFICATION PERFORMANCE CONSIDERING CULTURAL VARIABILITY

The basic emotional facial expressions are next classified using decision tree. In the first experiment, we generate a decision tree from a mixed-culture population. The sample space consists of 18 Indian subjects from C-DAC, Kolkata, database, 16 Japanese subjects from NICT, Japan, database, and 16 Euro-American subjects from Cohn–Kanade database. Thereby, the mixed population consists of a total of 50 subjects. This was divided into a training set of 30 subjects and a test set of 20 subjects. In the training set there are 10 subjects from each culture. In the test set, there were 8 Indians, 6 Japanese, and 6 Euro-Americans. For training the mixed-culture decision tree, four frames (neutral frame and three significant frames) are chosen from each subject's video sequence—starting from neutral to the apex level of the six basic expressions.

Optical flow from two successive significant frames is calculated to generate 540 ($30 \times 3 \times 6$) training cases. The 26 dimensional feature vectors $A_0, A_1, \ldots, A_{25}$ from the third column of Table 3.2 are used as the attributes of a decision tree. A mixed-culture decision tree is constructed using the algorithm C5.0. The decision tree and its subtrees are depicted in Figure 3.9. The leaf nodes of the tree are the six basic facial expressions: happiness (H), surprise (S_u), anger (A), fear (F), disgust (D), and sadness (S_a).

The rule base generated from the decision tree has been tested on a test set of 20 subjects of the created mixed-culture population. As before, four frames are chosen from each sequence for each subject, starting from the neutral to the apex of the six basic expressions. By generating optical flow vectors from successive significant frames, we have 360 ($20 \times 3 \times 6$) test cases. The performance over test case is expressed in terms of confusion matrix given in Table 3.5. The nonzero

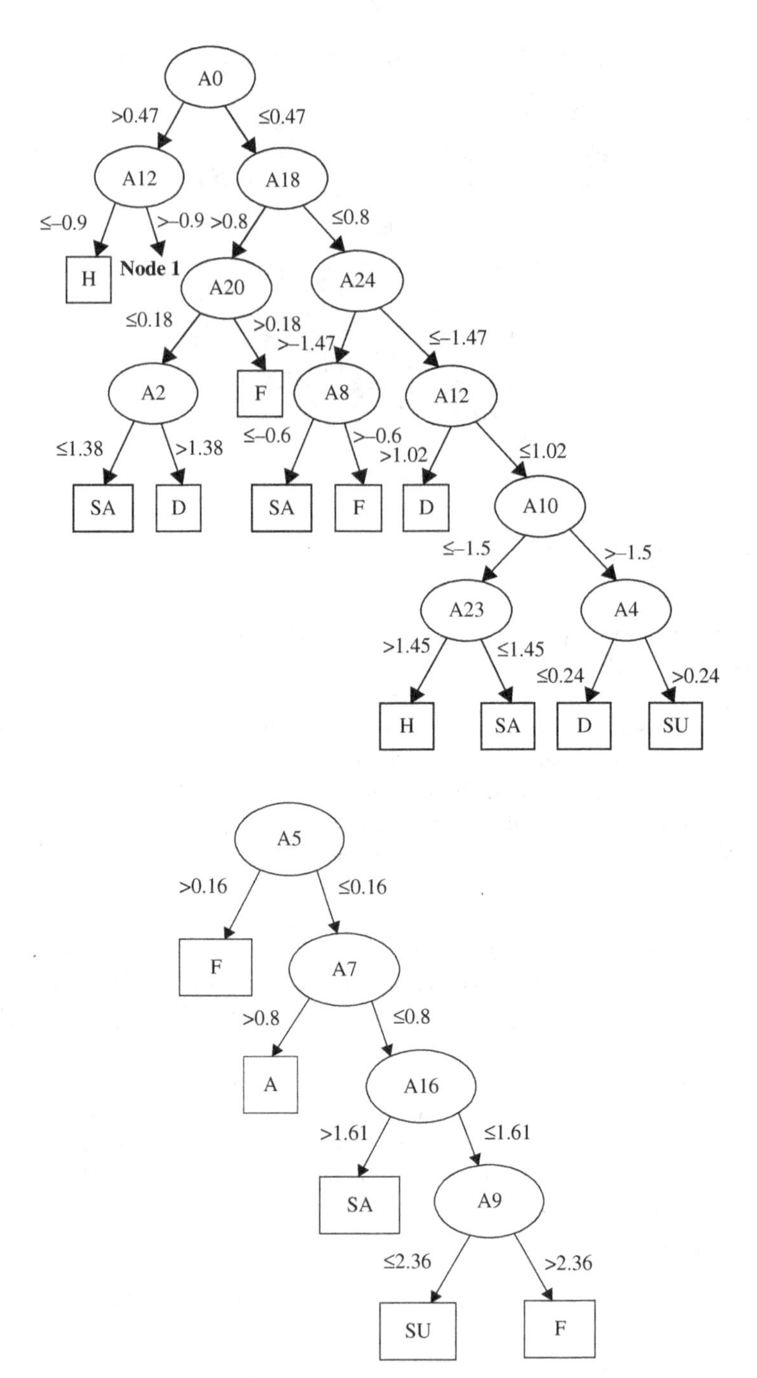

FIGURE 3.9 Decision tree for facial expression classification for mixed-culture population: (a) main tree, (b) subtree originated from node 1 of (a), (c) subtree originated from node 2 of (b), (d) subtree originated from node 3 of (c), (e) subtree originated from node 4 of (d).

TABLE 3.5 Evaluation of test data of mixed-culture population

	Happ	Surp	Ang	Fear	Disg	Sad
Happiness	20	15	7	9	8	1
Surprise	10	21	5	13	2	9
Anger	10	3	17	10	7	13
Fear	5	21	7	6	10	11
Disgust	8	7	23	5	8	9
Sadness	6	5	25	7	6	11

values in all the non-diagonal elements of the confusion matrix demonstrate the high misclassification rate.

In the next set of experiments, we generate culture-specific decision trees for each of the three cultures under consideration. The training and test sets are constructed as before. The tree for Indian subjects is depicted in Figures 3.10 and 3.11.

The attribute usage of the decision tree for Indian subjects is provided in Table 3.6.

The attribute A_{23} (standard deviation of flow projection in mouth) has the maximum entropy and hence the maximum discriminatory power. This signifies that the mouth region carries the maximum information to classify basic facial expressions. This supports the fact that the mouth has the maximum deformability and as a result suffers maximum deformation during expression display. The rule base generated from the decision tree depicted in Figure 3.10, has been tested on 20 Indian subjects of C-DAC, Kolkata, database displaying six basic emotional facial expressions. The result of evaluation on 360 test cases is expressed by the confusion matrix given in Table 3.7.

Most of the non-diagonal elements of the confusion matrix are either zero, or have a considerably lower value compared to the diagonal elements. This signifies that the misclassification rate is quite low. We also constructed decision tree and generated the rule base for the Japanese and Euro-American cultures. The tree generation has been done on training sets of 30 subjects of NICT, Japan, database and Cohn–Kanade database. The results of the test case evaluation on 20 test subjects, from these two databases are provided in Tables 3.8 and 3.9, respectively. From the confusion matrices, it is evident that the classification performance of the generated rule base improves when the decision tree becomes culture specific. This establishes the fact that there is indeed cultural variability in the display of basic facial expressions.

The performance of the decision tree resembles human recognition of facial expressions. Elfenbein *et al.* [23] conducted a meta-analysis of cross-cultural emotion recognition studies, and found that individuals were better at recognizing emotional expressions by members of their own cultural group. The within-group advantage is attributed to subtle variations in the style of expression displayed across cultures.

The classification errors for the different decision trees are summarized in Table 3.10.

It is clearly observed that the classification error significantly decreases when the decision tree is culture specific. The error is found to be higher for mixed-culture

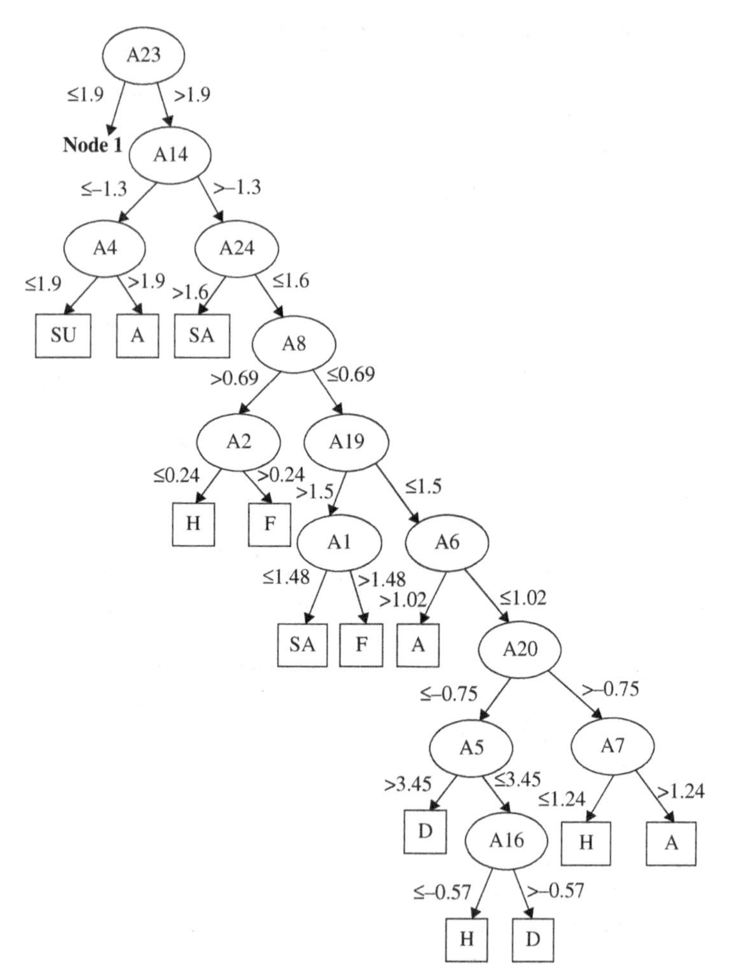

FIGURE 3.10 Decision tree for facial expression classification for Indian subjects—the main tree.

TABLE 3.6 Attribute usage of decision tree of Figure 3.10

Attribute	% usage	Attribute	% usage	Attribute	% usage
A_{23}	100%	A_{20}	76%	A_{14}	48%
A_{25}	46%	A_0	46%	A_{21}	42%
A_{24}	30%	A_3	27%	A_9	27%
A_{22}	27%	A_1	27%	A_8	26%
A_5	23%	A_7	23%	A_6	22%
A_{19}	21%	A_{10}	18%	A_{16}	17%
A_{12}	15%	A_{17}	13%	A_{17}	13%
A_4	10%	A_{15}	5%	A_{13}	5%
A_{18}	2%				

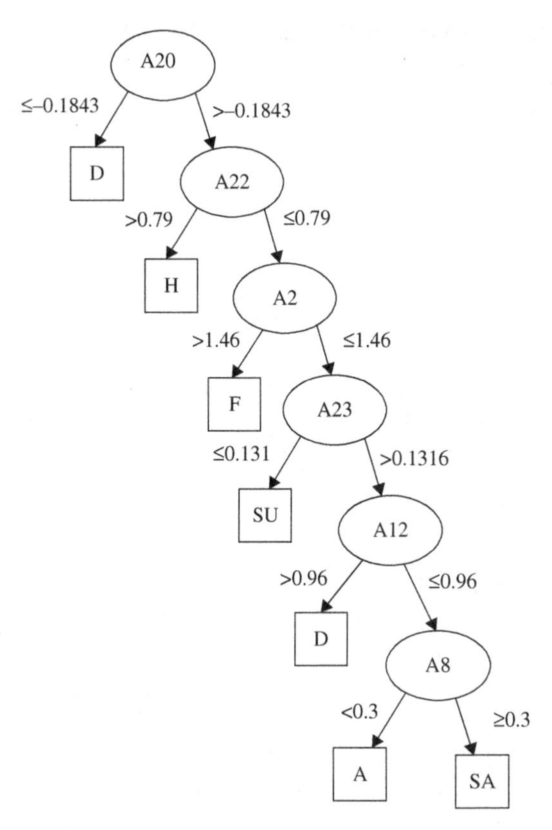

FIGURE 3.11 Subtree originated from node 1 of decision tree in Figure 3.10.

TABLE 3.7 Evaluation on test data for Indian subjects from C-DAC, Kolkata, database

	Happ	Surp	Ang	Fear	Disg	Sad
Happiness	56	4				
Surprise	2	51		7		
Anger			50	2	8	
Fear		19	2	39		
Disgust			3		53	4
Sadness			5		18	37

TABLE 3.8 Evaluation on test data for Japanese subjects from NICT, Japan, database

	Happ	Surp	Ang	Fear	Disg	Sad
Happiness	52	8				
Surprise	3	47		10		
Anger			50		10	
Fear		15		45		
Disgust			10		48	2
Sadness		3	10	6	2	30

TABLE 3.9 Evaluation on test data for Euro-American subjects from Cohn–Kanade database

	Happ	Surp	Ang	Fear	Disg	Sad
Happiness	56	4				
Surprise		57		3		
Anger			54		6	
Fear		13		47		
Disgust					50	10
Sadness			3		7	50

TABLE 3.10 Classification error of different decision tree classifiers

Culture	% classification error
Mixed culture	77
Indian	20
Japanese	25
Euro-American	14

decision tree as compared to the culture-specific ones. Another important finding from the confusion matrices is that the recognition rate for the expression happiness is the highest among all six basic facial expressions. According to Posamentier *et al.* [24] a happy smile should be easily recognizable. Facial expressions of negative emotions show a recognition deficit because they can be easily confused, such as mistaking fear for surprise, or anger for disgust, etc.

3.5 CONCLUSION

In this chapter, we have presented a cross-cultural study on facial expressions. In the present work, we have extracted the regions of prominent muscle movement using optical flow. Appropriate mapping has been established to convert these movements into action units. Occurrence of AUs across different cultures has been studied and it has been established that they possess inter-culture variations. Rule base has been generated using decision tree for classification of six basic facial expressions. It has been demonstrated that the performance of the rule-based classifier improves when the rule base becomes culture specific.

REFERENCES

[1] C. Darwin, *The Expression of Emotions in Man and Animals*, University of Chicago Press, 1965.

[2] A. Mehrabian, "Communication without words," *Psychol. Today*, 2(4): 53–56, 1968.

[3] V. Bruce, "What the human face tells the human mind: some challenges for the robot human interface," in Proceeding of the International Workshop on Robot and Human Communication, 1992, 44–51.

[4] A. Takeuchi and K. Nagao, *Communicative Facial Displace as a New Conversational Modality*, InterCHI'93, ACM Press, 1993, pp. 187–193.

[5] F. Hara and H. Kobayashi, "State of the art in component development for interactive communication with humans," *Adv. Robot.*, 11(6): 585–604, 1997.

[6] M. Pantic and L. J. M. Rothkrantz, "Automatic analysis of facial expressions: the state of the art," *IEEE Transactions on Pattern Analysis and Machine Intelligence*, 22(12): 1424–1445, 2000.

[7] B. Fasel and J. Luettin, "Automatic facial expression analysis: a survey," *Pattern Recogn.*, 36(1): 259–275, 2003.

[8] P. Ekman, W. V. Friesen, and P. Ellsworth, "*Emotion in the Human Face*," Oxford University Press, 1972.

[9] P. Ekman and W. V. Friesen, *Facial Action Coding System (FACS): Manual*, Consulting Psychologists Press, Pal Alto, CA, 1978.

[10] P. Ekman, "Universals and cultural differences in facial expressions of emotion," in *Nebraska Symposium on Motivation*, vol. 19 (ed J. Cole), University of Nebraska Press, Lincoln, 1972, pp. 207–283.

[11] J. I. Tsai, "Variation among european americans in emotional facial expression," *J. Cross Cult. Psychol.*, 34(6): 650–657, 2003.

[12] P. Ekman, W. V. Friesen, and J. C. Hager, "*The Facial Action Coding System: A Technique for the Measurement of Facial Movement*," Consulting Psychologist, San Francisco, 2002.

[13] Y. Zhang and Q. Ji, "Active and dynamic information fusion for facial expression understanding from image sequences," *IEEE Transactions on Pattern Analysis and Machine Intelligence*, 27(5): 699–714, 2005.

[14] FACS Manual, http://face-and-emotion.com/dataface/facs/manual/TOC.html

[15] P. Ekman and E. L. Rosenberg (eds), *What the Face Reveals: Basic and Applied Studies of Spontaneous Expression Using the Facial Action Coding System (FACS)*, Oxford University Press, 2005.

[16] B. K. P. Horn and B. G. Schunck, "Determining optical flow," *Artif. Intell.*, 17: 185–203, 1981.

[17] P. Viola and M. J. Jones, "Robust real-time face detection," *Int. J. Comput. Vis.*, 57(2): 137–154, 2004.

[18] S. Mitra and T. Acharya, *Data Mining: Multimedia, Soft Computing, and Bioinformatics*, John Wiley & Sons, New York, 2003, ISBN 0-471-46054-0.

[19] Z. Ruttkay, "Cultural dialects of real and synthetic emotional facial expressions," *AI Soc.*, 24(3): 307–315, 2009.

[20] J. Munkres, *Topology*, 2nd edition, Prentice Hall, 1999.

[21] T. Kanade, J. F. Cohn and Y. Tian, "Comprehensive database for facial expression analysis," Proceedings of the Fourth IEEE International Conference on Automatic Face and Gesture Recognition (FG'00), Grenoble, France, 2000, pp. 46–53.

[22] P. Lucey, J. F. Cohn, T. Kanade, J. Saragih, S. Ambadar, and I. Matthews, "The extended Cohn-Kanade dataset (CK+): A complete expression dataset for action unit and emotion-specified expression," in proceedings of the Third International Workshop on CVPR for

Human Communicative Behavior Analysis (CVPR4HB 2010), San Francisco, USA, 2010, pp. 94–101.

[23] A. H. Elfenbein and N. Ambady, "On the universality and cultural specificity of emotion recognition: a meta-analysis," *Psychol. Bull.*, 128: 203–235, 2002.

[24] M. T. Posamentier and H. Abdi, "Processing faces and facial expressions," *Neuropsychol. Rev.*, 13(3): 113–143, 2003.

AUTHOR BIOGRAPHIES

Chandrani Saha, who was a researcher in the Advanced Signal Processing Laboratory of the Centre for the Development of Advanced Computing (C-DAC), Kolkata, India, expired in December 2011.

Chandrani Saha received the B.Tech. degree from the Institute of Radio Physics and Electronics, University of Calcutta, India, in 2000 followed by M.Sc. in Physics from Jadavpur University, India, in 2004. Then, she was associated with the Centre for Development of Advanced Computing (C-DAC), Kolkata, India. She received the regional and national 1st runner up Young IT Professionals Award from Computer Society of India in 2005 and 2006. She had worked in different research and development projects on intelligent robotics, information security, etc. For the last few years she was working on Automatic Facial Expression Analysis in the Indo-Japan collaborative project "Facial feature extraction and human behavior analysis for a dialogue system." Her research interests included image processing, pattern recognition, and affective computing.

Washef Ahmed received his Bachelor of Engineering in Computer Science and Engineering from Burdwan University in 2004 and completed his Master of Technology in Computer Technology from Jadavpur University in 2007. He initially joined Wipro Technologies and served as a project engineer for 1 year. Afterward, he joined C-DAC, Kolkata, India, and since then he has been working as an engineer at Advanced Signal Processing Group, C-DAC, Kolkata, Ministry of Communication and Information Technology, Government of India.

His interests cover the domains of image processing and pattern recogntion, data structure, software engineering, and database management system. For the last 3 years he has been doing research and development on facial expression analysis.

Soma Mitra received the M.E. degree in Electronics and Telecommunication Engineering from Jadavpur University, India, in 1988. In 1990 she joined MAEP of DIT at Jadavpur University as a research engineer. In 1992 she joined C-DAC, Kolkata, where she is currently working as the Associate Director in the Advanced Signal Processing Group. She has visited the University of Rhode Island, USA, as an UNDP Fellow and NICT, Japan, under the JST fellowship.

Her research interests are image processing, pattern recognition, and soft computing. She has worked in different research and developmental projects sponsored by CFSL, DRDL, DIT, BRNS, etc. She is the coauthor of more than 10 papers in international journals and conferences. Her recent work areas are steganography, face recognition, and facial expression analysis.

Debasis Mazumdar received the M.Tech. degree in Electronics and Communication Engineering from the Institute of Radio Physics and Electronics, University of Calcutta, India, in 1990. In 1990 he was with the research and development division of Webel Telecommunication Industry, India. In 1991 he joined Central Electronics Engineering Research Institute, CSIR, Government of India. In 1993 he joined C-DAC, Kolkata, where he is currently working as the Joint Director in the Advanced Signal Processing Group. He has visited the University of Bremen, Germany, as an UNDP Fellow and NICT, Japan, under the JST fellowship. He has co-authored more than 20 papers in international journals and conferences.

His research interests include image processing, pattern recognition, speech processing, and human–computer interaction. As a senior researcher he has played a key role in developing industrial products like face recognition system, automatic bullet cartridge recognition system, real-time tracking software. Currently he is working on natural signal processing in the context of perception engineering and the statistical geometry of different feature space.

Sushmita Mitra is the Head of and a professor in Machine Intelligence Unit, Indian Statistical Institute, Kolkata, India. She is a fellow of the Indian National Academy of Engineering and the National Academy of Sciences, India. From 1992 to 1994 she was in the RWTH, Aachen, Germany, as a DAAD fellow. She was a visiting professor in the computer science departments of the University of Alberta, Edmonton, Canada, in 2004, 2007; Meiji University, Japan, in 1999, 2004, 2005, 2007; and Aalborg University, Esbjerg, Denmark, in 2002, 2003.

Dr. Mitra is a senior member of IEEE, and a fellow of the Indian National Academy of Engineering and the National Academy of Sciences, India. She is the author of the books *Neuro-Fuzzy Pattern Recognition: Methods in Soft Computing* and *Data Mining: Multimedia, Soft Computing, and Bioinformatics* published by John Wiley & Sons, and *Introduction to Machine Learning and Bioinformatics* by Chapman and Hall/CRC Press. Her research interests include pattern recognition, data mining, bioinformatics, soft computing, and multimedia.

4

A SUBJECT-DEPENDENT FACIAL EXPRESSION RECOGNITION SYSTEM

CHUAN-YU CHANG AND YAN-CHIANG HUANG

Department of Computer Science and Information Engineering, National Yunlin University of Science and Technology, Douliou, Taiwan

Facial expression recognition is one of the most popular topics in emotion analysis. Since facial expressions may be expressed differently by different people, results obtained using facial expression recognition systems based on general expression models are inaccurate. The proposed facial expression recognition system uses the facial features of an individual user. A radial basis function neural network is applied to classify seven emotions: neutral, happy, angry, surprised, sad, scared, and disgusted. Experimental results show that the proposed system can accurately identify emotions from facial expressions.

4.1 INTRODUCTION

Emotion is a neurobiological activity. Emotion recognition plays an important role in human interaction. In face-to-face communication, emotion manifests itself in several ways, such as through facial expression, paralanguage, and linguistic language.

Human emotional expressions are vital for interpersonal interaction. General expressions may include word choices, tone of voice, and body language, such as posture and physiological responses. The human face is regarded as a channel of communication that helps coordinate conversations in human–human interactions [1]. Ekman found that facial expressions are universally expressed and recognized by all humans [2]. Therefore, facial expressions are the most important

Emotion Recognition: A Pattern Analysis Approach, First Edition. Edited by Amit Konar and Aruna Chakraborty.
© 2015 John Wiley & Sons, Inc. Published 2015 by John Wiley & Sons, Inc.

information for emotion perception in face-to-face communication. Since facial expressions play a significant role in conveying human emotion, facial expression recognition is an important step in the development of a computer-facilitated human interaction system. A well-known facial expression model was given by Ekman and Friesen in Reference 3. They identified six basic human emotions, fear, surprise, sadness, anger, disgust, and happiness, and their associated facial expressions [4]. A seventh facial expression, neutral, was considered in Reference 5.

Extracting significant facial features is important in the design and implementation of automatic facial expression systems. Facial expression systems can be roughly categorized as either model- or image-based. In model-based approaches [6, 7], several feature points and landmarks are defined in the facial area. Changes of the points and landmarks are utilized for recognizing facial expressions. Although these approaches usually have low dimensionality, to ensure good performance, model-based approaches often require robust facial feature tracking techniques, which are often complex. In image-based approaches, features are extracted from the whole image of a face. These approaches are usually fast and simple, but their dimensionality is high. In order to reduce dimensionality, principal component analysis (PCA) [8–10] has been applied. Recently, Xiao *et al.* proposed a generalized framework for modeling and recognizing facial expressions on multiple manifolds [11]. In their method, each expression is modeled with an individual manifold. However, the computational cost is relatively high.

In the present study, features are extracted from color images. Although using gray scale images is faster and easier, the proposed method uses color images, which are transformed into various color spaces, to preserve information.

Besides choosing the proper facial features, accurate classification is also important. Modern classifiers, including Support Vector Machines (SVM)[12] and learning vector quantization neural networks (LVQNN)[8], have been used for facial expression recognition. In this chapter, a radial basis function neural network (RBFNN) is adopted due to its superior generalization ability [13].

Figure 4.1 shows the approach of a conventional facial expression recognition system, where facial expression is recognized using information of general expressions. Wang *et al.* proposed a person-independent facial expression space to analyze facial expressions based on supervised locality-preserving projections [14]. Matsugu *et al.* proposed a subject-independent facial expression recognition method, in which

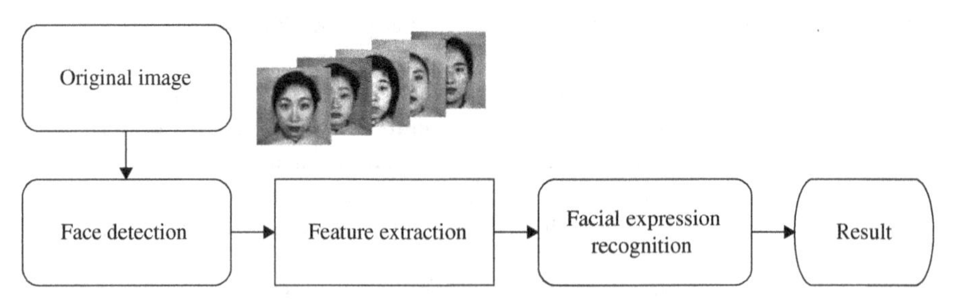

FIGURE 4.1 Conventional facial expression recognition system.

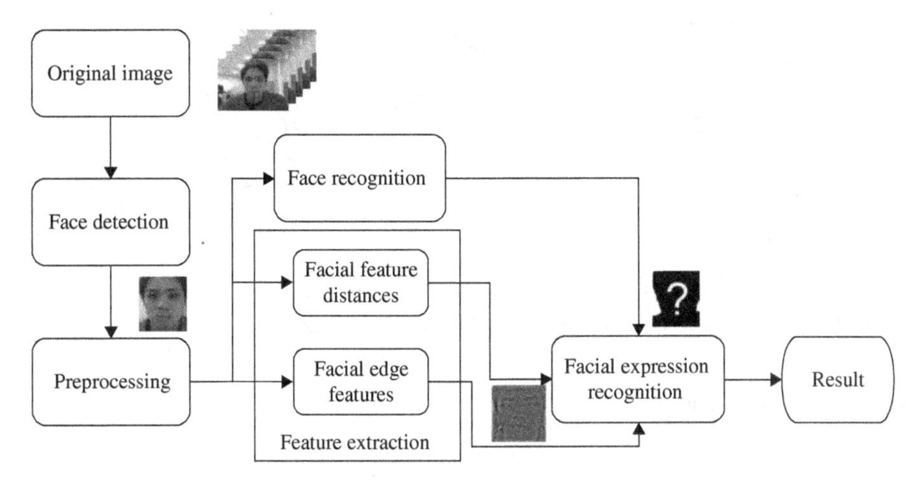

FIGURE 4.2 Subject-dependent facial expression recognition system.

a combination of a rule-based algorithm and a convolution neural network is applied to identify facial expressions [15]. Wang *et al.* proposed a face and facial expression recognition algorithm, which can simultaneously classify the given image into one of the basic facial expression categories and identify the person [5]. Their approach allows subject-independent facial expression recognition. However, facial expressions may vary with subject, so these approaches have low accuracy. In the present study, face recognition is incorporated into the facial expression system to improve performance. Figure 4.2 shows the proposed approach, where the face information of individual users is utilized in facial expression recognition. The original image is captured by a web camera. A face detection method is adopted to detect faces in the images. Then, facial features are extracted for further processing. A face recognition system is then used to recognize individual users. Based on the user's identity and facial features, an RBFNN classifier is used to recognize facial expressions.

The rest of this chapter is organized as follows. Section 4.2 presents the proposed method, including face detection, preprocessing, feature extraction, face recognition, and the facial expression recognition. Experiment results and comparisons with other methods are given in Section 4.3. Conclusions are given in Section 4.4.

4.2 PROPOSED METHOD

The proposed method consists of five major steps: face detection, preprocessing, face recognition, facial feature extraction, and facial expression recognition. Details of these processes are described below.

4.2.1 Face Detection

In this chapter, an adaptive color space switching approach is adopted for face detection [16]. In Reference 16, skin color distributions under five color spaces

(a)

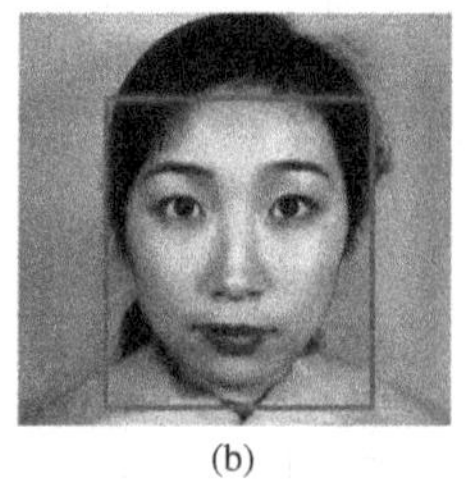
(b)

(c)

FIGURE 4.3 Samples of face detection (a) color image, (b) gray-scale image (printed with permission from JAFFE database) and (c) multiple-face image.

and Laws texture energy are used to represent facial features. The adaptive color space switching is used to select the most appropriate color space to detect faces. Faces are detected automatically under environments with complex backgrounds and varied illumination. The detected faces are extracted as face images, as shown by the red rectangles in Figure 4.3. The detected face images are utilized in subsequent processes.

4.2.2 Preprocessing

Preprocessing contains three major steps, namely, face image normalization, pupil detection, and the Gabor transform. Figure 4.4 shows a flowchart of the preprocessing procedure.

1. **Face image normalization.** Since the face detection method in Reference 16 may detect more than one face in an image, only the biggest face is selected for further processing. Then, the bilinear interpolation method is applied to normalize the size of the face image to 200×250 pixels.

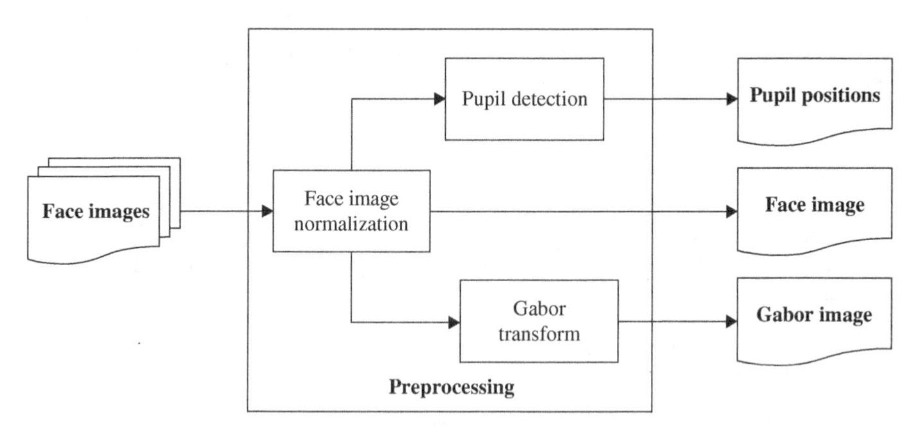

FIGURE 4.4 Preprocessing flowchart.

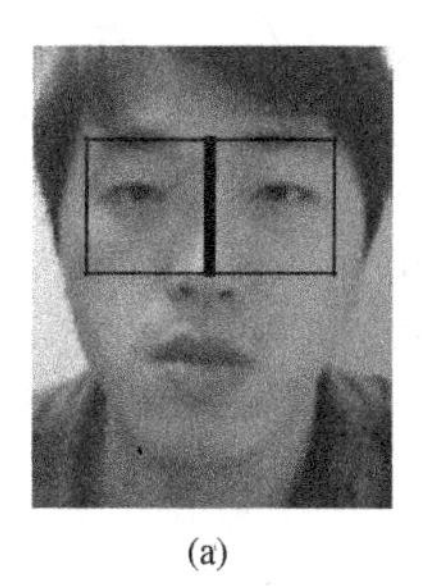
(a)

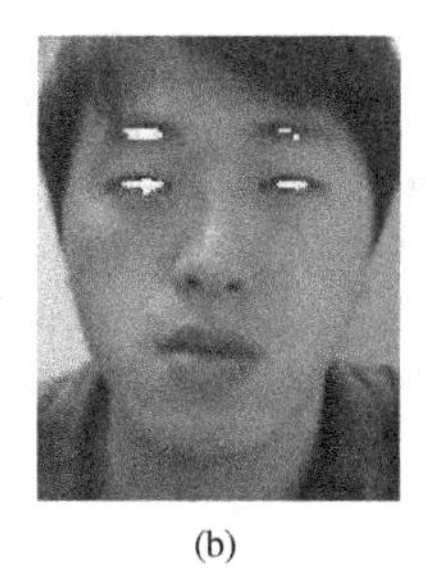
(b)

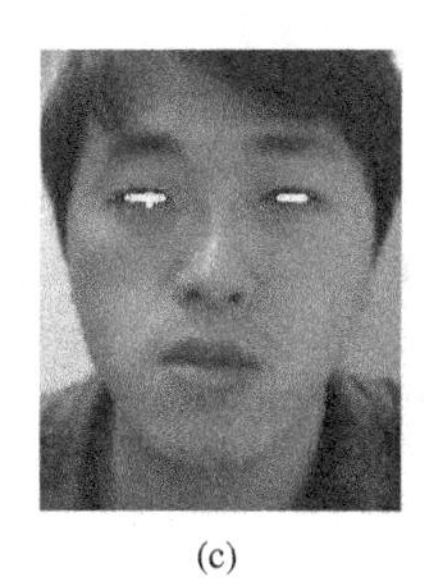
(c)

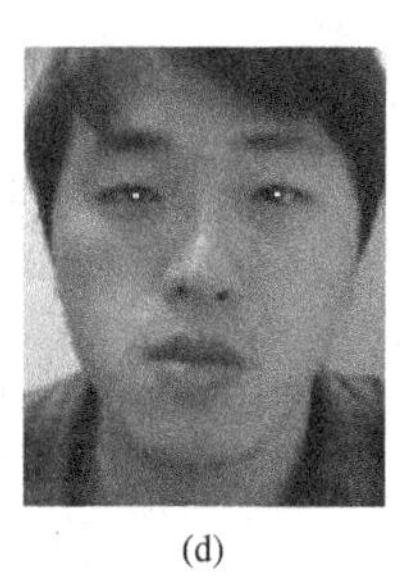
(d)

FIGURE 4.5 Pupil detection: (a) rough eye regions, (b) possible eyes, (c) detected eyes, and (d) detected pupil positions.

2. **Pupil detection.** The centers of eyes are the most important key points in a face. Therefore, pupil positions are first detected. Then, according to the pupil positions, other facial features are extracted. Based on the relationship of facial features, the rough eye regions are determined, as shown in Figure 4.5a. Possible eye regions are extracted from the YCbCr color space using

$$P_i(x,y) = \begin{cases} 1 & (Cr_i(x,y) > TCr_i) \; \& \; (Cb_i(x,y) > TCb_i) \\ 0 & \text{otherwise} \end{cases} \tag{4.1}$$

where $i = R, L$ represent the left and right eyes, respectively. TCr_i and TCb_i are the thresholds in the Cr and Cb color spaces of rough eye regions, respectively. In this study, the thresholds were set to 60% of the intensity in the Cr and Cb color spaces, respectively. Figure 4.5b shows the detected possible eye regions. In general, the eyes appear in the center of rough eye regions. Thus, the Euclidean distance between the possible eyes regions and the center of the rough eye regions is calculated.

$$\text{MeanDis}_i = \frac{\sum_{y=0}^{ph} \sum_{x=0}^{pw} P_i(x,y) \cdot \|(x,y) - (x_{c_i}, y_{c_i})\|}{\sum_{y=0}^{ph} \sum_{x=0}^{pw} P_i(x,y)} \tag{4.2}$$

Here (x_{c_i}, y_{c_i}) are the centers of two rough eye regions ($i = R, L$). ph and pw are the height and width of the ith rough eye region, respectively. The eye region can be detected using

$$\text{eye}_i(x,y) = \begin{cases} 1 & P_i(x,y) \cdot \|(x,y) - (x_{c_i}, y_{c_i})\| < \text{MeanDis}_i \\ 0 & \text{otherwise} \end{cases} \tag{4.3}$$

Finally, the pupil positions are obtained by calculating the centers of the detected eyes. Figure 4.5c shows the detected eye regions and Figure 4.5d shows the centers of the pupil positions.

3. **Gabor transform.** To decrease the influence of illumination variance in a face image, the Gabor transform method proposed in Reference 17 is adopted to obtain an average Gabor image from the color face image. The general form of the Gabor filter is given by

$$g_{s,\theta}(p,q) = \frac{k^2}{\sigma^2} \cdot e^{-\frac{k^2(p^2+q^2)}{2\sigma^2}} \cdot \left[e^{ik(p^2+q^2)-e^{-\frac{\sigma^2}{2}}}\right] \tag{4.4}$$

where k is the scale of a kernel wavelet, determined as $2^{-s/2} \times 2^{-\pi}$; σ is the standard deviation of the wavelet filter, often set to 2π; (p, q) are the coordinates of a pixel in the image; s is the scale parameter ($s = 0,1,2,3,4$ in this chapter); and θ is the orientation of the Gabor wavelet ($\theta = 0, \pi/8, 2\pi/8, \ldots, 7\pi/8$). Accordingly, there are 40 Gabor filters with different scales and orientations. The transformation is defined as the face image convolution with the Gabor filter with scale s and orientation θ:

$$\begin{aligned} g_{s,\theta}(x,y) &= [f(x,y) * g^{re}_{s,\theta}(p,q)]^2 + [f(x,y) * g^{im}_{s,\theta}(p,q)]^2 \\ &= \left[\sum_p \sum_q f(x+p,y+q) \cdot g^{re}_{s,\theta}(p,q)\right]^2 \\ &\quad + \left[\sum_p \sum_q f(x+p,y+q) \cdot g^{im}_{s,\theta}(p,q)\right] \end{aligned} \tag{4.5}$$

where $f(x,y)$ is a face image; the symbol $*$ denotes the convolution operator; and $g_{s,\theta}(p,q)$ is a Gabor filter with scale s and orientation θ. The superscripts *re* and *im* indicate the real part and the imaginary part, respectively. Once the 40 Gabor transformations are obtained, the Gabor face is obtained using

$$O(x,y) = \frac{1}{N} \sum_{s=0}^{4} \sum_{\theta=o}^{\frac{7\pi}{8}} G_{s,\theta}(x,y) \tag{4.6}$$

where $s = 0, 1, 2, 3, 4$, $\theta = 0, \pi/8, 2\pi/8, \ldots, 7\pi/8$, and $N = 40$. Figures 4.6a and 4.6c show two face images with different sizes taken under different illumination conditions. Figures 4.6b and 4.6 d show the Gabor faces of Figures 4.6a and 4.6c, respectively. The two Gabor faces have identical sizes and similar illuminations; that is, the Gabor transform normalizes scale and equalizes illumination.

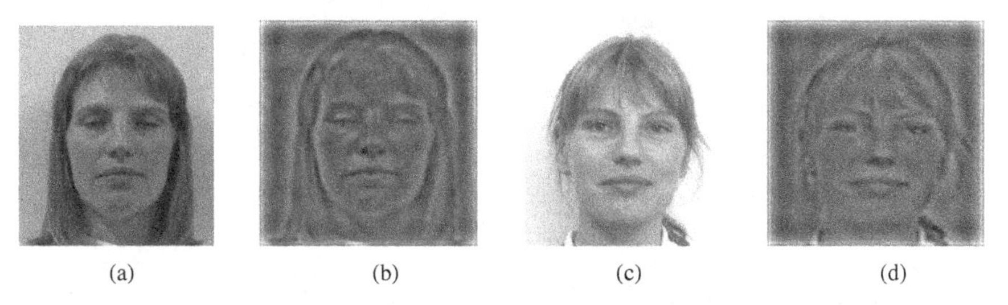

FIGURE 4.6 (a) Face image taken under dark illumination and (b) its Gabor face. (c) Face image taken under bright illumination and (d) its Gabor face.

4.2.3 Facial Feature Extraction

To recognize the expression of a face, the position changes of eyebrows, eyes, and mouth, and facial muscle movement need to be determined. In this chapter, two types of feature are extracted: facial feature distances and facial edge features.

1. **Facial feature distances.** Facial feature distances are used to describe the changes between facial feature points with and without expressions. The facial feature distance extraction procedure flowchart is shown in Figure 4.7. Based on the detected pupil positions, five regions are outlined, namely, two eye regions, two eyebrow regions, and a mouth region, as shown in Figure 4.8a Then, 16 facial feature points are extracted from these regions, as shown in Figure 4.8b The details of the 16 extracted facial features are given below.
   (i) **Eye feature extraction.** To extract accurate eye features, a rough iris region is outlined in the center of the eye region. The diameter of the

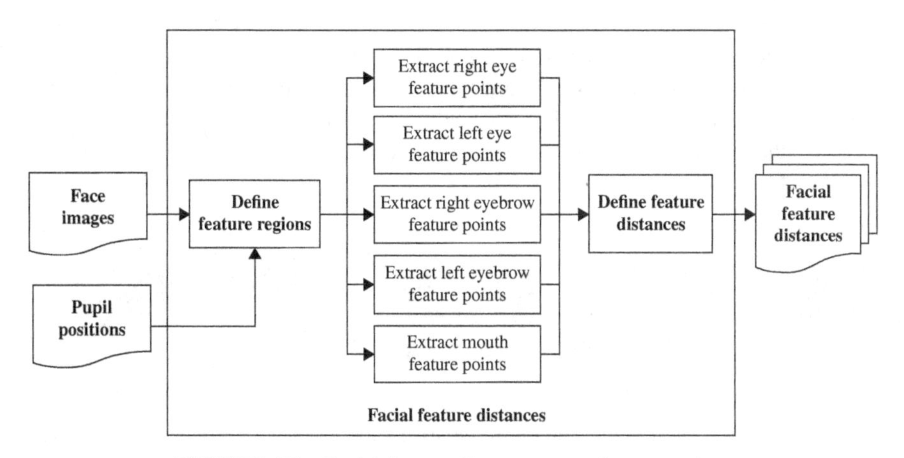

FIGURE 4.7 Facial feature distance extraction procedure.

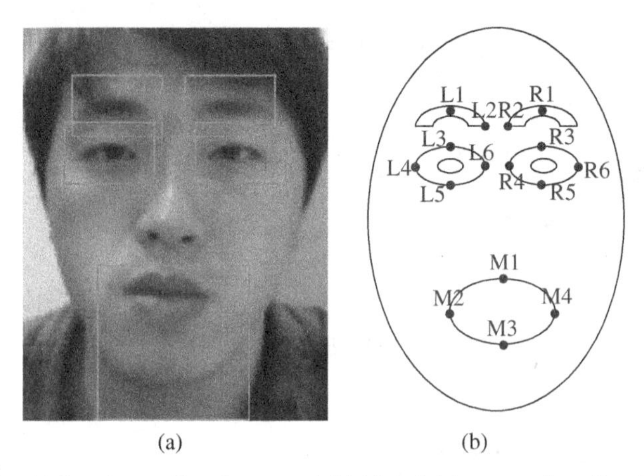

FIGURE 4.8 (a) 5 regions of face image and (b) 16 facial feature points. Part (a) is extracted from authors' own database.

circle is set to 1/3 × height of the eye region, as shown in Figure 4.9a. The iris region in the rough iris region is extracted from the YCbCr color space using

$$\text{iris}_i(x, y) = \begin{cases} 1 & (|Cr_{\text{pupil}_i}(x, y) - Cr_i(x, y)| < 5)\ \&\ (Cb_{\text{pupil}_i}(x, y) - Cb_i(x, y)| < 5) \\ 0 & \text{otherwise} \end{cases} \tag{4.7}$$

where Cb_{pupil_i} and Cr_{pupil_i} are the intensities in the *Cb* and *Cr* color spaces of pupil positions, respectively. Figure 4.9b shows the detected iris region. According to the detected pupil positions and the iris region, the rough eye region is determined, where the major radius and the minor radius of the ellipse are set to 1.5 × height of the iris region and the 2/3 width of the eye image, respectively. Figure 4.9c shows the outlined rough eye region. Then, the Sobel edge detection method is utilized to obtain the edge information in the rough eye region. The detected edges are shown in Figure 4.9d. Accordingly, four eye feature points are extracted from the top, bottom, leftmost region, and rightmost region, respectively, as show in Figure 4.9e.

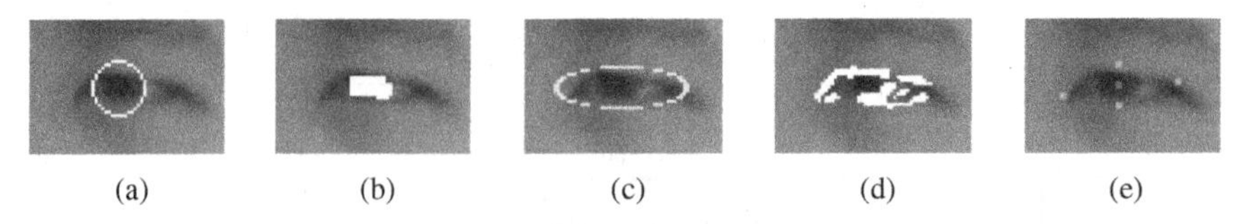

FIGURE 4.9 Eye feature extraction: (a) rough iris region, (b) detected iris region, (c) rough eye region, (d) detected edges, and (e) detected eye feature points.

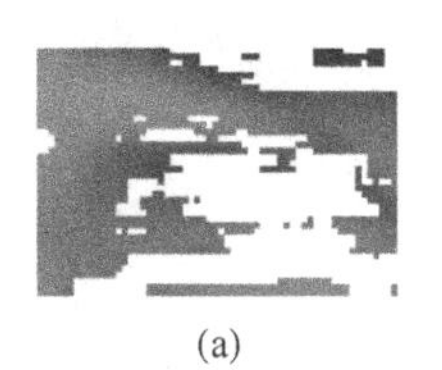
(a)

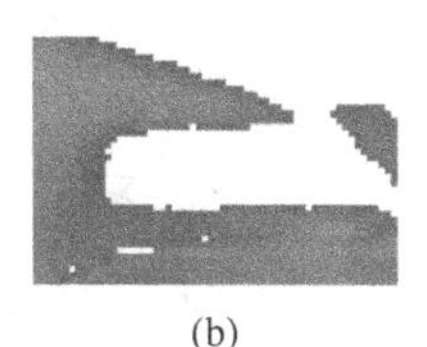
(b)

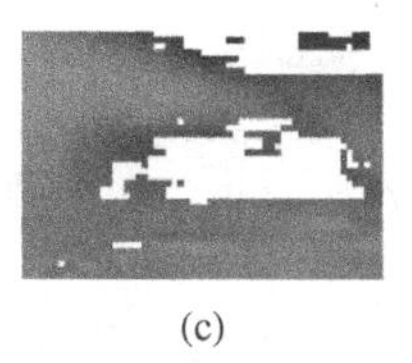
(c)

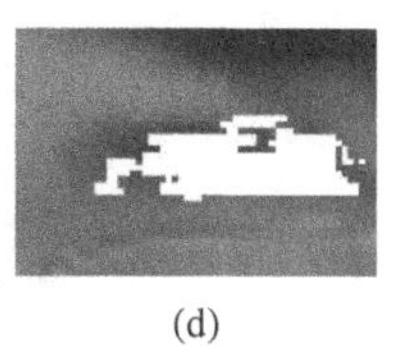
(d)

FIGURE 4.10 Eyebrow feature extraction: (a) probable eyebrow regions in *S* space, (b) probable eyebrow regions in *V* space, (c) rough eyebrow region, and (d) detected eyebrow.

(ii) **Eyebrow feature point extraction.** The rough eyebrow regions are extracted from the HSV color space using

$$SV_i(x,y) = \begin{cases} 1 & (S_i(x,y) < MeanS_i)\ \&\ (V_i(x,y) < MeanVi) \\ 0 & \text{otherwise} \end{cases} \tag{4.8}$$

where $S_i(x, y)$ and $V_i(x, y)$ are the intensity of the saturation and value of the *i*th eyebrow, respectively. $MeanS_i$ and $MeanV_i$ are the means of the intensity in the saturation and the value, respectively. An eight-neighbor labeling method is used to select the label with the maximum area. Figures 4.10a and 4.10b show the probable eyebrow regions in the saturation and value spaces, respectively. Figures 4.10c and 4.10 d show the probable eyebrow region and the detected eyebrow, respectively. The eyebrow feature points from the leftmost region and the center of the label are obtained.

(iii) **Mouth feature point extraction.** The method in Reference 18 is adopted to extract mouth features from the outlined mouth region. Then, four mouth feature points are extracted from the top, bottom, leftmost, and rightmost regions, respectively. Figure 4.11a shows the extracted 16 facial feature points. 17 feature distances are used to describe face changes, as shown in Figrue 4.11b. The 17 facial feature distances are referred to as D1, D2,..., D17, respectively.

2. **Facial edge features.** Facial edge features are extracted from the edges of a face image. The procedure used to extract facial edge features is shown in Figure 4.12.

   To obtain the facial edges, the Gabor face image is convoluted with a Sobel edge detection mask. Figure 4.13a shows an example of a facial edge image. On the edge image, the face is segmented into 16 blocks according to the center of the detected pupils, as shown in Figure 4.13b. The 16 blocks comprise the "inner face" region. The edge features of each block are calculated using the following equation:

$$F_i = \left(\sum_{y=0}^{bh} \sum_{x=0}^{bw} \left(\frac{B_i(x,y)}{bh \times bw} \right) \right) - \left(\sum_{i=1}^{Blocks} \sum_{y=0}^{bh} \sum_{x=0}^{bw} \left(\frac{B_i(x,y)}{bh \times bw \times Blocks} \right) \right) \tag{4.9}$$

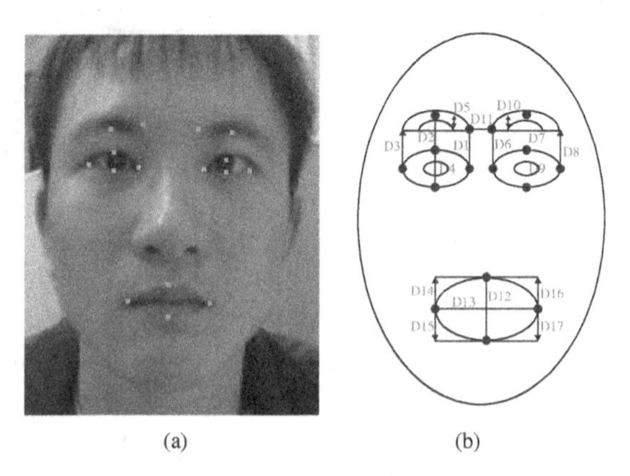

FIGURE 4.11 (a) 16 Facial feature points, and (b) 17 facial feature distances. Part (a) is extracted from authors' own database.

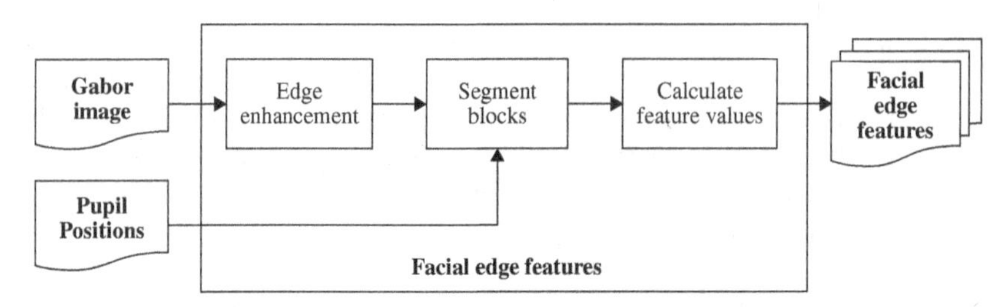

FIGURE 4.12 Facial edge feature extraction procedure.

where $B_i(x,y)$ is the gradient of point (x, y) in the ith block and bw and bh are the width and height of the block, respectively. *Blocks* is the number of blocks, which is set to 16 in this chapter.

4.2.4 Face Recognition

Subject-dependent facial expression recognition is achieved by identifying a user's face before expression recognition is conducted. In this chapter, the method in

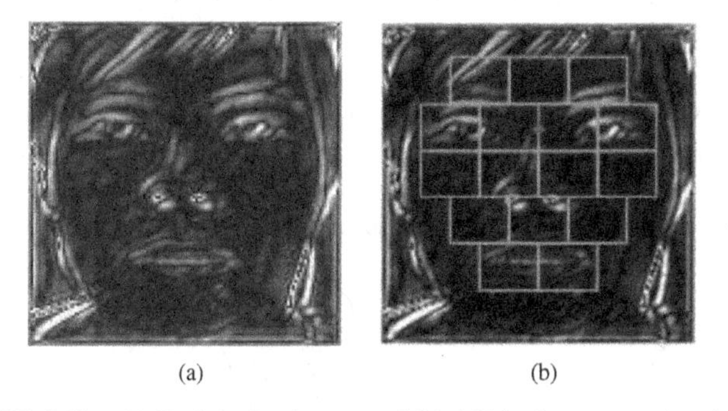

FIGURE 4.13 (a) Facial edge image and (b) 16 blocks segmenting the face.

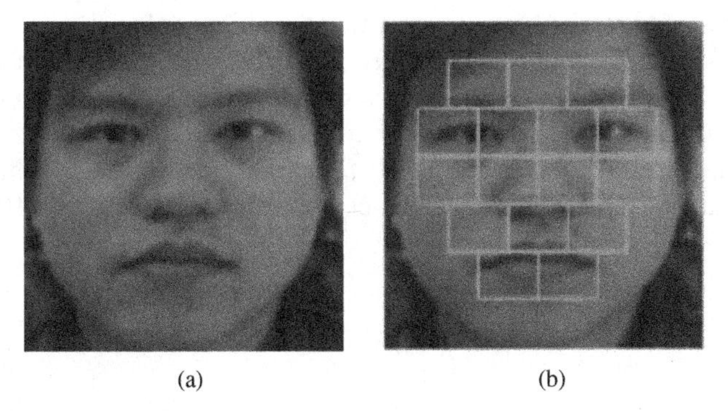

FIGURE 4.14 (a) Full face image and (b) "inner face" image. The blocks indicate the inner face region. Part (a) is extracted from authors' own database.

Reference 17 is adopted for this task. In Reference 17, PCA was applied to extracted features to select adequate centers for the classifier. An AdaBoost committee machine was used to improve the recognition. However, in Reference 17, a full face image, such as the extracted face image shown in Figure 4.14a, was used. Since the background and hair significantly affect recognition, only the "inner face" image is used to identify the user in the proposed method, as show in Figure 4.14b.

4.2.5 Facial Expression Recognition

A schematic diagram of the proposed subject-dependent facial expression recognition is shown in Figure 4.15. The three steps are personal expression model construction, feature normalization, and classification using an RBFNN.

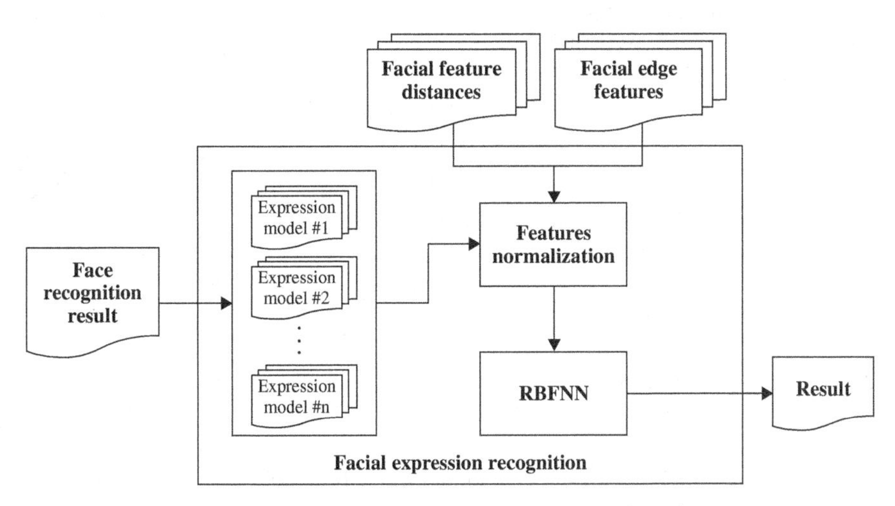

FIGURE 4.15 Facial expression recognition flow chart.

1. **Personal expression model construction.** Since facial expressions are expressed differently by different people, a personal expression model is constructed for each user. The facial features are then extracted from the neutral expression image of each user. The average features are calculated from the facial features of each user. The *i*th facial feature distance and the *i*th facial edge feature of the *m*th user can be expressed using Equations 4.10 and 4.11, respectively.

$$D_i^{\mathrm{Neutral}_m} = \frac{1}{p}\sum_{j=1}^{p} D_{i,j}^{\mathrm{Neutral}_m} \tag{4.10}$$

$$F_i^{\mathrm{Neutral}_m} = \frac{1}{p}\sum_{j=1}^{p} F_{i,j}^{\mathrm{Neutral}_m} \tag{4.11}$$

where $D_{i,j}^{\mathrm{Neutral}_m}$ and $F_{i,j}^{\mathrm{Neutral}_m}$ are the *i*th facial feature distance and the *i*th facial edge feature in the *j*th neutral expression image of the *m*th user, respectively. p is the number of neutral expressions of the *m*th user.

2. **Feature normalization.** The facial feature distances and the facial edge features are compared with the subject-dependent expression model to detect feature changes. The feature changes between the extracted facial features and the neutral expression of the *m*th user are calculated as follows:

$$\Delta D_i^m = D_i^m - D_i^{\mathrm{Natural}_m} \tag{4.12}$$

$$\Delta F_i^m = F_i^m - F_i^{\mathrm{Natural}_m} \tag{4.13}$$

where D_i^m represents the *i*th facial feature distances of the *m*th user, $i = 1, 2, \ldots, 17$. F_i^m represents the *i*th facial edge features of the *m*th user, $i = 1, 2, \ldots, 16$. Equations 4.14 and 4.15 are used to normalize the results in Equations 4.12 and 4.13, respectively:

$$D_i^{m\_nor} = \mathrm{Normalize}(\Delta D_i^m) \tag{4.14}$$

$$F_i^{m\_nor} = \mathrm{Normalize}(\Delta F_i^m) \tag{4.15}$$

The normalization function Normalize(·) is defined as:

$$f^{Nor} = \frac{f - f_L}{f_H - f_L} \times range \tag{4.16}$$

where f is the original feature, and f_L and f_H are the lower bound and upper bound of the feature f, respectively. *range* is the normalization scale, which is set to 1 in this chapter.

3. **Radial basis function neural network.** The RBFNN [13] is adopted as the classifier for recognition. An RBFNN comprises an input layer, a hidden layer,

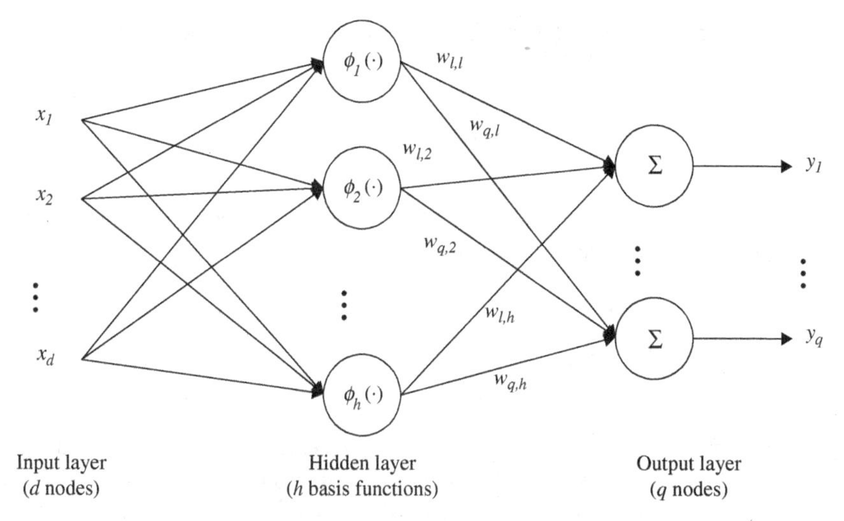

FIGURE 4.16 Architecture of a Radial Basis Function Neural Network.

and an output layer. The hidden layer has high dimensionality. A nonlinear transformation is applied from the input layer to the hidden layer. The higher the dimension of the hidden space, the more accurate the approximation will be. The architecture of the RBFNN is shown in Figure 4.16.

The 17 normalized facial feature distances ($D_i^{m_{nor}}$) and 16 normalized facial edge features ($F_i^{m_{nor}}$) are used to construct an input vector of the RBFNN:

$$\mathbf{x} = [x_1, x_2, \ldots, x_l, \ldots, x_{33}]^{\mathrm{T}} \tag{4.17}$$

where x_l is the lth feature in the feature vector. The output of the jth node in the output layer is obtained using

$$y_j(t) = \sum_{k=1}^{h} w_{j,k}(t)\phi(\mathbf{x}(t), \mathbf{c}_k(t), \sigma_k(t)) \tag{4.18}$$

where x$\in R^{n\times 1}$ is an input feature vector, $w_{j,k}$ is the weight vector between the jth output neuron and the kth hidden neuron, h is the number of neurons in the hidden layer, and $C_k \in R^{n\times 1}$ is the kth center node of the RBFNN. The radial basis function of a neuron in the hidden layer is given as

$$\phi(\mathbf{x}(t), \mathbf{c}_k(t), \sigma_k(t)) = exp\left(-\frac{\|\mathbf{x}(t) - \mathbf{c}_k(t)\|^2}{\sigma_k^2(t)}\right) \tag{4.19}$$

where the symbol $\|\cdot\|$ denotes the Euclidean norm and θ_k is the kth bandwidth of the Gaussian function. θ_k is initialized as

$$\sigma_k(0) = d_{\max}/\sqrt{h} \tag{4.20}$$

where $d_{\max}$ is the maximum Euclidean distance between the selected centers. Accordingly, Equation 4.18 can be rewritten as

$$y_j(t) = \sum_{k=1}^{h} w_{j,k}(t) \cdot exp\left(-\frac{\|\mathbf{x}(t) - \mathbf{c}_k(t)\|^2}{\sigma_k^2(t)}\right) \tag{4.21}$$

Thus, the output of the RBFNN can be represented in vector form as $\mathbf{y} = [y_1, y_2, \cdot, y_q]^{\mathrm{T}}$, where q is the number of classes of expressions. In RBFNN, a stochastic gradient-based supervised learning algorithm is used. The error cost function at the tth iteration is defined as

$$\begin{aligned} J(t) &= \frac{1}{2}\sum_{j=1}^{q} e(t)^2 = \frac{1}{2}\sum_{j=1}^{q}[y_{j,\text{desire}}(t) - y_j(t)] \\ &= \frac{1}{2}\sum_{j=1}^{q}\left[y_{j,\text{desire}}(t) - \sum_{k=1}^{h} w_{j,k}(t) \cdot exp\left(-\frac{\|\mathbf{x}(t) - \mathbf{c}_k(t)\|^2}{\sigma_k^2(t)}\right)\right]^2 \end{aligned} \tag{4.22}$$

where $y_{j,\text{desire}}(t) \in \{0, 1\}$ is the corresponding desired output of the jth node in the output layer. The update equations for the neural network parameters are

$$w_{j,k}(t+1) = w(t) + \mu_w e(t)\Psi(t) \tag{4.23}$$

$$\mathbf{c}_k(t+1) = \mathbf{c}_k(t) + \mu_c \frac{e(t)w_k(t)}{\sigma_k^2(t)} \cdot exp(-\|\mathbf{x}(t) - \mathbf{c}_k(t)\|^2/\sigma_k^2(t))[\mathbf{x}(t) - \mathbf{c}(t)] \tag{4.24}$$

$$\sigma_k(t+1) = \sigma_k(t) + \mu_\sigma \frac{e(t)w_k(t)}{\sigma_k^3(t)} \cdot exp(-\|\mathbf{x}(t) - \mathbf{c}_k(t)\|^2/\sigma_k^2(t))\|\mathbf{x}(t) - \mathbf{c}(t)\|^2 \tag{4.25}$$

where $\Psi(n) = [\phi\mathbf{x}(t), \mathbf{c}_l(t), \sigma_l(t), \ldots, \phi\mathbf{x}(t), \mathbf{c}_h(t), \sigma_h(t)]^T$, μ_w, μ_c, and μ_σ are appropriate learning rates. When the RBFNN has converged ($| e(t) | \leq$ *threshold*), the testing feature vector $[x_1, x_2, \ldots, x_3]$ is input to the trained RBFNN for classification. After the computation, the element with the minimum distance between the output nodes and the desire results represents the recognized class of the expression. That is, the expression category is determined as

$$m = argmin_{\forall j}[y_{j,\text{desire}}(t) - y_j(t)]^2 \qquad j = 1, 2, \ldots, q \tag{4.26}$$

4.3 EXPERIMENT RESULT

To evaluate the performance of the proposed method, the Taiwanese Facial Expression database (TWFE) and the Japanese Female Facial Expression database (JAFFE) were used [19]. TWFE has five expressions each for 30 individuals, including 23 males and 7 females, aged 22–26. The expressions are neutral, happy, angry, surprised, and sad. For each expression, there are 10 color full-frontal images. JAFFE contains 210 images of 10 females with 7 expressions, namely neutral, happy, angry, surprised, sad, disgusted, and scared. Figures 4.17 and 4.18 show samples of TWFE and JAFFE, respectively. Since the images in the JAFFE database are gray scale, the 16 facial feature points could not be extracted using the proposed method. The facial feature points were determined manually. Five experiments were performed: (1) parameter determination of RBFNN (number of hidden nodes and threshold); (2) comparison of facial features (two types were compared); (3) comparison of face recognition systems ("inner face" and full face were compared); (4) comparison of a subject-dependent

FIGURE 4.17 Samples from TWFE: (a) neutral, (b) happy, (c) angry, (d) surprised, and (e) sad. Extracted with permission from TWFE database.

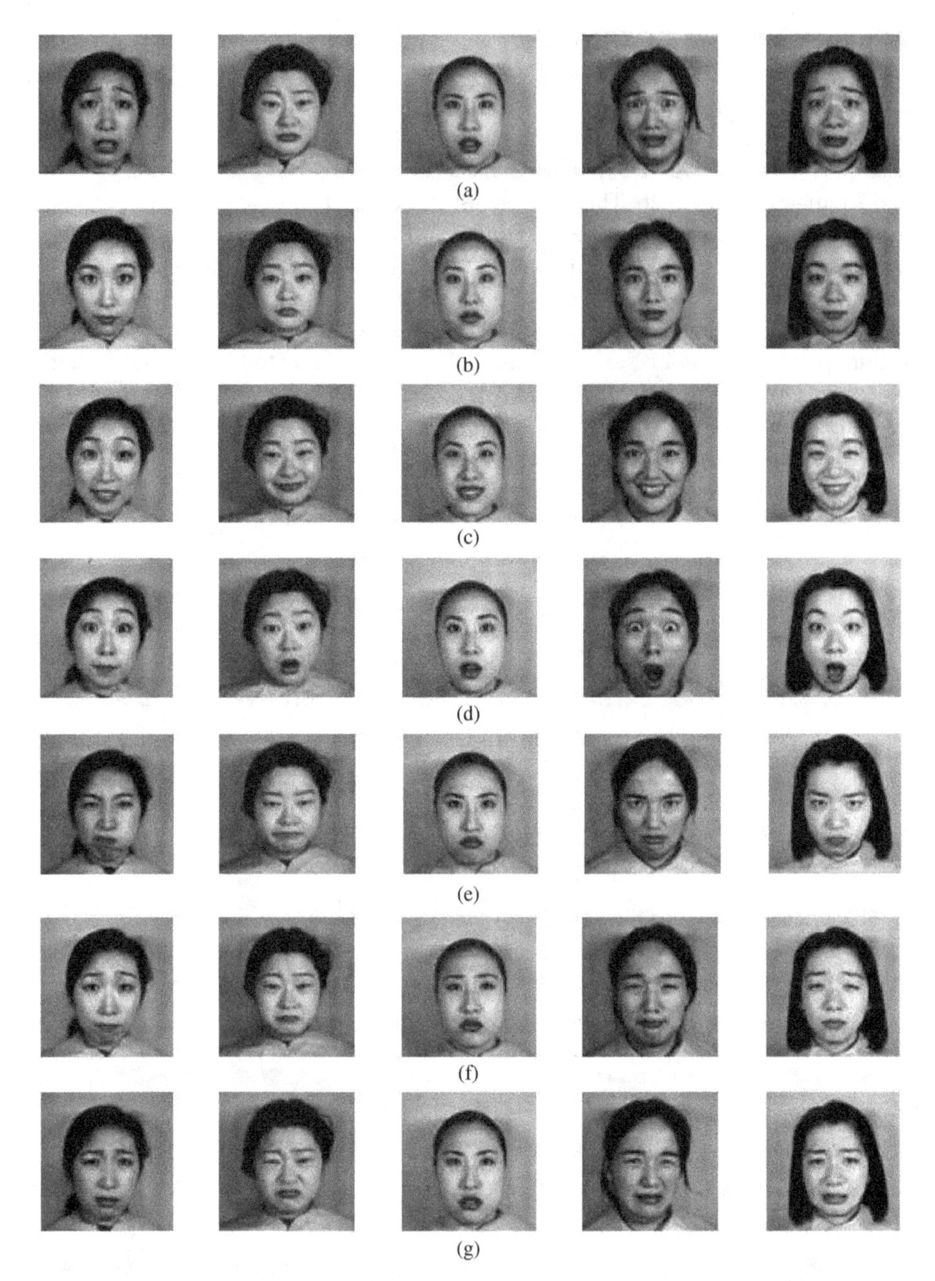

FIGURE 4.18 Samples from JAFFE: (a) scared, (b) neutral, (c) happy, (d) surprised, (e) angry, (f) sad, and (g) disgusted. Extracted with permission from JAFFE database [20].

TABLE 4.1 Recognition rates (FER, %) of the RBFNN for the TWFE database under various settings

	Threshold		
Hidden node	0.01	0.001	0.0001
1100	46.8%	95.7%	95.2%
1150	46.0%	95.9%	95.2%
1200	45.5%	95.6%	95.3%
1250	44.0%	95.5%	95.1%
1300	43.5%	95.5%	95.3%

and a subject-independent facial expression recognition systems (a conventional and the proposed expression recognition systems were compared); (5) comparison with other approaches (other approaches were compared with the proposed approach using the JAFFE database [20]).

4.3.1 Parameter Determination of the RBFNN

To obtain the best recognition rate using the RBFNN, the number of hidden nodes and a threshold must be properly determined. The two databases were used to obtain the best parameters. Both databases were equally divided into two subsets, one for training and the other for testing. The facial expression recognition rate is defined as

$$FER(\%) = \sum_{i=1}^{class\_num} n_i / img \tag{4.27}$$

where n_i is the number of correctly recognized images of the ith expression, img is the number of images in the database, and $class\_num$ is the number of expressions. $class\_num$ is 5 in the TWFE database and 7 in the JAFFE database. The recognition results for various setups for the TWFE database are shown in Table 4.1. The highest recognition rate was obtained when the number of hidden nodes was set to 1150 and the threshold was set to 0.001. The recognition results for the JAFFE database are shown in Table 4.2. The highest recognition rate was achieved when the number of hidden nodes was set to 195 and the threshold was set to 0.001.

The number of hidden nodes is determined via the training process. In practice, the number of hidden nodes can be estimated using

$$h = 1.5 \times P \times E \times N \tag{4.28}$$

where P is the number of people to be trained, E is the number of categories of expression for each person, and N is the number of expression images used for training for each person. For instance, TWFE has five expressions for 30 individuals. Each expression has 10 color full-frontal images. Assume that half of the images

TABLE 4.2 Recognition rates (FER, %) of the RBFNN for the JAFEE database under various settings

	Threshold			
Hidden node	0.01	0.001	0.0001	0.00001
180	85.85%	87.74%	87.74%	87.74%
195	87.26%	88.57%	87.74%	88.21%
210	87.74%	87.74%	87.74%	87.26%
225	85.85%	87.26%	86.79%	86.32%
240	85.85%	87.26%	86.79%	86.32%

are used for training; the estimated number of hidden neurons is thus $1.5 \times 30 \times 5 \times 5 = 1125$. JAFFE contains 210 images of 10 females with 7 expressions. Each expression has 3 gray-level full-frontal images. Assume that two expression images of each expression category for each person are used for training. Thus, the estimated number of hidden nodes is $1.5 \times 10 \times 7 \times 2 = 210$. A reasonable number of hidden nodes is the estimated number $\pm$ 15%. The confusion matrixes of the RBFNN with the selected parameters in the two databases are shown in Tables 4.3 and 4.4, respectively. In Table 4.3, 137 neutral expression facial images were correctly recognized as neutral expressions, with 1 image incorrectly recognized as happy, 2 images as angry, and 10 images as sad. The accuracy of each expression is defined as

$$Accuracy_i = n_i / img_i \tag{4.29}$$

where img_i is the number of the images of the ith expression. For the TWFE database, the accuracies for neutral, happy, angry, surprised, and sad expressions are 91.3%, 98.0%, 96.7%, 97.3%, and 96.0%, respectively. In Table 4.4, the accuracies for neutral, happy, angry, surprised, sad, scared, and disgusted expressions are 86.67%, 93.33%, 93.33%, 86.67%, 93.33%, 86.67%, and 80.00%, respectively. The accuracies are at least 91.3% and 80.00% for the TWFE and JAFFE databases, respectively. Therefore, the proposed method is effective.

TABLE 4.3 Confusion matrix obtained using optimal parameters for the TWFE database

Actual\system	Neutral	Happy	Angry	Surprised	Sad	Accuracy
Neutral	137	1	2	0	10	91.3%
Happy	0	147	2	0	1	98.0%
Angry	2	3	145	0	0	96.7%
Surprised	0	2	1	146	1	97.3%
Sad	0	2	4	0	144	96.0%
Overall accuracy						95.9%

TABLE 4.4 Confusion matrix obtained using optimal parameters for the JAFFE database

Actual\system	Neutral	Happy	Angry	Surprised	Sad	Scared	Disgusted	Accuracy
Neutral	13	1	0	0	1	0	0	86.67%
Happy	0	14	0	0	1	0	0	93.33%
Angry	0	0	14	0	1	0	0	93.33%
Surprised	0	1	0	13	0	1	0	86.67%
Sad	0	0	0	0	14	1	0	93.33%
Scared	0	0	1	0	1	13	0	86.67%
Disgusted	0	0	3	0	0	0	12	80.00%
Overall accuracy								88.57%

4.3.2 Comparison of Facial Features

Two types of facial features were extracted, namely, facial feature distances and facial edge features. To demonstrate the discriminability of these features, the recognition rates obtained using only facial feature distances, only facial edge features, and a combination of both features (the proposed method) were compared. Two databases were used in this experiment. In the TWFE database, 5 expressions were recognized, namely, neutral, happy, angry, surprised, and sad. In the JAFFE database, 7 expressions were recognized, namely, scared, neutral, happy, surprised, angry, sad, and disgusted. Both databases were equally divided into two subsets, one for training and the other for testing. The experiment results are shown in Table 4.5. Using only facial feature distances, the accuracies are 79.93% and 87.62% for the TWFE and JAFFE databases, respectively. Using only facial edge features, the accuracies are 86.83% and 63.21% for the TWFE and JAFFE databases, respectively. The proposed method, which uses both the facial distances and facial edge features, has accuracies of 95.87% and 88.21% for the TWFE and JAFFE databases, respectively. The results show that the proposed method is more effective than using only one type of facial feature. For TWFE, for which the facial feature points were automatically detected, using facial edge features only has a higher accuracy than using facial feature distances only. However, for JAFFEE, the reverse is true because the facial feature points in JAFFEE were determined manually. The highest recognition accuracy was achieved when both facial feature distances and facial edge features were considered.

TABLE 4.5 Comparison of the accuracy of facial expression recognition obtained using various features

	Facial feature distances only	Facial edge features only	Proposed method
TWFE	79.93%	86.73%	95.87%
JAFFE	87.62%	63.21%	88.21%

TABLE 4.6 Results of the face recognition obtained using inner face and full face images

	Accuracy
Full face images	73.58%
Inner face images	97.64%

4.3.3 Comparison of Face Recognition Using "Inner Face" and Full Face

The JAFFE database was used in this experiment. It was divided into two subsets, one for training and the other for testing. The full face images and "inner face" images were used to identify the user using the PCA-based method proposed in Reference 17. The experiment results are shown in Table 4.6. The accuracy for the inner face images is 97.64% and that for the full face images is 73.58%. Therefore, the accuracy can be improved by using inner face images.

4.3.4 Comparison of Subject-Dependent and Subject-Independent Facial Expression Recognition Systems

The proposed subject-dependent expression recognition system was compared with a conventional expression recognition system. In the conventional approach, the expression recognition is based on the general expression model, which is constructed from all members in a database. Two experiments were conducted on the TWFE and JAFFE databases to recognize 5 facial expressions (neutral, happy, angry, surprised, and sad). In the first experiment, the database was divided equally into the training set and the testing set. The second experiment used facial expression recognition with the leave-one-person-out method; that is, all but one person's images were used for training. The test was conducted on the images of the person that was left out. The two experiments are referred to as Test 1 and Test 2, respectively. The experiment results are shown in Table 4.7. In Test 1, the accuracies of the conventional facial expression recognition system are 93.1% and 92.7% for the TWFE and JAFFE databases, respectively. In Test 2, the accuracies are 58.5% and 66.9%, respectively. The accuracies of the proposed method in Test 1 are 95.9% and 96.9%, and those in Test 2 are 78.1% and 96.0%, respectively. The results show that the proposed approach outperforms conventional facial expression recognition.

TABLE 4.7 Comparison of a conventional expression recognition method with the proposed method

	TWFE database		JAFFE database	
	Test 1	Test 2	Test 1	Test 2
Conventional method	93.1%	58.5%	92.7%	66.9%
Proposed method	95.9%	78.1%	96.9%	96.0%

TABLE 4.8 Comparison with the method proposed in Reference 21

Approach	Test 1	Test 2
Horikaw [21]	97.30%	67.00%
Proposed method	98.61%	81.60%

TABLE 4.9 Comparison with the method proposed in Reference 6

Approach	Leave-one-image-out	Leave-one-person-out
Zheng [6]	98.36%	77.05%
Proposed method	98.88%	78.58%

4.3.5 Comparison with Other Approaches

In order to demonstrate the capability of the proposed approach, the proposed method was compared with other approaches using the JAFFE database. In the first test, 140 images were used for training and the remaining ones for testing. The other test used the leave-one-person-out method. The results are shown in Table 4.8. The method of Horikawa [21] was used to recognize 7 expressions. The accuracies for the proposed method are 98.61% and 81.60% for Test 1 and Test 2, respectively, which are better than those of Horikawa's [21] method. The method of Zheng [6] was used to recognize 6 expressions, namely, happy, angry, surprised, sad, scared, and disgusted. The results of the leave-one-person-out method and the leave-one-image-out method are shown in Table 4.9. The results show that the accuracies for the proposed method are 98.88% and 78.58%, respectively, which are better than those of the method of Zheng [6] (98.36% and 77.05%, respectively). The method of Bashyal [8] was used to recognize 7 expressions. 163 images were randomly selected for training and the remaining 50 images were used for testing (10 times). The results are shown in Table 4.10. For the training and testing, the proposed method obtained a lower error rate than that of Bashyal [8]. The overall accuracy (98.90% vs. 87.51%) was also better.

4.4 CONCLUSION

Since facial expressions are expressed differently by different people, face recognition was combined with facial expression recognition in this chapter. Facial features, including the changes in the positions of eyebrows, eyes, and mouth, and facial muscle

TABLE 4.10 Comparison with the method proposed in Reference 8

Approach	Training data error rate	Testing data error rate	Overall accuracy
Bashyal [8]	6.3 ± 1.1	32.4 ± 7.3	87.51%
Proposed method	0.0 ± 0.0	4.58 ± 2.2	98.90%

movement are extracted in the proposed approach. With identity information and the extracted facial features, an RBFNN is used to classify expressions. Experiment results show that the proposed approach is better than approaches that use the whole face image. The proposed method outperforms conventional expression systems. A few pointers of reference to relevant similar works for interested readers include [22–29].

ACKNOWLEDGMENT

This work was supported by the National Science Council, Taiwan, under grant NSC 98-2218-E-006-004.

REFERENCES

[1] A. Takeuchi and K. Nagao, "Communicative facial displays as a new conversational modality," in Proceedings of the ACM/IFIP Conference on Human Factors in Computing Systems, 1993, pp. 187–193.

[2] P. Ekman, "The argument and evidence about universals in facial expressions of emotion," in *Handbook of Social Psychophysiology*, 1999, pp. 143–164.

[3] P. Ekman and W. V. Friesen, *The Facial Action Coding System: A Technique for the Measurement of Facial Movement*, Consulting Psychologists Press, San Francisco, 1978.

[4] P. Ekman and W. V. Friesen, "Constants across cultures in the face and emotion," *J. Person. Soc. Psychol.*, 17(2): 124–129, 1971.

[5] H. Wang and N. Ahuja, "Facial expression decomposition," in Proceedings of the IEEE International Conference on Computer Vision, 2003, pp. 958–965.

[6] W. Zheng, X. Zhou, C. Zou, and C. Zhao, "Facial expression recognition using Kernel Canonical Correlation Analysis (KCCA)," *IEEE Transactions on Neural Networks*, 17: 233–238, 2006.

[7] C. Y. Chang, J. S. Tsai, C. J. Wang, and P. C. Chung, "Emotion recognition with consideration of facial expression and physiological signals," in Proceedings of the IEEE Symposium Series on Computational Intelligence, 2009, pp. 278–283.

[8] S. Bashyal and G. K. Venayagamoorthy, "Recognition of facial expressions using Gabor wavelets and learning vector quantization," *Eng. Appl. Artif. Intell.*, 21: 1056–1064, 2008.

[9] Y. Zilu, L. Jingwen, and Z. Youwei, "Facial expression recognition based on two dimensional feature extraction," in Proceedings of the 2008 International Conference on Software Process (ICSP 2008), 2008 pp. 1440–1444.

[10] S. M. Lajevardi and M. Lech, "Averaged gabor filter features for facial expression recognition," in Proceedings of the IEEE 2008, Digital Image Computing: Technique and Applications, 2008, pp. 71–76.

[11] R. Xiao, Q. Zhao, D. Zhang, and P. Shi, "Facial expression recognition on multiple manifolds," *Pattern Recogn.*, 44: 107–116, 2011.

[12] G. Littlewort, M. S. Bartlett, I. Fasel, J. Susskind, and J. Movellan, "Dynamics of facial expression extracted automatically from video," *Image Vis. Comput.*, 24: 615–625, 2006.

[13] S. Haykin, *Neural Network—A Comprehensive Foundation*, Prentice-Hall, Englewood Cliffs, NJ, 1999.

[14] H. Wang and K. Wang, "Affective interaction based on person-independent facial expression space," *Neurocomputing*, 71: 1889–1901, 2008.

[15] M. Matsugu, K. Mori, Y. Mitari, and Y. Kaneda, "Subject indepent facial expression recognition with robust face detection using a convolutional neural network," *Neural Netw.*, 16: 555–559, 2003.

[16] C. Y. Chang, Y. C. Tu, and H. H. Chang, "Adaptive color space switching based approach for face tracking," *Lect. Notes Comput. Sci.*, 4223: 244–252, 2006.

[17] C. Y. Chang and H. R. Hsu, "Application of principal component analysis to a radial basis function committee machine for face recognition," *Int. J. Innov. Comput. Inf. Control*, 5(11): 4145–4154, 2009.

[18] R. L. Hsu, A. M. Mohamed, and A. K. Jain, "Face detection in color images," *IEEE Transactions on Pattern Analysis and Machine Intelligence*, 24(5): 696–706, 2002.

[19] M. J. Lyons, S. Akamatsu, M. Kamachi, and J. Gyoba, "Coding facial expressions with Gabor wavelets," in Proceedings of the 3rd IEEE International Conference on Automatic Face and Gesture Recognition, 1998, pp. 200–205.

[20] M. J. Lyons, M. Kamachi, J. Gyoba, Japanese Female Facial Expressions (JAFFE) Database of digital images, 1997.

[21] Y. Horikawa, "Facial expression recognition using KCCA with combining correlation kernels and kansei information," in Proceedings of the Fifth International Conference on Computational Science and Applications, 2007, pp. 489–495.

[22] S. P. Aleksicand and K. A. Katsaggelos, "Automatic facial expression recognition using facial animation parameters and multistream HMMs," *IEEE Transactions on Information and Security*, 1(1): 3–11, 2006.

[23] S. Shiqian and Y. Baocai, "A robust face detection method," in Proceedings of the IEEE International Conference on Image and Graphics, 2004, pp. 302–305.

[24] C. Garcia, G. Zikos, and G. Tziritas, "Face detection in color images using wavelet packet analysis," in Proceedings of the IEEE International Conference on Multimedia Computing and Systems, vol. 1, 1999, pp. 703–708.

[25] E. Osuna, R. Freund, and F. Girosi, "Training support vector machines: an application to face detection," in Proceedings of the IEEE Conference on Computer Vision and Pattern Recognition, 1997, pp. 130–136.

[26] H.A. Rowley, S. Baluja, and T. Kanade, "Neural network-based face detection," *IEEE Transactions on Pattern Analysis and Machine Intelligence*, 20(1): 23–38, 1998.

[27] S. A. Billings, H. L. Wei, and M. A. Balikhin, "Generalized multiscale radial basis function networks," *Neural Netw.*, 20: 1081–1094, 2007.

[28] T. Xiang, M. K. H. Leung, and S. Y. Cho, "Expression recognition using fuzzy spatio-temporal modeling," *Pattern Recogn.*, 41: 204–216, 2008.

[29] Q. Zhang, C. Chen, C. Zhou, and X. Wei, "Independent component analysis of gabor features for facial expression recognition," in Proceedings of the 2008 International Symposium on Information Science and Engineering, 2008, pp. 84–87.

AUTHOR BIOGRAPHIES

Chuan-Yu Chang received the M.S. degree in Electrical Engineering from National Taiwan Ocean University, Keelung, Taiwan, in 1995, and the Ph.D. degree in Electrical Engineering from National Cheng Kung University, Tainan, Taiwan, in 2000.

From 2001 to 2002, he was with the Department of Computer Science and Information Engineering, Shu-Te University, Kaohsiung, Taiwan. From 2002 to 2006, he was with the Department of Electronic Engineering, National Yunlin University of Science and Technology, Yunlin, Taiwan, where since 2007, he has been with the Department of Computer and Communication Engineering (later Department of Computer Science and Information Engineering), and is currently a full professor and Dean of Research & Development. He is an associate editor of the *International Journal of Control Theory and Applications*. His current research interests include neural networks and their application to medical image processing, wafer defect inspection, digital watermarking, and pattern recognition. In the above areas, he has more than 150 publications in journals and conference proceedings.

Dr. Chang received the Excellent Paper Award of the Image Processing and Pattern Recognition Society of Taiwan in 1999, 2001, and 2009. He was also the recipient of the Best Paper Award in the International Computer Symposium in 1998 and 2010, the Best Paper Award in the Conference on Artificial Intelligence and Applications in 2001, 2006, 2007, and 2008, and the Best Paper Award in National Computer Symposium in 2005. He received the Best Paper Award from Symposium on Digital Life Technologies in 2010. He served as the Program CoChair of 2007 Conference on Artificial Intelligence and Applications, 2009 Chinese Image Processing and Pattern Recognition Society (IPPR) Conference on Computer Vision, Graphics, and Image Processing, and 2011 Workshop on Digital Life Technologies. He was a member of the organization and program committees more than 50 times, and he organized and chaired more than 30 technical sessions for many international conferences. He is a life member of IPPR and the Taiwanese Association for Artificial Intelligence (TAAI), senior member of IEEE, and is listed in Who's Who in the World, Who's Who in Science and Engineering, Who's Who in Asia, and Who's Who of Emerging Leaders.

Yan-Chiang Huang received the B.S. degree in Electronic Engineering in 2007 and the M.S. degree in Computer Science and Information Engineering in 2009 from the National Yunlin University of Science and Technology, Yunlin, Taiwan.

His research interests include neural networks and image processing.

5

FACIAL EXPRESSION RECOGNITION USING INDEPENDENT COMPONENT FEATURES AND HIDDEN MARKOV MODEL

Md. Zia Uddin

Electronic Engineering Department, Inha University, Incheon, Republic of Korea

Tae-Seong Kim

Bio-medical Engineering Department, Kyung Hee University, Seoul, Republic of Korea

Facial expression recognition (FER) from video is an essential research area in the field of human–computer interfaces (HCIs). In this work, we present a new method to recognize several facial expressions from time- sequential facial expression images. To produce robust facial expression features, Enhanced Independent Component Analysis (EICA) is utilized to extract independent component (IC) features which are further classified by Fisher Linear Discriminant Analysis (FLDA). Using these features, discrete Hidden Markov Models (HMMs) are utilized to model different facial expressions such as joy, anger, fear, and sadness. Performance of the proposed FER system is compared against four other conventional feature extraction approaches, including Principal Component Analysis (PCA), Independent Component Analysis (ICA), PCA-FLDA, and EICA, in conjunction with the same HMM scheme. Our preliminary results show that the proposed system yields much improved recognition rates reaching the mean recognition rate of 92.85%.

Emotion Recognition: A Pattern Analysis Approach, First Edition. Edited by Amit Konar and Aruna Chakraborty.
© 2015 John Wiley & Sons, Inc. Published 2015 by John Wiley & Sons, Inc.

5.1 INTRODUCTION

FER has been regarded as one of the fundamental technologies for HCI, which enables computers to interrelate with humans in a way to human to human interactions [1]. Since the FER technology provides computers a way of sensing user's emotional information, FER can contribute to a HCI system by responding to the expressive conditions of humans.

For the feature extraction, the early FER research works extracted useful features using PCA. PCA is a second-order statistical method to derive the orthogonal basis containing the maximum variability in an unsupervised manner that provides global image features. It is also commonly used for dimensionality reduction. Padgett and Cottrell [2] applied PCA on facial expression images to identify facial action units (FAUs) and to recognize facial expressions in which their best analysis was performed on the separated face region such as eyes and mouth. In References 3, 4, the authors also employed PCA as one of the feature extractors to solve FER with the Facial Action Coding System (FACS). However, the recent psychological research works show that facial expressions do not take place in separate ways, but the whole spatial relationships of the face can express the additional source of information in the facial perception of emotions [5]. Hence, most of the recent works such as References 6, 7 and 8 focused on the emotion-specified feature extraction rather than FAU. In some FER works, FLDA was applied to classify the PC features of the face images [6, 7]. Basically, FLDA is based on the class information that projects the data onto a subspace with the criterion that maximizes the between-class scatter and minimizes the within-class scatter of the projected data. Lately, Independent Component Analysis (ICA) has been extensively utilized for FER tasks due to its ability to extract local features [3,8–10]. As much of the information that distinguishes faces from different expressions stays in the higher-order statistics of the images [9], ICA is a better choice for FER than PCA, which follows the second-order statistics. ICA is a generalization of PCA that finds the independencies of the image features [11, 12]. Bartlett *et al.* [13] extracted the image representations for facial expression coding using ICA to classify 12 facial actions referred to FACS. Chao-Fa *et al.* [14] utilized ICA to extract the IC features of facial expression images to recognize the action units (AU) in the whole face as well as the lower and upper part of the faces. In addition, they encounter the limitation of AUs due to the fact that the separate facial actions do not directly draw the comparisons with human data [15]. Some researchers applied ICA for the emotion-specified FER where ICA was applied on the Japanese female facial expression data [16] to achieve good recognition accuracy. Bartlett *et al.* [9] again introduced ICA on the PCs (i.e., Enhanced ICA (EICA)) for face recognition where they showed that ICA outperforms PCA. Liu [17] applied EICA successfully for content-based face image retrieval using more than thousand frontal face images from FERET database [18]. Since FLDA is successful to classify the global PC features, it can be applied on the IC features of the time sequential face images to extract robust features for FER.

As for recognition techniques in FER, various distance measures such as Euclidian and Cosine distances were utilized [16]. Tian *et al.* [19] used neural network to extract FAUs by means of the Gabor wavelet-based features. Support Vector Machine (SVM) was also used on facial expression image regions in FER [14, 20]. Chao-Fa *et al.* [14] used SVM to extract the FAUs based on the IC features of facial expressions. Recently, HMM, a strong tool to decode time-sequential events, was popularly tested for FER [21–24]. Otsuka and Ohya [21] purposed HMMs to recognize the facial expressions of multiple persons utilizing time-sequential face features. Zhu *et al.* [22] applied HMM to recognize the emotions utilizing moment invariants. Cohen *et al.* [23] employed HMMs with the facial expression features extracted utilizing the Naïve-Bayes classifiers on the Cohn-Kanade database. More recently, Aleksic *et al.* [24] introduced a multi-stream HMM-based FER system utilizing Facial Animation Parameters (FAPs).Thus, HMM can be adopted successfully to model and decode the time-sequential features for robust FER.

In this chapter, we propose a novel approach for FER dealing with EICA, FLDA, and HMM. Here, local features are first extracted from the facial expression images through ICA over the PCs and further classified by FLDA that can be denoted as EICA-FLDA. These local features are then converted to discrete symbols or codes by means of vector quantization. Finally, discrete HMMs are applied to train and recognize the time-sequential code sequences. To compare the performance of our proposed approach, four comparison studies have been conducted: PCA, ICA, EICA, and FLDA over PCA (PCA-FLDA), as feature extractor in combination with HMM. The experimental results show that the proposed method significantly outperforms the conventional approaches.

The rest of the chapter is organized as follows. Section 5.2 presents the proposed system methodology from video preprocessing to expression training and recognition using HMMs. Section 5.3 represents the experimental results obtained using the proposed approach on the Cohn-Kanade expression database. Finally, we conclude this chapter in Section 5.4.

5.2 METHODOLOGY

Our FER system developed in this chapter consists of two major parts: EICA-FLDA as a feature extractor and HMMs as a recognizer. The major objectives is to find an efficient characterization of low-dimensional subspaces and ensure the temporally evolving features that reflects the changes of facial expressions representing one particular emotion. Due to its successful usage in pattern classification of consecutive events, HMMs are used as a recognizer for facial expression recognition where expressions are concatenated from neutral state to particular expression state. Figure 5.1 shows the overview of the proposed FER system, where H represents HMMs and L the likelihoods of HMMs.

5.2.1 Expression Image Preprocessing

In preprocessing of sequential images of facial expressions, image alignment is performed first to realign the common regions of the face. We utilized a face alignment approach as used in Reference 27 and manually matched the eyes and mouth of the faces in the designated coordinates. The face images are scaled and translated so that the sum of squared distance between the target coordinates and those of the transferred

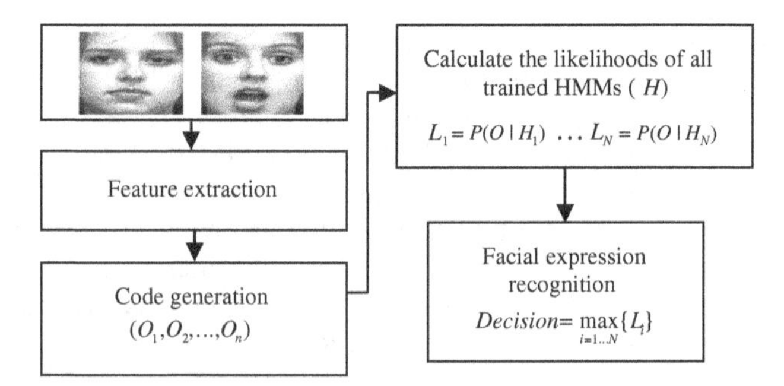

FIGURE 5.1 Overview of the proposed FER system.

features was minimized: the triangular shape from two eyes and one mouth is scaled and translated to fit the reference location. The typical realigned image consisted of 60×80 pixels. Afterward, the first frame of each input sequence is subtracted from the following frames to obtain the delta images [3] to obtain the facial expression change differences in the images over time. Figure 5.2 shows an exemplar set of surprise facial expression sequence and the corresponding delta images.

5.2.2 Feature Extraction

The key idea of the feature extraction in this method is the combination of EICA and FLDA. The purpose of this method is to find an optimal local representation of face images in a low-dimensional space and to lead the well-separated time-sequential features for robust training and recognition. The feature extraction consists of three fundamental stages: (1) PCA is performed first for dimensionality reduction, (2) ICA is applied on the reduced PCA subspace to find statistically independent basis images for corresponding facial expression image representation, and (3) FLDA is then employed to compress the same classes as close as possible and to separate the different classes as far as possible.

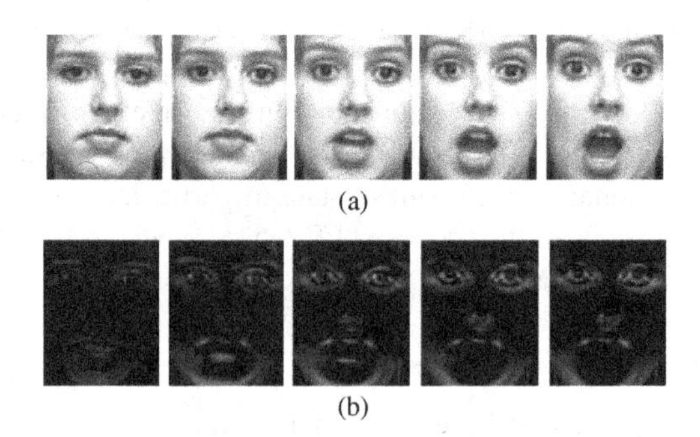

FIGURE 5.2 A sequence of surprise facial expression images: (a) the aligned video image sequence [25, 26] and (b) the corresponding delta image sequence. Extracted with permission from Cohn–Kanade database.

5.2.2.1 Principal Component Analysis As a first stage of the feature extraction, we initially apply PCA to the dataset to reduce the data dimension. PCA is a popular method that transforms the high-dimensional space to a reduced space capturing the maximum variability. Its fundamental is to compute the eigenvectors of the covariance data matrix Y and then the approximation is done using a linear combination of a few top eigenvectors. The covariance matrix of the sample training image vectors can be calculated as

$$P^T(YY^T)P = \Lambda \tag{5.1}$$

where P represents the matrix of orthonormal eigenvectors and Λ the diagonal matrix of the eigenvalues. The eigenvector associated with the largest eigenvalue indicates the axis of maximum variance and the following eigenvector with the second largest eigenvalue indicates the axis of the second largest variance. Thus, a certain number of eigenvectors chosen according to the highest eigenvalues defines the subspace. Besides, P reflects the original coordinate system onto the eigenvectors where the eigenvector corresponding to the largest eigenvalue indicates the axis of largest variance and the next largest one is the orthogonal axis of the largest one indicating the second largest variance, and so on. Usually, the eigenvalues that are close to zero carry negligible variance and hence can be excluded. Figure 5.3 shows the top 100 eigenvalues corresponding to the first 100 eigenvectors after PCA on the facial expression images.

Here, let us denote the dataset of n images with zero mean image vectors by $X = (x_1, x_2, \dots, x_n)^T$ and its corresponding PC projection vectors with low dimensions by $V = (v_1, v_2, \dots, v_n)^T$. Thus, the principal component of an expression image vector can be represented as follows:

$$V = XP_m \tag{5.2}$$

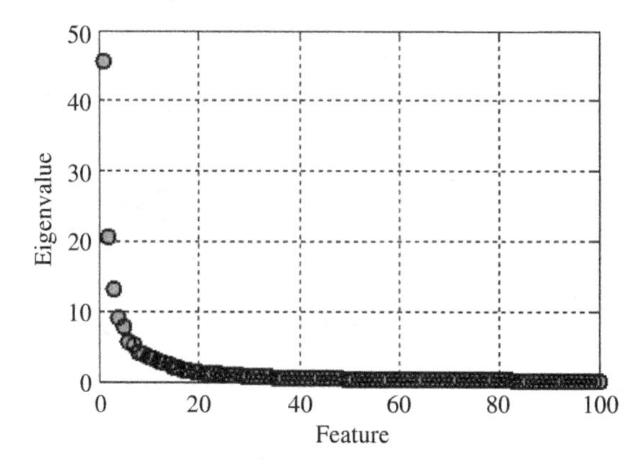

FIGURE 5.3 Top 100 eigenvalues corresponding to the eigenvectors.

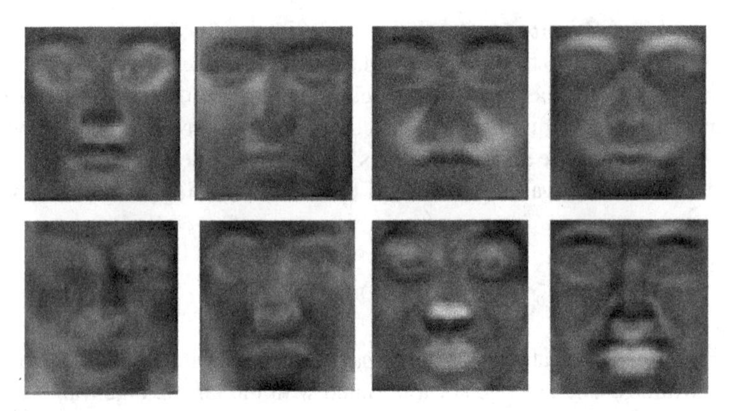

FIGURE 5.4 Eight basis images after applying PCA on the face images of the six expressions.

where P_m represents the first m number of chosen eigenvectors. Figure 5.4 shows eight basis images after applying PCA on the six facial expression images, namely joy, anger, disgust, sad, fear, and surprise.

5.2.2.2 Independent Component Analysis ICA is introduced to solve a blind source separation problem where the objective is to decompose mixture of observed signals into a linear combination of some unknown independent signals and their mixing matrix. The ICA algorithm finds the statistically independent basis images. If S and X are the collection of the basis and input images, respectively, then the relation between X and S is modeled as

$$X = AS \tag{5.3}$$

where A is an unknown linear mixing matrix of full rank. The sources are basically independent of each other and the mixing matrix is invertible. However, the ICA algorithm tries to find the mixing matrix A or the separating matrix so that

$$U = WX \tag{5.4}$$

$$U = WAS \tag{5.5}$$

where U represents an estimation of the independent sources. The estimation problem of finding A can be represented by prewhitening of the observed vectors X. Therefore, X is transformed into its prewhitened form Y such that

$$Y = KX \tag{5.6}$$

$$K = \Lambda^{-\frac{1}{2}}Z \tag{5.7}$$

where the correlation of Y is a unit matrix. Λ substitutes the diagonal matrix of the eigenvalues and Z orthonormal eigenvectors of the covariance matrix of X. After

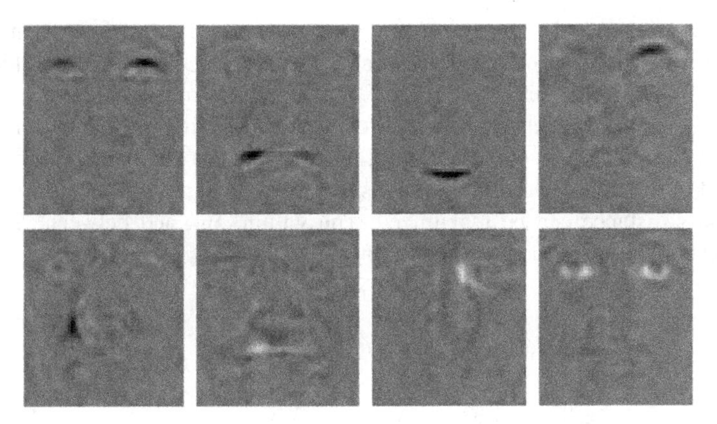

FIGURE 5.5 Eight basis images after applying ICA on the all face images of the six different facial expressions.

transforming the sample vectors X to Y, the ICA algorithm is applied on Y. Thus, the relation between Y and S becomes as

$$Y = BS \tag{5.8}$$

where B is the estimation of the unknown mixing matrix.

Before applying ICA, PCA is usually used to reduce the dimension of the total training image data which is also known as EICA [17]. In general, the ICA basis images focus on the local feature information unlike the global features in the PC basis images. Figure 5.5 shows eight ICA basis images for all the expressions where high-contrast parts represent local human face components such as eyebrows and lips used frequently in all expressions. The ICA algorithm is performed on P_m^T and thus m independent basis images in the rows of S are produced:

$$S = WP_m^T \tag{5.9}$$

$$P_m^T = W^{-1}S \tag{5.10}$$

$$X_r = VW^{-1}S \tag{5.11}$$

where $V = XP_m$ is the projection of the images X on P_m and X_r is the reconstructed original facial expression images.

Therefore, the enhanced independent component representation of expression images can be represented as

$$R = VW^{-1} \tag{5.12}$$

5.2.2.3 Fisher Linear Discriminant Analysis For the final step, FLDA is performed on the enhanced IC features, which can be called together as EICA-FLDA.

FLDA, also known as the fisherface method, is a supervised classification approach that utilizes the class-specific information maximizing the ratio of the within- and between-class scatter information. It looks for the vectors in the underlying space to create the best discrimination among different classes. It is well known for feature extraction and dimension reduction. In order to obtain maximum discrimination, it projects data onto the lower-dimensional space so that the ratio of the between- and within-class distance can be maximized. The within, S_{W}, and between, S_{B}, class scattering comparison is done by the following equations:

$$S_{\mathrm{B}} = \sum_{i=1}^{C} I_i(\bar{m}_i - \bar{\bar{m}})(\bar{m}_i - \bar{\bar{m}})^T \tag{5.13}$$

$$S_{\mathrm{W}} = \sum_{i=1}^{C} \sum_{m_k \in C_i} (m_k - \bar{m}_i)(m_k - \bar{m}_i)^T \tag{5.14}$$

where I_i is the number of vectors in the ith class C_i. c is the number of classes, and in our case, it represents the number of expressions. $\bar{\bar{m}}$ represents the mean of all vectors, $\bar{m}_i$ the mean of the class C_i, and m_k the vector of a specific class.

The optimal discrimination matrix is chosen by maximizing the ratio of the determinant of the between- and within-class scatter matrix as

$$D_{\mathrm{FLDA}} = \arg\max_{D} \frac{|D^T S_{\mathrm{B}} D|}{|D^T S_{\mathrm{W}} D|} \tag{5.15}$$

where D_{FLDA} is the set of discriminant vectors of S_{W} and S_{B} corresponding to the $(c-1)$ largest generalized eigenvalues λ and can be obtained via solving (5.16). The size of D_{FLDA} is $t \times r$, where $t \leq r$ and r is the number of elements in a vector.

$$S_{\mathrm{B}} d_i = \lambda_i S_{\mathrm{W}} d_i, \quad i = 1, 2, \ldots, (c-1) \tag{5.16}$$

where the rank of S_{B} is $(c-1)$ or less.

However, FLDA produces an optimal linear discriminant function that maps the input into the classification space on which the class identification of the samples is decided. The feature vectors using EICA-FLDA can be represented as

$$G = R D_{\mathrm{FLDA}}^T \tag{5.17}$$

Figure 5.6 shows an exemplar 3D EICA-FLDA feature plot where only four out of six expressions are shown for the clarity of presentation after applying FLDA on three ICs that are chosen on the basis of the top kurtosis values. In this plot, using those ICs, the features seem to be well separable.

Now, it seems clear that through the FLDA algorithm we can linearly classify the prototypes of different classes with the help of a low-dimensional subspace.

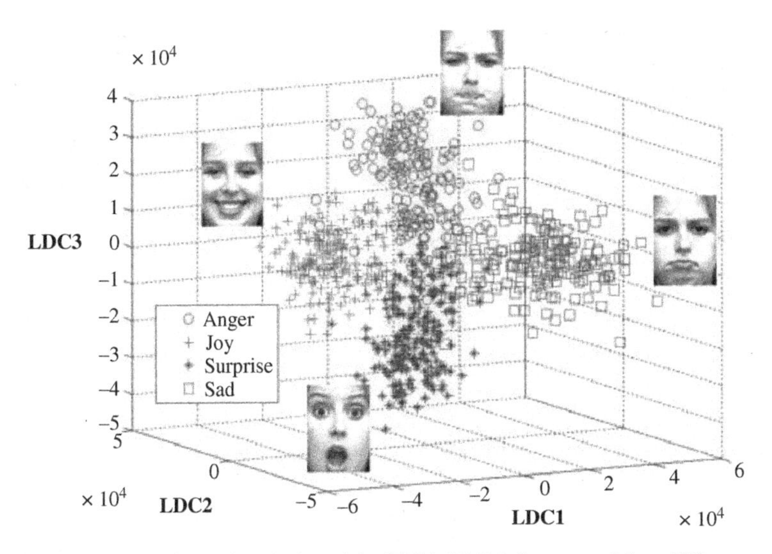

FIGURE 5.6 Three-dimensional plot of the EICA-FLDA features of four different expression images.

5.2.3 Codebook and Code Generation

As discrete HMMs are normally trained with symbols or codes of sequential data, the feature vectors obtained using EICA-FLDA are needed to be symbolized. The trained feature vectors formulate a codebook that is a set of temporal code signature of sequential dataset and the codebook is then regarded as a reference for recognizing the expressions. To obtain the codebook, vector quantization algorithm is performed on the feature vectors from training dataset. In our chapter, we utilize the Linde, Buzo, and Gray (LBG)'s clustering algorithm for vector quantization [28]. The LBG approach selects the first initial centroids and splits the centroids of the whole dataset. Then, it continues to split the dataset according to the codeword size, and the optimization is performed to reduce the distortion. After vector quantization is done, the index numbers are regarded as codes of the feature vectors to be implemented with HMMs.

As long as a feature vector is available, the index number of the closest code vector of the codebook from the feature is the code for that replace. Hence, every facial expression image is going to be assigned a code. If there are K image sequences of n length, then there will be K sequences of n length codes. The codes are the observations as denoted by O. Figure 5.7 shows the codebook generation and code selection from the codebook utilizing EICA-FLDA features.

5.2.4 Expression Modeling and Training Using HMM

An HMM is a collection of finite states connected by transitions [29]. Every state of an HMM can be described by two types of probabilities, namely transition probability

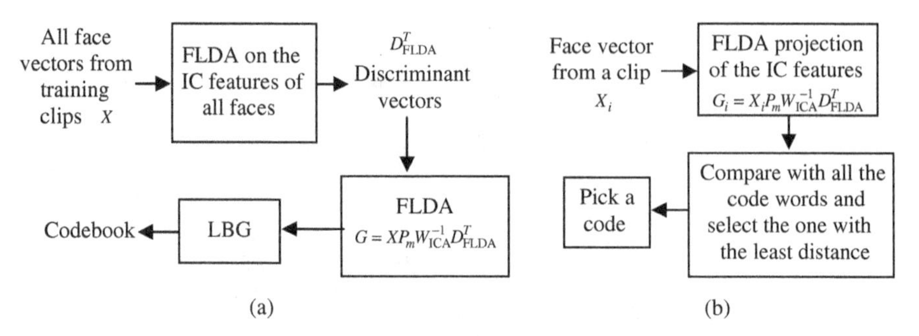

FIGURE 5.7 (a) Codebook generation and (b) code selection using EICA-FLDA features.

and symbol or code observation probability. A generic HMM can be expressed as $H = \{\Xi, \pi, A, B\}$, where Ξ denotes possible states, π the initial probability of the states, A the transition probability matrix between hidden states where state transition probability a_{ij} represents the probability of a changing state from i to j and B observation codes' probability from every state where the probability $b_j(d)$ indicates the probability of observing the code d from state j. If the number of expressions is N, then there will be a dictionary $(H_1, H_2, \ldots, H_N)$ of N trained models. We used the Baum–Welch algorithm [29] for HMM parameter estimation as follows:

$$\alpha_1(i) = \pi_i b_i(O_1) \qquad 1 \le i \le s \tag{5.18}$$

$$\alpha_{t+1}(j) = \sum_{i=1}^{s} \alpha_t(i) a_{ij} b_j(O_{t+1}) \quad t = 1, 2, \ldots, (n-1) \tag{5.19}$$

$$\beta_n(i) = 1 \qquad 1 \le i \le s \tag{5.20}$$

$$\beta_t(i) = \sum_{j=1}^{s} a_{ij} b_j(O_{t+1}) \beta_{t+1}(i) \qquad t = (n-1), (n-2), \ldots, 1 \tag{5.21}$$

$$\xi_t(i,j) = \frac{\alpha_t(i) a_{ij} b_j(O_{t+1}) \beta_{t+1}(j)}{\sum_{i=1}^{s} \sum_{j=1}^{s} \alpha_t(i) a_{ij} b_j(O_{t+1}) \beta_{t+1}(j)} \tag{5.22}$$

$$\hat{\pi}_i = \gamma_1(i) \qquad 1 \le i \le s \tag{5.23}$$

$$\gamma_t(i) = \sum_{j=1}^{s} \xi_t(i,j) \tag{5.24}$$

$$\hat{a}_{ij} = \frac{\sum_{t=1}^{n-1} \xi_t(i,j)}{\sum_{t=1}^{n-1} \gamma_t(i)} \tag{5.25}$$

$$\hat{b}_j(o) = \frac{\sum_{t=1,\, O_t=o}^{n-1} \gamma_t(j)}{\sum_{t=1}^{n-1} \gamma_t(i)} \tag{5.26}$$

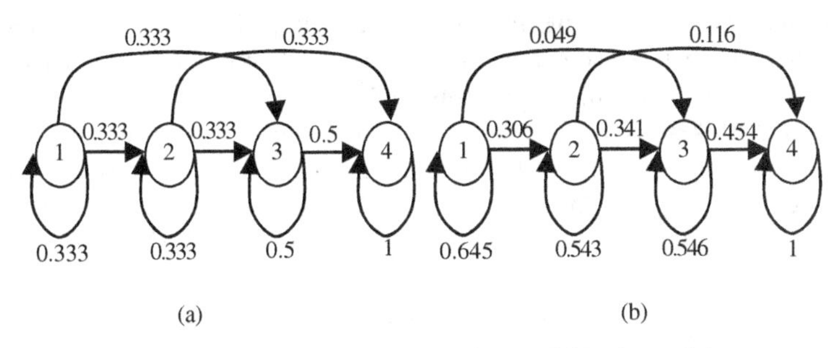

FIGURE 5.8 A sad HMM (a) before and (b) after training.

where n represents each expression image sequence length and s the number of states used in the models. α and β are the forward and backward variables, respectively, that are calculated from transition and observation matrices. $\xi_t(i,j)$ represents the probability of staying in a state i at time t and a state j at time $(t+1)$. $\gamma_t(i)$ is the probability of staying in the state i at time t. $\hat{a}_{ij}$ is the estimated transition probability from the state i to the state j and $\hat{b}_j(o)$ the estimated observation probability of a code o from the state j.

Figure 5.8 shows the structure and transition probabilities of a sad HMM before and after training with the codebook size of 64. In order to test a facial expression code sequence O, we find the appropriate HMM representing the maximum likelihood. Thus, to test an expression code sequence O (i.e., $O_1, O_2, \ldots, O_n$) with length n, the appropriate expression HMM can be found as

$$L = \arg \max_{i=1}^{N}(P(O|H_i)) \tag{5.27}$$

where L represents the maximum likelihood after applying O on all the trained expression HMMs H.

5.3 EXPERIMENTAL RESULTS

A set of experiments were performed with the proposed FER techniques in comparison to different feature extraction methods such as PCA, generic ICA, PCA-FLDA, EICA, and EICA-FLDA. To compare performance, we used the well-known facial expression dataset, the Cohn–Kanade facial expression database obtained from Carnegie Mellon University (CMU) [30]. We recognized six universal expressions such as anger, joy, sadness, disgust, fear, and surprise. A total of 288 sequences were used and 8 images which only displayed the frontal view of the facial expressions were contained in each sequence. A total of 25 sequences of anger, 35 of joy, 30 of sadness, 25 of disgust, 30 of fear, and 35 of surprise sequences were used in training and for the testing purpose, 11 of anger, 19 of joy, 13 of sadness,12 of disgust, 12 of fear, and 20 of surprise subsets were used. For empirical settings, we differentiated the number of feature selected in the PCA step to test the performance. We empirically selected 120 eigenvectors and codebook with the size 64 for our experiments.

The total mean of recognition rate from EICA representation of facial expression images with HMMs was 65.47% which was higher than that of the generic ICA and PCA recognition rates. Moreover, the best recognition rate from our experiment was obtained from our proposed method presenting 92.85% in comparison to the performance of 82.72% from the best conventional approach PCA-FLDA. The expressions such as surprise, joy, and sadness were recognized with very high accuracy (i.e., 93.75–100%). Tables 5.1–5.5 summarize the recognition results of each method.

TABLE 5.1 Person independent confusion matrix using PCA (unit: %)

Expression	Anger	Joy	Sadness	Surprise	Fear	Disgust
Anger	**30**	0	20	0	10	40
Joy	4	**48**	8	8	28	4
Sadness	0	6.06	**81.82**	12.12	0	0
Surprise	0	0	0	**68.75**	12.50	18.75
Fear	0	8.33	50	8.33	**33.33**	0
Disgust	0	8.33	25	0	0	**66.67**
Mean			**54.76**			

TABLE 5.2 Person-independent confusion matrix using ICA (unit : %)

Expression	Anger	Joy	Sadness	Surprise	Fear	Disgust
Anger	**30**	0	10	30	10	20
Joy	4	**60**	0	0	36	0
Sadness	0	6.06	**87.88**	6.06	0	0
Surprise	0	0	12.50	**81.25**	0	6.25
Fear	0	25	25	8.33	**33.33**	8.33
Disgust	0	8.33	25	0	0	**66.67**
Mean			**59.86**			

TABLE 5.3 Person-independent confusion matrix using EICA (unit : %)

Expression	Anger	Joy	Sadness	Surprise	Fear	Disgust
Anger	**60**	0	0	0	20	20
Joy	4	**72**	8	4	12	0
Sadness	0	6.06	**87.88**	6.06	0	0
Surprise	0	0	12.50	**81.25**	0	6.25
Fear	0	0	16.67	8.33	**50**	8.33
Disgust	25	8.33	25	0	0	**41.67**
Mean			**65.47**			

TABLE 5.4 Person-independent confusion matrix using PCA-FLDA (unit : %)

Expression	Anger	Joy	Sadness	Surprise	Fear	Disgust
Anger	**60**	0	10	0	0	30
Joy	0	**88**	0	0	8	4
Sadness	0	6.06	**87.88**	6.06	0	0
Surprise	0	0	0	**93.75**	6.25	0
Fear	0	8.33	8.33	8.33	**75**	0
Disgust	0	0	0	0	8.33	**91.67**
Mean			**82.72**			

TABLE 5.5 Person-independent confusion matrix using EICA-FLDA (unit : %)

Expression	Anger	Joy	Sadness	Surprise	Fear	Disgust
Anger	**80**	0	0	0	0	20
Joy	0	**96**	0	0	4	0
Sadness	0	0	**93.75**	0	6.25	0
Surprise	0	0	0	**100**	0	0
Fear	0	8.33	0	0	**91.67**	0
Disgust	0	0	0	0	8.33	**91.67**
Mean			**92.85**			

5.4 CONCLUSION

In this chapter, we have presented a novel system for FER utilizing EICA-FLDA for facial expression feature extraction and HMM for recognition. We have illustrated the performance of the proposed method applied on sequential facial expression datasets for six facial expressions. The result shows that EICA-FLDA can improve the facial expression feature extraction task. Furthermore, HMM dealing with EICA-FLDA processed sequential data can provide the higher recognition rate of 92.85% in comparison to the results from the conventional feature extracting methods with HMM. We conclude that the holistic face features can be used for the HMM method rather than usage of action units for HMM to recognize the partial face activities.

ACKNOWLEDGMENTS

This research was supported by the MKE (The Ministry of Knowledge Economy), Korea, under the ITRC (Information Technology Research Center) support program supervised by the NIPA (National IT Industry Promotion Agency) (NIPA-2011-C1090-1121-0003).

REFERENCES

[1] S. Mitra and T. Acharya, "Gesture recognition: a survey," *IEEE Transactions on Systems, Man, and Cybernetics*, 37(3): 311–324, 2007.

[2] C. Padgett and G. Cottrell, "Representing face images for emotion classification," *Advances in Neural Information Processing Systems*, vol. 9, MIT Press, Cambridge, MA, 1997.

[3] G. Donato, M. S. Bartlett, J. C. Hagar, P. Ekman, and T. J. Sejnowski, "Classifying facial actions," *IEEE Transactions on Pattern Analysis and Machine Intelligence*, 21(10): 974–989, 1999.

[4] P. Ekman and W. V. Priesen, *Facial Action Coding System: A Technique for the Measurement of Facial Movement*, Consulting Psychologists Press, Palo Alto, CA, 1978.

[5] M. Meulders, P. De Boeck, I. Van Mechelen, and A. Gelman, "Probabilistic feature analysis of facial perception of emotions," *Appl. Stat.*, 54(4): 781–793, 2005.

[6] A. J. Calder, A. M. Burton, P. Miller, A. W. Young, and S. Akamatsu, "A principal component analysis of facial expressions," *Vis. Res.*, 41: 1179–1208, 2001.

[7] S. Dubuisson, F. Davoine, and M. Masson, "A solution for facial expression representation and recognition," *Signal Process Image Commun.*, 17: 657–673, 2002.

[8] I. Buciu, C. Kotropoulos, and I. Pitas, "ICA and gabor representation for facial expression recognition," in Proceedings of the IEEE, 2003, pp. 855–858.

[9] M. S. Bartlett, J. R. Movellan, and T. J. Sejnowski, "Face recognition by independent component analysis," *IEEE Transactions on Neural Networks*, 13(6): 1450–1464, 2002.

[10] F. Chen and K. Kotani, "Facial expression recognition by supervised independent component analysis using MAP estimation," *IEICE Trans. Info. Syst.*, E91-D(2): 341–350, 2008.

[11] A. Hyvarinen, J. Karhunen, and E. Oja, *Independent Component Analysis*, John Wiley & Sons, 2001.

[12] Y. Karklin and M. S. Lewicki, "Learning higher-order structures in natural images," *Netw. Comput. Neural Syst.*, 14: 483–499, 2003.

[13] M. S. Bartlett, G. Donato, J. R. Movellan, J. C. Hager, P. Ekman, and T. J. Sejnowski, "Face image analysis for expression measurement and detection of deceit," in 6th Joint Symposium on Neural Computation Proceedings, 1999, pp. 8–15.

[14] C. Chao-Fa and F. Y. Shin, "Recognizing facial action units using independent component analysis and support vector machine," *Pattern Recogn.*, 39: 1795–1798, 2006.

[15] A. J. Calder, A. W. Young, and J. Keane, "Configural information in facial expression perception," *J. Exp. Psychol. Hum. Percept. Perform.*, 26(2): 527–551, 2000.

[16] M. J. Lyons, S. Akamatsu, M. Kamachi, and J. Gyoba, "Coding facial expressions with Gabor wavelets," in Proceedings of the Third IEEE International Conference on Automatic Face and Gesture Recognition, 1998, pp. 200–205.

[17] C. Liu, "Enhanced independent component analysis and its application to content based face image retrieval," *IEEE Transactions on Systems, Man, and Cybernetics*, 34(2): 1117–1127, 2004.

[18] P. J. Phillips, H. Wechsler, J. Huang, and P. Rauss, "The FERET database and evaluation procedure for face-recognition algorithms," *Image Vis. Comput.*, 16: 295–306, 1998.

[19] Y. Tian, T. Kanade, and J. Cohn, "Evaluation of Gabor wavelet based facial action unit recognition in image sequences of increasing complexity," in Proceedings of the 5th

IEEE International Conference on Automatic Face and Gesture Recognition, 2002, pp. 229–234.

[20] I. Kotsia and I. Pitas, "Facial expression recognition in image sequences using geometric deformation features and support vector machine," *IEEE Transactions on Image Processing*, 16: 172–187, 2007.

[21] T. Otsuka and J. Ohya, "Recognizing multiple person's facial expressions using HMM based on automatic extraction of significant frames from image sequences," in IEEE International Conference on Image Processing, 1997, pp. 546–549.

[22] Y. Zhu, L. C. De Silva, and C. C. Ko, "Using moment invariants and HMM in facial expression recognition," *Pattern Recogn. Lett.*, 23: 83–91, 2002.

[23] I. Cohen, N. Sebe, A. Garg, L. S. Chen, and T. S. Huang, "Facial expression recognition from video sequences: temporal and static modeling," *Comput. Vis. Image Underst.*, 91: 160–187, 2003.

[24] P. S. Aleksic and A. K. Katsaggelos, "Automatic facial expression recognition using facial animation parameters and multistream HMMs," *IEEE Transactions on Information and Security*, 1: 3–11, 2006.

[25] T. Kanade, J. F. Cohn, and Y. Tian, "Comprehensive database for facial expression analysis," in Proceedings of the Fourth IEEE International Conference on Automatic Face and Gesture Recognition (FG00), Grenoble, France, 2000, pp. 46–53.

[26] P. Lucey, J. F. Cohn, T. Kanade, J. Saragih, S. Ambadar, and I. Matthews, "The extended Cohn-Kanade Dataset (DK+): a complete expression dataset for action unit and emotion specified expression," in Proceedings of the Third International Workshop on CVPR for Human Communicative Behavior Analysis (CVPR 4HB 2010), San Francisco, 2010, pp. 94–101.

[27] L. Zhang and G. W. Cottrell, "When holistic processing is not enough: local features save the day," in Proceedings of the Twenty-sixth Annual Cognitive Science Society Conference, 2004.

[28] Y. Linde, A. Buzo, and R. Gray, "An algorithm for vector quantizer design," *IEEE Transaction on Communications*, 28(1): 84–94, 1980.

[29] L. R. Rabiner, "A tutorial on hidden Markov models and selected applications in speech recognition," *Proceedings of the IEEE*, 77(2): 257–286, 1989.

[30] J. F. Cohn, A. Zlochower, J. Lien, and T. Kanade, "Automated face analysis by feature point tracking has high concurrent validity with manual FACS coding," *Psychophysiology*, 36: 35–43, 1999.

AUTHOR BIOGRAPHIES

Md. Zia Uddin received his B.Sc. degree in Computer Science and Engineering from International Islamic University Chittagong, Bangladesh, in 2004. He has completed his M.S. leading to his Ph.D. degree from the Department of Biomedical Engineering, Kyung Hee University, Republic of Korea, in 2011. He is working now as a faculty member of Electronic Engineering Department, Inha University in Republic of Korea. His research interest includes pattern recognition, image processing, computer vision, and machine learning.

Tae-Seong Kim received the B.S. degree in Biomedical Engineering from the University of Southern California (USC) in 1991, the M.S. degrees in Biomedical and Electrical Engineering from USC in 1993 and 1998, respectively, and Ph.D. in Biomedical Engineering from USC in 1999. After his postdoctoral work in cognitive sciences at the University of California, Irvine, in 2000, he joined the Alfred E. Mann Institute for Biomedical Engineering and Department of Biomedical Engineering at USC as a research scientist and research assistant professor. In 2004, he moved to Kyung Hee University, Korea, where he is currently an associate professor in the Biomedical Engineering Department. His research interests have spanned various areas of biomedical imaging including magnetic resonance imaging (MRI), functional MRI, E/MEG imaging, DT-MRI, transmission ultrasonic CT, and magnetic resonance electrical impedance imaging. Lately, he has started research work in proactive computing at the u-Lifecare Research Center and the Center for Sustainable Healthy Buildings. Dr. Kim has published more than 60 peer-reviewed papers and 120 papers in proceedings, and holds 3 international patents. He is a member of IEEE, KOSOMBE, and Tau Beta Pi, and listed in Who's Who in the World '09–'11 and Who's Who in Science and Engineering '11–'12.

6

FEATURE SELECTION FOR FACIAL EXPRESSION BASED ON ROUGH SET THEORY

YONG YANG AND GUOYIN WANG

School of Computer Science and Technology, Chongqing University of Posts and Telecommunications, Chongqing, China

The study of emotion is a hot research topic in artificial intelligence. Although great progress has been made, several unsolved issues still exist. For example, which features are the most important for emotion. It is a research topic that was seldom studied in computer science; however, related research works have been conducted in cognitive psychology. In this chapter, feature selection for facial expression recognition is studied based on rough set theory, and a self-learning attribute reduction algorithm is proposed based on rough set and domain-oriented, data-driven, data-mining theories. Experimental results show that important features for facial expression can be found based on the rough set reduction algorithms compared with traditional feature selection methods, such as genetic algorithm. Furthermore, important and useful features can be identified by the proposed method with higher recognition rate. It is found that the features concerning the mouth are the most important ones in geometrical features for facial emotion recognition.

6.1 INTRODUCTION

In recent years, human-centered science and technology have been widely researched. Among the research subjects related with humans, emotion is an important one. There are many research branches related to emotion, such as affective computing [1–3],

Emotion Recognition: A Pattern Analysis Approach, First Edition. Edited by Amit Konar and Aruna Chakraborty.
© 2015 John Wiley & Sons, Inc. Published 2015 by John Wiley & Sons, Inc.

emotion intelligence [4], artificial psychology, and artificial emotion [5,6]. There are several achievements in this field in recent years, such as AIBO, the popular robot dog produced by Sony.

Emotion recognition is fundamental and important in the study of emotion. Usually, it is studied based on the facial or audio features by the methods such as Artificial Neural Network (ANN), fuzzy set, Support Vector Machine (SVM), and Hidden Markov Model (HMM), and the recognition rate often arrives at 64–98% [7].

Until now, research on emotion recognition is mainly according to the basic type of emotion, such as happiness, sadness, surprise, anger, disgust, fear, and neutral. Some applications focus on the particular emotion states, for example, sleepy is focused in a driver monitor system. It is very difficult for research on recognizing the complicated and mixed emotion such as bittersweet, since it can not be taken as a simple classification problem for such emotion, and it is hardly solved based on traditional classification methods. As for the complicated and mixed emotion, we prefer to model how it can be mixed by basic emotion states and how it can be changed over time, but it is beyond the scope of this chapter. In this chapter, basic emotion types are taken for emotion recognition.

Although some progress has been made in emotion recognition, there is a long way to go before a computer can act as a human in emotion recognition since there are many problems unsolved in psychology and cognitive theories, for example, how does emotion come into being, what is the essence of emotion, and what is the feature of emotion. Among these problems, which features are important and essential for emotion and which features are crucial for emotion recognition are open questions. It is a research topic that was seldom studied in computer science. However, related research works have been conducted in cognitive psychology [8–10].

There has been several research works related to the important features for emotion in cognitive psychology. On the basis of the results of psychological experiments, Sui *et al.* argue that the information conveyed by different facial parts have diverse effects on the facial expression recognition and the eyes play the most important role [8]. Wang *et al.* argue that the low spatial frequency information is important for emotion [9]. White argues that edge-based facial information is used for expression recognition [10].

In our previous works of emotion recognition in References 11–14, attribute reduction algorithms based on classical rough set are taken as the feature selection method for facial emotional, and SVM is taken as the classifiers. Some useful features concerning eyes and mouth are found. Based on these features, high correct recognition rates are achieved. However, classical rough set theory is based on equivalence relation. There must be a process of discretization in equivalence relation since the measured facial features are continuous values. Information might be lost or changed in the discretization process, thereby the result would be affected. To solve this problem, some research works have been taken. Shang *et al.* propose a new attribute algorithm, which integrates the discretion and reduction using information entropy-based uncertainty measures and evolutionary computation [15]. Jensen *et al.* propose a fuzzy-rough attribute reduction algorithm and an attribute reduction algorithm based on tolerance relation [16]. Although these research works can avoid the discretization process, the parameters in these methods should be given according to prior experience of domain experts, for example, the fuzzy set membership function in Jensen's method and the population amount in Shang's method. If there is no experience of domain experts, these methods will be useless in some extent. In this chapter, a self-learning attribute reduction algorithm is proposed based on rough set and domain-oriented, data-driven, data-mining (3DM) theories, which can avoid the process of discretization and find a suitable threshold for tolerance relation automatically.

It is difficult to get inner features without any interfere with the human subjects. Further, to ensure that a human subject is in natural state is also difficult. Research on emotion recognition is always taken based on facial or speech features. In this chapter, facial emotion recognition is researched based on the basic emotion types. Furthermore, feature selection for emotion recognition based on rough set theory is highlighted. Classical rough-set reduction algorithm, the proposed feature selection algorithm, and classical feature selection method, namely genetic algorithm, are all used for facial emotion feature selection. Experimental results show that important and useful features for emotion recognition can be found by the proposed method with a high recognition rate. Furthermore, the features concerning mouth are found to be the most important ones in geometrical features for facial emotion recognition. The rest of this chapter is organized as follows. In Section 6.2, a novel feature selection method for emotion recognition based on rough set theory is introduced. Feature selection for emotion recognition is discussed in Section 6.3. Finally, conclusions and future works are presented in Section 6.4.

6.2 FEATURE SELECTION FOR EMOTION RECOGNITION BASED ON ROUGH SET THEORY

6.2.1 Basic Concepts of Rough Set Theory

Rough set (RS) is a valid mathematical theory for dealing with imprecise, uncertain, and vague information, and it was developed by Z. Pawlak in 1980s [17, 18]. RS has been successfully used in many domains such as machine learning, pattern recognition, intelligent data analyzing, and control algorithm acquiring [19–21]. The advantage of RS is its ability for attribute reduction (knowledge reduction, feature selection). Some basic concepts of rough set theory are introduced here for the convenience of the following discussion.

Definition 6.1. A decision information system is defined as a quadruple $S = (U, C \cup D, V, f)$, where U is a finite set of objects, C is the condition attribute set, and $D = \{d\}$ is the decision attribute set.$\forall c \in C$, with every attribute $a \in C \cup D$, a set of its values Va is associated. Each attribute a determines a function $fa : U \longrightarrow Va$.

Definition 6.2. For a subset of attributes $B \subseteq A$, an indiscernability relation is defined by $Ind(B)=\{(x, y) \in U \times U\text{: } \forall_{a\in B}(a_x = a_y)\}$.

The indiscernability relation defined in this way is an equivalence relation. Obviously, $Ind(B) = \cap_{b\in B} Ind(\{b\})$. By $U/Ind(B)$, we mean the set of all equivalence classes in the relation *Ind*(*B*). The classical rough set theory is based on an observation that objects may be indiscernable due to limited available information, and the indiscernability relation defined in this way is an equivalence relation indeed. The intuition behind the notion of an indiscernability relation is that selecting a set of attribute $B \subseteq A$ effectively defines a partition of the universe into sets of objects that cannot be discerned using the attributes in B only. The equivalence classes $E_i \in U/Ind(B)$, induced by a set of attributes $B \subseteq A$, are referred to as object classes or simply classes. The classes resulted from *Ind*(*A*) and *Ind*(*D*) are called condition classes and decision classes, respectively.

Definition 6.3. A decision information system is a continuous value information system and it is defined as a quadruple $S = (U, C \cup D, V, f)$, where U is a finite set of objects, C is the condition attribute set, and $D = \{d\}$ is the decision attribute set, and c is the continuous value attribute.

A facial expression information system is a continuous value information system according to Definition 6.3.

If a condition attribute value is a continuous value, indiscernability relation cannot be used directly since it requires that the condition attribute values of two different samples are equal, which is difficult to satisfy. Consequently, a process of discretization must be taken, in which information may be lost or changed. The result of attribute reduction would be affected. Since all measured facial attributes are continuous value and imprecise to some extent, the process of discretization may affect the result of emotion recognition. We argue that two attribute values are taken as equal if they are similar in some range in a continuous value information systems. Then, it is important to find the suitable extent to take the different values as equal. Based on this idea, a method based on tolerance relation that avoids the process of discretization is proposed in this chapter.

Definition 6.4. A binary relation $R(x, y)$ defined on an attribute set B is called a tolerance relation if it satisfies

1. Symmetrical: $\forall_{x,y \in U}(R(x, y) = R(y, x))$.
2. Reflexive: $\forall_{x \in U}(R(x, x) = 1)$.

From the standpoint of a continuous value information system, a relation could be set up for a continuous value information system as follows.

Definition 6.5. Let an information system $S = (U, C \cup D, V, f)$ be a continuous value information system. A relation $R(x, y)$ is defined as:

$$R(x, y) = \{(x, y) | x \in U \wedge y \in U \wedge \forall_{a \in C}(|a_x - a_y| \leq \varepsilon, 0 \leq \varepsilon \leq 1)\}$$

Apparently, $R(x, y)$ is a tolerance relation according to Definition 4 since $R(x, y)$ is symmetrical and reflexive. In classical rough set theory, an equivalence relation constitutes a partition of U, but a tolerance relation constitutes a cover of U, and equivalence relation is a particular type of tolerance relation.

Definition 6.6. Let $R(x, y)$ be a tolerance relation based on Definition 6.5, $n_R(x_i) = \{x_j | x_j \in U \wedge \forall_{a \in C}(|a_{x_i} - a_{x_j}| \leq \varepsilon)\}$ is called a tolerance class of x_i, and $|n_R(x_i)| = |\{x_j | x_j \in n_R(x_i), 1 \leq j \leq U\}|$ is the cardinal number of the tolerance class of x_i.

According to Definition 6.6, $\forall x \in U$, the bigger the tolerance class of x is, the more uncertainty it will be and the less knowledge it will contain. On the contrary, the smaller the tolerance class of x is, the less uncertainty it will be and the more knowledge it will contain. Accordingly, the concept of knowledge entropy and conditional entropy could be defined as follows.

Definition 6.7. Let $U = \{x_1, x_2, \cdots, x_{|U|}\}$, $R(x, y)$ be a tolerance relation, the knowledge entropy $E(R)$ of relation R is defined as $E(R) = -\frac{1}{|U|}\sum_{i=1}^{|U|} \log_2 \frac{|n_R(x_i)|}{|U|}$.

Definition 6.8. Let R and Q be tolerance relations defined on U, a relation satisfying R and Q simultaneously can be taken as $R \cup Q$, and it is also a tolerance relation. $\forall_{x_i} \in U$, $n_{R \cup Q} = n_R(x_i) \cap n_Q(x_i)$, therefore, the knowledge entropy of $R \cup Q$ can be defined as $E(R \cup Q) = -\frac{1}{|U|}\sum_{i=1}^{|U|} \log_2 \frac{|n_{R \cup Q}(x_i)|}{|U|}$.

Definition 6.9. Let R and Q be tolerance relations defined on U, the conditional entropy of R with respect to Q is defined as $E(Q|R) = E(R \cup Q) - E(R)$.

Let $S = (U, C \cup D, V, f)$ be a continuous value information system, relation K be a tolerance relation defined on its condition attribute set C, and relation L be an equivalence relation (a special tolerance relation) defined on its decision attribute set D. According to Definition 6.7, Definition 6.8, and Definition 6.9, we can get $E(D|C) = E(L|K) = E(K \cup L) - E(K) = -\frac{1}{|U|}\sum_{i=1}^{|U|} \log_2 \frac{|n_{K \cup L}(x_i)|}{|U|} - (-\frac{1}{|U|}\sum_{i=1}^{|U|} \log_2 \frac{|n_K(x_i)|}{|U|}) = -\frac{1}{|U|}\sum_{i=1}^{|U|} \log_2 \frac{|n_{K \cup L}(x_i)|}{|n_K(x_i)|}$. The conditional entropy $E(D|C)$ has a clear meaning, that is, it is a ratio between the knowledge of all attributes (condition attribute set plus decision attribute set) and the knowledge of the condition attribute set.

6.2.2 Feature Selection Based on Rough Set and Domain-Oriented Data-Driven Data Mining Theories

In this section, a novel attribute reduction algorithm is proposed based on rough set and domain-oriented, data-driven, data-mining (3DM) theories [22, 23].

3DM is a data-mining theory proposed by Guoyin Wang [22, 23]. According to the theory, knowledge could be expressed in different ways, that is, some relationship exists between the different formats of the same knowledge. In order to keep the knowledge unchanged in a data-mining process, the properties of the knowledge should remain unchanged during the knowledge transformation process [24]. Otherwise, mistake may occur in the process of knowledge transformation. Based on this understanding, knowledge reduction can be seen as a process of knowledge transformation, in which properties of the knowledge should be remained.

In the application of emotion recognition, neither faces nor emotions are entirely same. For any two different emotion samples, there must be some different features in the samples. Accordingly, an emotion sample belongs to an emotion state according to its features which are different to the others. From this standpoint, we argue that the discernability of the condition attribute set with respect to the decision attribute set can be taken as an important property of knowledge in the course of knowledge acquisition in emotion recognition. Based on the idea of 3DM, the discernability should be unchanged in the process of knowledge acquisition and attribute reduction.

Definition 6.10. Let $S = (U, C \cup D, V, f)$ be a continuous value information system. If $\forall_{x_i, x_j \in U}(d_{x_i} \neq d_{x_j} \rightarrow \exists_{a \in C}(a_{x_i} \neq a_{x_j}))$, it is certainly discernable for the continuous value information system S.

The discernability is taken as a fundamental ability that a continuous information system has in this chapter. According to 3DM, the discernability should be unchanged if feature selection is done for a continuous value information system. From Definition 6.10, we can have $\forall_{x_i, x_j \in U}(d_{x_i} \neq d_{x_j} \rightarrow \exists_{a \in C}(|a_{x_i} - a_{x_j}| > \varepsilon))$. Therefore, according to Definition 6.6, we can have $\forall_{x_i, x_j \in U}(x_j \notin n_R(x_i) \wedge x_i \notin n_R(x_j) \rightarrow n_R(x_i) \neq n_R(x_j))$. Accordingly, the discernability of a tolerance relation can be defined as follows.

Definition 6.11. Let $R(x, y)$ be a tolerance relation according to Definition 6.5, if $\forall_{x_i, x_j \in U}(d_{x_i} \neq d_{x_j} \rightarrow n_R(x_i) \neq n_R(x_j), R(x, y))$ has the certain discernability.

If $R(x, y)$ has certain discernability, according to Definition 6.11, $\forall_{x_i, x_j \in U}(n_R(x_i) = n_R(x_j) \rightarrow d_{x_i} = d_{x_j})$, therefore, $\forall_{x_i, x_j \in U}(x_i, x_j \in n_R(x_i) \rightarrow d_{x_i} = d_{x_j})$.

Theorem 6.1. $E(D|C) = 0$ is a necessary and sufficient condition of that there is certain discernability for the condition attribute set with respect to the decision attribute set in tolerance relation.

Proof. Let $S = (U, R, V, f)$ be a continuous value information system, relation K be a tolerance relation defined on the condition attribute set C, relation L be an equivalence relation (a special tolerance relation) defined on decision attribute set D.

(Necessity). If there is certain discernability for the condition attribute set with respect to the decision attribute set in tolerance relation, according to Definition 6.11, $\forall_{x_i, x_j \in U}(x_i, x_j \in n_K(x_i) \rightarrow d_{x_i} = d_{x_j})$, then, $n_K(x_i) \subseteq n_L(x_i)$, $n_{K \cup L}(x_i) = n_K(x_i)$ $\quad |n_{K \cup L}(x_i)| = |n_K(x_i)|$, $\quad E(D|C) = E(L|K) = -\frac{1}{|U|}\sum_{i=1}^{|U|} \log_2 \frac{|n_{K \cup L}(x_i)|}{|n_K(x_i)|} = -\frac{1}{|U|}\sum_{i=1}^{|U|} \log_2 1 = 0$.

(Sufficiency). $\forall x_i \in U$, we can have $n_{K \cup L}(x_i) \subseteq n_K(x_i)$, $|n_{K \cup L}(x_i)| \leq |n_K(x_i)|$. Since $E(D|C) = E(L|K) = -\frac{1}{|U|}\sum_{i=1}^{|U|} \log_2 \frac{|n_{K \cup L}(x_i)|}{|n_K(x_i)|} = 0$, we can have $\forall x_i \in U$, $|n_{K \cup L}(x_i)| = |n_K(x_i)|$, that is, $n_{K \cup L}(x_i) = n_K(x_i)$. Therefore, the decision values should be equal for the different samples included in the same tolerance class. Accordingly, we can have $\forall_{x_i, x_j \in U}(x_i, x_j \in n_R(x_i) \rightarrow d_{x_i} = d_{x_j})$, therefore, $\forall_{x_i, x_j \in U}(d_{x_i} \neq d_{x_j} \rightarrow \exists_{a \in C}(a_{x_i} \neq a_{x_j}))$, and there is certain discernability for the condition attribute set with respect to decision attribute set in tolerance relation. This completes the proof.

It is apparent from Theorem 6.1 that $E(D|C) = 0$, if $R(x, y)$ has certain discernability.

For a given continuous value information system S, there could be many different tolerance relations according to different threshold ε under the condition $E(D|C) = 0$, but the biggest granular and the best generalization are always required for knowledge acquisition. According to the principle, we can have the following results.

1. If the threshold ε in tolerance relation is 0, then the tolerance class $n_R(x_i)$ of an instance x_i contains x_i itself only, we can have $n_{R \cup Q}(x_i) = n_R(x_i) = \{x_i\}$, and $E(D|C) = 0$. It is the smallest tolerance class for the tolerance relation, the smallest knowledge granular and the smallest generalization.
2. If the threshold ε in tolerance relation is increased from 0, both $n_R(x_i)$ and $n_{R \cup Q}(x_i)$ are increased. If $n_R(x) \subseteq n_Q(x)$, then, $n_{R \cup Q}(x_i) = n_R(x_i)$, $|n_{R \cup Q}(x_i)| = |n_R(x_i)|$, $E(D|C) = 0$, and the granular of knowledge is increased.
3. If the threshold ε in tolerance relation is increased to a critical point named ε_{opt}. both $n_R(x_i)$ and $n_{R \cup Q}(x_i)$ are increased, and $n_{R \cup Q}(x_i) = n_R(x_i)$, $|n_{R \cup Q}(x_i)| = |n_R(x_i)|$, $E(D|C) = 0$, and the granular of knowledge is the biggest under the condition that certain discernability of the condition attribute set with respect to the decision attribute set in tolerance relation is unchanged.
4. If the threshold ε in tolerance relation is increased from ε_{opt} and $\varepsilon < 1$, then $n_{R \cup Q}(x_i) \neq n_R(x_i)$, $|n_{R \cup Q}(x_i)| \neq |n_R(x_i)|$, $E(D|C) \neq 0$, then the certain discernability is changed. If $\forall_{x_i \in U}(n_Q(x_i) \subset n_R(x_i))$, then $n_{R \cup Q}(x_i) = n_Q(x_i)$, $|n_{R \cup Q}(x_i)| = |n_Q(x_i)|$, and $|n_Q(x_i)| < |n_R(x_i)|$. Therefore, $E(D|C) = E(Q|R) = -\frac{1}{|U|}\sum_{i=1}^{|U|} \log_2 \frac{|n_{R \cup Q}(x_i)|}{|n_R(x_i)|} = -\frac{1}{|U|}\sum_{i=1}^{|U|} \log_2 \frac{|n_Q(x_i)|}{|n_R(x_i)|} > 0$. Since $|n_Q(x_i)|$ is held and $|n_R(x_i)|$ is increased with the threshold of ε increase, $E(D|C)$ is increased.
5. If the threshold ε in tolerance relation is increased to $\varepsilon = 1$, then $n_R(x_i) = U$ and $\forall_{x_i \in U}(n_Q(x_i) \subseteq n_R(x_i))$, $n_{R \cup Q}(x_i) = n_Q(x_i)$, $|n_{R \cup Q}(x_i)| = |n_Q(x_i)|$, so, $E(D|C) = E(Q|R) = -\frac{1}{|U|}\sum_{i=1}^{|U|} \log_2 \frac{|n_{R \cup Q}(x_i)|}{|n_R(x_i)|} = -\frac{1}{|U|}\sum_{i=1}^{|U|} \log_2 \frac{|n_Q(x_i)|}{|U|}$. Since the equivalence class of Q is held, $E(D|C)$ is constant.

The relationship between entropy, condition entropy, and ε can be shown in Figure 6.1.

From Figure 6.1 and the discussion above, if the threshold value of ε takes ε_{opt}, it will make $E(D|C) = 0$; therefore, the certain classification ability of the condition attribute set with respect to the decision attribute set will be unchanged. At the same time, the tolerance class of x is the biggest. In a sense, the knowledge granular is the biggest in ε_{opt}, then the generalization should be the best.

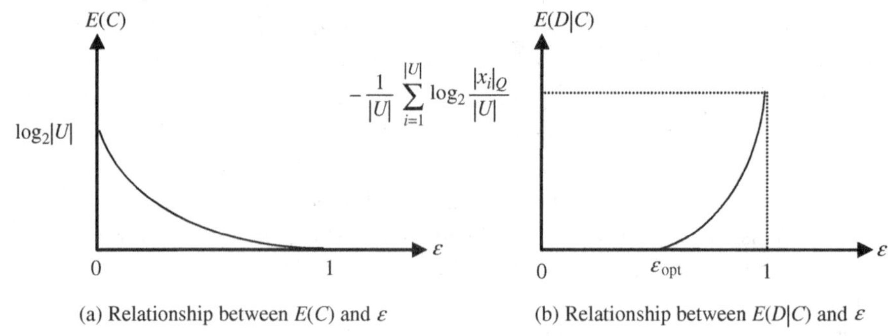

(a) Relationship between $E(C)$ and ε (b) Relationship between $E(D|C)$ and ε

FIGURE 6.1 Relationship between entropy, condition entropy, and ε.

In summary, parameter selection of ε is discussed, based on 3DM, a suitable threshold value of ε, ε_{opt}, is found. The choice of $\varepsilon = \varepsilon_{opt}$ helps improving the classification ability of the condition attribute set with respect to the decision attribute set. Also at the same time, it helps attainly maximal generalization. It is predominant for the course of finding ε_{opt} since the method is based on data only and dose not need experiences of domain experts. Therefore, the method is more robustness.

In this chapter, the threshold of ε_{opt} is searched in [0,1] based on binary search algorithm.

6.2.3 Attribute Reduction for Emotion Recognition

The discernability of the condition attribute set with respect to the decision attribute set in tolerance relation is a fundamental feature of knowledge of a continuous value information system. The discernability should be unchanged according to 3DM. Since $E(D|C) = 0$ is a necessary and sufficient condition for keeping the discernability of the condition attribute set with respect to the decision attribute set in tolerance relation. Therefore, a self-learning attribute reduction algorithm (SARA) is proposed for continuous value information systems as follows.

Algorithm 6.1 Self-learning Attribute Reduction Algorithm (SARA)

Input: a decision table $S = (U, C \cup D, V, f)$ of a continuous information system, where U is a finite set of objects, C is the condition attribute set, and $D = \{d\}$ is the decision attribute set.

Output: a relative reduction B of S.

Step1: Compute ε_{opt}, then set up a tolerance relation on the condition attribute set C.

Step2: Compute condition entropy $E(D|C)$.

Step3: $\forall a_i \in C$, compute $E(D|\{a_i\})$. Sort a_i according to $E(D|\{a_i\})$ descendant.

Step4: Let $B = C$, deal with each a_i as below.

Step4.1: Compute $E(D|B - \{a_i\})$;

Step4.2: If $E(D|C) = E(D|B - \{a_i\})$, attribute a_i should be reduced, and $B = B - \{a_i\}$, otherwise, a_i could not be reduced, B is holding.

Let $|U| = n$, $|C| = m$. The time complexity of Step1 is $O(n)$, the time complexity of Step2 is $O(mn^2)$, the time complexity of Step3 is $O(mn^2)$, and the time complexity of Step4 is $O(m^2n^2)$; therefore, the time complexity of the algorithm is $O(m^2n^2)$.

6.3 EXPERIMENT RESULTS AND DISCUSSION

In this section, feature selection for emotion recognition is discussed. First, experiment condition is introduced. Second, different feature selection methods will be used for facial emotion feature selection. In the end, important features for emotion recognition will be discussed.

6.3.1 Experiment Condition

Since there are a few open facial emotional datasets including all the races, sex, and ages, three facial emotional datasets are identified, selected, and used in the experiments. The first dataset comes from the Cohn-Kanade AU-Coded Facial Expression (CKACFE) database [25–27] and the dataset is a representation of Western people in some extent. The second one is the Japanese Female Facial Expression (JAFFE) database [28] and it is a representation of Eastern women in some extent. The third one named CQUPTE [29] and it is collected from eight graduate students in Chongqing University of Posts and Communications, China, in which four are females and four are males. Details of the datasets are listed in Table 6.1

Some samples are shown in Figure 6.2. In each dataset, the samples are happiness, sadness, fear, disgust, surprise, and angry from left to right in Figure 6.2.

Facial expression of human being is expressed by the shape and position of facial components such as eyebrows, eyes, mouth, and nose. The geometric features, appearance features, wavelet features, and mixture features of facial expression are more popular for emotion recognition in recent years. Among all the facial features, the geometric facial features represent the shape and locations of facial components. Since geometric facial features are intuitionistic for the facial expression, it is taken as facial feature and discussed in the experiments.

The geometric facial features are the distance between two different feature points. The MPEG-4 standard is a popular standard for feature point selection. It extends Facial Action Coding System (FACS) to derive facial definition parameters (FDPs) and facial animation parameters (FAPs). There are 68 FAPs, in which 66 low parameters are defined according to FDPs to describe the motion of a human face. The FDP and low-level FAP can constitute a concise representation of a face, and they are

TABLE 6.1 Three facial emotional datasets

Dataset name	Samples	People	Emotion classes
CKACFE	405	97	Happiness, Sadness, Surprise, Anger, Disgust, Fear, Neutral
JAFFE	213	10	Happiness, Sadness, Surprise, Anger, Disgust, Fear, Neutral
CQUPTE	652	8	Happiness, Sadness, Surprise, Anger, Disgust, Fear, Neutral

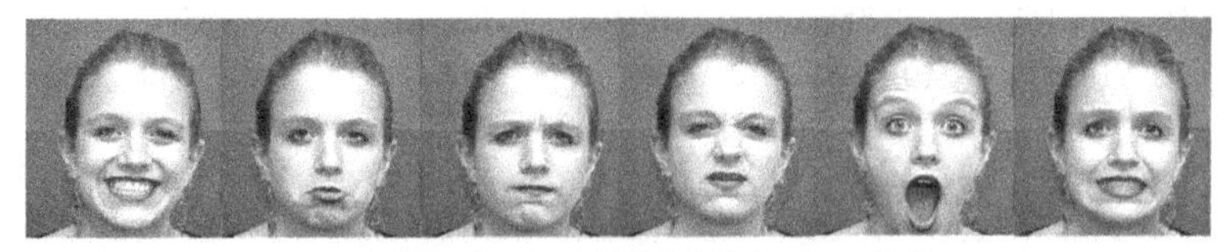

(a) Some images of CKACFE database [25]

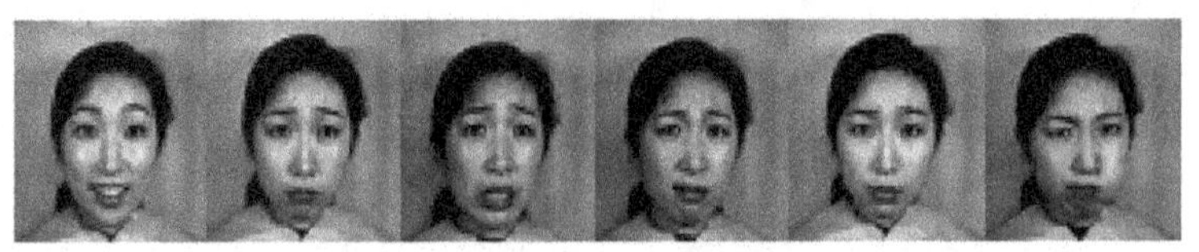

(b) Some images of JAFFE database [26]

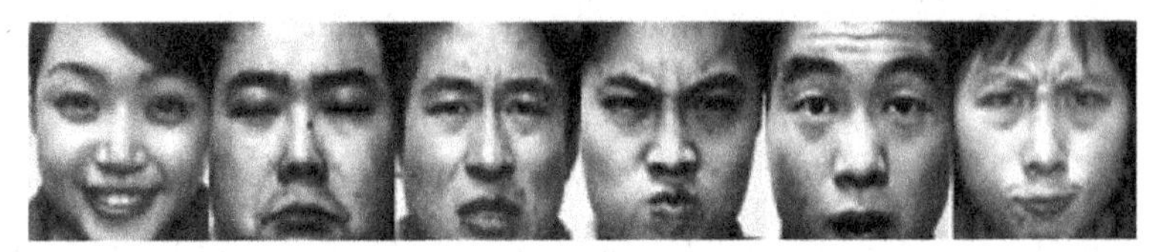

(c) Some images of CQUPTE database [27]

FIGURE 6.2 Facial emotion samples. Extracted with permission from the databases.

adequate for basic emotion recognition because of the varieties of expressive parameter. In the experiments, 52 low FAPs are chosen to represent emotion because some FAPs have little effect on facial expression. For example, the FAP named raise_l_ear denotes the vertical displacement of the left ear. Thus, a feature point set including 52 feature points is defined as shown in Figure 6.3. Based on the feature points, 33 facial features are extracted for emotion recognition according to References 11–13 and listed in Table 6.2. The 33 facial features can be divided into three groups. There

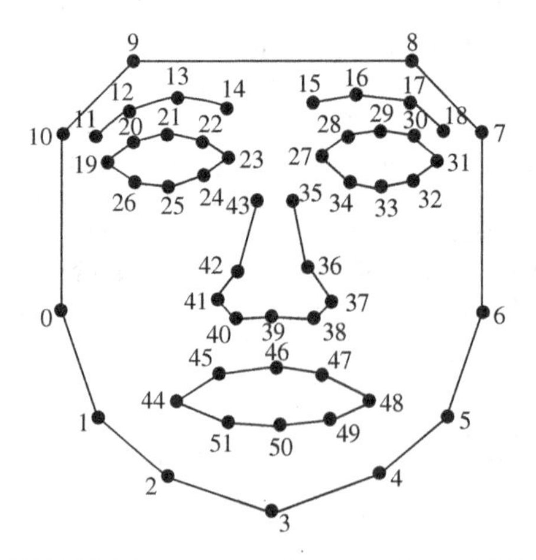

FIGURE 6.3 Feature points according to FAPs.

TABLE 6.2 Thirty-three features defined on 52 feature points

Feature	Description	Feature	Description	Feature	Description
d0	dis(11,19)	d11	dis(39,44)	d22	dis(44,48)/2
d1	dis(18,31)	d12	dis(39,48)	d23	dis(45,51)
d2	dis(21,25)	d13	dis(44,48)	d24	dis(47,49)
d3	dis(20,26)	d14	dis(46,50)	d25	dis(14,23)
d4	dis(22,24)	d15	dis(39,3)	d26	dis(15,27)
d5	dis(29,33)	d16	dis(21,A)	d27	dis(19,23)/2
d6	dis(28,34)	d17	dis(A,25)	d28	dis(27,31)/2
d7	dis(30,32)	d18	hei(A,44)	d29	(wid(19,23)+wid(27,31))/2
d8	dis(39,46)	d19	dis(29,B)	d30	(hei(11,39)+hei(18,39))/2
d9	dis(23,44)	d20	dis(B,33)	d31	(hei(14,39)+hei(15,39))/2
d10	dis(27,48)	d21	hei(B,48)	d32	(hei(44,39)+hei(48,39))/2

are 17 features in the first group that concern eyes and consists of d_0, d_1, d_2, d_3, d_4, d_5, d_6, d_7, d_{16}, d_{17}, d_{19}, d_{20}, d_{25}, d_{26}, d_{27}, d_{28}, and d_{29}; there are six features in the second group that concern cheek and consists of d_9, d_{10}, d_{18}, d_{21}, d_{30}, and d_{31}; there are 10 features in the third group that concern mouth and consists of d_8, d_{11}, d_{12}, d_{13}, d_{14}, d_{15}, d_{22}, d_{23}, d_{24}, and d_{32}. In Table 6.2, A is the midpoint of points 19 and 23, and B is the midpoint of points 27 and 31. dis(i, j) denotes the Euclid distance between points i and j; hei(i, j) denotes the horizontal distance between points i and j; wid(i, j) denotes the vertical distance between points i and j. Since dis(23, 27) is suitable for all kinds of expression, we normalize the distance features as follows. First, $x_i{}' = d_i/\text{dis}(23, 27)$. Second, the normalized distance is continually normalized in [0, 1] as follows:$x_i = \frac{x_i' - \min(x_i')}{\max(x_i') - \min(x_i')}$, $x_i \in [0, 1]$.

6.3.2 Experiments for Feature Selection Method for Emotion Recognition

In this section, there are five comparative experiments related with feature selection for emotion recognition.

In the first experiment, the proposed method SARA is taken as the method of feature selection for emotion recognition. In the second one, a classical attribution reduction algorithm named CEBARKNC [30] is taken as a method of feature selection for emotion recognition. CEBRKNC is selected in this comparative experiment since it is an attribute reduction algorithm based on conditional entropy in equivalence relation. In this experiment, a greedy algorithm proposed by Nguyen [31] is taken as a discretization method and it is done on the platform RIDAS [32]. In the third experiment, another classical attribute reduction algorithm named MIBARK [33] is taken as a method of feature selection. It is a reduction algorithm based on mutual information as the measure of importance of attribute. Similarly, the greedy algorithm proposed by Nguyen [31] is taken as a discretization method and it is done on the platform RIDAS also. In the fourth experiment, a traditional feature selection method, Genetic Algorithm (GA) [34], is used as the feature selection method for emotion

TABLE 6.3 The results of comparative experiments on the three datasets

	SARA+SVM		CEBARKNC+ SVM		MIBARK+ SVM		GA+SVM		SVM	
Database	CRR	RAN	CRR	RAN	CRR	RAN	CRR	RAN	CRR	RAN
CKACFE	76.01	11.25	73.07	12.5	75.05	17.75	73.09	14.25	79.80	33
JAFFE	69.37	11.5	63.17	11	63.98	14.5	55.89	14.25	74.46	33
CQUPTE	92.45	14	78.83	13.5	87.90	13.5	88.95	14.75	93.86	33
average	79.28	12.25	71.69	12.33	75.64	15.25	72.64	14.42	82.71	33

CRR, percentage of the correct recognition rate; RAN, amount of attributes after feature selection.

recognition. This experiment is done on WEKA [35], a famous machine learning tool, and CfsSubsetEval is taken as the evaluator for feature selection in WEKA. In the fifth experiment, all the 33 features are used for emotion recognition and the feature selection course is omitted.

SVM is a new machine learning method, and it is famous for its great ability for small-sample applications. Therefore, SVM is taken as a classifier for all the comparative experiments, and the same parameters are used in all the experiments. Meanwhile, fourfold cross-validation is taken for all the experiments.

The results of the comparative experiments are shown in Table 6.3.

From the experiment results of SARA+SVM and SVM shown in Table 6.3, we can find that SARA use nearly one-third features and get nearly the same correct recognition rate; therefore, SARA can be taken as a useful feature selection method for emotion recognition.

When we compare the experimental results of SARA+SVM and CEBARKNC+ SVM shown in Table 6.3, we can find SARA selects as much features as CEBARKNC, but SARA gets a better correct recognition rate than CEBARKNC. Further, from the comparative experiment results between SARA+SVM and MIBARK+SVM, or experimental results between SARA+SVM and GA+SVM shown in Table 6.3, we can find SARA can use fewer features than MIBARK or GA, but get higher recognition rate. Therefore, SARA can be taken as an effective feature selection method for emotion recognition than CEBARKNC, MIBARK, and GA, since the features selected by SARA have better discernability in emotion recognition.

Important features selected by the four feature selection methods are listed in Table 6.4.

From Table 6.4, we can find that the four feature selection algorithms can select different features for emotion recognition. Among all the experiment results, SARA selects three important features, x_{13}, x_{14}, and x_{15} for all the three emotion datasets, meanwhile CEBARKNC selects two important features, x_8 and x_{13}, MIBARK selects two important features, x_1 and x_{32}; however, GA cannot find any important feature for all the three datasets. Since better correct recognition rate can be achieved if SARA is used as a method of feature selection for emotion recognition. Therefore, x_{13}, x_{14}, and x_{15} can be seen more important for emotion recognition. Although the features of x_{13}, x_{14}, and x_{15} are normalized features, the importance of original features of d_{13}, d_{14}, and d_{15} are also evident. Since the feature of d_{13}, d_{14}, and d_{15} are all concerning mouth, we can draw a conclusion that the geometrical features concerning mouth are

TABLE 6.4 The important features in the three datasets

Database	SARA	CEBARKNC	MIBARK	GA
CKACFE	x_{13}, x_{14}, x_{15}	$x_1, x_8, x_{13}, x_{16}, x_{21}$	$x_1, x_8, x_{16}, x_{17}, x_{22}, x_{28}, x_{30}, x_{31}, x_{32}$	x_{14}, x_{25}
JAFFE	$x_5, x_{13}, x_{14}, x_{15}, x_{26}$	x_0, x_8, x_{13}	$x_1, x_{15}, x_{24}, x_{26}, x_{32}$	x_7, x_{25}, x_{32}
CQUPTE	$x_7, x_{13}, x_{14}, x_{15}, x_{24}, x_{30}, x_{31}$	$x_0, x_1, x_4, x_8, x_{13}, x_{26}, x_{27}, x_{32}$	$x_0, x_1, x_4, x_8, x_{13}, x_{23}, x_{24}, x_{32}$	$x_{11}, x_{14}, x_{22}, x_{23}, x_{24}, x_{32}$
Important features	x_{13}, x_{14}, x_{15}	x_8, x_{13}	x_1, x_{32}	–

the most important features for emotion recognition. The original selected features of SARA, CEBARKNC, and MIBARK are shown in Figure 6.4.

6.3.3 Experiments for the Features Concerning Mouth for Emotion Recognition

From the last section, we draw a conclusion that the geometrical features concerning mouth are important for emotion recognition. In this section, four experiments are proposed for the purpose of testing the importance of the geometrical feature concerning mouth for emotion recognition. In the first experiment, all the 33 facial features are used for emotion recognition. In the second experiment, only the features selected by SARA are used for emotion recognition. In the third experiment, all the features concerning mouth are deleted and 19 features are used for emotion recognition, in which there are 17 features concerning eyes d_0, d_1, d_2, d_3, d_4, d_5, d_6, d_7, d_{16}, d_{17}, d_{19}, d_{20}, d_{25}, d_{26}, d_{27}, d_{28}, and d_{29} and two features d_{30} and d_{31} concerning cheek but not mouth. In the fourth experiment, all the features concerning eyes are deleted and 12 features are used for emotion recognition, in which there are 10 features concerning mouth d_8, d_{11}, d_{12}, d_{13}, d_{14}, d_{15}, d_{22}, d_{23}, d_{24}, and d_{32} and two features d_{30} and d_{31} concerning cheek but not eyes. SVM is taken as aclassifier in the four experiments and is given the same parameters. Experiment results are listed in Table 6.5.

From Table 6.5, we can find that the correct recognition rate is decreased greatly without features related to mouth. Therefore, it is concluded that the features concerning mouth are the most important geometrical features for emotion recognition. On the other hand, we can find that the correct recognition rate does not decrease so much if there are no features concerning eyes. Therefore, the geometrical features concerning eyes do not play an important role in emotion recognition. But from the psychological experiments of Reference 8, Sui *et al.* found that the eyes play an important role in emotion; therefore, we may draw a conclusion that the geometrical features concerning mouth are the most important geometrical features for emotion recognition, and the geometrical features concerning eyes are not so important. Furthermore, the important features concerning eyes for emotion recognition should be discovered and used in emotion recognition in the further work. Meanwhile, we can find that the correct recognition rate is decreased in CKACFE more than in JAFFE and CQUPTE. Therefore, we can draw a conclusion that the geometrical features

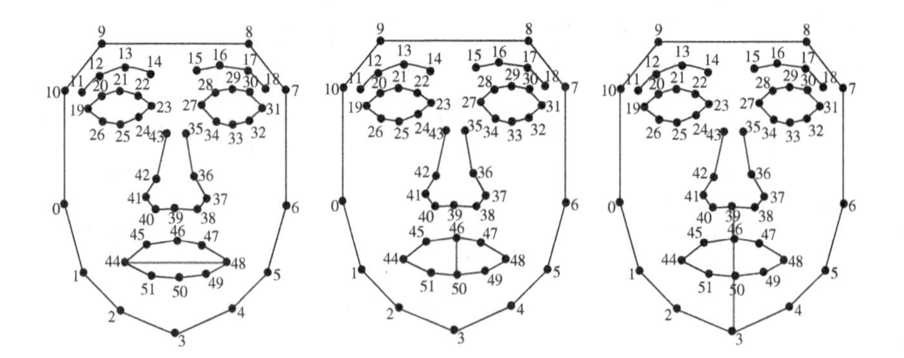

(a) Important features selected by SARA

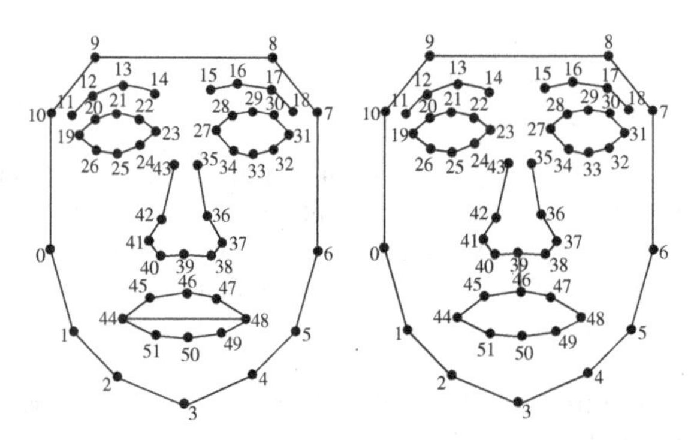

(b) Important features selected by CEBARKNC

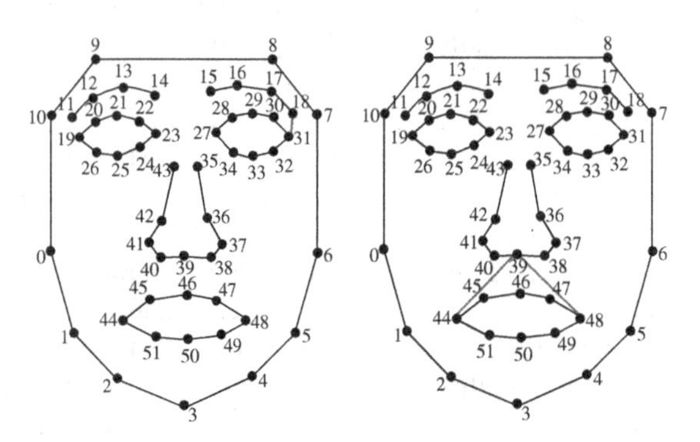

(c) Important features selected by MIBARK

FIGURE 6.4 Important features.

TABLE 6.5 The results of comparative experiments on the three datases

	SARA reserved		All features		No mouth		No eyes	
Dataset	CRR	RAN	CRR	RAN	CRR	RAN	CRR	RAN
CKACFE	76.01	11.25	79.80	33	45.18	19	75.15	12
JAFFE	69.37	11.5	74.46	33	53.32	19	58.29	12
CQUPTE	92.45	14	93.86	33	69.02	19	89.72	12
Average	79.28	12.25	82.71	33	55.84	19	74.39	12

concerning mouth are more important for emotion expression for the Caucasian than the Eastern people.

6.4 CONCLUSION

In this chapter, feature selection for emotion recognition is discussed. A self-learning attribute reduction algorithm is proposed based on rough set and domain-oriented, data-driven, data-mining theories. It is proved to be a good feature selection method for emotion recognition. Based on the comparative of feature selection for emotion recognition, rough set methods can be found better than Genetic Algorithm, and the geometrical features concerning mouth are found to be the most important geometrical features for emotion recognition.

ACKNOWLEDGMENTS

The work is supported by National Natural Science Foundation of China under Grant No. 60773113; Natural Science Foundation of Chongqing under Grant No. 2008BA2041, No.2008BA2017, and No.2007BB2445; Science & Technology Research Program of Chongqing Education Commission under Grant No. KJ110522; Chongqing Key Lab of Computer Network and Communication Technology Foundation under Grant No.CY-CNCL-2009-02; and Chongqing University science foundation under Grant No. A2009-26 and No.JK-Y-2010002.

REFERENCES

[1] R. W. Picard, *Affective Computing*, MIT Press, Cambridge, 1997.

[2] R. W. Picard, "Affective computing: challenges," *Int. J. Hum. Comput. Stud.*, 59(1–2): 55–64, 2003.

[3] R. W. Picard, E. Vyzas, and J. Healey, "Toward machine emotional intelligence: analysis of affective physiological state," in IEEE Transactions on Pattern Analysis and Machine Intelligence, 2001.

[4] A. Chakraborty and A. Konar, *Emotional Intelligence: A Cybernetic Approach*, Springer, 2009.

[5] Z. L. Wang, *Artificial Psychology*, Machinery Industry Press, China, 2006.

[6] Z. L. Wang, "Artificial psychology and artificial emotion," in CAAI Transactions on Intelligent Systems, China, 2006.

[7] Z. H. Zeng, M. Pantic, G. I. Roisman, and T. S. Huang, "A survey of affect recognition methods: audio, visual, and spontaneous expressions," *IEEE Transactions on Pattern Analysis and Machine Intelligence*, 31(1): 39–58, 2009.

[8] X. Sui and Y. T. Ren, "Online processing of facial expression recognition," *Acta Psychologica Sinica*, 39(1): 64–70, 2007 [Chinese].

[9] Y. M. Wang and X. L. Fu, "Recognizing facial expression and facial identity parallel processing or interactive processing," in Advances in Psychological Science, China, 2005.

[10] M. White, "Effect of photographic negation on matching the expressions and identities of faces," *Perception*, 30(8): 969–981, 2001.

[11] Y. Yang, G. Y. Wang, P. J. Chen, *et al.*, "Feature selection in audiovisual emotion recognition based on Rough set theory," in Transaction on Rough Set VII, LNCS 4400: 283–294, 2007.

[12] Y. Yang, G. Y. Wang, and P. J. Chen, "An emotion recognition system based on rough set theory," in Proceedings of the 2006 Conference on Advances in Intelligent IT: Active Media Technology, 2006.

[13] P. J. Chen, G. Y. Wang, Y. Yang, and J. Zhou, "Facial expression recognition based on rough set theory and SVM," in Proceedings of the RSKT2006, 2006.

[14] J. Zhou, G. Y. Wang, Y. Yang, and P. Chen, "Speech emotion recognition based on rough set and SVM," in Proceedings of the ICCI2006, 2006.

[15] L. Shang, Q. Wan, W. Yao, *et al.*, "An approach for reduction of continuous-valued attributes," *J. Comput. Res. Dev.*, 2005.

[16] R. Jensen and Q. Shen, "Tolerance-based and fuzzy-rough feature selection," in Proceedings of the 16th International Conference on Fuzzy Systems, 2007.

[17] Z. Pawlak, "Rough sets," *Int. J. Comp. Inform. Sci.*, 11: 341–356, 1982.

[18] Z. Pawlak, "Rough classification," *Int. J. Man Mach. Stud.*, 20: 469–483, 1984.

[19] Z. Pawlak, "Rough set theory and its applications to data analysis," *Cybern. Syst.*, 29(1): 661–688, 1998.

[20] R. W. Swiniarski and A. Skowron, "Rough set methods in feature selection and recognition," *Pattern Recogn. Lett.*, 24: 833–849, 2003.

[21] Z. Ning and A. Skowron, "A rough set-based knowledge discovery process," *Int. J. Appl. Math. Comput. Sci.*, 11(3): 603–619, 2001.

[22] G. Y. Wang and Y. Wang, "3DM: domain-oriented data-driven data mining," *Fundam. Inform.*, 90(4): 395–426, 2009.

[23] G. Y. Wang, "Introduction to 3DM: domain-oriented data-driven data mining," in Proceedings of the RSKT2008, 2008.

[24] S. Ohsuga, "Knowledge discovery as translation," in *Foundations of Data Mining and Knowledge Discovery*, (eds T. Y. Lin and X. Hu) Springer, 2005.

[25] The Cohn-Kanade AU-Coded Facial Expression Database, http://vasc.ri.cmu.edu/idb/html/face/facial_expression/index.html. Last accessed on Sept. 18, 2014.

[26] T. Kanade., J. F. Cohn, and Y. Tian, "Comprehensive database for facial expression analysis," in Proceedings of the Fourth IEEE International Conference on Automatic Face and Gesture Recognition (FG00), Grenoble, France, 2000, pp. 46–53.

[27] P. Lucey, J. F. Cohn, T. Kanade, J. Saragih, S. Ambadar, and I. Matthews, "The extended Cohn-Kanade Dataset (DK+): a complete expression dataset for action unit and emotion specified expression," in Proceedings of the Third International Workshop on CVPR for Human Communicative Behavior Analysis (CVPR 4HB 2010), San Fransisco, 2010, pp. 94–101.

[28] The Japanese Female Facial Expression (JAFFE) Database, http://www.kasrl.org/jaffe.html. Last accessed on Sept. 18, 2014.

[29] Chongqing University of Posts and Telecommunications Emotional Database (CQUPTE), http://cs.cqupt.edu.cn/users/904/docs/9317-1.rar. Last accessed on Sept. 16, 2014.

[30] G. Y. Wang, H. Yu, and D. C. Yang, "Decision table. reduction based on conditional information entropy," *J. Comput.*, 25: 759–766, 2002.

[31] H. S. Nugyen and A. Skowron, "Quantization of real value attributes: rough set and boolean reasoning approach," in Proceedings of the Second Joint Annual Conference on Information Science, Wrightsville Beach, NC, September 28–October 1, 1995.

[32] G. Y. Wang, Z. Zheng, and Y. Zhang, "RIDAS—a rough set based intelligent data analysis system," in The First International Conference on Machine Learning and Cybernetics (ICMLC2002), 2002.

[33] D. Q. Miao and G. R. Hu, *Journal of Computer Research and Development*, 1999.

[34] D. E. Goldberg, *Genetic Algorithms in Search, Optimization and Machine Learning*, Addison-Wesley Longman Publishing, Boston, MA, 1989.

[35] G. Holmes, A. Donkin, and I. H. Witten, "WEKA: a machine learning workbench," in Second Australian and New Zealand Conference on Intelligent Information Systems, 1994.

AUTHOR BIOGRAPHIES

Yong Yang was born in Heqing, China, in 1976. He received the B.Sc. degree in Telecommunication Engineering in 1997 and the M.Sc. degree in Computer Science in 2003 from the Chongqing University of Posts and Telecommunications, Chongqing, China, and the Ph.D. degree in Computer Science from Southwest Jiaotong University, Chengdu, China, in 2009. His research interests include affective computing, human–computer interaction, cloud computing, and parallel data mining.

Guoyin Wang was born in Chongqing, China, in 1970. He received the B.Sc., M.Sc., and Ph.D. degrees in Computer Science from Xi'an Jiaotong University, Xi'an, China, in 1992, 1994, and 1996, respectively. He is Steering Committee Chair of International Rough Set Society and Chairman of the Rough Set and Soft Computation Society, Chinese Association for Artificial Intelligence. His research interests include rough set theory, granular computing, and data mining.

7

EMOTION RECOGNITION FROM FACIAL EXPRESSIONS USING TYPE-2 FUZZY SETS

ANISHA HALDER AND AMIT KONAR

Artificial Intelligence Laboratory, Department of Electronics and Telecommunication Engineering, Jadavpur University, Kolkata, India

ARUNA CHAKRABORTY

Department of Computer Science & Engineering, St. Thomas' College of Engineering and Technology, Kolkata, India

ATULYA K. NAGAR

Mathematics and Computer Science Department, Liverpool Hope University, Liverpool, UK

Facial expression of a person representative of similar emotions is not always unique. Naturally, the facial features of a subject taken from different instances of the same emotion have wider variations. In the presence of two or more facial features, the variation of the attributes together makes the emotion recognition problem more complicated. This variation is the main source of uncertainty in the emotion recognition problem, which has been addressed here in two steps using type-2 fuzzy sets. First, a type-2 fuzzy face-space is constructed with the background knowledge of facial features of different subjects for different emotions. Second, the emotion of an unknown facial expression is determined based on the consensus of the measured facial features with the fuzzy face-space. Both interval and general type-2 fuzzy sets have been used separately to model the fuzzy face-space. The interval type-2 fuzzy set involves primary membership functions for m facial features obtained from n subjects, each

Emotion Recognition: A Pattern Analysis Approach, First Edition. Edited by Amit Konar and Aruna Chakraborty.
© 2015 John Wiley & Sons, Inc. Published 2015 by John Wiley & Sons, Inc.

having l instances of facial expressions for a given emotion. Besides, the general type-2 fuzzy set employing the primary membership functions mentioned earlier also involves the secondary membership functions for individual primary membership curve, which has been obtained here by formulating and solving an optimization problem. The optimization problem attempts to minimize the difference between two decoded signals: (1) the type-1 defuzzification of the average primary membership functions obtained from n subjects and (2) the type-2 defuzzified signal for a given primary membership function with secondary memberships as unknown. The uncertainty management policy adopted using general type-2 fuzzy set has resulted in a classification accuracy of 97.328% in comparison to 90.88% obtained by its interval type-2 counterpart.

7.1 INTRODUCTION

Emotion recognition is currently gaining importance for its increasing scope of applications in human–computer interactive systems. Several modalities of emotion recognition, including facial expression, voice, gesture, and posture, have been studied in the literature. However, irrespective of the modality, emotion recognition comprises two fundamental steps involving feature extraction and classification [1]. Feature extraction refers to determining a set of features/attributes, preferably independent, which together represents a given emotional expression. Classification aims at mapping emotional features into one of several emotion classes.

Performance of an emotion recognition system greatly depends on feature selection and classifier design. A good classification algorithm sometimes cannot yield high classification accuracy for poorly selected features. On the other hand, even after the selection of a complete set of features, describing an emotion, we occasionally fail to recognize the emotion correctly for the selection of a poor classifier. Most commonly used techniques for feature selection in the emotion recognition problem include principal component analysis [2], independent component analysis [3], rough sets [4], Gabor filter [5], and Fourier descriptors [6]. Among the popularly used techniques for emotion classification, neural net-based mapping [7,8], fuzzy relational approach [9], linear discriminate analysis [3], and hidden Markov model [2,3] need special mention. A brief overview of the existing research on emotion recognition is given next.

Ekman and Friesen took an early attempt to recognize facial expression from the movements of cheek, chin, and wrinkles [10]. Their experiments confirmed the existence of a good correlation between basic movements of the facial action units and facial expressions. Kobayashi and Hara [11–13] designed a scheme for the recognition of human facial expressions using the well-known back-propagation neural algorithms [14,15]. Their scheme is capable of recognizing six common facial expressions depicting happiness, sadness, fear, anger, surprise, and disgust. Yamada proposed an alternative method of emotion recognition through classification of visual information [16].

Fernandez-Dols *et al.* proposed a scheme for decoding emotions from facial expressions and content [17]. Kawakami *et al.* [14] designed a method for the construction of emotion space using neural network. Busso and Narayanan [18] analyzed the scope of facial expressions, speech, and multimodal information in emotion recognition. Cohen *et al.* [19] developed a scheme for emotion recognition from the temporal variations in facial expressions obtained from the live video sequence of the subjects. They used hidden Markov model to automatically segment and recognize facial expression. Gao *et al.* presented a scheme for facial expression recognition from a single facial image using line-based caricatures [20]. Among other significant contributions in emotion recognition, the works presented in References 6, 10–13, 21–30, 47 need special mention. For a more complete literature survey, which cannot be given here for space restriction, readers may refer to two outstanding papers by Pantic *et al.* [30, 31].

Emotional features greatly depend on the psychological states of the subjects. For example, facial expressions of a subject, while experiencing the same emotion, have wider variations, resulting in parametric changes in individual feature. Further, different subjects experiencing the same emotion have difference in their facial features. Repeated experiments with a large number of subjects, each having multiple instances of similar emotional experience, reveal that apparently there exists a small but random variations of facial features around specific fixed points [32]. The variation in different instances of facial expression for similar emotive experience of an individual can be regarded as an intrapersonal-level uncertainty [33]. On the other hand, the variation in facial expression of individuals for similar emotional experience can be treated as interpersonal-level uncertainty [33].

The random variations in features can be modeled with fuzzy sets. Classical (type-1) fuzzy sets, pioneered by Zadeh [34], has widely been used over the last five decades for modeling uncertainty of ill-defined systems. Type-1 fuzzy sets employ a single membership function to represent the degree of uncertainty in measurements of a given feature. So, it can capture the variation in measurements of a given feature for different instances of a specific emotion experienced by a subject. In Reference 9, the authors considered a fixed membership function to model the uncertainty involved in a feature for a given emotion, disregarding the possibility of variation in the membership curves for different subjects.

This chapter, however, models the above form of interpersonal level uncertainty by Interval Type-2 Fuzzy Sets (IT2FS). IT2FS employs an upper membership function (UMF) and a lower membership function (LMF) to capture the uncertainty involved in a given measurement of a feature within the bounds of its two membership curves at the point of the measurement. However, the degree of correct assignment of membership for each membership curve embedded between the UMF and LMF in IT2FS is treated as unity, which always is not correct. General Type-2 Fuzzy Sets (GT2FS) can overcome the above problem by considering a secondary membership grade that represents the correctness in (primary) membership assignment at each measurement points [35]. Naturally, GT2FS is expected to give us better results in emotion classification for its representational advantage over IT2FS.

One fundamental problem in GT2FS that limits its application in classification problems, perhaps, is due to users' inability to correctly specify the secondary memberships. In this chapter, we determine the secondary memberships by extracting certain knowledge from the individual primary assignments for each feature of a given emotion for a subject. The knowledge extracted is encoded as an optimization problem with secondary memberships as unknown. The solution of the optimization problem, carried out off-line, provides the secondary grades. The secondary grades are later used to reduce the uncertainty involved in primary memberships of individual feature for all subjects at a given measurement point.

The chapter provides two alternative approaches to emotion recognition from an unknown facial expression, when the emotion class of individual facial expression of a large number of experimental subjects is available. The first approach deals with IT2FS to construct a fuzzy face-space based on the measurements of a set of features from a given set of facial expressions carrying different emotions. An unknown facial expression is classified into one of several emotion classes by determining the maximum support of an emotion class to a given set of measurements of a facial expression. The class having the maximum support is declared as the emotion of the unknown facial expression. In spirit, this is similar to how a fuzzy rule-based system for classification works.

The second approach employs GT2FS to construct a fuzzy face-space, comprising both primary and secondary membership functions, obtained from known facial expressions of several subjects containing multiple instances of the same emotion for each subject. The emotion class of an unknown facial expression is determined by obtaining maximum support of each class to the given facial expression. The class with the maximum support is the winner. The maximum support evaluation here employs both primary and secondary memberships and thus is slightly different from the IT2FS-based classification.

Experiments reveal that the classification accuracy of emotion of an unknown person by the GT2FS-based scheme is as high as 97%. When secondary memberships are ignored, and classification is performed with IT2FS, the classification accuracy falls by a margin of 7%. The additional 7% classification accuracy obtained by GT2FS, however, has to pay a price for additional complexity of $(m \in n \in k)$ multiplications, where m, n, and k denote the number of features, the number of subjects, and the number of emotion classes, respectively.

The chapter is divided into seven sections. Section 7.2 provides fundamental definitions associated with type-2 fuzzy sets, which will be required in the rest of the chapter. In Section 7.3, we propose the principle of uncertainty management in fuzzy face-space for emotion recognition. Section 7.4 deals with secondary membership evaluation procedure for a given type-2 primary membership function. Experimental details are given in Section 7.5 and three methods of performance analysis are undertaken in Section 7.6. Conclusions are listed in Section 7.7.

7.2 PRELIMINARIES ON TYPE-2 FUZZY SETS

7.2.1 Type-2 Fuzzy Sets

In this section, we define some terminologies related to Type-1 (T1) and Type-2 (T2) fuzzy sets. These definitions will be used throughout the chapter.

Definition 7.1. Given a universe of discourse X, a conventional *type-1 fuzzy set A* defined on X is given by a two-dimensional membership function, also called type-1 membership function. The *membership function*, denoted by $\mu_A(x)$, is a crisp number in [0, 1] for a generic element $x \in X$. Usually, the fuzzy set A is expressed as two tuples [36] and is given by

$$A = (x, \mu_A(x))|\forall x \in X \tag{7.1}$$

An alternative representation of the fuzzy set A is also found in the literature [37] and is given by

$$A = \int_{x \in X} \mu_A(x)|x \tag{7.2}$$

where $\int$ denotes union of all admissible x.

Definition 7.2. A type-2 fuzzy set $\widetilde{A}$ is characterized by a three-dimensional membership function, also called type-2 membership function, which itself is fuzzy. The type-2 membership function is usually denoted by $\mu_{\widetilde{A}}(x, u)$, where $x \in X$ and $u \in J_x \subseteq [0, 1]$. Usually, the fuzzy set $\widetilde{A}$ is expressed as two tuples:

$$\widetilde{A} = ((x, u), \mu_{\widetilde{A}}(x, u))|x \in X, u \in J_x \subseteq [0, 1]. \tag{7.3}$$

An alternative form of representation of the type-2 fuzzy set is given in (7.4).

$$\widetilde{A} = \int_{x \in X} \int_{u \in J_x} \mu_{\widetilde{A}}(x, u)|(x, u), J_x \subseteq [0, 1] \tag{7.4}$$

$$= \int_{x \in X} \left[\int_{u \in J_x} f_x(u)/u \right] /x, J_x \subseteq [0, 1] \tag{7.5}$$

where $f_x(u) = \mu_{\widetilde{A}}(x, u) \in [0.1]$. The $\int \int$ denotes union over all admissible x and u.

Definition 7.3. At each point of x, say $x = x^/$, the two-dimensional plane containing axes u and $\mu(x^/, u)$ is called the *vertical slice* of $\mu_{\widetilde{A}}(x, u)$ [38]. A *secondary membership function* is a vertical slice of $\mu_{\widetilde{A}}(x, u)$. Symbolically, it is given by $\mu_{\widetilde{A}}(x, u)$ at $x = x^/$ for $x^/ \in X$ and $\forall u \in J_{x^/} \subseteq [0, 1]$.

$$\mu_{\widetilde{A}}(x, u) = \int_{u \in J_{x^/}} f_{x^/}(u)|u, J_{x^/} \subseteq [0, 1] \tag{7.6}$$

where $0 \leq f_{x^/}(u) \leq 1$. The amplitude of a secondary membership function is called secondary grade [38]. In (7.6), $J_{x^/}$ is the primary membership of $x^/$.

Definition 7.4. Uncertainty in the primary membership of a type-2 fuzzy set $\widetilde{A}$ is represented by a bounded region, called *footprint of uncertainty* (FOU), which is the defined as the union of all primary memberships, that is,

$$FOU(\widetilde{A}) = \cup_{x \in U} J_x \tag{7.7}$$

In Reference 38, $FOU(\widetilde{A})$ is also defined as $D\widetilde{A}$. Thus,

$$D\widetilde{A}(x) = J_x \forall x \in X \tag{7.8}$$

Thus, (7.8) reduces to

$$\widetilde{A} = \int \int_{(x,u) \in D\widetilde{A}} \mu_{\widetilde{A}}(x, u)/(x, u) \tag{7.9}$$

If all the secondary grades of a type-2 fuzzy set $\widetilde{A}$ are equal to 1, that is,

$$\mu_{\widetilde{A}}(x, u) = 1; \forall x \in X, \forall u \in J_x \subseteq [0, 1] \tag{7.10}$$

then $\widetilde{A}$ is called IT2FS. The FOU is bounded by two curves, called LMF and UMF, denoted by $\underline{\mu}_{\widetilde{A}}(x)$ and $\overline{\mu}_{\widetilde{A}}(x)$, respectively, where $\underline{\mu}_{\widetilde{A}}(x)$ and $\overline{\mu}_{\widetilde{A}}(x)$ at all x, respectively, take up the minimum and the maximum of the membership functions of the embedded type-1 fuzzy sets [15] in the FOU.

7.3 UNCERTAINTY MANAGEMENT IN FUZZY-SPACE FOR EMOTION RECOGNITION

This section provides a general overview of the proposed scheme for emotion recognition using type-2 fuzzy sets. Here, the emotion recognition problem is considered as uncertainty management in fuzzy space after encoding the measured facial attributes by type-2 fuzzy sets.

Let $F = f_1, f_2, \ldots, f_m$ be the set of m facial features. Let $\mu_{\widetilde{A}}(f_i)$ be the primary membership in [0, 1] of the feature f_i to be a member of set $\widetilde{A}$, and $\mu(f_i, \mu_{\widetilde{A}}(f_i))$ be the secondary membership of the measured variable f_i in [0, 1]. If the measurement of a facial feature, f_i, is performed p times on the same subject experiencing the same emotion, and the measurements are quantized into q intervals of equal size, we can evaluate the frequency of occurrence of the measured variable f_i in q quantized intervals. The interval containing the highest frequency of occurrence then can be identified, and its center, m_i, approximately represents the mode of the measurement variable f_i. The second moment, σ_i, around m_i is determined, and an exponential bell-shaped (Gaussian) membership function centered on m_i and with a spread σ_i is used to represent the membership function of the random variable f_i. This function represents the membership of f_i to be CLOSE-TO the central value, m_i. It may be noted that a bell-shaped (Gaussian-like) membership curve would have a peak at the center with a membership value one, indicating that membership at this point is the largest for an obvious reason of having the highest frequency of f_i at the center.

On repetition of the above experiment for variable f_i on n subjects, each experiencing the same emotion, we obtain n such membership functions, each one for one individual subject. Naturally, the measurement variable f_i now has both intra- and inter-personal level uncertainties. The intralevel uncertainty occurs due to the preassumption of the bell shape (Gaussian function) of the primary membership function, and the interlevel uncertainty occurs due to multiplicity of the membership function for n subjects. Thus, a new measurement for an unknown facial expression can be encoded using all the n-membership curves, giving n possible membership values, thereby giving rise to uncertainty in the fuzzy space.

The uncertainty involved in the present problem has been addressed here by two distinctive approaches: (i) IT2FS and (ii) GT2FS. Naturally, the former approach is simple, but more error-prone as it ignores the intralevel uncertainty. The second approach is more robust and is capable to take care of both the uncertainties. However, it needs additional complexity as it involves more computation. Although both the approaches have many common steps, we briefly outline the IT2FS-based one for its simplicity, and then explain the other without repeating the common steps further.

7.3.1 Principles Used in the IT2FS Approach

The primary membership functions for a given feature f_i corresponding to a particular emotion taken from n subjects together forms a IT2FS $\tilde{A}$, whose FOU is bounded by a lower and an upper membership curves $\underline{\mu}_{\tilde{A}}(f_i))$ and $\underline{\mu}_{\tilde{A}}(f_i))$, respectively, where

$$\underline{\mu}_{\tilde{A}}(f_i)) = \text{Min}\{\mu^1_{\tilde{A}}(f_i), \mu^2_{\tilde{A}}(f_i), \ldots, \mu^n_{\tilde{A}}(f_i)\} \tag{7.11}$$

$$\overline{\mu}_{\tilde{A}}(f_i)) = \text{Max}\{\mu^1_{\tilde{A}}(f_i), \mu^2_{\tilde{A}}(f_i), \ldots, \mu^n_{\tilde{A}}(f_i)\} \tag{7.12}$$

are evaluated for all f_i, and $\mu^j_{\tilde{A}}(f_i), 1 \leq j \leq n$ denotes the primary membership function of feature f_i for subject j in IT2FS $\tilde{A}$.

Figure 7.1 provides the FOU for a given feature f_i. Now, for a given measurement $f_i^/$, we obtain an interval $[\underline{\mu}_{\tilde{A}}(f_i^/), \overline{\mu}_{\tilde{A}}(f_i^/)]$, representing the entire span of uncertainty

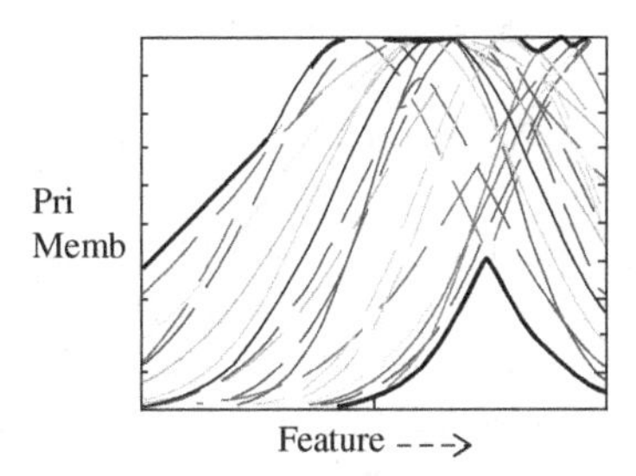

FIGURE 7.1 FOU for feature Mouth-Opening.

of the measurement variable $f_i^{/}$ in the fuzzy space, induced by n primary membership functions: $\mu_{\tilde{A}}^{j}(f_i)$, $1 \leq j \leq n$. The interval $[\underline{\mu}_{\tilde{A}}(f_i^{/}), \overline{\mu}_{\tilde{A}}(f_i^{/})]$ is evaluated by replacing f_i by $f_i^{/}$ in (7.11) and (7.12).

If there exist m different facial features, then for each feature we would have such an interval, and consequently we obtain m such intervals given by $[\underline{\mu}_{\tilde{A}}(f_1^{/}), \overline{\mu}_{\tilde{A}}(f_1^{/})], [\underline{\mu}_{\tilde{A}}(f_2^{/}), \overline{\mu}_{\tilde{A}}(f_2^{/})], \ldots\ldots, [\underline{\mu}_{\tilde{A}}(f_m^{/}), \overline{\mu}_{\tilde{A}}(f_m^{/})]$.

The proposed IT2FS reasoning system employs a particular format of rules, commonly used in fuzzy classification problems [39]. Consider, for instance, a fuzzy rule, given by Rc: if f_1 is $\widetilde{A_1}$ AND f_2 is $\widetilde{A_2}$ ….. AND f_m is $\widetilde{A_m}$ then emotion class is c.

Here, f_i for $i = 1$ to m denote m measurements (feature values) for the IT2FS $\widetilde{A_1}, \widetilde{A_2}, \ldots \widetilde{A_m}$, respectively, where

$$\widetilde{A}_i = [\underline{\mu}_{\tilde{A}_i}(f_i), \overline{\mu}_{\tilde{A}_i}(f_i)], \forall i. \tag{7.13}$$

In the present formulation, we selected $\widetilde{A_1} = \widetilde{A_2} = \ldots = \widetilde{A_m} = \widetilde{A}$ (say). Since an emotion is characterized by all of these m features, to find the overall support of the m features (m measurements made for the unknown subject) to the emotion class c represented by the n primary memberships, we use the fuzzy meet operation

$$s_c^{\min} = \mathrm{Min}\{\underline{\mu}_{\tilde{A}}(f_1^{/}), \underline{\mu}_{\tilde{A}}(f_2^{/}), \ldots\ldots, \underline{\mu}_{\tilde{A}}(f_m^{/})\} \tag{7.14}$$

$$s_c^{\max} = \mathrm{Min}\{\overline{\mu}_{\tilde{A}}(f_1^{/}), \overline{\mu}_{\tilde{A}}(f_2^{/}), \ldots\ldots, \overline{\mu}_{\tilde{A}}(f_m^{/})\} \tag{7.15}$$

Thus, we can say that the unknown subject is experiencing the emotion class c at least to the extent $s_c^{\min}$, and at most to the extent $s_c^{\max}$.

To reduce the nonspecificity associated with the interval S_{c-i}, different approaches can be taken. For example, the most conservative approach would be to use lower bound, while the most liberal view would to use the upper bound of the interval as the support for the class c. In absence of any additional information, a balanced approach would be to use center of the interval as the support for the class c by the n primary memberships to the unknown subject. This idea is supported by Mendel [40] and Lee [41]. We compute the center, S_c of the interval S_{c-i},

$$S_c = (s_c^{\min} + s_c^{\max})/2 \tag{7.16}$$

Thus, S_c is the degree of support that the unknown facial expression is in emotion class c.

Now to predict the emotion of a person from his/her facial expression, we determine S_c for each emotion class. Presuming that there exist k emotion classes, let us denote them by $S_1, S_2, \ldots, S_k$ for emotion classes $1, 2, \ldots, k$, respectively. Since a given facial expression may convey different emotions with different degrees, we resolve the conflict by ranking the S_i for $i = 1$ to k, and thus determine the emotion class r, for which $S_r \geq S_i$ for all i following the Rule R_c.

To make the algorithm robust, we consider association of fuzzy-encoded measurements with emotion class by considering the weakest reliability of the joint occurrence of the fuzzy measurements and identify the winning emotion class having this measure of reliability superseding the same of other emotion classes.

7.3.2 Principles Used in the GT2FS Approach

The previous approach employs a reasoning mechanism to compute the degree of support of k emotion classes induced by m features for each class to an unknown facial expression using a set of $k \times m$ IT2FS. The GT2FS-based reasoning realized with measurements taken from n subjects, however, requires $k \times m \times n$ GT2FS to determine the emotion class of an unknown facial expression. The current approach tunes the primary membership values for the given measurements using the secondary memberships of the same measurement, and thus reduces the degree of intralevel uncertainty of the primary membership functions. The reduction in the degree of uncertainty helps in improving the classification accuracy of emotion at the cost of additional complexity required to evaluate type-2 secondary membership functions and also to reason with $k \times m \times n$ fuzzy sets.

Let f_i' be the measurement of the ith feature for an unknown subject. Now, by consulting the n primary membership functions obtained from n subjects for a given emotion, we obtain n primary membership values for the measurement f_i' given by $\mu^1_{\tilde{A}}(f_i'), \mu^2_{\tilde{A}}(f_i'), \ldots\ldots, \mu^n_{\tilde{A}}(f_i')$. Let the secondary membership values for each primary membership, respectively, be $\mu(f_i', \mu^1_{\tilde{A}}(f_i')), \mu(f_i', \mu^2_{\tilde{A}}(f_i')), \ldots\ldots \mu(f_i', \mu^n_{\tilde{A}}(f_i'))$. Since the secondary memberships denote the degree of accuracy of the primary memberships, the uncertainty in a primary membership function can be reduced by multiplying each primary membership value by its secondary membership. Thus, the modified primary membership values are given by

$$\begin{aligned} {}^{\text{mod}}\mu^1_{\tilde{A}}(f_i') &= \mu^1_{\tilde{A}}(f_i') \times \mu(f_i', \mu^1_{\tilde{A}}(f_i')) \\ {}^{\text{mod}}\mu^2_{\tilde{A}}(f_i') &= \mu^2_{\tilde{A}}(f_i') \times \mu(f_i', \mu^2_{\tilde{A}}(f_i')) \\ &\cdots\cdots\cdots\cdots\cdots\cdots \\ {}^{\text{mod}}\mu^n_{\tilde{A}}(f_i') &= \mu^n_{\tilde{A}}(f_i') \times \mu(f_i', \mu^n_{\tilde{A}}(f_i')) \end{aligned} \tag{7.17}$$

where ${}^{\text{mod}}\mu^j_{\tilde{A}}(f_i')$ denotes the modified primary membership value for jth subject. The next step is to determine the range of ${}^{\text{mod}}\mu^j_{\tilde{A}}(f_i')$ for $j = 1 - n$, comprising the minimum and the maximum given by $[{}^{\text{mod}}\underline{\mu}_{\tilde{A}}(f_i'), {}^{\text{mod}}\overline{\mu}_{\tilde{A}}(f_i')]$, where

$${}^{\text{mod}}\underline{\mu}_{\tilde{A}}(f_i') = \text{Min}\left\{{}^{\text{mod}}\mu^1_{\tilde{A}}(f_i'), {}^{\text{mod}}\mu^2_{\tilde{A}}(f_i'), \ldots\ldots, {}^{\text{mod}}\mu^n_{\tilde{A}}(f_i')\right\}$$

$${}^{\text{mod}}\overline{\mu}_{\tilde{A}}(f_i') = \text{Max}\left\{{}^{\text{mod}}\mu^1_{\tilde{A}}(f_i'), {}^{\text{mod}}\mu^2_{\tilde{A}}(f_i'), \ldots\ldots, {}^{\text{mod}}\mu^n_{\tilde{A}}(f_i')\right\}$$

Now for m features, the rule-based type-2 classification is performed in a similar manner as in the previous section with the replacement of $\underline{\mu}_{\tilde{A}}(f_i^/)$ and $\overline{\mu}_{\tilde{A}}(f_i^/)$ by ${}^{\text{mod}}\underline{\mu}_{\tilde{A}}(f_i^/)$ and ${}^{\text{mod}}\overline{\mu}_{\tilde{A}}(f_i^/)$, respectively.

7.3.3 Methodology

We briefly discuss the main steps involved in fuzzy face-space construction based on the measurements of m facial features for n-subjects, each having l instances of facial expression for a particular emotion. We need to classify an facial expression of an unknown person into one of k emotion classes.

IT2FS-Based Emotion Recognition

1. We extract m facial features for n subjects, each having l instances of facial expression for a particular emotion. The above features are extracted for k emotion classes.
2. We construct a fuzzy face-space for each emotion class separately. The fuzzy face-space for an emotion class comprises a set of n primary membership functions for each feature. Thus, we have m groups of n primary membership functions. Each primary membership curve is constructed from l facial instances of a subject attempted to exhibit a particular emotion in her facial expression by acting.
3. For a given set of features $f_1^/, f_2^/, \ldots, f_m^/$ obtained from an unknown facial expression, we determine the range of membership for feature $f_i^/$, given by $[\underline{\mu}_{\tilde{A}}(f_i^/), \overline{\mu}_{\tilde{A}}(f_i^/)]$, where $\tilde{A}$ is a given interval type-2 fuzzy set with a primary membership function defined as CLOSE-TO-centre-value-m of the respective membership function.
4. Now for an emotion class j, we take fuzzy meet operation over the ranges for each feature to evaluate the range of uncertainty for individual emotion class. The meet operation here is computed by taking cumulative t-norm of $\underline{\mu}_{\tilde{A}}(f_i^/)$ and $\overline{\mu}_{\tilde{A}}(f_i^/)$ separately for $i = 1$ to m, and thus obtaining $S_j^{\min}$ and $S_j^{\max}$, respectively.
5. The support of the jth emotion class to the measurements is evaluated by taking average of $S_j^{\min}$ and $S_j^{\max}$, and defining the result by S_j.
6. Now by using classifier rule, we determine the maximum support offered by all the k emotion classes and declare the unknown facial expression to have emotion r, if $S_r > S_i$ for all emotion classes $i = 1$ to k.

GT2FS-Based Emotion Recognition

1. This step is the same as that for IT2FS-based emotion recognition.
2. The construction for primary membership functions here is the same as given in step 2 of IT2FS-based recognition scheme. In addition, we need to construct

secondary membership functions for individual primary membership curves. The procedure for the construction of secondary membership curves will be discussed in Section 7.4.

3. For a given feature $f_i^/$, we consult each primary and secondary membership curve under a given emotion class and take the product of primary and secondary memberships at $f_i = f_i^/$. The resulting membership value obtained for the membership curves for the subject w is given by

$$^{\text{mod}}\mu_{\tilde{A}}^{w}(f_i^/) = \mu_{\tilde{A}}^{w}(f_i^/) \times \mu(f_i^/, \mu_{\tilde{A}}^{w}(f_i^/)) \tag{7.18}$$

where the parameters have their usual significance.

Now, for $w = 1$ to n, we evaluate $\mu_{\tilde{A}}^{w}(f_i^/)$ and thus obtain the minimum and the maximum values of $\mu_{\tilde{A}}^{w}(f_i^/)$, to obtain a range of uncertainty $[^{\text{mod}}\underline{\mu}_{\tilde{A}}(f_i^/), {}^{\text{mod}}\overline{\mu}_{\tilde{A}}(f_i^/)]$. This is repeated for all features under each emotion class.

4. Step 4 is the same as that in the IT2FS-based recognition scheme with the replacement of $\underline{\mu}_{\tilde{A}}(f_i^/)$ and $\overline{\mu}_{\tilde{A}}(f_i^/)$, respectively, by $^{mod}\underline{\mu}_{\tilde{A}}(f_i^/)$ and $^{mod}\overline{\mu}_{\tilde{A}}(f_i^/)$.
5. Steps 5 and 6 are exactly similar to those in the IT2FS-based recognition scheme.

7.4 FUZZY TYPE-2 MEMBERSHIP EVALUATION

This section provides an introduction to type-2 membership evaluation [15,35,42]. Classical fuzzy sets consider primary membership function $\mu_A(x)$ of a fuzzy linguistic variable x, in a fuzzy set A, where $x \in X$and $A \subseteq X$. Usually $\mu_A(x)$ is defined by an expert in a specialized domain or extracted from data. However, it has been found that different experts assign different membership functions for the variable x. Similarly, different datasets for the same problem may also lead to different membership functions. Consider, for example, a fuzzy linguistic variable age in universe AGE, where age belongs to 0–120 years. Let YOUNG be a fuzzy set in the universe AGE. Now, the membership function $\mu_{\text{YOUNG}}(\text{age})$ can be assigned by different people in different forms. Considering a Gaussian-type function, for $\mu_{\text{YOUNG}}(\text{age})$, we note that the peak of the function may be fixed at 18, 20, or 22 years depending on the relative preference of the individual. Such memberships suffer from two distinct types of uncertainty. The first type of uncertainty, called the intralevel uncertainty, is due to approximate values of membership within a given function assigned by an expert. The interlevel uncertainty refers to relative variations among the membership values for a given value of age.

Although theoretically very sound, type-2 fuzzy set has limitedly been used over the last two decades because of the users' inadequate knowledge to correctly assign the secondary memberships. This chapter, however, overcomes this problem by extracting type-2 membership function from its type-1 counterpart by an evolutionary algorithm. A brief outline to the construction of secondary membership function is given in this section.

Intuitively, when an expert assigns a grade of membership, he/she is relatively more certain to determine the location of the peaks and the minima of the function, but may not have enough background to correctly assign the membership values in the rest. Presuming that the (secondary) membership values at the peak and the minima are close to 1, we attempt to compute secondary memberships at the remaining part of the secondary membership function. The following assumptions are used to construct an objective function, which needs to be minimized to obtain the solution of the problem.

1. Let $x = x_p$ and $x = x_q$ be two successive optima (peak/minimum) on the primary membership function $\mu_A(x)$. Then, at any point x, lying between x_p and x_q, the secondary membership $\mu(x, \mu_A(x))$ will be smaller than both $\mu(x_p, \mu_A(x_p))$ and $\mu(x_q, \mu_A(x_q))$.
2. The falloff in secondary membership at a point x away from its value at a peak/minimum $\mu(x_p, \mu_A(x_p))$ is exponential and is given by

$$\mu(x, \mu_A(x)) = \mu(x_p, \mu_A(x_p)).\exp(-|x - x_p|) \tag{7.19}$$

   The secondary membership at any point x between two consecutive optima at $x = x_p$ and $x = x_q$ in the primary membership is selected from the range $[\alpha, \beta]$, where

$$\begin{aligned} \alpha &= \mu(x_p, \mu_A(x_p)).\exp(-|x - x_p|) \\ \beta &= \mu(x_q, \mu_A(x_q)).\exp(-|x - x_q|) \end{aligned} \tag{7.20}$$

3. The defuzzified signal obtained on averaging the primary memberships at all x from n sources is equal to the defuzzified signal obtained from type-2 primary and secondary functions for individual sources. The logic behind this, though apparent, is briefly explained here. The interlevel uncertainty involved in the averaged membership function is the minimum with respect to individual primary function, as the uncertainty is also averaged at a given value of x, and is thus reduced.
4. The unknown secondary membership at two values of x separated by a small positive δ should have a small difference. This is required to avoid sharp changes in the secondary grade.

Let the primary membership functions for feature $f_i = x$ from n sources be $\mu^1_{\tilde{A}}(x), \mu^2_{\tilde{A}}(x), \ldots \mu^n_{\tilde{A}}(x)$. Then, the average membership function that represents a special form of fuzzy aggregation is given by

$$\mu_{\tilde{A}}(x) = \frac{\sum_{i=1}^{n} \mu^i_{\tilde{A}}(x)}{n}, \forall x \tag{7.21}$$

that is, at each position of $x = x_j$, the above membership aggregation is employed to evaluate a new composite membership profile $\mu_{\tilde{A}}(x)$. The defuzzified signal obtained by the centroid method [36] from the averaged primary membership function is given by

$$\bar{\bar{c}} = \frac{\sum_{\forall x} x.\mu_{\tilde{A}}(x)}{\sum_{\forall x} \mu_{\tilde{A}}(x)} \tag{7.22}$$

Further, the type-2 centroidal defuzzified signal obtained from the ith primary and secondary membership functions is defined as

$$\overline{c_i} = \frac{\sum_{\forall x} x.\mu^i_{\tilde{A}}(x).\mu(x, \mu^i_{\tilde{A}}(x))}{\sum_{\forall x} \mu^i_{\tilde{A}}(x).\mu(x, \mu^i_{\tilde{A}}(x))} \tag{7.23}$$

Using assumptions 3 and 4, we construct a performance index J_i to compute secondary membership for the ith subject for a given emotion.

$$J_i = (\overline{c_i} - \bar{\bar{c}})^2 + \sum_{x=x_1}^{x_{R-1}} \{\mu((x+\sigma), \mu^i_{\tilde{A}}(x+\sigma)) - \mu(x, \mu^i_{\tilde{A}}(x))\}^2 \tag{7.24}$$

The second term in (7.24) acts as a regularizing term to prevent abrupt changes in the membership function. In (7.24), x_1 and x_R are the smallest and the largest values of a given feature considered over R sample points of $\mu^i_{\tilde{A}}(x)$, respectively. In (7.24), $\sigma = (x_R - x_1)/(R-1)$ and $x_k = x_1 + (k-1)\sigma$ for $k = 1$ to R. The secondary membership evaluation problem, now, transforms to minimization of J_i by selecting $\mu(x, \mu^i_{\tilde{A}}(x))$ from a given range $[\alpha, \beta]$, where α and β are the secondary memberships at the two optima in secondary membership around the point x. Expressions (7.20) are used to compute α and β for each x separately. Note that, for each subject carrying individual emotion, we have to define (7.23) and (7.24) and find the optimal secondary membership functions.

Any derivative-free optimization algorithm can be used to minimize J_i and to obtain $\mu(x, \mu^i_{\tilde{A}}(x))$ at each x except the optima on the secondary membership. Differential Evolution (DE) [43] is one such derivative-free optimization algorithm, which has fewer control parameters, and has outperformed the well-known binary-coded Genetic Algorithm [44] and Particle Swarm Optimization algorithms [45] with respect to standard benchmark functions [46]. Further, DE is simple and involves only a few lines of code, which motivated us to employ it to solve the above optimization problem.

An iteration of the classical DE [43] algorithm consists of the four basic steps—initialization of a population of vectors, mutation, crossover or recombination, and, finally, selection. The main steps of classical DE are given below:

Set the generation number $t = 0$ and randomly initialize a population of NP individuals $\vec{P}_t = \{\vec{X}_1(t), \vec{X}_2(t), \ldots, \vec{X}_{NP}(t)\}$ with $\vec{X}_i(t) = \{x_{i,1}(t), x_{i,2}(t), \ldots, x_{i,D}(t)\}$ and each individual uniformly distributed in the range $[\vec{X}_{\min}, \vec{X}_{\max}]$, where

$\vec{X}_{\min} = \{x_{\min,1}, x_{\min,2}, \ldots, x_{\min,D}\}$ and $\vec{X}_{\max} = \{x_{\max,1}, x_{\max,2}, \ldots, x_{\max,D}\}$ with $i = [1, 2, \ldots, NP]$, where NP = population size.

while stopping criterion is not reached, **do**
for *1* to *NP*

1. **Mutation:**
Generate a donor vector $\vec{V}(t) = \{v_{i,1}(t), v_{i,2}(t), \ldots, v_{i,D}(t)\}$ corresponding to the ith target vector $\vec{X}_1(t)$ by the following scheme $\vec{V}_1(t) = \vec{X}_{r1}(t) + F * (\vec{X}_{r2}(t) - \vec{X}_{r3}(t))$, where $r1$, $r2$, and $r3$ are mutually exclusive random integers in the range $[1, NP]$
2. **Crossover:**
Generate trial vector $\vec{U}_i(t) = \{u_{i,1}(t), u_{i,2}(t), \ldots, u_{i,D}(t)\}$ for the ith target vector $\vec{X}_1(t)$ by binomial crossover as

$$\vec{u}_{i,j}(t) = \vec{v}_{i,j}(t) \text{ if } rand(0,1) < C_r$$
$$= \vec{x}_{i,j}(t) \text{ otherwise.}$$

3. **Selection:**
Evaluate the trial vector $\vec{U}_i(t)$
if $f(\vec{U}_i(t)) \le f(\vec{X}_i(t))$,
then $\vec{X}_i(t+1) = \vec{U}_i(t)$
$f(\vec{X}_i(t+1)) = f(\vec{U}_i(t))$
end if
end for
4. Increase the counter value $t = t + 1$.
end while.

The parameters used in the algorithm, namely scaling factor "F" and crossover rate "C_r," should be initialized before calling the "while" loop. The terminate condition can be defined in many ways, a few of which include: (i) fixing the number of iterations N, (ii) when best fitness of population does not change appreciably over successive iterations, and (iii) either of (i) and (ii), whichever occurs earlier.

An algorithm to compute the secondary membership function of a type-2 fuzzy set from its primary counterpart using DE is outlined below.

1. Obtain the averaged primary membership function $\mu_A(x)$ from the primary membership functions $\mu^i_{\tilde{A}}(x)$ obtained from n sources, that is, $1, \ldots, n$. Evaluate $\overline{\overline{c}}$ and also $\overline{c_i}$ for a selected primary membership function $\mu^i_{\tilde{A}}(x)$ using (7.22) and (7.23), respectively.
2. Find the optima on $\mu^j_{\tilde{A}}(x)$ for a given j. Let the set of x corresponding to the optima be S. Set the secondary membership $\mu(x, \mu^j_{\tilde{A}}(x))$ to 0.99 (close to 1) for all $x \in S$.

3. For each $x \in X$, where $x \notin S$, identify the optima closest around x from S. Let they be located at $x = x_p$ and $x = x_q$, where $x_p < x < x_q$. Determine α and β for each x, given by (7.20).
4. For each x, where $\mu(x, \mu^j_{\tilde{A}}(x))$ lies in $[\alpha, \beta]$, minimize J_j by DE.
5. Obtain $\mu(x, \mu^j_{\tilde{A}}(x))$ for all x after the DE converges.
6. Repeat step 2 onwards for all j.

For a Gaussian primary membership function, the minimum occurs at infinity, but the minimum value is practically 0 when x is $m \pm 4\sigma$, where μ and σ are mean and standard deviation of x, respectively. In Step 2, the minimum is taken as $m \pm 4\sigma$ and we obtain x by dividing the range $[(m - 4\sigma), (m + 4\sigma)]$ into equal intervals of the same length.

An illustrative plot of secondary membership function for a given primary is given in Figure 7.2.

7.5 EXPERIMENTAL DETAILS

In this section, we present the experimental details of emotion recognition using the principles introduced in Sections 7.3 and 7.4. Here, we consider five emotion classes (i.e., $k = 5$), including anger, fear, disgust, happiness, and relaxation. The experiment is conducted with two sets of subjects: (a) the first set of n (= 10) subjects is considered for designing the fuzzy face-space and (b) the other set of 30 facial expressions taken from six unknown subjects are considered to validate the result of the proposed emotion classification scheme. Five facial features (i.e., $m = 5$) have been used here to design the type-2 fuzzy face-space.

We now briefly overview the main steps of feature extraction followed by fuzzy face-space construction and emotion recognition of an unknown subject using the preconstructed face-space.

7.5.1 Feature Extraction

Feature extraction is a fundamental step in emotion recognition. This chapter considers extraction of features from emotionally rich facial expressions synthesized by the subjects by acting. Existing research results [9,47] reveal that the most important facial regions responsible for the manifestation of emotion are the eyes and the lips. This motivated us to select the following features: Left Eye Opening (EO_L), Right Eye Opening (EO_R), Distance between the Lower Eyelid to the Eyebrow for the Left Eye (LEE_L), Distance between the Lower Eyelid to Eyebrow for the Right Eye (LEE_R), and the Maximum Mouth opening (MO) including the lower and the upper lips. Figure 7.3 explains the above facial features on a selected facial image.

For extraction of any of the features mentioned earlier, the first step that needs to be carried out is to separate out the skin and the non-skin regions of the image.

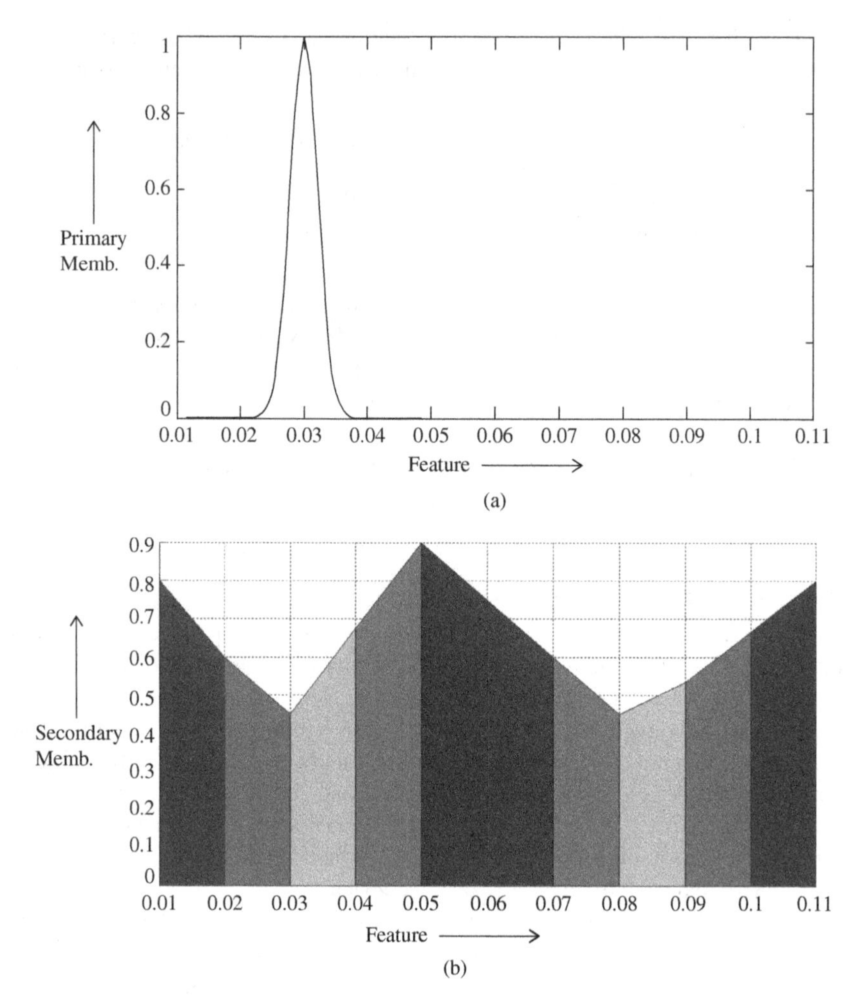

FIGURE 7.2 (a) Primary membership function for a given feature and (b) its corresponding secondary membership function obtained by minimizing J_i.

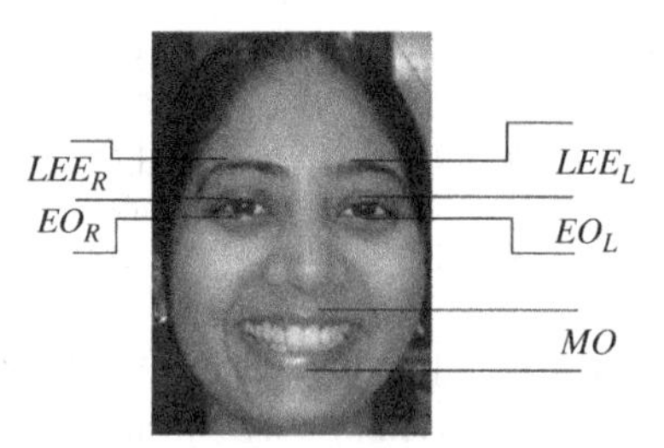

FIGURE 7.3 Facial features of an image. Taken with permission from Indian Women database (Jadavpur University).

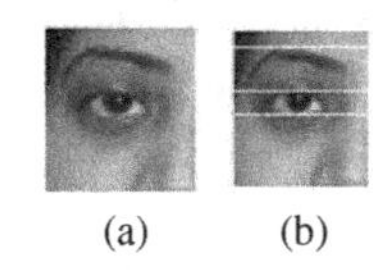

(a) (b)

FIGURE 7.4 (a) Localized eye search region. (b) Detection of eye features.

Estimation of eye features (EO_L, LEE_L, EO_R, and LEE_R): To compute the eye features, we first localize the eye region as shown in Figure 7.4a. A look at Figure 7.4 reveals that there is a sharp change in color while moving from the forehead region to the eyebrow region. Thus, to detect the location of the eyebrow, we take the average intensity (in three primary color planes) over each row of the image from the top and identify the row with a maximum dip in all the three planes. This row indicates the top of the eyebrow region (Figure 7.4b). Similarly, we detect the lower eyelid by identifying the row with a sharp dip in intensity in all the three planes, while scanning the face up from the bottommost row. The location of the top eyelid region is identified by scanning the face up from the marked lower eyelid until a dip in the three color planes are noted together.

Estimation of Mouth Opening (MO): In order to estimate the mouth opening, we first localize the mouth region as shown in Figure 7.5a. Then, a conversion from R-G-B to perceptually uniform $L^* - a^* - b^*$ color space is undertaken in order to represent the perceptual difference in color by Euclidean distance [48]. The k-means clustering algorithm is applied next on this image to get three clusters. The three clusters are the skin, lip, and teeth regions. The cluster with the highest intensity variance in $L^* - a^* - b^*$ color space is declared as the lip region. Thus, we select the lip cluster (Figure 7.5b) to determine the mouth opening. To obtain the mouth opening, we take the average intensity of three primary pixel colors and plot the row average of such value contributed by each pixel against the row number (Figure 7.5c). It is observed that the width of the zero-crossing zone in Figure 7.5c provides a measure of mouth opening.

Experiments are undertaken both on colored image database such as the Indian Women (Jadavpur University) database and gray-scale images including Japanese Female Facial Expression (JAFFE) and Cohn–Kanade databases. The principles of feature extraction introduced above is used for color image analysis, while for gray-scale images, the processing needed for feature extraction is done manually. The gray image processing is required to compare relative performance of our techniques with the existing ones tested on gray-scale images. Selective images from three facial

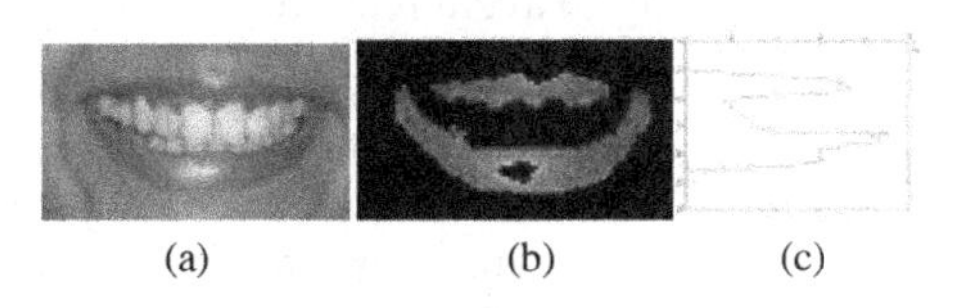

(a) (b) (c)

FIGURE 7.5 (a) Mouth search area. (b) Lip cluster. (c) Graph of average intensity over each row against the row position.

TABLE 7.1 Experiment done on different databases: JAFFE [49], Indian Women Database (Jadavpur University) [9], and Cohn–Kanade Database [48, 50]

	Emotion				
Databases	Anger	Disgust	Fear	Happiness	Sadness
JAFFE	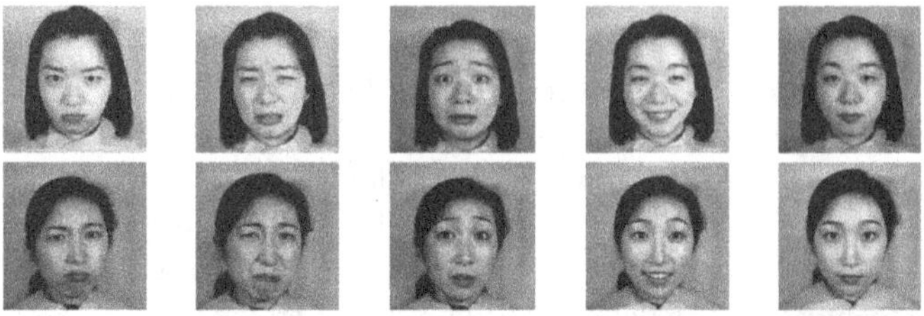				
Indian	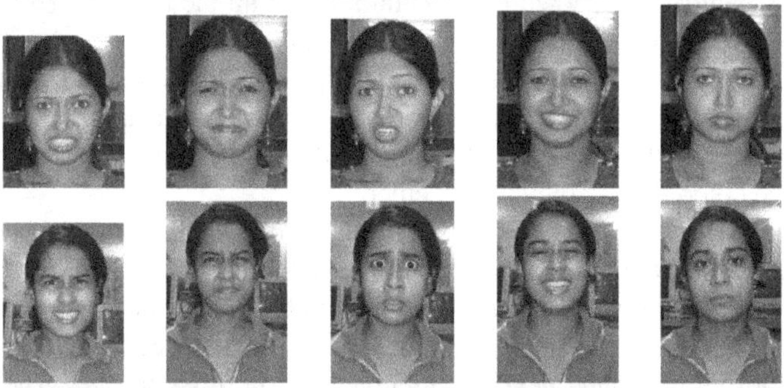				
Cohn-Kanade	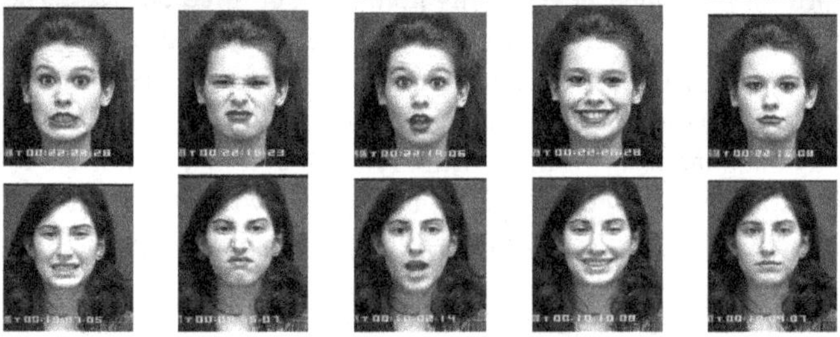				

Source: Reprinted with permission from JAFFE, Indian Women, and Cohn–Kanade databases

expression databases are given in Table 7.1. Training and test image data partition for three experimental databases is given in Table 7.2. The training data include y instances of z distinct emotions from x subjects.

The following explanation in this section is given with respect to the Indian Women Database (Jadavpur University) only to avoid repetition of the same issues.

7.5.2 Creating the Type-2 Fuzzy Face-Space

The interval type-2 fuzzy face-space contains only the primary membership functions for each facial feature. Since we have five facial features and the experiment includes five distinct emotions of 10 subjects, we obtain $10 \times 5 \times 5 = 250$ primary membership

TABLE 7.2 Training and test data for three databases: x = no. of subjects, y = no. of instances, and z = no. of emotions

Databases used	Training images ($x \times y \times z$)	Test images selected at random emotions
JAFFE	$5 \times 3 \times 5$	40
Indian Women (Jadavpur University)	$10 \times 10 \times 5$	40
Cohn–Kanade	$10 \times 5 \times 5$	40

curves. To compute primary memberships, 10 instances of a given emotion are used. These 250 membership curves are grouped into 25 heads, each containing 10 membership curves of 10 subjects for a specific feature for a given emotion. Figure 7.6 gives an illustration of one such group of 10 membership functions for the feature EO_L for the emotion: disgust.

For each primary membership function, we have a corresponding secondary membership function. Thus, we obtain 250 secondary membership functions. One illustrative type-2 secondary membership for subject 1 for the feature EO_L for the emotion disgust is given in Figure 7.7. The axes in the figure represent feature (EO_L), primary, and secondary membership values as indicated.

7.5.3 Emotion Recognition of an Unknown Facial Expression

The emotion recognition problem addressed here attempts to determine the emotion from facial expression of an unknown subject. To keep the measurements in an emotional expression normalized and free from distance variation from the camera focal plane, we construct a bounding box, covering only the face region, and the reciprocal

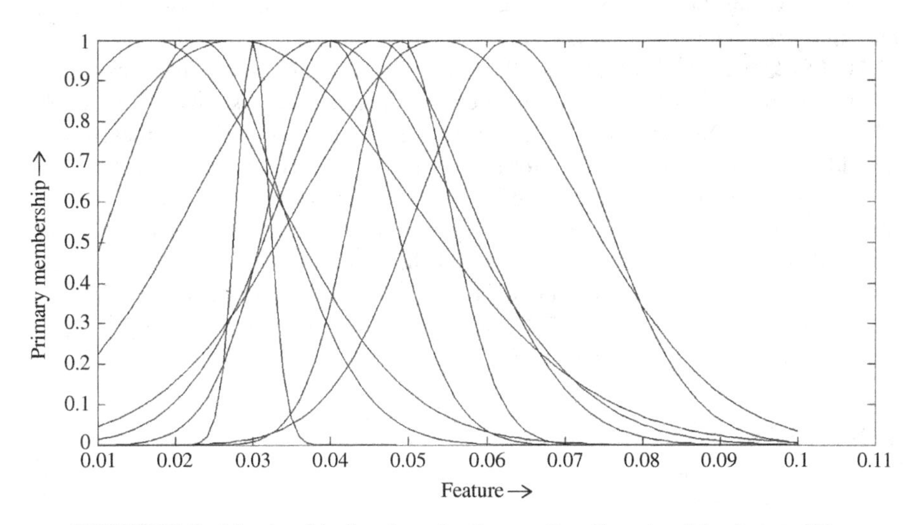

FIGURE 7.6 Membership functions for the emotion disgust and the feature EO_L.

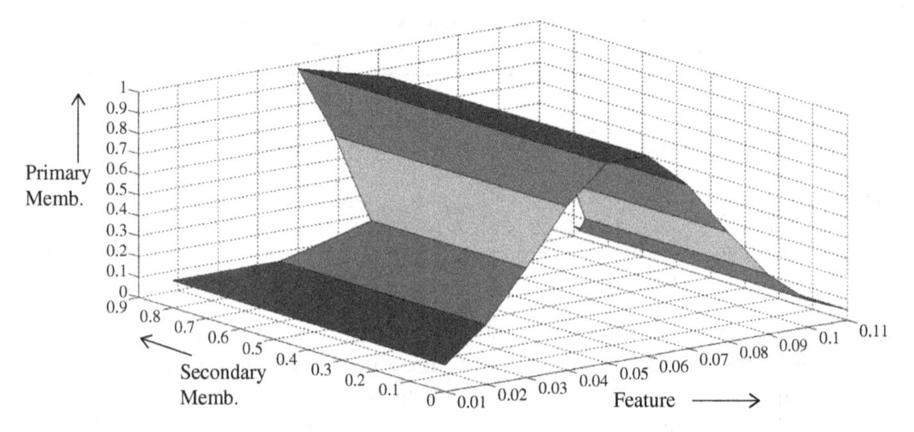

FIGURE 7.7 Secondary membership curve of a subject for emotion disgust.

FIGURE 7.8 Facial image of an unknown subject.

of the diagonal of the bounding box is used as a scale factor for normalization of the measurements. The normalized features obtained from Figure 7.8 are enlisted in Table 7.3. We now briefly explain the experimental results obtained by two alternative reasoning methodologies incorporating IT2FS and GT2FS.

IT2FS-Based Recognition: The IT2FS-based recognition scheme considers a fuzzy face-space of 5 sets of 10 primary membership functions as shown in Figure 7.6, where each set refers to one particular feature obtained from 10 sources for an individual emotion. Consequently, for five distinct emotions we have 25 such sets of primary membership functions. Table 7.4 provides the evaluation of type-2 primary membership values for a feature: EO_L consulting 10 primary membership functions obtained from 10 subjects, representative of the facial expression for disgust. The range of these memberships is given in the last column of Table 7.4. For each feature, we obtain five tables like Table 7.4, each one for a given emotion. Thus, for five features, we would have altogether 25 such tables.

TABLE 7.3 Extracted features of Figure 7.8

EO_L	EO_R	MO	LEE_L	LEE_R
0.026	0.026	0.135	0.115	0.115

TABLE 7.4 Calculated type-2 primary membership values for the feature: EO_L under the emotion disgust

Feature	Primary memberships (μ_{pri})	Range ($\min\{\mu_{pri}\}, \max\{\mu_{pri}\}$)
	0.65	
	0.10	
	0.15	
	0.45	
EO_L(pri)	0.18	0.08–0.65
	0.55	
	0.08	
	0.41	
	0.16	
	0.12	

Table 7.5 provides the results of individual range in primary membership for each feature experimented under different emotional conditions. For example, the entry (0–0.12) corresponding to the row Anger and column EO_L gives an idea about the extent of the EO_L for the unknown subject matches with known subjects from the emotion class Anger. The results of computing fuzzy meet operation over the range of individual features taken from facial expressions of the subjects under same emotional condition are given in Table 7.5. The average of the ranges along with its center value is also given in Table 7.5. It is observed that the center has the largest value ($=0.3115$) for the emotion Happiness.

GT2FS-Based Recognition: We now briefly illustrate the GT2FS-based reasoning for emotion classification. Here, the secondary membership function corresponding to the individual primary membership function of five features obtained from facial expressions carrying five distinct emotions for 10 different subjects are determined using curves as that shown in Figure 7.7.

Table 7.6 provides the summary of the primary and secondary memberships obtained for EO_L for the emotion disgust. The range computation for the feature EO_L is also shown in the last column of Table 7.6 with its center value. The same computations are repeated for all emotions, and the range evaluated in the last column of Table 7.7 indicates that the center of this range here too has the largest value ($=0.2775$) for the emotion happiness.

7.6 PERFORMANCE ANALYSIS

Performance analysis for emotion recognition itself is an open-ended research problem, as there is a dearth of literature on this topic. Most of the current literature on emotion research claims their algorithm to have outperformed others simply by comparing the classification accuracy, irrespective of the features used and the (facial) databases employed to validate the performance of individual algorithms. Naturally, because of non-uniformity in selected features and the training data (images) obtained

TABLE 7.5 Calculated feature ranges and center value for each emotion

Emotion	Range of Primary Membership for Features					Range S_j^c after fuzzy meet operation (center)
	EO_L	EO_R	MO	LEE_L	LEE_R	
Anger	0–0.12	0–0.133	0.37–0.864	0.001–0.425	0.003–0.434	0–0.12 (0.06)
Disgust	0.08–0.65	0.079–0.67	0–0.474	0.03–0.54	0.03–0.47	0–0.47 (0.235)
Fear	0.001–0.0422	0–0.053	0.284–0.826	0.027–0.554	0.031-0.629	0–0.0422 (0.0211)
Happiness	0–0.623	0–0.649	0.593–0.926	0.32–0.735	0.284–0.86	0–0.623 (0.3115)
Relaxed	0–0.274	0–0.272	0–0.004	0.057–0.528	0.051–0.562	0–0.004 (0.002)

TABLE 7.6 Calculated type-2 membership values for the feature: EO_L under the emotion disgust

Feature	Primary memberships (μ_{pri})	Secondary memberships (μ_{sec})	$\mu^{\text{mod}} = \mu_{\text{pri}} \times \mu_{\text{sec}}$	Range ($\min\{\mu^{\text{mod}}\}, \max\{\mu^{\text{mod}}\}$)
	0.65	0.72	0.468	
	0.10	0.55	0.055	
	0.15	0.58	0.087	
	0.45	0.68	0.306	
EO_L	0.18	0.56	0.1008	0.044–0.468
	0.55	0.68	0.374	
	0.08	0.55	0.044	
	0.41	0.63	0.2583	
	0.16	0.53	0.0848	
	0.12	0.59	0.0708	

from different databases, the comparison apparently has no pragmatic values. This chapter, however, compares the relative performance of the proposed GT2FS algorithms with five traditional emotion recognition algorithms/techniques and the IT2FS-based scheme introduced here, considering a common framework of their features and databases. The algorithms used for comparison include Linear Support Vector Machine (SVM) Classifier [47], (Type-1) Fuzzy Relational Approach [9], Principal Component Analysis (PCA) [51], Multilayer Perceptron (MLP) [52,53], and Radial Basis Function Network (RBFN) [52,53] and IT2FS.

Table 7.8 shows the classification accuracy of the two algorithms we proposed using three facial image databases, that is, the Japanese Female Facial Expression (JAFFE) Database, Indian Women Database (Jadavpur University), and Cohn–Kanade database. Experimental classification accuracy obtained for different other algorithms mentioned earlier using the three databases are given in Table 7.10.

Two statistical tests called McNemar's test [54] and Friedman test [55] and one new test called root mean square (RMS) error test are undertaken to analyze the relative performance of the proposed algorithms over existing ones.

7.6.1 The McNemar's Test

Let f_{A} and f_{B} be two classifier outputs obtained by algorithms A and B, respectively, when both the algorithms have a common training set R. We now define a null hypothesis [54]:

$$Pr_{R,x}[f_{\text{A}}(x) = f(x)] = Pr_{R,x}[f_{\text{B}}(x) = f(x)] \tag{7.26}$$

where $f(x)$ be the experimentally induced target function to map any data point x onto one of k -specific emotion classes.

Let n_{01} be the number of examples misclassified by f_{A} but not by f_{B}, and n_{10} be the number of examples misclassified by f_{B} but not by f_{A}.

TABLE 7.7 Calculated feature ranges and center value for each emotion

	Range of Primary Membership for Features					
Emotion	EO_L	EO_R	MO	LEE_L	LEE_R	Range S_j^c after fuzzy meet operation (center)
Anger	0–0.14	0–0.17	0.39–0.964	0.0034–0.814	0.0029–0.781	0–0.14 (0.07)
Disgust	0.044–0.468	0.041–0.531	0–0	0–0.39	0–0.283	0–0 (0)
Fear	0–0.298	0–0.275	0.04–0.742	0.054–0.473	0.057–0.511	0–0.275 (0.1375)
Happiness	0–0.555	0–0.604	0.573–0.910	0.133–0.851	0.3214–0.7213	0–0.555 (0.2775)
Relaxed	0–0.132	0–0.221	0–0	0.013–0.458	0.046–0.552	0–0 (0)

TABLE 7.8 Percentage accuracy of our proposed methods over three databases

	JAFFE (%)	Indian Women (Jadavpur University) (%)	Cohn–Kanade (%)	Average accuracy (%)
IT2FS	90	90	92.5	90.88
GT2FS	97.5	97.5	97.5	97.328

Then following Reference 54, we define a statistic

$$Z = \frac{(|n_{01} - n_{10}| - 1)^2}{n_{01} + n_{10}} \tag{7.27}$$

Let A be the proposed GT2FS algorithm and B is one of six typical algorithms. We thus evaluate $Z = Z_1$ through Z_6, where Z_j denotes the comparator statistic of misclassification between the GT2FS (Algorithm: A) and the jth of six algorithms (Algorithm: B), where the suffix j refers to the algorithm in row number j of Table 7.9.

Table 7.9 is evaluated to obtain Z_1 through Z_6 and the hypothesis has been rejected, when

$$Z_j > \chi^2_{1,0.95} = 3.841459$$

The last inequality indicates that the probability of the null hypothesis is correct only to a level of 5% and so, we reject it. If the hypothesis is not rejected, we consider its acceptance. The decision about acceptance or rejection is also included in Table 7.9.

7.6.2 Friedman Test

The Friedman test [55] ranks the algorithms for each dataset separately. The best performing algorithm gets rank 1. In case of ties, average ranks are assigned.

Let r_i^j be the rank of jth of k (= 6) algorithms on the ith of N (= 3) datasets. The average rank of algorithm j then is evaluated by

$$R_j = \frac{1}{N} \sum_{\forall i} r_i^j \tag{7.29}$$

The null hypothesis here states that all the algorithms are equivalent, so their individual ranks R_j should be equal. The Friedman statistic χ_F^2 is distributed following (7.29) with $k - 1$ degrees of freedom. Here, $k = 7$ and $N = 3$. A larger N of course is desirable; however, emotion databases being fewer, finding large N is not feasible. Here, we consider percentage accuracy of classification as the basis of rank. Table 7.10 provides the percentage accuracy of classification with respect to three databases, JAFFE, Indian Women (Jadavpur University), and Cohn–Kanade, and the respective ranks of the algorithm.

$$\chi_F^2 = \frac{12N}{k(k+1)} \left[\sum_j R_j^2 - \frac{k(k+1)^2}{4} \right] \tag{7.30}$$

TABLE 7.9 Statistical comparison of performance using McNemar's test with three databases. Reference Algorithm A = GT2FS

	JAFFE		Indian Women (Jadavpur University)		Cohn–Kanade	
Classifier algorithm B used for comparison	Z_j	Comments on acceptance/ rejection of hypothesis	Z_j	Comments on acceptance/ rejection of hypothesis	Z_j	Comments on acceptance/ rejection of hypothesis
IT2FS	1.333	Accept	0.5	Accept	0.5	Accept
SVM	0.5	Accept	0	Accept	0	Accept
Fuzzy Relational Approach	2.25	Accept	1.333	Accept	1.333	Accept
PCA	0	Accept	2.25	Accept	2.25	Accept
MLP	8.1	Reject	8.1	Reject	8.1	Reject
RBFN	11.077	Reject	10.083	Reject	10.083	Reject

TABLE 7.10 Average ranking of classification algorithms by Friedman test

	Classification Accuracy			Ranks obtained			
CA	JAFFE	Indian	Cohn–Kanade	JAFFE	Indian	Cohn–Kanade	Average rank (R_j)
A	97.5	97.5	97.5	1	1	1	1
B_2	92.5	95	95	3	2	2	2.33
B_3	90	90	92.5	4	4	3	3.66
B_4	87.5	92.5	90	5	3	4	4
B_5	95	87.5	87.5	2	5	5	4
B_6	72.5	75	72.5	6	6	6	6
B_7	65	67.5	67.5	7	7	7	7

CA = Classifier Algorithm, A = GT2FS, B_1 = SVM, B_2 = IT2FS, B_3 = Fuzzy Relational Approach, B_4 = PCA, B_5 = MLP, B_6 = RBFN

Now, from Table 7.10, we obtain R_js and evaluate χ_F^2 using (7.29), with $N = 3$, $k = 7$.

$$\chi_F^2 = \frac{12N}{k(k+1)}\left[\sum_j R_j^2 - \frac{k(k+1)^2}{4}\right] \\ = 14.9586 > \chi^2_{6,0.95}(12.592) \tag{7.31}$$

So, the null hypothesis, claiming that all the algorithms are equivalent, is wrong and, therefore, the performances of the algorithms are determined by their ranks only. The order of ranking of the algorithm is apparent from their average ranks. The smaller the average rank, the better is the algorithm. Let > be a comparator of relative ranks, where $x > y$ means the algorithm x is better in rank than the algorithm y. Table 7.10 indicates that the relative order of ranking of the algorithm by Friedman test as GT2FS> SVM> IT2FS> Fuzzy Relational Approach> PCA> MLP> RBFN. It is clear from Table 7.10 that the average rank of GT2FS is 1 and average rank of IT2FS is 3, claiming GT2FS outperforms all the algorithms by Friedman test.

7.6.3 The Confusion Matrix-Based RMS Error

The statistical tests given earlier compare individual algorithms based on their success in classification and failures in misclassification for all classes together, and ignore the percentage classification of two algorithms under each emotion class. Confusion matrices [9, 24, 51] provide the classification accuracy of an algorithm in individual classes. Naturally, a metric may be defined to compare the percentage classification accuracy of two algorithms for each emotion class. Let A and B be two classification algorithms, and d_{ii}^{A} and d_{ii}^{B} denote the diagonal entries of the two confusion matrices obtained by algorithms A and B, respectively. Then, the RMS error norm, representing

TABLE 7.11 RMS error of other tested algorithms with respect to reference algorithm A = GT2FS

	RMS error obtained for different algorithm B				
Classification algorithm B	JAFFE	Indian Women (Jadavpur University)	Cohn–Kanade	Average RMS error	Rank obtained
SVM	14.529	11.056	11.056	12.2137	2
Fuzzy Relational Approach	24.031	14.489	21.3396	19.9532	5
IT2FS	19.717	19.717	14.529	17.9877	4
PCA	6.665	14.529	21.3396	14.1778	3
MLP	49.1	50.6577	49.1	49.6192	6
RBFN	69.7286	67.9	67.9	68.5095	7

a comparator of two algorithms A and B induced by the respective six diagonals in two confusion matrices, is given by

$$\mathrm{Error}_{\mathrm{RMS}}(\mathrm{A}, \mathrm{B}) = \left[\sum_{i=1}^{5} (d_{ii}^{A} - d_{ii}^{B})^2 \right]^{1/2} \tag{7.32}$$

Extensive experiments with seven different algorithms including GT2FS and IT2FS reveal that GT2FS outperforms the rest in percentage classification of accuracy in each emotion class. This motivated us to select A = reference algorithm and B = any one of six other algorithms already introduced earlier to determine the RMS error norm between algorithms A and B.

It is clear from (7.31) that smaller the RMS error norm, the closer is the performance of algorithm B with respect to algorithm A. A look at Table 7.11 reveals that the average of the RMS errors obtained by three databases is the smallest for SVM, indicating that SVM is the next best algorithm when comparison with GT2FS. The ranking of the six algorithms based on the measure of the average of the RMS errors obtained from experiments with three databases, considering GT2FS as the reference algorithm (with rank = 1) is also shown in Table 7.11.

The order of ranks of the seven algorithms is given below using ">" operator: GT2FS > SVM> PCA > IT2FS>Fuzzy Relational Approach > MLP > RBFN. The order of ranking indicates that GT2FS and IT2FS have relative ranking 1 and 4, respectively.

A difference in order of ranking obtained by Friedman test and the RMS error norm metric is due to the difference in the ranking criteria. The Friedman test compares all algorithms with respect to overall classification accuracy of individual algorithm, whereas the RMS error norm considers relative ranking of the algorithms from the individual class performance.

Irrespective of the three tests, the GT2FS outperforms all the six classifier algorithms, whereas IT2FS has a rank 3 in Friedman test and 4 in RMS error-based test.

7.7 CONCLUSION

The chapter employs IT2FS- and GT2FS-based automatic emotion recognition of an unknown facial expression, when the background knowledge about a large face database with known emotion class is available. The GT2FS-based recognition scheme requires type-2 secondary membership functions, a computation of which by an evolutionary approach is also provided. Both the schemes first construct a fuzzy face-space, and then infer the emotion class of the unknown facial expression by determining the maximum support of the individual emotion classes using the pre-constructed fuzzy face-space. The class with the highest support is regarded as the emotion of the unknown facial expression.

The IT2FS-based recognition scheme takes care of the intersubject-level uncertainty in computing the maximum support of individual emotion class. The GT2FS-based recognition scheme, however, takes care of both the intersubject- and intrasubject-level uncertainties, and thus offers a higher classification accuracy for the same set of features. Experimental analysis confirms that the classification accuracy of emotion by employing GT2FS is 97.328% and by IT2FS is 90.88%.

Our proposed method is different from classical type-2 (type-1) based recognition as the type-1 fuzzy set does not involve subjective opinion of multiple sources and employs relational matrices, which themselves play a major role to determine the classification accuracy. Since relational matrices are designed by subject expert, it is often difficult to obtain unique relational matrices for a given recognition system. The proposed scheme, however, does not involve relational matrix and thus it is free from subjective bias on classification accuracy.

The more the number of subjects used for constructing the fuzzy facespace, the better would be the fuzzy facespace, and thus better would be the classification accuracy. Since the fuzzy facespace is created off-line, the online computational load to recognize emotion is insignificantly small in IT2FS. The computational load in GT2FS, however, is large as it includes an optimization procedure to determine the secondary membership for each emotion and for each subject. This additional complexity in GT2FS, however, offers approximately 7% improvement in classification accuracy in comparison to the same by IT2FS. Statistical test employed and RMS error norm based comparator clearly indicate that GT2FS outperforms six selected algorithms, and to the best of our knowledge has the best performance over all the six algorithms.

REFERENCES

[1] A. Chakraborty, A. Konar, P. Bhowmik, and A. K. Nagar, "Stability, chaos and limit cycles in recurrent cognitive reasoning systems," in *The Handbook on Reasoning-Based Intelligent Systems* (eds K. Nakamatsu and L. C. Jain), 2013.

[2] D. Datcu and L. J. M. Rothkrantz, "Emotion recognition using bimodal data fusion," in Proceedings of the 12th International Conference on Computer Systems and Technologies, 2011, pp. 122–128.

[3] Md. Zia Uddin, J. J. Lee, and T. S. Kim. "Independent shape component-based human activity recognition via Hidden Markov Model," *Appl. Intell.* 33(2): 193–206, 2010.

[4] Y. Yang, Z. Chen, Z. Liang, and G. Wang, "Attribute reduction for massive data based on rough set theory and MapReduce," in Proceedings of the RSKT, 2010, pp. 672–678.

[5] C.-Y. Chang, S.-C. Li, P.-C. Chung, J.-Y. Kuo, and Y.-C. Tu, "Automatic facial skin defect detection system," in Proceedings of the BWCCA, 2010, pp. 527–532.

[6] O. A. Uwechue and S. A. Pandya, *Human Face Recognition Using Third-Order Synthetic Neural Networks*, Kluwer, Boston, MA, 1997.

[7] A. Bhavsar and H. M. Patel, "Facial expression recognition using neural classifier and fuzzy mapping," in Proceedings of the IEEE INDICON, Chennai, India, 2005, pp. 383–387.

[8] Y. Guo and H. Gao, "Emotion recognition system in images based on fuzzy neural network and HMM," in Proceedings of the Fifth IEEE ICCI, 2006, pp. 73–78.

[9] A. Chakraborty, A. Konar, U. K. Chakraborty, and A. Chatterjee, "Emotion recognition from facial expressions and its control using fuzzy logic," *IEEE Transactions on Systems, Man, and Cybernetics, Part A: Systems and Humans*, 39(4): 726–743, 2009.

[10] P. Ekman and W. V. Friesen, *Unmasking the Face: A Guide to Recognizing Emotions from Facial Clues*, Prentice-Hall, Englewood Cliffs, NJ, 1975.

[11] H. Kobayashi and F. Hara, "Measurement of the strength of six basic facial expressions by neural network," *Trans. Jpn. Soc. Mech. Eng. C*, 59(567): 177–183, 1993.

[12] H. Kobayashi and F. Hara, "Recognition of mixed facial expressions by neural network," *Trans. Jpn. Soc. Mech. Eng. C*, 59(567): 184–189, 1993.

[13] H. Kobayashi and F. Hara, "The recognition of basic facial expressions by neural network," *Trans. Soc. Instrum. Control. Eng.*, 29(1):112–118, 1993.

[14] F. Kawakami, S. Morishima, H. Yamada, and H. Harashima, "Construction of 3-D emotion space using neural network," in Proceedings of the Third International Conference on Fuzzy Logic, Neural Nets and Soft Computing, Iizuka, Japan, 1994, pp. 309–310.

[15] J. M. Mendel and D. Wu, *Perceptual Computing*, IEEE Press, Piscataway, NJ, 2010.

[16] H. Yamada, "Visual information for categorizing facial expression of emotion," *Appl. Cogn. Psychol.*, 7(3): 257–270, 1993.

[17] J. M. Fernandez-Dols, H. Wallbotl, and F. Sanchez, "Emotion category accessibility and the decoding of emotion from facial expression and context," *J. Nonverbal Behav.*, 15(2): 107–123, 1991.

[18] C. Busso and S. Narayanan, "Interaction between speech and facial gestures in emotional utterances: a single subject study," *IEEE Transactions on Audio, Speech, and Language Processing*, 15(8): 2331–2347, 2007.

[19] I. Cohen, N. Sebe, A. Garg, L. S. Chen, and T. S. Huang, "Facial expression recognition from video sequences: temporal and static modeling," *Comput. Vis. Image Underst.*, 91(1/2): 160–187, 2003.

[20] Y. Gao, M. K. H. Leung, S. C. Hui, and M. W. Tananda, "Facial expression recognition from line-based caricatures," *IEEE Transactions on Systems, Man, and Cybernetics, Part A: Systems and Humans*, 33(3): 407–412, 2003.

[21] M. Rizon, M. Karthigayan, S. Yaacob, and R. Nagarajan, "Japanese face emotions classification using lip features," in Proceedings of the GMAI, 2007, pp. 140–144.

[22] H. Zhao, Z. Wang, and J. Men, "Facial complex expression recognition based on fuzzy kernel clustering and support vector machines," in Proceedings of the Third ICNC, 2007, pp. 562–566.

[23] Y. Lee, C. W. Han, and J. Shim, "Fuzzy neural networks and fuzzy integral approach to curvature-based component range facial recognition," in Proceedings of the International Conference on Convergence Information Technology, 2007, pp. 1334–1339.

[24] G. U. Kharat and S. V. Dudul, "Neural network classifier for human emotion recognition from facial expressions using discrete cosine transform," in Proceedings of the 1st International Conference on Emerging Trends in Engineering and Technology, 2008, pp. 653–658.

[25] J. M. Sun, X. S. Pei, and S. S. Zhou, "Facial emotion recognition in modern distant system using SVM," in Proceedings of the Seventh International Conference on Machine Learning and Cybernetics, Kunming, China, 2008, pp. 3545–3548.

[26] H. Tsai, Y. Lai, and Y. Zhang, "Using SVM to design facial expression recognition for shape and texture features," in Proceedings of the Ninth International Conference

on Machine Learning and Cybernetics, Qingdao, China, July 11–14, 2010, pp. 2697–2704.

[27] B. Biswas, A. K. Mukherjee, and A. Konar, "Matching of digital images using fuzzy logic," *AMSE Publ.*, 35(2): 7–11, 1995.

[28] M. Paleari, R. Chellali, and B. Huet, "Features for multimodal emotion recognition: an extensive study," in Proceedings of the CIS, 2010, pp. 90–95.

[29] R. W. Picard, E. Vyzas, and J. Healey, "Toward machine emotional intelligence: analysis of affective physiological state," *IEEE Transactions on Pattern Analysis and Machine Intelligence*, 23(10): 1175–1191, 2001.

[30] M. Pantic and I. Patras, "Dynamics of facial expression: recognition of facial actions and their temporal segments from face profile image sequences," *IEEE Transactions on Systems, Man and Cybernetics, B (Cybernetics)*, 36(2): 433–449, 2006.

[31] M. Pantic and L. J. M. Rothkrantz, "Automatic analysis of facial expressions: the state of the art," *IEEE Transactions on Pattern Analysis and Machine Intelligence*, 22(12): 1424–1445, 2000.

[32] H. Zhang, D. Liu, and Z. Wang, *Controlling Chaos: Suppression, Synchronization and Chaotification*, Springer, New York, 2009.

[33] J. M. Mendel and D. Wu, *Perceptual Computing: Aiding People in Making Subjective Judgments*, IEEE-Wiley Press.

[34] L. A. Zadeh, "Fuzzy sets," *Inf. Control*, 8(3): 338–353, 1965.

[35] J. M. Mendel and H. Wu, "Type-2 fuzzistics for symmetric interval type-2 fuzzy sets: Part 1, forward problems," *IEEE Transactions on Fuzzy Systems*, 14(6): 718–792, 2006.

[36] H. J. Zimmermann, *Fuzzy Set Theory and Its Applications*, Springer-Verlag, Berlin, Germany, 2001.

[37] J. R. Agero and A. Vargas, "Calculating functions of interval type-2 fuzzy numbers for fault current analysis," *IEEE Transactions on Fuzzy Systems*, 15(1): 31–40, 2007.

[38] H. Wu, Y. Wu, and J. Luo, "An interval type-2 fuzzy rough set model for attribute reduction," *IEEE Transactions on Fuzzy Systems*, 17(2): 301–315, 2009.

[39] O. Cordón, M. J. del Jesus, and F. Herrera, "A proposal on reasoning methods in fuzzy rule-based classification systems," *Int. J. Approx. Reason.*, 20(1): 21–45, 1999.

[40] J. M. Mendel, "On the importance of interval sets in type-2 fuzzy logic systems," in Proceedings of the Joint 9th IFSA World Congress and 20th NAFIPS International Conference, Vancouver, BC, Canada, July 25–28, 2001, pp. 1647–1652.

[41] C.-H. Lee, J.-L. Hong, Y.-C. Lin, and W.-Y. Lai, "Type-2 fuzzy neural network systems and learning," *Int. J. Comput. Cogn.*, 1(4): 79–90, 2003.

[42] J. M. Mendel and R. I. John, "Type-2 fuzzy sets made simple," *IEEE Transactions on Fuzzy Systems*, 10(2): 117–127, 2002.

[43] S. Das, A. Abraham, U. K. Chakraborty, and A. Konar, "Differential evolution using a neighbourhood-based mutation operator," *IEEE Transactions on Evolutionary Computation*, 13(3): 526–553, 2009.

[44] Z. Michalewicz, *Genetic Algorithms + Data Structure = Evolution Programs*, Springer-Verlag, Berlin, Germany, 1992.

[45] P. Engelbrecht, *Fundamentals of Computational Swarm Intelligence*, John Wiley & Sons, New York, 2005.

[46] P. N. Suganthan, N. Hansen, J. J. Liang, K. Deb, Y.-P. Chen, A. Auger, and S. Tiwari, "Problem definitions and evaluation criteria for the CEC 2005 special session on real-parameter optimization," Technical Report, Nanyang Technological University, Singapore, 2005.

[47] S. Das, A. Halder, P. Bhowmik, A. Chakraborty, A. Konar, and A. K. Nagar, "Voice and facial expression based classification of emotion using linear support vector," in Proceedings of the Second International Conference on Developments in Systems Engineering, 2009, pp. 377–384.

[48] P. Lucey, J. F. Cohn, T. Kanade, J. Saragih, S. Ambadar, and I. Matthews, "The extended Cohn–Kanade Dataset (DK+): a complete expression dataset for action unit and emotion specified expression," in Proceedings of the Third International Workshop on CVPR for Human Communicative Behavior Analysis (CVPR 4HB 2010), San Fransisco, CA, 2010, pp. 94–101.

[49] M. J. Lyons, M. Kamachi, and J. Gyoba, "Japanese Female Facial Expressions (JAFFE)," in Database of Digital Images, 1997.

[50] T. Kanade, J. F. Cohn, and Y. Tian, "Comprehensive database for facial expression analysis," in Proceedings of the Fourth IEEE International Conference on Automatic Face and Gesture Recognition (FG'00), Grenoble, France, 2000, pp. 46–53.

[51] A. Chakraborty and A. Konar, *Emotional Intelligence: A Cybernetic Approach*, Springer-Verlag, Heidelberg, Germany, 2009.

[52] M. Gargesha and P. Kuchi, "Facial expression recognition using a neural network," *Artif. Neural Comput. Syst.*, 1–6, 2002.

[53] J. M. Mendel, "Fuzzy sets for words: a new beginning," in Proceedings of the IEEE International Conference on Fuzzy System, 2003, pp. 37–42.

[54] J. Demsar, "Statistical comparisons of classifiers over multiple data sets," *J. Mach. Learn. Res.*, 7: 1–30, 2006.

[55] T. G. Dietterich, "Approximate statistical tests for comparing supervised classification learning algorithms," *Neural Comput.*, 10(7): 1895–1923, 1998.

[56] K. V. Price, R. M. Storn, and J. A. Lampinen (eds), *Differential Evolution: A Practical Approach*, Springer, New York, 2005.

[57] D. J. Kim, S. W. Lee, and Z. Bien, "Facial emotional expression recognition with soft computing techniques," in Proceedings of the 14th IEEE International Conference on Fuzzy System, 2005, pp. 737–742.

[58] R. Cowie, E. Douglas-Cowie, J. G. Taylor, S. Ioannou, M. Wallace, and S. Kollias, "An intelligent system for facial emotion recognition," in Proceedings of IEEE ICME, Amsterdam, The Netherlands, July 2005, pp. 904–907.

[59] M. H. Bindu, P. Gupta, and U. S. Tiwary, "Cognitive model based emotion recognition from facial expressions for live human computer interaction," in Proceedings of the IEEE Symposium on CIISP, 2007, pp. 351–356.

[60] N. Esau, E. Wetzel, L. Kleinjohann, and B. Kleinjohann, "Real-time facial expression recognition using a fuzzy emotion model," in Proceedings of the IEEE International Conference on Fuzzy System, 2007, pp. 1–6.

[61] A. Khanam, M. Z. Shafiq, and M. U. Akram, "Fuzzy based facial expression recognition," in Proceedings of the Congress on Image and Signal Processing, 2008, pp. 598–602.

[62] H. Zhao and Z. Wang, "Facial action units recognition based on fuzzy kernel clustering," in Proceedings of the Fifth International Conference on Fuzzy Systems and Knowledge Discovery, 2008, pp. 168–172.

[63] Y. Sun, Z. Li, and C. Tang, "An evolving neural network for authentic emotion classification," in Proceedings of the Fifth International Conference on Natural Computation, 2009, pp. 109–113.

[64] C. Chang, Y. Huang, and C. Yang, "Personalized facial expression recognition in color image," in Proceedings of the Fourth International Conference on Innovative Computing, Information and Control, 2009, pp. 1164–167.

[65] M. Fu, Y. Kuo, and K. Lee, "Fusing remote control usages and facial expression for emotion recognition," in Proceedings of the Fourth International Conference on Innovative Computing, Information and Control, 2009, pp. 132–135.

[66] C. Huang and Y. Huang, "Facial expression recognition using model-based feature extraction and action parameters classification," *J. Vis. Commun. Image Represent.*, 8(3): 278–290, 1997.

[67] K. Huang, S. Huang, and Y. Kuo, "Emotion recognition based on a novel triangular facial feature extraction method," in *Proceedings of IJCNN*, 2010, pp. 1–6.

[68] Y. Zhang and Q. Ji, "Active and dynamic information fusion for facial expression understanding from image sequences," *IEEE Transactions on Pattern Analysis and Machine Intelligence*, 27(5): 699–714, 2005.

[69] C. Liu and H. Wechsler, "Probabilistic reasoning models for face recognition," in Proceedings of the IEEE Conference on Computer Vision and Pattern Recognition, 1998, pp. 827–832.

[70] G. U. Kharat and S. V. Dudul, "Emotion recognition from facial expression using neural networks," in Conference on Human System Interactions, 2009, pp. 207–219.

[71] A. Halder, A. Chakraborty, A. Konar, and A. K. Nagar, "Computing with words model for emotion recognition by facial expression analysis using interval type-2 fuzzy sets," in Proceedings of FUZZ-IEEE, Hyderabad, India, 2013.

[72] J. M. Mendel, R. I. John, and F. Liu, "Interval type-2 fuzzy logic systems made simple," *IEEE Transactions on Fuzzy Systems*, 14(6): 808–821, 2006.

AUTHOR BIOGRAPHIES

Anisha Halder received the B.Tech. degree in Electronics and Communication Engineering from the Haldia Institute of Technology, Midnapore, India, and the M.E. degree in Control Engineering from Electronics and Telecommunication Engineering Department, Jadavpur University, Kolkata, India, in 2007 and 2009, respectively. She is currently pursuing her Ph.D. degree in facial expression analysis for emotion recognition jointly under the guidance of Prof. Amit Konar and Dr. Aruna Chakraborty at Jadavpur University, India.

Her principal research interests include artificial intelligence, pattern recognition, cognitive science, and human–computer interaction. She is the author of over 25 papers published in top international journals and conference proceedings.

Amit Konar received the B.E. degree from Bengal Engineering and Science University, Howrah, India, in 1983 and the M.E., M.Phil., and Ph.D. (Engineering) degrees from Jadavpur University, Calcutta, India, in 1985, 1988, and 1994, respectively. In 2006, he was a visiting professor with the University of Missouri, St. Louis. He is currently a professor with the Department of Electronics and Telecommunication Engineering (ETCE), Jadavpur University, where he is the Founding Coordinator of the M.Tech. program in Intelligent Automation and Robotics. He has supervised fifteen Ph.D. theses. He has over 200 publications in international journals and conference proceedings.

He is the author of eight books, including two popular texts *Artificial Intelligence and Soft Computing* (CRC Press, 2000) and *Computational Intelligence: Principles, Techniques and Applications* (Springer, 2005). He serves as the Associate Editor of *IEEE Transactions on Systems, Man and Cybernetics, Part A*, and *IEEE Transactions on Fuzzy Systems*.

His research areas include the study of computational intelligence algorithms and their applications to the domains of electrical engineering and computer science. Specifically, he has worked on fuzzy sets and logic, neurocomputing, evolutionary algorithms, Dempster–Shafer theory, and Kalman filtering, and has applied the principles of computational intelligence in image understanding, VLSI design, mobile robotics, pattern recognition, brain–computer interfacing, and computational biology. He was the recipient of All India Council for Technical Education (AICTE) accredited 1997–2000 Career Award for Young Teachers for his significant contribution in teaching and research.

Aruna Chakraborty received the M.A. degree in Cognitive Science and the Ph.D. degree in Emotional Intelligence and Human–Computer Interactions from Jadavpur University, Kolkata, India, in 2000 and 2005, respectively. She is currently an associate professor in the Department of Computer Science & Engineering, St. Thomas' College of Engineering & Technology, Kolkata. She is also a visiting faculty of Jadavpur University, where she offers graduate-level courses on intelligent automation and robotics and cognitive science.

She, with her teacher Prof. A. Konar, has written a book titled Emotional *Intelligence: A Cybernetic Approach* (Springer, Heidelberg, 2009). She serves as an Editor to the *International Journal of Artificial Intelligence and Soft Computing*, Inderscience, UK. Her current research interests include artificial intelligence, emotion modeling, and their applications in next-generation human–machine interactive systems. She is a nature lover and loves music and painting.

Atulya K. Nagar holds the Foundation Chair, as professor of computer and mathematical sciences, at Liverpool Hope University and is Head of the Department of Mathematics and Computer Science. A mathematician by training, Prof. Nagar possesses multidisciplinary expertise in natural computing, bioinformatics, operations research, and systems engineering. He has an extensive background and experience of working in universities in the United Kingdom and India. He has been an expert reviewer for the Biotechnology and Biological Sciences Research Council (BBSRC) Grants Peer-Review Committee for Bioinformatics Panel and serves on Peer-Review College of the Arts and Humanities Research Council (AHRC) as a scientific expert member. He has coedited volumes on Intelligent Systems and Applied Mathematics; he is the Editor-in-Chief of the *International Journal of Artificial Intelligence and Soft Computing* (IJAISC) and serves on editorial boards for a number of prestigious journals as well as on the International Programme Committee (IPC) for several international conferences.

He received a prestigious Commonwealth Fellowship for pursuing his Doctorate in Applied Non-Linear Mathematics, which he earned from the University of York in 1996. He holds B.Sc. (Hons.), M.Sc., and M.Phil. (with Distinction) from MDS University of Ajmer, India. Prior to joining Liverpool Hope, Prof. Nagar was with the Department of Mathematical Sciences, and later at the Department of Systems Engineering, Brunel University, London.

8

EMOTION RECOGNITION FROM NON-FRONTAL FACIAL IMAGES

WENMING ZHENG

Key Laboratory of Child Development and Learning Science, Ministry of Education, Southeast University, Nanjing, P. R. China

HAO TANG

HP Labs, Palo Alto, CA, USA

THOMAS S. HUANG

Beckman Institute for Advanced Science and Technology, University of Illinois at Urbana-Champaign, Urbana, IL, USA

Emotion recognition from facial images is a very active research topic in human–computer interaction (HCI). Earlier research on human emotions can be traced back to Darwin's pioneer work. The study of the relationship between facial expression and emotion by Ekman and Friesen in the 1970s paved the way for the facial emotion recognition research and has since attracted a lot of researchers to this area. According to Ekman and his colleague, there are six basic emotions that are universal to human beings, namely, anger (AN), disgust (DI), fear (FE), happiness (HA), sadness (SA), and surprise (SU). During the past several decades, various methods have been proposed to recognize the emotion categories based on human facial images. However, most of the previous research focuses on the frontal or nearly frontal facial images. In contrast to the frontal or nearly frontal facial images, emotion recognition from non-frontal facial images is much more challenging yet of more practical utility. In this chapter, we survey the recent advances of the current non-frontal facial emotion recognition methods, including the three-dimensional (3D) facial expression recognition and multiview facial expression recognition. We will especially be focusing our

Emotion Recognition: A Pattern Analysis Approach, First Edition. Edited by Amit Konar and Aruna Chakraborty.
© 2015 John Wiley & Sons, Inc. Published 2015 by John Wiley & Sons, Inc.

attention on the facial emotion recognition issue from the non-frontal facial images. Discussions and conclusions are presented at the end of this chapter.

8.1 INTRODUCTION

Human emotion are the result of many psychological and physiological reactions, which play a crucial role in human–human communication. Although the study of emotion has a long history, there is still little agreement on the emotion mechanism. The earlier research of emotion can be traced back to Darwin [1], who studied emotion expressions from an ethological perspective and showed that emotional expressions are closely related to survival [1,2]. James [3] and Lange [4] independently developed the James–Lange theory of emotion, which states that emotions are the effective results of human physiological changes. Another emotion theory that is contrary to the James–Lange theory was developed by Cannon [5], who believed that emotions are prior to the physical behaviors. Emotions, despite their different mechanisms, can be generally described in two different ways [2,6]. The first one is to classify emotions into a set of discrete statuses, for example, the six basic emotions introduced by Ekman and Friesen [7], and the second one is to use the so-called emotional dimension approach [6] to describe emotions. The emotional dimension method emphasizes that there are close relationships among different emotions. A typical example is the 2D emotion describing approach, in which emotions are described by valence and arousal [8,9]. The valence dimension represents the degree of how positive or negative the emotions are, whereas the arousal dimension represents the degree of how excited or apathetic the emotions are [6].

It is well known that human individuals often express their emotion information through a broad range of verbal and nonverbal cues. Speech and audio are two verbal cues, while facial expressions, body gestures, EEG signals, and physiological signals are the typical nonverbal cues. Figure 8.1 enumerates some of the popular

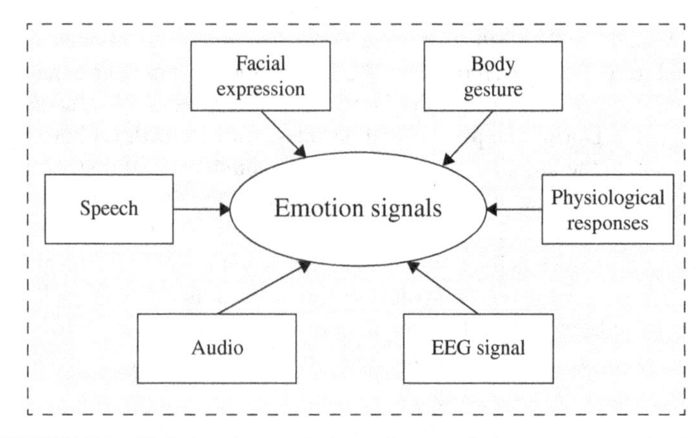

FIGURE 8.1 Verbal and nonverbal emotion cues in human communications.

emotion cues, which provide useful information to deal with the emotion recognition problem.

Over the last several decades, emotion recognition has been a very active research topic, which has attracted more and more researchers and scientists from psychology, physiology, cognitive science, computer science, and behavioral science. For instance, to tackle the emotion-related computing issues, Picard [10] proposed a research direction called "Affective Computing," which was regarded as the computing that relates to, arises from, or deliberately influences emotions. Of late, the emotion computing research has begun to gain application in behavioral science, security, animation, education, and HCI [11].

Overall, emotion recognition methods can be basically categorized into two: unimodal-based methods and multimodal-based methods [2, 6, 12, 13, 90]. The former deals with the emotion recognition issue using a single modality and tackles the emotion recognition of each modality in an independent way. Most of the previous emotion recognition methods in the earlier literature belong to the unimodal-based methods. Some of the most famous methods are listed as follows: emotion recognition from speech and audio [14–16], emotion recognition from body gesture [17], emotion recognition from facial expression [18, 19], emotion recognition from physiological parameters [20, 21], and emotion recognition from EEG signals [22–24]. While more and more researchers focus their research on unimodal emotion recognition, psychological studies show that human individuals tend to use different emotion cues rather than a single one in human–human communications [25, 26]. Moreover, it is notable that there are situations in which emotion signals of a single modality may be influenced by others and hence might not obtain the desired result [2]. In contrast to unimodal-based emotion recognition approaches, multimodal-based emotion approaches make use of more emotion channels to boost the emotion recognition performance. Some of the most famous multimodal emotion recognition methods are bimodal emotion recognition from facial expression and speech [27], bimodal emotion recognition from facial expression and body gesture [28–30], and emotion recognition from facial expression, speech, and body gesture [12, 13, 89].

Among the various emotional cues, facial expression and speech are two important ones. According to Mehrabian [32], human speech accounts for 38% of a communication message while facial expression accounts for 55%. Due to the importance of speech and facial expression in human emotion communications, most of the current emotion recognition research focuses on the two emotion modalities. The earlier work of automatic emotion recognition using facial expression can be traced back to the 1970s when Suwa *et al.* presented the very first work in Reference 33. However, the earliest research on human emotion can be traced back to the Darwin's pioneer work [1], who pointed out that emotion expressions are universal to human beings and animals. In the 1970s, Ekman and Friesen developed the famous Facial Action Coding System (FACS) to code human facial expressions, which has become one of the most widely used and versatile methods for describing facial behaviors [31, 34, 35]. The original FACS consisted of 44 action units (AUs) and 30 of them are related to the contraction of a specific set of facial muscles [19]. The FACS was updated by Ekman, Friesen, and Hager in 2002 and is still known

TABLE 8.1 Description of AUs in FACS

AU#	Descriptions	AU#	Descriptions	AU#	Descriptions
1	Inner brow raiser	16	Lower lip depressor	30[a]	Jaw sideways
2	Outer brow raiser	17	Chin raiser	31	Jaw clencher
4	Brow lower	18	Lip pucker	32[a]	Bite
5	Upper lid raiser	19	Tongue show	33[a*]	Blow
6	Cheek raiser and lid compressor	20	Lip stretcher	34[a]	Puff
7	Lid tightener	21	Neck tightener	35[a]	Suck
9	Nose wrinkler	22	Lip funneler	36[a]	Bulge
10	Upper lip raiser	23	Lip tightener	37[a]	Lip wipe
11	Nasolabial furrow	24	Lip presser	38	Nostril dilator
12	Lip corner puller	25	Lips part	39	Nostril compressor
13	Sharp lip puller	26	Jaw drop	43	Eye closure
14	Dimpler	27	Mouth stretch	45	Blink
15	Lip corner depressor	28	Lips suck	46	Wink
		29[a]	Jaw thrust		

[a]Action Descriptor.
Source: From Reference 31.

as FACS. The new version of FACS defines 32 AUs, which are anatomically related to the contraction or relaxation of one or more muscles. The authors also define a number of action descriptors. These action descriptors differ from the AUs in that they neither have a specific muscular basis nor have specific distinguishing behaviors. Table 8.1 summarizes the facial behaviors associated with the AUs or action descriptors [31].

Moreover, according to Ekman and Friesen [7], there are six basic emotions that are universal to human beings, namely anger (AN), disgust (DI), fear (FE), happiness (HA), sadness (SA), and surprise (SU). These basic emotions can be coded via the emotional facial action coding system (EMFACS) [36]. Figure 8.2 shows the prototypical facial expression images corresponding to the six basic emotions.

Of late, the recognition of these six basic emotions from human facial expressions has become a very active research topic in HCI. Various methods have been proposed

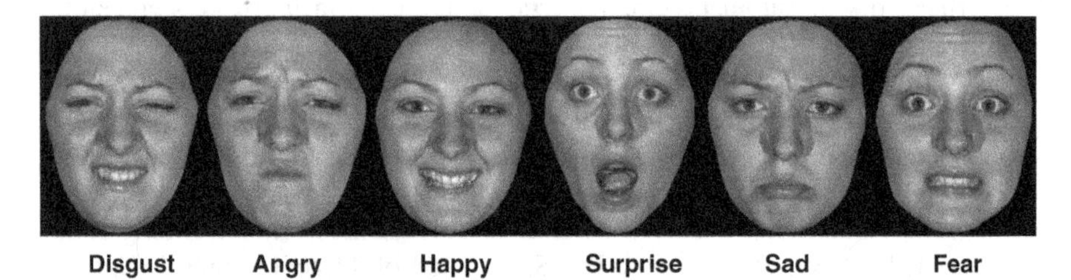

FIGURE 8.2 Six prototypical facial expressions defined by Ekman and Friesen. Extracted with permission from BU-3DFE database [37].

to this end during the past decades. One may refer to References 12, 13, 18, 19, 38, and 39 for a survey.

Although automatic emotion-specified expression recognition has been extensively explored in the past several decades, most of the previous approaches are restricted to the frontal- or nearly frontal-view facial images. For many real-world applications, the frontal view face images are not available, as the subjects move freely when emotion is captured. Thus restricting emotion analysis using only frontal images is not realistic. On the other hand, the previous experimental findings on both psychology [40] and computer vision [41] in face recognition show that the non-frontal facial images can achieve better recognition performance than the frontal ones. Motivated by these new findings, more and more researchers have turned their research interest from the frontal facial emotion recognition topic to the non-frontal one. In contrast to the frontal or near-frontal facial emotion recognition research, however, recognizing non-frontal facial emotions is more challenging. The major challenges are mainly because of the large variations in pose views, which result in the difficulties of facial feature extraction due to the face occlusion and a more complex classifier design. Moreover, both the pose estimation [42] and face alignment [43] to non-frontal facial images in real scenarios are challenging work. In addition, the lack of publicly available 3D or multiview facial expression database may also impede research development. Nevertheless, there are still a few papers presented recently to address the non-frontal facial emotion recognition issue [44–56]. The goal of this chapter is to survey the recent advances in non-frontal view facial emotion recognition research and discuss the future development in this field.

The remainder of this chapter is organized as follows: In Section 8.2, we give a brief introduction of the facial emotion recognition methods. In Section 8.3, we briefly review the facial expression databases that are commonly used for the non-frontal facial emotion recognition researches. Section 8.4 presents the recent advances of non-frontal facial emotion recognition methods. Discussion and conclusions are presented in Section 8.5.

8.2 A BRIEF REVIEW OF AUTOMATIC EMOTIONAL EXPRESSION RECOGNITION

8.2.1 Framework of Automatic Facial Emotion Recognition System

Automatic emotion recognition from facial images has been a very hot research topic in computer vision and pattern recognition over the past three decades due to its technical challenge and the potential wide-ranging applications across many fields. For literature surveys, see References 13, 18, 19, 38, and 57. The major goal of facial emotion recognition is to classify each facial image into one of the six basic emotion-specified expression categories (i.e., anger, disgust, fear, happiness, sadness, and surprise). Figure 8.3 illustrates the general framework of an automatic facial emotion recognition system, where the facial emotion recognition system consists of three parts.

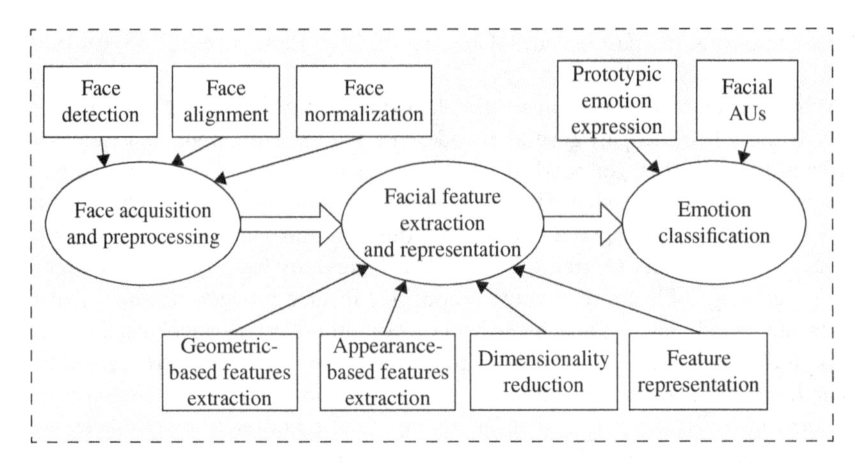

FIGURE 8.3 The general framework of an automatic facial emotion recognition system.

The first part is to deal with facial image acquisition, which involves face detection, face alignment, and face normalization. The second part is to tackle facial feature extraction and feature representation, including the facial feature extraction, the dimensionality reduction of feature vectors, and the feature representation. The feature extraction operations could be used to extract either geometric features or appearance features or both of them. The dimensionality reduction operation is optional depending on the number of features and the number of training samples. The last part focuses on the facial emotion classification. Generally, two different ways could be used for this purpose. One is to choose a classifier to classify each test facial image into one of the basic emotion-specified expressions, and the other one is to perform the AU classification [19] and then make the emotion-specified expression decision based on the EMFACS rule [36].

Although a lot of facial emotion recognition methods have been proposed over the last 40 years, they can be classified into several categories in terms of different classification rules. For instance, we can classify them into static image-based approaches and dynamic image-based approaches in terms of whether the temporal information is used or not. We can also classify them into 2D facial image-based approaches and 3D facial image-based approaches according to whether 2D or 3D images are used. Furthermore, based on the features we extract during emotion recognition, we can also classify emotion recognition methods into geometric feature-based approaches and appearance feature-based approaches [19].

Geometric features and appearance features provide complementary discriminative information for emotion recognition. Geometric features are usually extracted from the areas around the mouth, eyes, brows, and nose, and used to represent the shapes or locations of the facial components. Appearance features aim to capture emotion information from skin textures, which can be extracted from either the whole face image or some specific face regions. Comparisons between geometric features and appearance features in facial emotion recognition have been presented in the literature [58,59]. In the following subsections, we will overview some of the frequently used facial feature extraction methods in facial emotion recognition.

8.2.2 Extraction of Geometric Features

Perhaps the straight and effective way of extracting geometric features for emotion recognition is to utilize FACS. Although FACS is mainly used for coding facial actions, it also provides a very effective way for us to locate the landmark points that convey the emotion information. In FACS, the AUs are divided into upper- and lower-face categories because the AUs of the lower face have little influence on the ones in the upper face and vice versa. In Reference 60, Tian *et al.* used the geometric features for AU recognition, where a set of landmark points located on the positions of AU regions were used for geometric feature representation. Another commonly used way of extracting geometric features is to locate the landmark points according to the definition of the facial animation parameters (FAPs) presented in ISO MPEG-4 standard [61]. The FAPs contain 68 parameters associated with 10 groups. These parameters may be used to describe the visemes and expressions, or the displacement of the feature points as shown in Figure 8.4. The use of FAPs as the geometric features

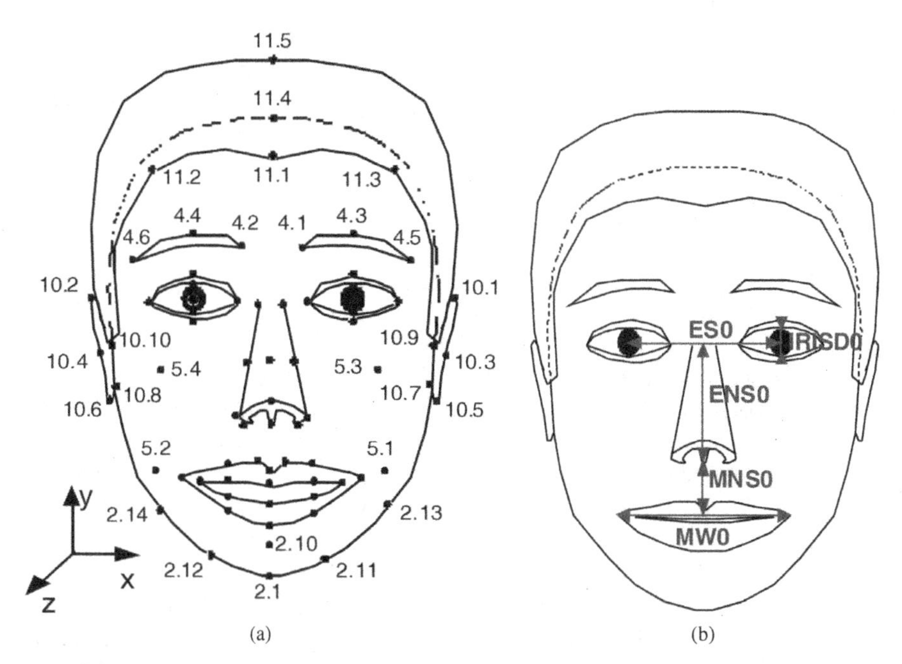

FIGURE 8.4 (a) Few feature points of FAPs in ISO MPEG-4 standard; (b) A face model in its neutral state on which the FAPUs are defined by the fractions of distances between the marked key facial feature points. The five FAPUs are as follows: IRISD0, Iris diameter; ES0, Eye separation; ENS0, Eye–nose separation; MNS0, Mouth–nose separation; MW0, Mouth width. From Reference 61.

for facial emotion recognition has appeared in many previous papers, especially for 3D facial expression recognition [52,53,59,62,63].

Besides the aforementioned two popular methods for extracting the geometric feature points, there are other methods for extracting the geometric features. For instance, Zhang *et al.* [58, 64, 65] used the coordinates of 34 fiducial points manually located on the face as geometric features. Chang *et al.* [66] used the coordinates of 58 feature points located by active shape model as geometric features. For more details about the extraction of geometric features, please also refer to References 18, 19, and 38.

8.2.3 Extraction of Appearance Features

To obtain appearance features, one of the simplest ways is to use the gray-scale values of the face region. However, the extracted features may be less discriminative for emotion recognition and sensitive to the variations of illuminations, face image scales, and head pose views. More efficient appearance features have been presented in the literature during the past several years. Among them, the Gabor wavelet features [67], the scale invariant feature transform (SIFT) features [68, 69], and the local binary pattern (LBP) features [70] are the most popular ones used in facial emotion recognition.

1. Gabor features have been proven to be very effective for both face and facial emotion recognition; they are extracted by convolving the face image with a set of Gabor filters, and the coefficients of the convolution are used as the Gabor features. A typical Gabor kernel is defined as the product of a Gaussian envelope with a harmonic oscillation function [58, 71], that is,

$$\Psi_{u,v}(\mathbf{z}) = \frac{\|\mathbf{k}_{u,v}\|^2}{\sigma^2} \exp\left(-\frac{\|\mathbf{k}_{u,v}\|^2\|\mathbf{z}\|^2}{2\sigma^2}\right)\left[\exp(i\mathbf{k}_{u,v}\cdot\mathbf{z}) - \exp\left(-\frac{\sigma^2}{2}\right)\right]$$

   where $\mathbf{z} = (x, y)$ denotes the spatial location, $\mathbf{k}_{u,v} = \mathbf{k}_v \exp(i\phi_u)$, $\mathbf{k}_v = \pi/2^v$, $\phi_u = \pi u/N$ $(u = 0, 1, \ldots, N-1)$, and $\sigma = \pi$, where u and v define the orientation and scale of the Gabor kernels, respectively. Figure 8.5 shows an example of a set of Gabor filters under five scales $(v = 1, 2, \ldots, 5)$ and six orientations $(u = 0, 1, \ldots, 5)$, where Figure 8.5a depicts the real parts of the Gabor filters, Figure 8.5b depicts the imaginary parts of the Gabor filters, and Figure 8.5c depicts the magnitude parts of the Gabor filters.
2. The LBP feature was originally proposed by Ojala *et al.* [70] as a gray scale invariant texture descriptor and showed excellent performance in the classification of various kinds of textures. LBP aims at capturing the local image information and has evidence of being used as a feature descriptor [72, 73] for facial emotion description. Suppose that g_c is the gray value of a center pixel located at $c = (x, y)$ and g_s $(s = 0, \ldots, S-1)$ represents the gray values of S

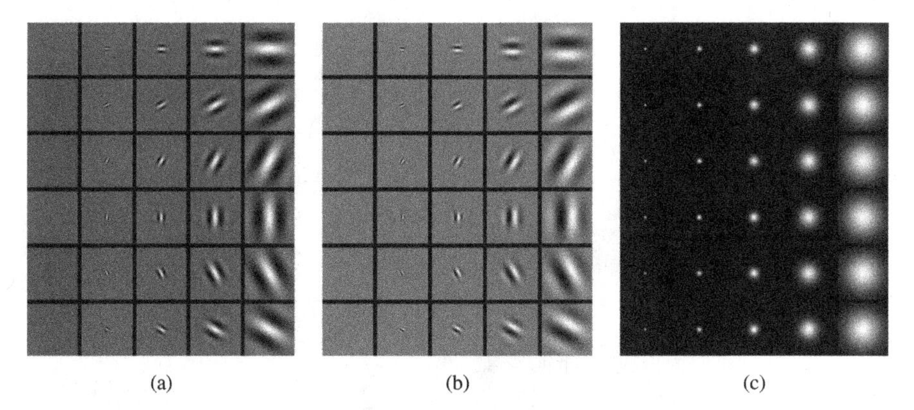

(a) (b) (c)

FIGURE 8.5 A set of Gabor filters with five scales and six orientations: (a) the real parts; (b) the imaginary parts; (c) the magnitude parts.

equally spaced pixels on a circle of radius R around this center pixel c, then the LBP operator at the center position c is defined as

$$\mathrm{LBP}_{S,R} = \sum_{s=0}^{S-1} f(g_s - g_c)2^s, \text{ where } f(g_s - g_c) = \begin{cases} 1, & g_s - g_c \geq 0 \\ 0, & g_s - g_c < 0 \end{cases}$$

In Reference 73, Zhao *et al.* proposed the LBP-TOP descriptor as an extension of the LBP operator to handle the spatiotemporal feature extraction. They showed excellent performance of the LBP-TOP features in the facial expression recognition of image sequences [73]. An example of using the LBP-TOP operator on a set of face patches of image sequences is illustrated in Figure 8.6.

3. The SIFT feature extraction method was originally proposed by David Lowe for the detection of local features and matching in images [68,69]. The SIFT features are invariant to image translation, scaling, rotation, and partially invariant to illumination changes. They are also robust to local geometric distortion. The SIFT feature extraction procedures usually can be divided into four steps [69]: scale-space extrema detection, keypoint localization, orientation assignment, and keypoint descriptor. However, some of the steps can be omitted in practical applications. Figure 8.7 illustrates an example of extracting the SIFT features from a set of landmark points of the face.

8.3 DATABASES FOR NON-FRONTAL FACIAL EMOTION RECOGNITION

The research of non-frontal facial emotion recognition largely depends on the availability of multiview facial image database. The lack of the databases will impede

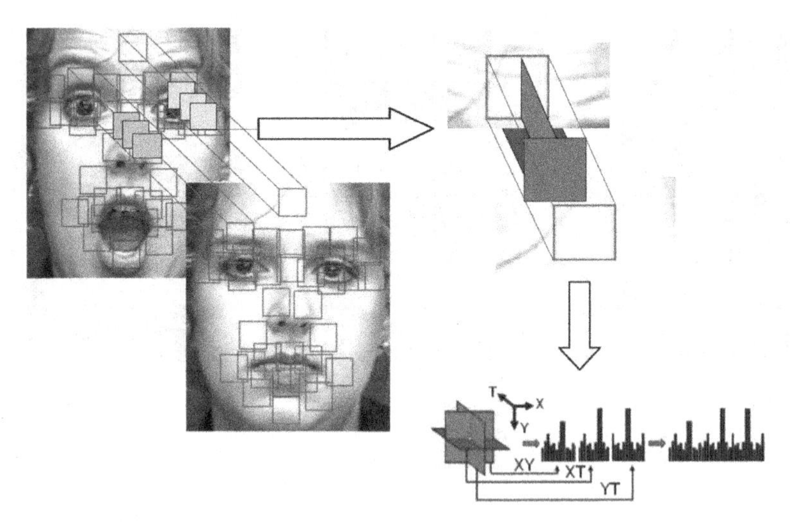

FIGURE 8.6 Example of using LBP-TOP operator on the face image sequences, where the LBP operator is applied in local image patches. From Reference 74.

the development in this field. In recent years, several multiview facial expression databases have been built and are now available for researchers, in which the BU-3DFE database [37] and its dynamic version BU-4DFE [75], the Multi-PIE database [76], and the Bosphorus 3D database [77] are three popular ones for dealing with the non-frontal-view facial emotion recognition. In what follows, we briefly introduce the three databases. For more details, please refer to References 37, 75, and 76.

8.3.1 BU-3DFE Database

The BU-3DFE database was recently developed by Yin *et.al.* at Binghamton University, which is designed to sample 3D facial behaviors with different prototypical

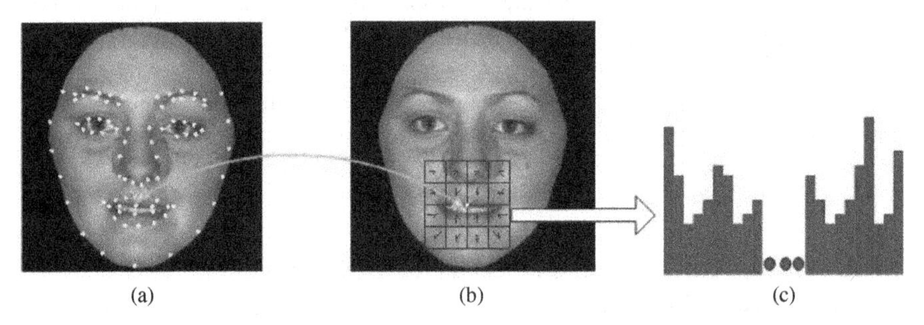

FIGURE 8.7 Illustration of extracting the SIFT features from a set of landmark points of the face. (a) The facial image was annotated by a set of landmark points; (b) construction of a SIFT descriptor for a specific landmark points; (c) construction of a histogram for describing the associated landmark points.

emotional states at various levels of intensities. It consists of a total of 100 subjects, among which 56 are females and 44 are males. The subjects are well distributed across different ethnic or racial ancestries, including White, Black, East Asian, Middle-East Asian, Hispanic Latino, and others. The database is recorded with a 3D face imaging system using random light pattern projection in the speckle projection flash environment [37]. During the record, each subject is asked to perform the six basic facial expressions plus the neutral facial expression, and each facial expression has four different levels of intensities (low, middle, high, and the highest), except for the neutral facial expression which has only one intensity level. As a result, each subject contains 25 3D facial expression models, resulting in a total of 2500 3D facial expression models in the database. In the database, associated with each 3D facial expression model, there is a raw 3D face mesh model, a cropped 3D face mesh model, a pair of texture images with two-angle views (about +45° and –45° away from the face frontal normal), a frontal-view texture image, a set of 83 facial feature points, and a facial pose vector, which give a complete 3D description of a face under a specific facial expression. A more detailed description about the BU-3DFE database can be found in Reference 37. Figure 8.8 shows some examples of the 3D facial expression models in the database, where four different levels of intensities are shown in each of the six basic facial expressions.

Additionally, as the number of 2D facial images provided by the BU-3DFE database is relatively small (only a pair of texture images with two-angle views and a frontal-view texture image), it may not be enough to handle the research of the non-frontal facial emotion recognition problem. An effective way of solving this problem is to project each 3D face model onto more different projection views using computer software, for example, OpenGL [78]. Figure 8.9 illustrates an example of projecting a 3D face model onto five projection views (0°, 30°, 45°, 60°, 90°) using the OpenGL software, resulting in five yaw views of 2D facial images.

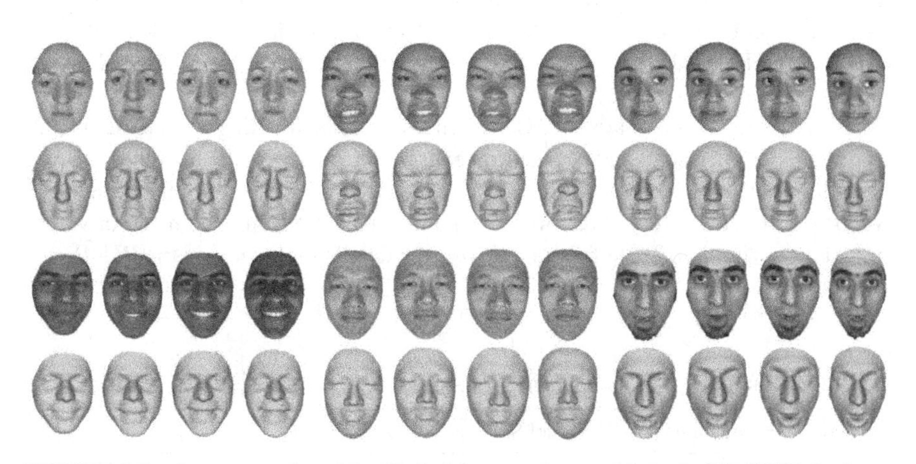

FIGURE 8.8 Some examples of the 3D facial expression models in the BU-3DFE database, where each subject shows four different levels of intensities of each of the six basic facial expressions. Extracted with permission from BU-3DFE database.

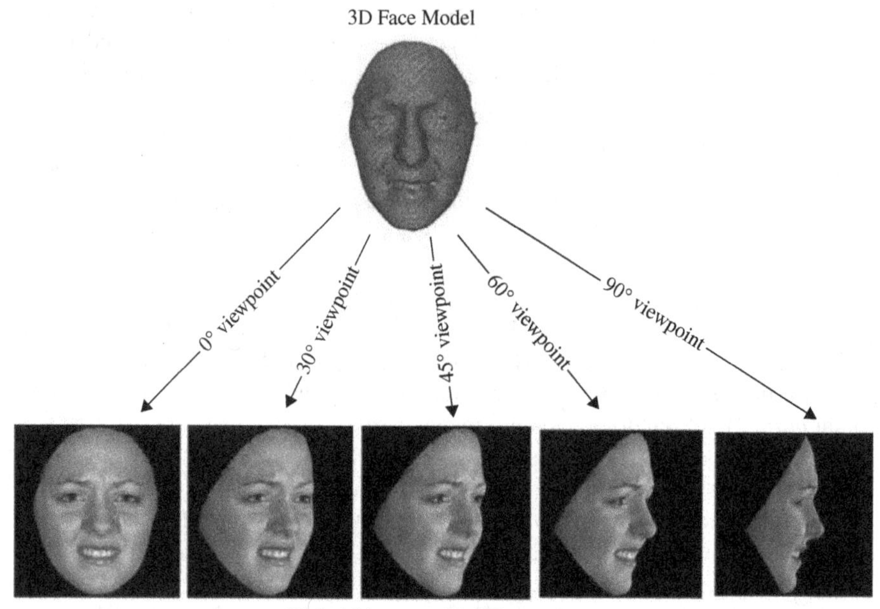

FIGURE 8.9 Example of projecting a 3D face model onto five projection views, resulting in five views of 2D facial images. Taken with permission from BU-3DFE database.

8.3.2 BU-4DFE Database

The BU-3DFE database is a static facial expression database and hence is difficult to use for dynamic facial emotion recognition. To overcome this limitation, Yin *et al.* at Binghamton University built a dynamic 3D facial expression (also widely referred to as BU-4DFE) database. The BU-4DFE database consists of 606 3D facial video sequences from 101 subjects.

For each subject, there are six 3D sequences associated with six universal facial expressions. In each sequence, the neutral expression appears in the beginning and at the end. For each expression sequence, three videos are captured, among which are two gray scale videos (upper and lower) for range data creation and one color video for texture generation. Table 8.2 summarizes the details of the data in the BU-4DFE database.

TABLE 8.2 Summary of the BU-4DFE database

Subject#	Expression#	3D Model Sequences#	2D Texture Videos#	3D Models#
101	6	606	606	60,600

Source: From Reference 75.

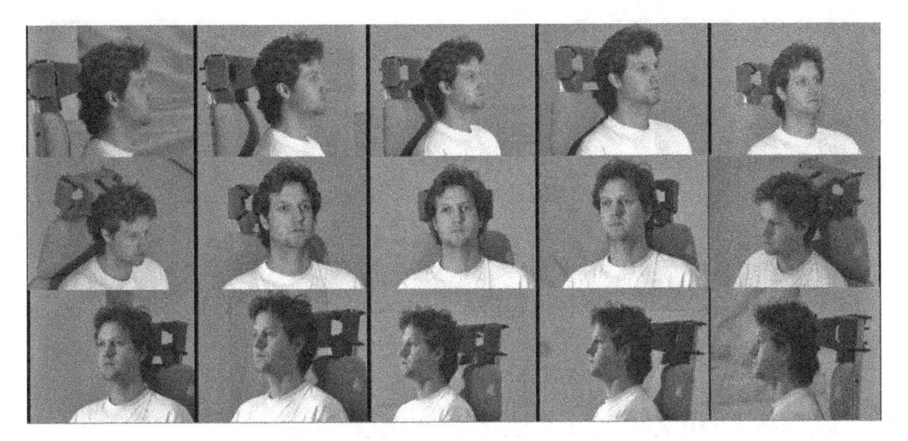

FIGURE 8.10 Examples of the facial images with 15 pose views. Taken with permission from Multi-PIE database [24].

8.3.3 CMU Multi-PIE Database

The Multi-PIE facial expression database, which was recently built by the research group at Carnegie Mellon University (CMU), consists of 755,370 facial images from 337 different subjects, in which 60% are European American, 35% are Asian, 3% are African American, and 2% others. The average age of the subjects in the database is 27.9 years. Among the 337 subjects, 235 are males and 102 are females. The facial images are captured under 15 view points and 19 illumination conditions in up to four recording sessions over a span of 6 months. Examples of the facial images captured under the 15 view points are shown in Figure 8.10. For more details about the database, please refer to Reference 76.

8.3.4 Bosphorus 3D Database

The aforementioned BU-3DFE, BU-4DFE, and Multi-PIE databases are mainly used for prototypic emotion-specific, expression-based emotion recognition. To explore the AU recognition or the AU-based emotion recognition research, a new 3D facial expression database called the Bosphorus database has been built recently. The Bosphorus database consists of 105 subjects (60 males and 45 females) covering various poses, expressions, and occlusion conditions. In the database, there are 4652 face scans that are captured from a structured-light 3D digitizer device. Each scan has been manually labeled with 24 facial landmark points. The database also contains up to 35 expressions per subject, FACS scoring (includes intensity and asymmetry codes for each AU), systematic head poses (13 yaw and pitch rotations), and varieties of face occlusions (beard and moustache, hair, hands, eyeglasses). Figure 8.11 shows some facial examples of the six basic prototypic expressions in the Bosphorus database.

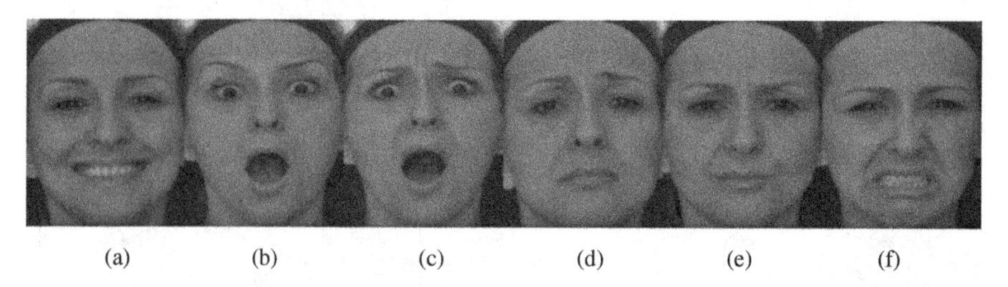

(a) (b) (c) (d) (e) (f)

FIGURE 8.11 Examples of six basic prototypic expressions: (a) happiness, (b) surprise, (c) fear, (d) sadness, (e) anger, (f) disgust. Taken with permission from Bosphorus database [77].

8.4 RECENT ADVANCES OF EMOTION RECOGNITION FROM NON-FRONTAL FACIAL IMAGES

In the aforementioned sections, we briefly introduced the facial expression recognition methods and pointed out that most of the previous facial emotion recognition research work focuses on frontal or nearly frontal facial images. Moreover, we also pointed out the necessity for and advantage of conducting non-frontal facial emotion recognition. For these reasons, more and more researchers have been attracted to the non-frontal facial emotion recognition field in recent years [11,44,45,49,50,52,53,55,56,62,79–81].

In this section, we will survey the recent research advances in this field. Specifically, we categorize the facial emotion methods as 3D facial model-based approaches and non-frontal 2D facial image-based approaches. Noting that Fang *et al.* [82] have presented a survey on 3D facial expression recognition, we focus the most parts of our survey on facial emotion recognition from non-frontal 2D facial images.

8.4.1 Emotion Recognition from 3D Facial Models

The method of using of 3D face model for face recognition has been proven to achieve better recognition results than 2D facial images due to its robustness to the poses, scales, and lighting variations [77]. The success of using 3D face model in face recognition has attracted many researchers to investigate the facial emotion recognition based on 3D face models. Over the last 5 years, many facial emotion recognition methods based on 3D face models have appeared in the literature [52, 53,62,63,79,81,83–86]. Most of the methods used the geometric features extracted from the 3D models with a neutral 3D model as reference to describe the facial emotion changes. To perform the emotion classification, we may choose a classifier, for example, the support vector machine or Adaboost [52,53], and then classify each 3D face model into one of the basic emotion categories based on the geometric features. For the recent advances in 3D facial emotion recognitions, the readers may refer to the recent survey presented by Fang *et al.* [82].

TABLE 8.3 Recent advances in multiview 2D facial image-based emotion recognition methods

			Database	
Authors	Features	Classifier	BU-3DFE	Multi-PIE
Zheng *et al.* [55]	Sparse SIFT	k-NN	√	
Hu *et al.* [44]	Geometric Features	SVM	√	
Hu *et al.* [45]	HoG , LBP and Sparse SIFT	SVM	√	
Moor and Bowden [47]	LBP	SVM		√
Moor and Bowden [87]	LBP	SVM	√	√
Zheng *et al.* [56]	Dense SIFT	Linear Classifier	√	
Tang *et al.* [54]	Dense SIFT	HMM+k-NN	√	
Rudovic *et al.* [49]	Geometric Features	SVM		√
Rudovic *et al.* [50]	Geometric Features	SVM	√	√

8.4.2 Emotion Recognition from Non-frontal 2D Facial Images

Although 3D face model-based methods are able to deal with pose variation problem in emotion recognition, the need for the 3D face model would limit its practical applications in the cases when the 3D face models are not available. However, as an alternative approach, the use of multiview 2D facial images to deal with the non-frontal facial emotion recognition problem has received an increasing interest from researchers. Table 8.3 lists some of the recent advances in non-frontal facial emotion recognition methods.

8.4.2.1 Geometric Feature-Based Approaches The first attempt to handle the non-frontal facial emotion recognition problem was conducted by Hu *et al.* [44], in which the authors aimed to explore the issue of whether non-frontal facial images could achieve better recognition performance than the frontal or nearly frontal ones in expression recognition. The authors investigated the facial emotion recognition problem on a set of images with five yaw views, that is, 0°, 30°, 45°, 60°, and 90°, generated from the BU-3DFE database. Figure 8.12 shows some examples of the 2D facial images rendered from the 3D face models of BU-3DFE database. In this method, the authors extracted a set of normalized geometric features by calculating the 2D displacement of the facial feature points around the eyes, eyebrows, and mouth between the six basic emotional expressions and the neutral expression of the same person in the same pose views. Experiments were carried out using the subject-independent cross-validation strategy (80 of the 100 subjects were used as the training data and the remaining 20 subjects were used as the test data), where several classifiers (Linear Bayes normal classifier, quadratic Bayes normal classifier, Parzen classifier, support vector machine, and *k*-nearest neighbor classifier) were used. The experiments were repeated five times and the lowest overall error rate was 33.5% when the SVM classifier was used. Moreover, the experimental results show

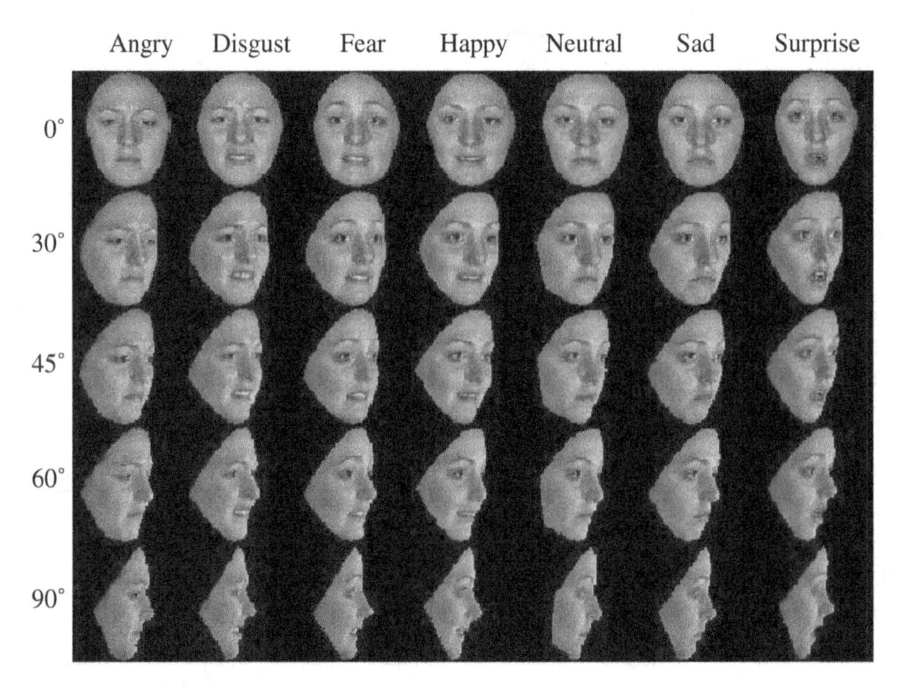

FIGURE 8.12 Some facial images of different expressions obtained by projecting 3D face models in BU-3DFE database onto different views. Taken with permission from BU-3DFE database [55].

that, when the facial view is fixed at 45°, the lowest average error rate (= 28.5%) is achieved. This suggests that the frontal view is not the best projection direction for recognizing facial expressions.

In References 49 and 50, Rudovic *et al.* proposed another method of using the geometric features to investigate pose-invariant facial expression recognition using regression approaches. A three-step procedure was proposed for this purpose. Specifically, in Reference 49, the authors used four regression approaches, namely, linear regression (LR), support vector regression (SVR), relevance vector regression (RVR), and Gaussian process regression (GPR), to investigate the regression-based facial expression recognition approaches. The pose estimation was first carried out and was assigned to one of the previously known poses (±15°, ±30°, ±45° pan angles and ±15°, ±30° tilt angles for the BU-3DFE database, 0°, −15°, −30°, −45° pan angles for the Multi-PIE database). Then, 39 landmark points were manually located on the 2D facial image. The pose normalization based on regression methods was applied to map the geometric features of the estimated pose view onto the frontal one. Finally, the emotion recognition experiments on the Multi-PIE database using 10-fold person-independent cross-validation were conducted, where the SVM classifier trained on the frontal view was used for emotion classification. The experimental results on

Multi-PIE database show that better recognition performance could be achieved when the RVR method with frontal classifier (FC) is used. In Reference 50, the same authors proposed to use the coupled GPR (CGPR) method for the same purpose. In this method, the multiclass LDA method was first applied to reduce the dimensionality of the input feature vectors consisting of 39 facial landmark points. Then, the pose estimation and normalization operations were performed based on the given input geometric features, in which the CGPR method was utilized for pose normalization. The SVM classifier was also used for classification in this method. Experiments on both BU-3DFE and Multi-PIE databases were carried out to evaluate the recognition performance of the method using fivefold cross-validation strategy, where seven classes of emotion categories (the six basic emotion categories plus the neutral emotion) were involved in the BU-3DFE database and only four facial emotion categories, that is, disgust, happy, surprise, and neutral, were involved in the experiments. For the BU-3DFE database, the average test error rates on the frontal pose, non-frontal pose, and the unknown pose facial images were 25.2%, 25.7%, and 27.2%, respectively. For the Multi-PIE database, the average test error rates on the frontal pose and the non-frontal pose facial images were 4.8% and 11.7%, respectively.

8.4.2.2 Sparse SIFT Feature-Based Approaches Hu *et al.* [45] used SIFT features, LBP features, and the histogram of Gaussian (HOG) to investigate the multiview (0°, 30°, 45°, 60°, and 90°) facial expression recognition. Three subspace analysis approaches, namely, locality preserving projection (LPP), principal component analysis (PCA), and linear discriminant analysis (LDA), were used to reduce the dimensionality of the feature vectors. Then, two recognition schemes, namely, the pose + emotion cascade scheme and the view × emotion composite scheme, were proposed to carry out the recognition of emotion expression and pose views for an input image. In the experiments, the SVM classifier was adopted for the final emotion classification. The experimental results suggest that the SIFT + LPP method achieves the best performance with 26.94% average error rate. Especially, the lowest error rate (26.13%) and the highest error rate (28.55%) were achieved at the 30° and the 90° pose views, respectively.

Zheng *et al.* [55] also used the SIFT feature to explore the non-frontal-view facial emotion recognition. Similar to Reference 45, 2D facial images with five pan pose views (0°, 30°, 45°, 60°, and 90°) were generated from the 3D face model of the BU-3DFE database, which resulted in a set of 12,000 facial images. However, different from Reference 45, the SIFT features in Reference 55 were extracted directly from 83 landmark points provided by the BU-3DFE database, without the key points detection step. Hence, each facial image corresponded to a 10,624-dimensional feature vector. To reduce the dimensionality of the feature vectors, the authors proposed a new discriminant analysis method based on a minimal Bayes error estimation. They conducted experiments with 10 independent trials to evaluate the recognition performance of their method. In each trial, they randomly partitioned the 100 subjects into two groups. One group contained face images of 80 subjects of all six expressions, four intensities, and five views and comprised a training set of 9600 face images. The other group contained face images of 20 subjects and comprised a test set of 2400

TABLE 8.4 Average error rate (%) of different emotions versus different views

Emotions	0°	30°	45°	60°	90°	Ave.
Happiness	12.50	12.50	12.87	11.88	16.87	13.32
Sadness	27.87	23.00	22.00	21.13	24.25	23.65
Angust	19.75	19.12	18.75	19.63	27.12	20.88
Fear	38.25	38.38	38.12	35.75	39.75	38.05
Surprise	8.88	9.12	9.00	8.87	10.50	+9.27
Disgust	25.00	23.25	23.00	25.50	26.75	24.70
Average	22.04	20.90	20.62	20.46	24.21	21.65

Source: From Reference 55.

face images. Each trial of an experiment involved a different random partition of the 100 subjects into a training set and a test set, and the results of the 10 independent trials were averaged. The overall error rate achieved is as low as 21.65%. Table 8.4 summarizes the average error rate of different emotions versus different views, from which one can see that the best results are achieved when the facial views are between 30° and 60°.

8.4.2.3 Dense SIFT Feature-Based Approaches It is notable that the landmark points used for SIFT feature extraction in References 44, 45, and 55 are sparse and should be labeled manually. Moreover, the effectiveness of those methods are only evaluated in five yaw views. In practical facial emotion recognition applications, however, one may encounter much more complex pose views. Instead of using the dense SIFT features, Zheng *et al.* [56] and Tang and Huang [54] proposed to extract dense SIFT features to overcome the drawbacks of the spare SIFT-based facial emotion recognition methods. In both papers, the authors first generated 35 facial images corresponding to 35 pose views from each 3D face model of the BU-3DFE database, where the 35 pose views were seven yaw angles (−45°, −30°, −15°, 0,° +15°, +30°, and +45°) and five pitch angles (−30°, −15°, 0°, +15°, and +30°). As a consequence, they obtained $100 \times 6 \times 5 \times 7 = 21{,}000$ facial images in total for the experiments. Figure 8.13 shows some examples of the generated multiview face images.

The concept of the dense SIFT feature description method is illustrated in Figure 8.14, which needs neither the face image alignment nor the facial landmark points localization.

To extract the dense SIFT features, they placed a dense grid on a facial image, and extracted a 128-dimensional SIFT descriptor at each node of the grid with a fixed scale and orientation. The SIFT descriptor was formed from the histogram of intensity gradients within a neighborhood window of the grid node. Hence, it is a distinctive feature for representing the texture variation in this local region. Among the extracted SIFT feature vectors, those SIFT features with extremely small magnitudes corresponding to the SIFT features extracted from the black background in the images as well as those extracted from the low contrast portion on the face were abandoned. Figure 8.15 shows an example of extracting the dense SIFT feature from a face image, where Figure 8.15a shows the positions of SIFT features to be

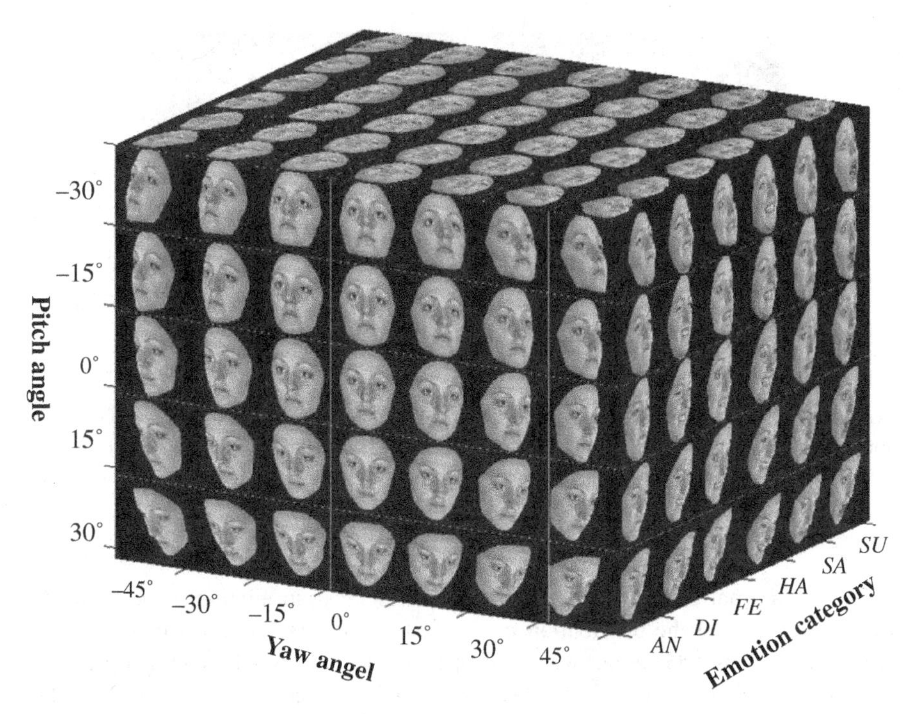

FIGURE 8.13 Some facial images rendered from the BU-3DFE database, covering the facial images of six basic emotions, seven yaw angles, and five pitch angles. From Reference 56.

extracted and Figure 8.15b shows the final positions after abandoning the positions with black background or low contrast portion on the face.

In favor of the non-frontal facial emotion recognition, Zheng *et al.* [56] proposed to use the regional covariance matrix (RCM) [88] method to represent the facial image. The entries of each RCM are then concatenated into a high dimensional feature vector for emotion recognition, in which PCA is further used to reduce the dimensionality of the feature vectors. Moreover, to further extract the discriminative features for emotion recognition, the authors developed a new discriminant analysis method

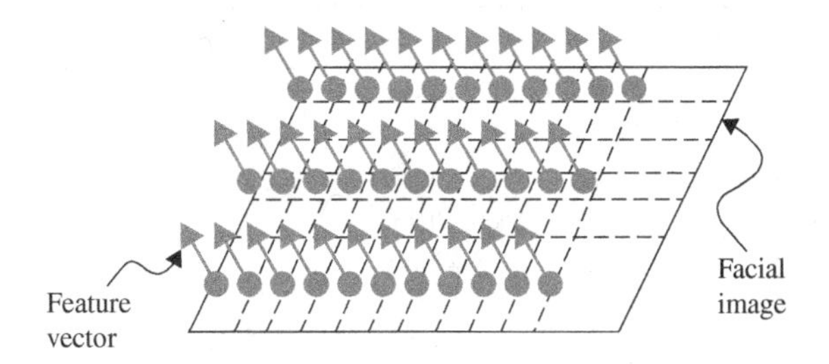

FIGURE 8.14 The whole facial region is divided into some patches, and each patch produces a SIFT feature vector. From Reference 56.

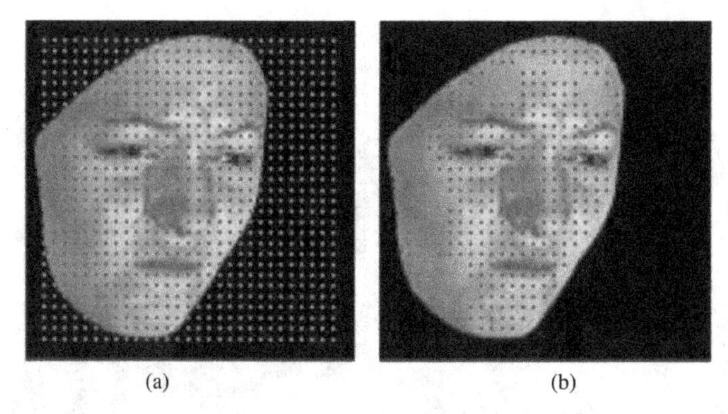

FIGURE 8.15 (a) The SIFT feature vectors are extracted from the grid nodes, shown as red dots; (b) abandon the SIFT feature vectors with extremely small magnitudes, which correspond to the SIFT feature vectors extracted from the black background in the images as well as those extracted from the low-contrast portion on the face. From Reference 54.

based on minimal Bayes error estimation, where Gaussian mixture model (GMM) was used to model the distribution of the feature vectors. Mixtures of Gaussians rather than a single Gaussian was used in their method. Finally, they adopted the k-nearest neighbor classifier for the facial emotion classification. Experiments on the BU-3DFE database were carried out using fivefold cross-validation strategy. They used different numbers of Gaussian mixture components, namely, 16, 32, 64, 128, and 256, in the experiments and found that the best recognition result was achieved when 256 Gaussian mixture components were used. Table 8.5 shows the average error rates (%) of different emotions versus different views, where 256 Gaussian mixture components are used.

Tang *et al.* [54] proposed to use the supervector representation method to represent each facial image, where the supervector was obtained based on the ergodic hidden Markov model (EHMM). The diagram of the method is illustrated in Figure 8.16.

TABLE 8.5 Average error rates (%) of different emotions versus different views using our method, where 256 Gaussian mixture components are used

Views	−30°	−15°	−0°	+15	+30°	Ave.
−45°	39.67	35.67	31.00	33.00	43.00	36.47
−30°	30.67	28.33	27.67	28.50	38.50	30.73
−15°	28.33	29.17	25.83	25.83	33.17	28.47
0°	30.83	27.83	25.17	25.67	31.83	28.27
+15°	32.33	29.33	26.33	28.50	32.00	29.70
+30°	32.33	29.33	29.33	32.67	35.50	31.83
+45°	40.17	33.50	31.33	35.83	43.67	36.90
Ave.	33.48	30.45	28.10	30.00	36.81	31.77

Source: From Reference 56.

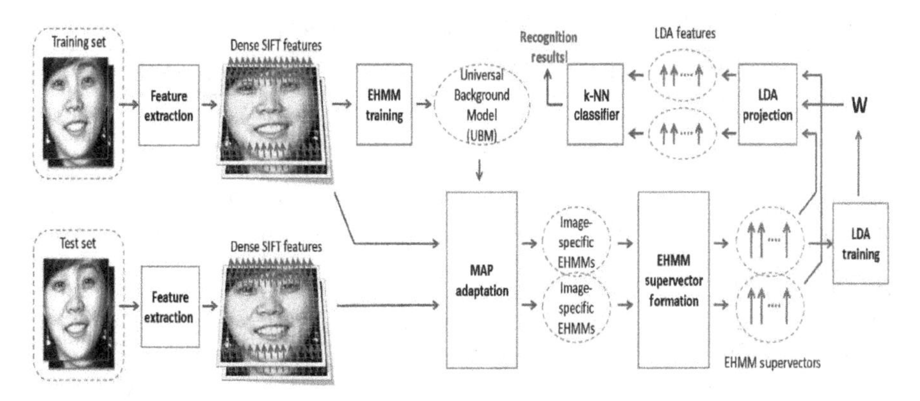

FIGURE 8.16 The schematic overview of the proposed approach. From Reference 54.

Specifically, the dense SIFT feature vectors were first extracted from each facial image. Then, a universal background model (UBM) was developed by training an EHMM over all facial images of the training set. The UBM was then maximum a posteriori (MAP) adapted to each facial image in the training and test sets to produce the image-specific EHMMs. Based on these EHMMs, the supervector representation of the facial images can be obtained by means of an upper bound approximation of the Kullback–Leibler divergence (KLD) rate between two EHMMs. Finally, facial emotion recognition was performed based on k-NN classifier in a linear discriminant subspace of the EHMM supervectors (defined by a linear projection W). In the experiment, the fivefold cross-validation strategy was adopted to evaluate the performance of the method. At each fold, four out of the five groups were chosen as the training data to train the UBM (3 states, 64-component Gaussian mixture emission densities) and the remaining group was used as the test data. The UBM was then MAP adapted to every facial image in both the training and test sets to produce the image-specific EHMMs, which were used to construct the corresponding EHMM supervectors. Then, linear discriminant analysis (LDA) was used to transform the the original EHMM supervectors from both the training and test sets onto a low dimension feature space, in which the k-NN classifier was employed to perform the facial emotion recognition. The overall error rate, as well as the average error rates for different views and the average error rates for different facial expressions, are reported in Table 8.6. The rightmost column represents the average recognition error rates for different views (a total of 35 views), the bottom row represents the average recognition error rates for different facial expressions (a total of six universal facial expressions), and the bottom-right corner cell represents the average overall recognition error rate.

8.4.2.4 LBP Feature-Based Approaches The use of LBP features for non-frontal facial emotion recognition was first investigated by Hu *et al.* [45]. A more extensive study on the use of LBP to investigate the non-frontal facial emotion recognition was recently proposed by Moore and Bowden [47, 87]. In References 47 and 87, the

TABLE 8.6 Recognition error rates with respect to different pose views

%	AN	DI	FE	HA	SA	SU	Ave.
−45, −30	21.0	34.0	48.0	10.0	45.0	5.0	27.2
−45, −15	23.0	21.0	41.0	7.0	47.0	10.0	24.8
−45, +0	34.0	19.0	35.0	17.0	32.0	7.0	24.0
−45, +15	34.0	12.0	37.0	19.0	40.0	4.0	24.3
−45, +30	40.0	18.0	30.0	14.0	41.0	9.0	25.3
−30, −30	25.0	20.0	33.0	14.0	31.0	8.0	21.8
−30, −15	30.0	24.0	54.0	22.0	37.0	8.0	29.2
−30, +0	26.0	20.0	53.0	11.0	38.0	7.0	25.8
−30, +15	20.0	19.0	41.0	7.0	43.0	6.0	22.7
−30, +30	37.0	12.0	35.0	7.0	37.0	5.0	22.2
−15, −30	37.0	13.0	43.0	11.0	44.0	4.0	25.3
−15, −15	28.0	19.0	35.0	13.0	42.0	6.0	23.8
−15, +0	25.0	18.0	37.0	14.0	38.0	8.0	23.3
−15, +15	33.0	22.0	52.0	19.0	41.0	5.0	28.7
−15, +30	28.0	29.0	38.0	12.0	33.0	6.0	24.3
+0, −30	26.0	21.0	42.0	11.0	35.0	6.0	23.5
+0, −15	36.0	17.0	46.0	11.0	32.0	6.0	24.7
+0, +0	39.0	18.0	38.0	18.0	34.0	3.0	25.0
+0, +15	27.0	15.0	32.0	10.0	42.0	5.0	21.8
+0, +30	30.0	18.0	38.0	10.0	40.0	6.0	23.7
+15, −30	35.0	23.0	45.0	12.0	33.0	6.0	25.7
+15, −15	37.0	22.0	49.0	13.0	26.0	6.0	25.5
+15, +0	37.0	22.0	49.0	11.0	30.0	9.0	26.3
+15, +15	43.0	14.0	41.0	13.0	22.0	7.0	23.3
+15, +30	37.0	13.0	31.0	8.0	41.0	5.0	22.5
+30, −30	37.0	12.0	30.0	10.0	39.0	6.0	22.3
+30, −15	29.0	16.0	39.0	12.0	35.0	4.0	22.5
+30, +0	27.0	24.0	46.0	14.0	42.0	6.0	26.5
+30, +15	32.0	31.0	43.0	15.0	28.0	6.0	25.8
+30, +30	47.0	17.0	45.0	7.0	25.0	6.0	24.5
+45, −30	45.0	14.0	28.0	6.0	32.0	5.0	21.7
+45, −15	36.0	23.0	34.0	18.0	41.0	9.0	26.8
+45, +0	42.0	21.0	38.0	11.0	41.0	9.0	27.0
+45, +15	28.0	13.0	44.0	10.0	42.0	5.0	23.7
+45, +30	24.0	33.0	44.0	15.0	46.0	9.0	28.5
Ave.	32.4	19.6	40.4	12.3	37.0	6.3	24.7

Source: From Reference 54.

authors used not only the traditional LBP features but also the variations of the LBP features for the facial expression description. Specifically, they explored six kinds of LBP variations as shown in Table 8.7 to evaluate the performance of LBP features in the facial emotion recognition. The BU-3DFE database and the Multi-PIE database were, respectively, used to evaluate the performance of the the various LBP features, where multiclass SVM was used as the classifier.

TABLE 8.7 Variations of LBP features used in Reference 87

Features	Description of the LBP features
LBP^{riu2}	Uniform rotation invariant LBP
LBP^{ri}	Rotation invariant LBP
LBP^{gm}	Uniform LBP obtained from gradient magnitude image
LBP^{u2}	Standard local uniform binary patterns with a neighborhood of 8 pixels and a radius of 1 pixel
LBP^{ms}	Multi-scale LBP where radius varies from 1 to 8 pixels
LGBP	LBP features are extracted from Gabor images, where 40 different Gabor images are composed from applying Gabor kernels at different scales and orientations

In the experiments, the authors used a set of non-frontal facial images to evaluate their method. For the BU-3DFE database, they generated five different poses (0°, 30°, 45°, 60°, and 90°) via OpenGL software from each 3D face model, whereas for the Multi-PIE database, the facial images corresponding to seven pose views (0°, 15°, 30°, 45°, 60°, 75°, and 90°) were chosen. The experiments were carried out using the 10-fold cross-validation strategy. For the BU-3DFE database, the 100 subjects were randomly partitioned into a subset of 90 subjects as the training data and a subset of 10 subjects as the test data. The best recognition accuracy (=77.67%) on the BU-3DFE database was achieved when the LGBP features were used in the experiments. For the CMU Multi-PIE database, the training and test sets were divided by 80% and 20%, respectively, and the recognition accuracies were 80.17% when the LGBP features were used.

8.5 DISCUSSIONS AND CONCLUSIONS

In this chapter, we have surveyed the recent advances of the non-frontal facial emotion recognition methods, including the current commonly used 3D facial database and multiview 2D facial image database. Although significant advances have been achieved in the non-frontal facial emotion recognition fields, as we can see from this chapter, there is still a long way to go before the non-frontal facial emotion recognition methods become applicable in real scenarios. Below are some of the major issues that should be considered for future research:

1. Most of the existing non-frontal facial emotion recognition methods are developed based on posed facial images. However, in real applications, what we face are spontaneous emotions. Consequently, more attention should be paid to spontaneous emotion recognition problems.
2. The existing non-frontal facial emotion recognition methods have focused on recognition of prototypic expression of the basic emotions. Few of them consider the emotion intensities. More efforts should be paid to this field.

3. Most of the methods dealing with the emotion recognition problem only utilize the discriminative information extracted from a single view of a 2D facial image. It is notable that the fusion of multiple pose views would result in improvement in the recognition accuracy of the emotion.
4. Most methods dealing with emotion recognition are based on non-frontal prototypic expression images. The investigation of the emotion recognition problem based on AU classification is less explored.

ACKNOWLEDGMENTS

This work was partly supported by Natural Science Foundations of China under grants 61073137 and 60872160, and partly by the Jiangsu Natural Science Foundations under Grant BK2010243.

REFERENCES

[1] C. Darwin, *The Expression of the Emotions in Man and Animals*, John Murray, London, 1872.

[2] N. Sebe, I. Cohen, and T. S. Huang, "Multimodal emotion recognition," in *Handbook of Pattern Recognition and Computer Vision*, 2005, pp. 981–256.

[3] W. James, *The Principles of Psychology*, vol. I, Henry Holt, New York, 1890.

[4] C. G. Lange and W. James, *The Emotions*, vol. 1, Williams & Wilkins, 1922.

[5] W. B. Cannon, "The James–Lang theory of emotion: a critical examination and an alternative theory," *Am. J. Psychol.*, 39: 106–124, 1927.

[6] H. Gunes and M. Pantic, "Automatic, dimensional and continuous emotion recognition," *Int. J. Synth. Emot.*, 1(1): 68–99, 2010.

[7] P. Ekman and W. V. Friesen, "Constants across cultures in the face and emotion," *J. Personal Soc. Psychol.*, 17(2): 124–129, 1971.

[8] J. A. Russell, "A circumplex model of affect," *J. Personal Soc. Psychol.*, 39(6): 1161–1178, 1980.

[9] H. Schlosberg, "Three dimensions of emotion," *Psychol. Rev.*, 61(2): 81–88, 1954.

[10] R. W. Picard, *Affective Computing*, MIT Press, Cambridge, MA, 1997.

[11] S. Taheri, P. Turaga, and R. Chellappa, "Towards view-invariant expression analysis using analytic shape manifolds," in IEEE International Conference on Automatic Face & Gesture Recognition and Workshops, 2011.

[12] M. Pantic and L. J. M. Rothkrantz, "Toward an affect-sensitive multimodal human–computer interaction," *Proceedings of IEEE*, 91(9): 1370–1390, 2003.

[13] Z. Zeng, M. Pantic, G. I. Roisman, and T. S. Huang, "A survey of affect recognition methods: audio, visual, and spontaneous expressions," *IEEE Transactions on Pattern Analysis and Machine Intelligence*, 31(1): 39–58, 2009.

[14] M. E. Ayadi, M. S. Kamel, and F. Karray, "Survey on speech emotion recognition: features, classification schemes, and databases," *Pattern Recognit.*, 44: 572–587, 2011.

[15] I. Luengo, E. Navas, and I. Hernaez, "Feature analysis and evaluation for automatic emotion identification in speech," *IEEE Transactions on Multimedia*, 12(6): 490–501, 2010.

[16] Z. Zeng, J. Tu, P. Pianfetti, M. Liu, T. Zhang, Z. Zhang, T. S. Huang, and S. Levinson, "Audio-visual affect recognition through multi-stream fused HMM for HCI," in International Conference on Computer Vision and Pattern Recognition, 2005, pp. 967–972.

[17] A. Camurri, I. Lagerlöf, and G. Volpe, "Recognizing emotion from dance movement: comparison of spectator recognition and automated techniques," *Int. J. Hum. Comput. Stud.*, 59(1–2): 213–225, 2003.

[18] B. Fasel and J. Luettin, "Automatic facial expression analysis: a survey," *Pattern Recognit.*, 36: 259–275, 2008.

[19] Y. L. Tian, T. Kanade, and J. F. Cohn, "Facial expression analysis," in *Handbook of Facial Recognition* (eds S. Z. Li and A. K. Jain), Springer-Verlag, 2004.

[20] K. H. Kim, S. W. Bang, and S. R. Kim, "Emotion recognition system using short-term monitoring of physiological signals," *Med. Biol. Eng. Comput.*, 42: 419–427, 2004.

[21] R. W. Picard, E. Vyzas, and J. Healey, "Toward machine emotional intelligence: analysis of affective physiological state," *IEEE Transactions on Pattern Analysis and Machine Intelligence*, 23(10): 1175–1191, 2001.

[22] D. O. Bos, "EEG-based emotion recognition using hybrid filtering and higher order crossings," 2009.

[23] M. Li, Q. Chai, T. Kaixiang, A. Wahab, and H. Abut, "EEG emotion recognition," in *In-Vehicle Corpus and Signal Processing for Driver Behavior* (eds K. Takeda *et al.*), 2009.

[24] P. C. Petrantonakis and L. J. Hadjileontiadis, "EEG-based emotion recognition using hybrid filtering and higher order crossings," in The 3rd International Conference on Affective Computing and Intelligent Interaction and Workshops, 2009, pp. 1–6.

[25] K. R. Scherer and H. G. Wallbott, "Analysis of nonverbal behavior," in *Handbook of Discourse Analysis*, vol. 2, 1985, pp. 199–230.

[26] K. R. Scherer and H. Ellgring, "Multimodal expression of emotion: affect programs or component appraisal patterns?" *Emotion*, 7(1): 2007.

[27] C. Busso, Z. Deng, S. Yildirim, M. Bulut, C. M. Lee, A. Kazemzadeh, S. Lee, U. Neumann, and S. Narayanan, "Analysis of emotion recognition using facial expressions speech and multimodal information," in International Conference on Multimodal Interfaces, 2004, pp. 205–211.

[28] H. Gunes and M. Piccardi, "Bi-modal emotion recognition from expressive face and body gestures," *J. Netw. Comput. Appl.*, 30(4): 1334–1345, 2007.

[29] H. Gunes and M. Piccardi, "Automatic temporal segment detection and affect recognition from face and body display," *IEEE Transactions on Systems, Man, and Cybernetics, Part B*, 39(1): 64–84, 2009.

[30] C. Shan, S. Gong, and P. W. McOwan, "Beyond facial expressions: learning human emotion from body gestures," in Proceedings of the British Machine Vision Conference, 2007.

[31] P. Ekman, W. V. Friesen, and J. C. Hager, "Facial Action Coding System," The Manual on CD ROM, Research Nexus Division of Network Information Research Corporation, Salt Lake City, UT 2002.

[32] A. Mehrabian, "Communication without words," *Psychol. Today*, 2(4): 53–56, 1968.

[33] M. Suwa, N. Sugie, and K. Fujimora, "A preliminary note on pattern recognition of human emotional expression," in Proceedings of the Fourth International Joint Conference on Pattern Recognition, 1978, pp. 408–410.

[34] P. Ekman and W. V. Friesen, "Pictures of facial affect," Human Interaction Laboratory, University of California Medical Center, San Francisco, CA, 1976.

[35] P. Ekman, "Facial expression and emotion," *Am. Psychol.*, 48(4): 384–392, 1993.

[36] W. V. Friesen and P. Ekman, "EMFACS-7: Emotional Facial Action Coding System," unpublished manuscript, University of California at San Francisco, 1983.

[37] L. Yin, X. Wei, Y. Sun, J. Wang, and M. J. Rosato, "A 3D facial expression database for facial behavior research," in Proceedings of Seventh International Conference on Automatic Face and Gesture Recognition, 2006, pp. 211–216.

[38] M. Pantic and L. J. M. Rothkrantz, "Automatic analysis of facial expressions: the state of the art," *IEEE Transactions on Pattern Analysis and Machine Intelligence*, 22(12): 1424–1445, 2000.

[39] A. Vinciarelli, M. Pantic, and H. Bourlard, "Social signal processing: survey of an emerging domain," *Image Vis. Comput.*, 27: 1743–1759, 2009.

[40] V. Bruce, T. Valentine, and A. Baddeley, "The basis of the 3/4 view advantage in face recognition," *Appl. Cogn. Psychol.*, 1: 109–120, 1987.

[41] X. Liu, J. Rittscher, and T. Chen, "Optimal pose for face recognition," in 2006 IEEE Computer Society Conference on Computer Vision and Pattern Recognition, vol. 2, 2006, pp. 1439–1446.

[42] E. Murphy-Chutorian and M. M. Trivedi, "Head pose estimation in computer vision: a survey," *IEEE Transactions on Pattern Analysis and Machine Intelligence*, 31(4): 607–626, 2009.

[43] T. F. Cootes, C. J. Taylor, D. H. Cooper, and J. Graham, "Active shape models–their training and application," *Comput. Vis. Image Underst.*, 38–59, 1995.

[44] Y. Hu, Z. Zeng, L. Yin, X. Wei, J. Tu, and T. S. Huang, "A study of non-frontal-view facial expressions recognition," in Proceedings of ICPR, 2008, pp. 1–4.

[45] Y. Hu, Z. Zeng, L. Yin, X. Wei, X. Zhou, and T. S. Huang, "Multi-view facial expression recognition," in International Conference on Automatic Face and Gesture Recognition, 2008.

[46] S. Kumano, K. Otsuka, J. Yamato, E. Maeda, and Y. Sato, "Pose-invariant facial expression recognition using variable-intensity templates," *Int. J. Comput. Vis.*, 83(2): 178–194, 2009.

[47] S. Moore and R. Bowden, "The effect of pose on facial expression recognition," in British Machine Vision Conference, 2009.

[48] M. Pantic and I. Patras, "Dynamics of facial expression: recognition of facial actions and their temporal segments from face profile image sequences," *IEEE Transactions on Systems, Man and Cybernetics, Part B*, 36(2): 433–449, 2006.

[49] O. Rudovic, I. Patras, and M. Pantic, "Regression-based multi-view facial expression recognition," in Proceedings of International Conference on Pattern Recognition (ICPR), 2010.

[50] O. Rudovic, I. Patras, and M. Pantic, "Coupled Gaussian process regression for pose-invariant facial expression recognition," *European Conference on Computer Vision (ECCV)*, 350–363, 2010.

[51] A. O. Sajama, "Supervised dimensionality reduction using mixture models," in International Conference on Machine Learning, 2005, pp. 768–775.

[52] H. Tang and T. S. Huang, "3D facial expression recognition based on automatically selected features," in CVPR 2008 Workshop on 3D Face Processing (CVPR-3DFP'08), Anchorage, Alaska, June 2008.

[53] H. Tang and T. S. Huang, "3D facial expression recognition based on properties of line segments connecting facial feature points," in IEEE International Conference on Automatic Face and Gesture Recognition (FG'08), Amsterdam, The Netherlands, September 2008.

[54] H. Tang, M. Hasegawa-Johnson, and T. S. Huang, "Non-frontal view facial expression recognition based on ergodic hidden Markov model supervectors," in IEEE Conference on Multimedia & Expo (ICME), 2010, pp. 1202–1207.

[55] W. Zheng, H. Tang, Z. Lin, and T. S. Huang, "A novel approach to expression recognition from non-frontal face images," in Proceedings of IEEE ICCV, 2009, pp. 1901–1908.

[56] W. Zheng, H. Tang, Z. Lin, and T. S. Huang, "Emotion recognition from arbitrary view face images," in Proceedings of European Conference on Computer Vision (ECCV2010), 2010, pp. 490–503.

[57] A. Samal and P. Iyengar, "Automatic recognition and analysis of human faces and facial expressions: a survey," *Pattern Recognit.*, 25(1): 65–77, 1992.

[58] Z. Zhang, M. Lyons, M. Schuster, and S. Akamatsu, "Comparison between geometry-based and Gabor wavelets-based facial expression recognition using multi-layer perception," in Proceedings of Third IEEE International Conference on Automatic Face and Gesture Recognition, Nara, Japan, 1998, pp. 454–459.

[59] L. Zhang, D. W. Tjondronegoro, and V. Chandran, "Evaluation of texture and geometry for dimensional facial expression recognition," in 2011 International Conference on Digital Image Computing: Techniques and Applications (DICTA2011), 2011.

[60] Y.-L. Tian, T. Kanade, and J. F. Cohn, "Recognizing action units for facial expression analysis," *IEEE Transactions on Pattern Analysis and Machine Intelligence*, 23(2): 1–19, 2001.

[61] ISO/IEC 14496-2, 1999.

[62] H. Soyel and H. Demirel, "Facial expression recognition using 3D facial feature distances," in Image Analysis and Recognition, Volume 4633 of Lecture Notes in Computer Science, 2007, pp. 831–838.

[63] H. Soyel and H. Demirel, "3D facial expression recognition with geometrically localized facial features," in 23rd International Symposium on Computer and Information Sciences (ISCIS2008), 2008, pp. 1–4.

[64] W. Zheng, X. Zhou, C. Zou, and L. Zhao, "Facial expression recognition using kernel canonical correlation analysis (KCCA)," *IEEE Transactions on Neural Networks*, 17(1): 233–238, 2006.

[65] G. Guo and C. R. Dyer, "Learning from examples in the small sample case: face expression recognition," *IEEE Transactions on Systems, Man, and Cybernetics, Part B: Cybernetics*, 35(3): 477–488, 2005.

[66] Y. Chang, C. Hu, and M. Turk, "Manifold of facial expression," in IEEE International Workshop on Analysis and Modeling of Faces and Gestures, 2003, pp. 28–35.

[67] J. G. Daugman, "Complete discrete 2-D Gabor transforms by neural networks for image analysis and compression," *IEEE Transactions on Acoustics, Speech and Signal Processing*, 36(7): 1169–1179, 1988.

[68] G. G. Lowe, "Object recognition from local scale-invariant features," in Proceedings of the Seventh IEEE International Conference on Computer Vision, 1999, pp. 1150–1157.

[69] D. G. Lowe, "Distinctive image features from scale-invariant keypoints," *Int. J. Comput. Vis.*, 60(2): 91–110, 2004.

[70] T. Ojala, M. Pietikäinen, and T. Mäenpää, "Multiresolution gray-scale and rotation invariant texture classification with local binary patterns," *IEEE Transactions on Pattern Analysis and Machine Intelligence*, 971–987, 2002.

[71] M. Lyons, J. Budynek, and S. Akamatsu, "Automatic classification of single facial images," *IEEE Transactions on Pattern Analysis and Machine Intelligence*, 21(12): 1357–1362, 1999.

[72] X. Feng, "Facial expression recognition based on local binary patterns and coarse-to-fine classification," 2004.

[73] G. Zhao and M. Pietikäinen, "Dynamic texture recognition using local binary pattern with an application to facial expressions," *IEEE Transactions on Pattern Analysis and Machine Intelligence*, 29(6): 915–928, 2007.

[74] X. Huang, G. Zhao, M. Pietikäinen, and W. Zheng, "Dynamic facial expression recognition using boosted component-based spatiotemporal features and multi-classifier fusion," in Advanced Concepts for Intelligent Vision Systems (ACIVS). Lecture Notes in Computer Science, vol. 6475, 2010, pp. 312–322.

[75] L. Yin, X. Chen, Y. Sun, T. Worm, and M. Reale, "A high-resolution 3D dynamic facial expression database," in The Eighth International Conference on Automatic Face and Gesture Recognition (FGR08), 2008.

[76] R. Gross, I. Matthews, J. Cohn, T. Kanade, and S. Baker, "Multi-PIE," *Image Vis. Comput.*, 28: 807–813, 2010.

[77] A. Savran, N. Alyüz, H. Dibeklioğlu, O. Çeliktutan, B. Gökberk, B. Sankur, and L. Akarun, "Bosphorus database for 3D face analysis," in Biometrics and Identity Management, 2008, pp. 47–56.

[78] OpenGL: The Industry's Foundation for High Performance Graphics, http://www.opengl.org/. Last accessed on Sept. 18, 2014.

[79] S. Berretti, A. D. Bimbo, P. Pala, B. B. Amor, and M. Daoudi, "A set of selected SIFT features for 3D facial expression recognition," in 2010 International Conference on Pattern Recognition, 2010, pp. 4125–4128.

[80] I. Mpiperis, S. Malassiotis, and M. G. Strintzis, "Bilinear models for 3D face and facial expression recognition," *IEEE Transactions on Informatation Forensics and Security*, 3(3): 498–511, 2008.

[81] J. Wang, L. Yin, X. Wei, and Y. Sun, "3D facial expression recognition based on primitive surface feature distribution," in Proceedings of the 2006 IEEE Conference on Computer Vision and Pattern Recognition (CVPR2006), 2006.

[82] T. Fang, X. Zhao, O. Ocegueda, S. K. Shah, and I. A. Kakadiaris, "3D facial expression recognition: a perspective on promises and challenges," in 2011 IEEE International

Conference on Automatic Face & Gesture Recognition and Workshops (FG 2011), 2011, pp. 603–610.

[83] A. M. Bronstein, M. M. Bronstein, and R. Kimmel, "Three-dimensional face recognition," *Int. J. Comput. Vis.*, 64(1): 5–30, 2005.

[84] I. A. Kakadiaris, G. Passalis, G. Toderici, N. Murtuza, N. Karampatziakis, and T. Theoharis, "3D face recognition in the presence of facial expressions: an annotated deformable model approach," *Transactions on Pattern Analysis and Machine Intelligence (PAMI)*, 13(12), 2007.

[85] Y. Sun and L. Yin, "Facial expression recognition based on 3D dynamic range model sequences," in Proceedings of the 10th European Conference on Computer Vision: Part II, 2008, pp. 58–71.

[86] J. Sung and D. Kim, "Pose-robust facial expression recognition using view-based 2D+3D AAM," *IEEE Transactions on Systems, Man, and Cybernetics, Part A: Systems and Humans*, 38: 852–866, 2008.

[87] S. Moore and R. Bowden, "Local binary patterns for multi-view facial expression recognition," *Comput. Vis. Image Underst.*, 115: 541–558, 2011.

[88] O. Tuzel, F. Porikli, and P. Meer, "Region covariance: a fast descriptor for detection and classification," in Proceedings of the ECCV, 2006, pp. 589–600.

[89] G. Castellano, L. Kessous, and G. Caridakis, "Multimodal emotion recognition from expressive faces, body gestures and speech," in Proceedings of the Second International Conference on Affective Computing and Intelligent Interaction, 2007.

[90] R. Cowie, E. Douglas-Cowie, N. Tsapatsoulis, G. Votsis, S. Kollias, W. Fellenz, and J. G. Taylor, "Emotion recognition in human–computer interaction," *IEEE Signal Processing Magazine*, 18(1): 32–80, 2001.

AUTHOR BIOGRAPHIES

Wenming Zheng is a professor of the Research Center for Learning Science (RCLS) at Southeast University, Nanjing, China. He received the B.S. degree in Computer Science from Fuzhou University, Fuzhou, Fujian, China, in 1997, the M.S. degree in Computer Science from Huaqiao University, Quanzhou, Fujian, China, in 2001, and the Ph.D. degree in Signal and Information Processing from Southeast University in 2004. He served as a visiting scholar from 2008 to 2009 at the Image Formation and Processing Group, Beckman Institute, University of Illinois, Urbana-Champaign. His research interests include pattern recognition, computer vision, neural computation, and machine learning, from theory to application. Dr. Zheng received the Microsoft Professorship Award in 2006, the Natural Science Award (2nd prize) from Ministry of Education, China, in 2008, and the Science and Technology Progress Award (2nd prize) of Jiangsu province in 2009.

Hao Tang is a researcher at Hewlett-Packard Laboratories, Palo Alto, CA. He received the Ph.D. degree in Electrical Engineering from the University of Illinois at Urbana-Champaign in 2010, the M.S. degree in Electrical Engineering from Rutgers University in 2005, the M.E. and B.E. degrees, both in Electrical Engineering, from the University of Science and Technology of China, Hefei, China, in 2003 and 1998, respectively. His broad research interests include statistical pattern recognition, machine learning, computer vision, speech and multimedia signal processing. From 1998 to 2003, he was on the faculty in the Department of Electronic Engineering and Information Science at the University of Science and Technology of China, where he also served as a senior software engineer, principal researcher, research director, and research manager at Anhui USTC iFlyTEK Co., Ltd. He received the Science and Technology Progress Award (1st Prize) from the Government of Anhui Province, China, in 2000, and the National Science and Technology Progress Award (2nd Prize) from the State Council of China in 2003. He was a co-recipient of Best Student Paper Award at the International Conference on Pattern Recognition in 2008, a recipient of IBM T.J. Watson Emerging Leader in Multimedia Award in 2009, and a corecipient of the Excellence Award of the 12th China Patent Awards in 2010.

Thomas Huang received the B.S. degree in Electrical Engineering from National Taiwan University, Taipei, Taiwan, China, and the M.S. and Sc.D. degrees in Electrical Engineering from the Massachusetts Institute of Technology (MIT), Cambridge. He was on the faculty of the Department of Electrical Engineering at MIT from 1963 to 1973, and on the faculty of the School of Electrical Engineering and Director of its Laboratory for Information and Signal Processing at Purdue University from 1973 to 1980. In 1980, he joined the University of Illinois at Urbana-Champaign, where he is now the William L. Everitt Distinguished Professor of Electrical and Computer Engineering, a research professor at the Coordinated Science Laboratory, the Head of the Image Formation and Processing Group at the Beckman Institute for Advanced Science and Technology, and cochair of the Institute's major research theme: human–computer intelligent interaction. Dr. Huang's professional interests lie in the broad area of information technology, especially the transmission and processing of multidimensional signals. He has published 20 books and over 500 papers in network theory, digital filtering, image processing, and computer vision. He is a member of the National Academy of Engineering, a foreign member of the Chinese Academies of Engineering and Science, and a fellow of the International Association of Pattern Recognition, IEEE, and the Optical Society of America, and has received a Guggenheim Fellowship, an A.von Humboldt Foundation Senior US

Scientist Award, and a Fellowship from the Japan Association for the Promotion of Science. He received the IEEE Signal Processing Society's Technical Achievement Award in 1987 and the Society Award in 1991. He was awarded the IEEE Third Millennium Medal in 2000. Also in 2000, he received the Honda Lifetime Achievement Award for "contributions to motion analysis." In 2001, he received the IEEE Jack S. Kilby Medal. In 2002, he received the King-Sun Fu Prize, International Association of Pattern Recognition, and the Pan Wen-Yuan Outstanding Research Award. He is a founding editor of the International Journal of Computer Vision, Graphics, and Image Processing and an editor of the Springer Series in Information Sciences, published by Springer.

9

MAXIMUM A POSTERIORI BASED FUSION METHOD FOR SPEECH EMOTION RECOGNITION

LING CEN

*Human Language Technology Department, Institute for Infocomm Research (I^2R), A*STAR, Singapore*

ZHU LIANG YU

College of Automation Science and Engineering, South China University of Technology, Guangzhou, China

WEE SER

Centre for Signal Processing, Nanyang Technological University, Singapore

With the increasing demand for spoken language interfaces in human–computer interactions, automatic recognition of emotional states from human speech has become increasingly important. In our previous work, we have proposed a hybrid scheme that combines the Probabilistic Neural Network (PNN) and the Universal Background Model-Gaussian Mixture Model (UBM-GMM) for speech emotion recognition. In this chapter, we extend the hybrid scheme into a more general Maximum A Posteriori (MAP) based fusion method. The proposed fusion method is capable of effectively combining the strengths of several (two or more) classification methods for recognition of emotional states in speech signals. In order to illustrate the effectiveness of the proposed method, PNN, UBM-GMM, and k-Nearest Neighbor (k-NN) are used as base classifiers in the numerical experiments presented in this chapter. Numerical results show that higher accuracies can be achieved compared with those obtained using the base classifiers alone in the classification of 15 emotional states for the samples extracted from the LDC database. It is also shown from the

Emotion Recognition: A Pattern Analysis Approach, First Edition. Edited by Amit Konar and Aruna Chakraborty.
© 2015 John Wiley & Sons, Inc. Published 2015 by John Wiley & Sons, Inc.

experiment results that the proposed MAP-based method can work well with a small training dataset.

9.1 INTRODUCTION

The technology for human–computer interaction has advanced rapidly over the recent decade. To realize communications between humans and computers, the computers need to recognize, understand, and response to human speech. The recognition of human speech is one of the most important components in human–computer conservational communication. In human speech, besides linguistic information, the paralinguistic information that refers to the implicit messages such as emotional states of the speaker is conveyed too. The human emotions are the mental and physiological states, which reflect not only the mood but also the human personality. They are associated with the feelings, thoughts, and behaviors of humans. Although the emotional states conveyed in speech do not alter the linguistic content, they carry important information on the speaker such as his/her desire, intent, and response to the outside world [1,2]. The phonologically identical utterances, for example, can be expressed in different emotions and convey various meanings. To realize natural and smooth interaction between humans and computers, as such, it is important that a computer is able to recognize not only the linguistic contents but also the emotional states conveyed in human speech so that personalized responses can be delivered accordingly in human–computer interaction applications.

Automatic detection of human emotions has found a lot of applications in many areas, such as learning, security, medicine, and entertainment [3,4]. With the wide use of Internet and rapid development of e-learning applications, automatic tutoring systems are expectedly able to interactively adjust the content of teaching and speed of delivery according to the users' response that are expressed implicitly in their speech. In security of audio surveillance, detection of abnormal emotions, for example, stress, fear and nervousness, helps identifying liars or suspicious human subjects. Similarly, automatically recognizing emotions from patients could also be helpful in clinical studies as well as in psychosis monitoring and diagnosis assistance. As for commercial applications, emotion classification has been applied in, for example, customer services and call centers, where machines could adjust their responses automatically by distinguishing satisfaction or dissatisfaction conveyed in speech of customers. Interactive games have become most popular in the family of computer games. Emotion recognition could help bring in more fun features and improve the adaptive interaction between computers and human players. The demand for natural human-like machines has motivated a lot of research attentions and interests in automatic detection of emotional states of human subjects.

Automatic recognition of speech emotion is, however, a very challenging task because there lacks of a precise definition and model of emotions [5]. Figure 9.1 shows a simple structure of the speech emotion recognition system [6], which can be broken down into three modules, namely preprocessing, feature extraction, and

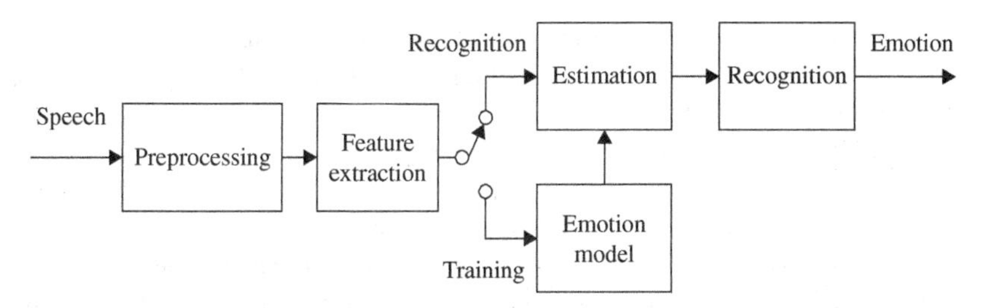

FIGURE 9.1 Typical structure of a speech emotion recognition system.

classification. The major works of earlier research on speech emotion recognition are on feature extraction, selection, and the use of a single classifier for emotion recognition [7].

The behavior of the acoustic features in various emotions has been investigated in many studies [1, 8–12]. The pitch, the intensity, and the rate of speech as well as the duration ratio of voiced and unvoiced regions have been commonly used in speech emotion recognition [1, 5, 13–15]. Besides these features directly extracted from speech signals, the features derived from mathematical transformation of basic acoustic features, for example, Mel-frequency cepstral coefficients (MFCCs) [16, 17] and linear prediction-based cepstral coefficients (LPCCs) [16], have also been used for detecting emotions from speech. For those features that are calculated on a frame basis, instead of using the raw features, the feature statistics, such as mean, median, range, standard deviation, maximum, minimum, and linear regression coefficients, are usually used [1,5]. Even though many studies have tried to investigate the acoustic features that are correlated to emotions, however, there is still no conclusive evidence to show which set of features can provide the best recognition accuracy [18].

Many classification methods have been applied in speech emotion recognition in the literature. In Reference 13, k-NN classifier and majority voting of subspace specialists were employed for the recognition of sadness, anger, happiness, and fear and the best accuracy achieved was 79.5%. Neural network (NN) was used to recognize eight emotions including happiness, teasing, fear, sadness, disgust, anger, surprise, and neutral in Reference 19 with an accuracy of 50%. In Reference 5, the acoustic information was combined with the lexical and discourse information for speech emotion recognition. The Linear Discriminant Classifier with Gaussian class-conditional probability and k-NN were used for recognizing negative and non-negative emotions from speech signals and a maximum accuracy of 89.06% was achieved. In Reference 20, a real-time emotion recognizer using NN was developed for call center applications, and achieved a classification accuracy of 77% in recognizing agitation and calm emotions using eight features chosen by a feature selection algorithm. In Reference 21, the Support Vector Machines (SVMs) were used to detect anger, happiness, sadness, and neutral with an average accuracy of 73%. In Reference 18, a rough set theory was used for feature selection and the SVMs were employed for classifying six different emotions, that is, happiness, anger, surprise, sadness, fear, and neutral state. An accuracy of 74.75% was achieved with 13 selected features. In Reference 22, the existence of a universal psychobiological mechanism of emotions in speech was explored by studying the recognition of fear, joy, sadness, anger, and disgust in nine languages, obtaining 66% of overall accuracy.

Although several papers have been published on speech emotion recognition, very few have considered hybrid classification methods [23]. Decision-level fusion is an approach to improve recognition performance by combining multiple base classifiers. There are two key issues in decision fusion, one is the optimum combination of base decisions, and the other is the design of base classifiers. In Reference 23], two hybrid schemes, stacked generalization [24] and the unweighted vote, were applied for emotion recognition. These two schemes achieved accuracies of 72.18 % and 70.54 %, respectively, when they were used to recognize anger, disgust, fear, happiness, sadness, and surprise. In Reference 7, a hybrid classification method that combines the SVM and the Decision Tree (DT) was proposed. The highest accuracy for classifying neutral, anger, and loud was reported to be 72.4%. How effective the decisions from various base classifiers can be combined to generate the overall class decision is a critical factor that determines the accuracy of the classification. A fusion scheme that combines the Probabilistic Neural Network (PNN) and the Universal Background Model-Gaussian Mixture Model (UBM-GMM) has been presented by the authors in Reference 25. In this paper, the class-conditional probability is calculated by counting the corresponding data samples. As a result, the method requires a large training dataset for training all possible combinations of local decisions from the base classifiers. In this chapter, we extend the above method into a more general MAP-based fusion method.

The improvement achieved by the proposed method can be briefly concluded below.

1. This fusion method is able to effectively combine the strengths of several (two or more) classification methods at the decision level.
2. The implementation of the method is simple and it does not require tuning of any prespecified parameters.
3. More importantly, it can work with a small training set. This is achieved by calculating the posterior probability with Bayes' theorem under the assumption of mutual stochastic independence of local decisions given a true class.

To illustrate the effectiveness of the proposed method, the PNN, UBM-GMM, and k-NN methods are used as the base classifiers for the numerical experiments presented in this chapter. The results from experiments have shown that the proposed method can achieve higher recognition accuracy compared with those obtained using base classifiers alone.

The remaining part of this chapter is organized as follows. The acoustic feature extraction process is discussed in Section 9.2. In Section 9.3, the proposed MAP-based fusion classification method is presented. Numerical results are shown in Section 9.4 and the concluding remarks are given in Section 9.5.

9.2 ACOUSTIC FEATURE EXTRACTION FOR EMOTION RECOGNITION

Feature extraction is an important process in the automatic recognition of emotional states from human speeches. Prior to feature extraction, speech signals have to be first preprocessed including pre-emphasis, framing, and windowing processes.

1. Pre-emphasis process
   In order to emphasize important frequency components in the signal, the speech data sampled with a rate of 16 kHZ are first high-pass filtered by a Finite Impulse Response (FIR) filter called pre-emphasis filter given by

$$H(z) = 1 - 0.9375z^{-1} \tag{9.1}$$

2. Framing process
   Signal frames of length 25 ms are then extracted from the filtered speech signal at an interval of 10 ms based on the assumption that the signal within a frame is stationary or quasi-stationary.
3. Windowing process
   A Hamming window is applied to each signal frame to reduce signal discontinuity in order to avoid spectral leakage.

Three short-time cepstral features are extracted as shown in References 3 and 4, which are LPCCs, PLP cepstral coefficients, and MFCCs. For easy reference of readers, we briefly introduce these features which are also given in Reference 4.

1. Linear prediction-based cepstral coefficients
   Linear Prediction (LP) analysis based on the source-filter model has been one of the most important speech analysis technologies. It models the vocal tract transfer function by an all-pole filter with a transfer function given by

$$H(z) = \frac{1}{1 - \sum_{i=1}^{p} a_i z^{-i}} \tag{9.2}$$

   where a_i is the filter coefficients. The speech signal, S_t, assumed to be stationary over the analysis frame is approximated as a linear combination of the past p samples, given as

$$\hat{S}_t = \sum_{i=1}^{p} a_i S_{t-i} \tag{9.3}$$

   In (9.3), a_i can be found by minimizing the mean square filter prediction error between $\hat{S}_t$ and S_t. The cepstral coefficients is considered to be more reliable and robust than the LP filter coefficients. It can be computed directly from the LP filter coefficients using the recursion given as

$$\hat{c}_k = a_k + \sum_{i=1}^{k-1} \left(\frac{i}{k}\right) c_i a_{k-i} \qquad 0 < k \leq p \tag{9.4}$$

where c_k is the cepstral coefficients.

2. Perceptual linear prediction cepstral coefficients
   PLP, proposed by Hermansky [26], is based on three concepts from the psychophysics of hearing to derive an estimate of the auditory spectrum:
   (a) critical-band spectral resolution;
   (b) equal-loudness curve;
   (c) intensity-loudness power law.
   It combines the Discrete Fourier Transform (DFT) and LP technique. In PLP analysis, the speech signal is processed based on hearing perceptual properties prior to LP analysis. The auditory spectrum of speech is then approximated by the autoregressive all-pole model. The calculation of PLP cepstral coefficients involves six steps as shown in Figure 9.2.
   (a) Spectral analysis
   The short-time power spectrum is achieved for each speech frame.
   (b) Critical-band spectral resolution
   The power spectrum is warped onto a Bark scale and convolved with the power spectral of the critical band filter, in order to simulate the frequency resolution of the ear which is approximately constant on the Bark scale.

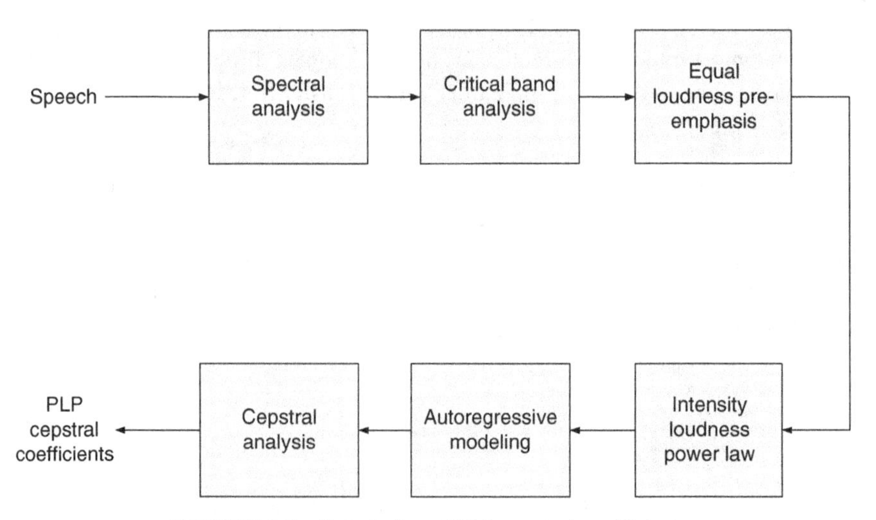

FIGURE 9.2 Calculation of PLP cepstral coefficients.

(c) Equal-loudness pre-emphasis
An equal-loudness curve is used to compensate for the non-equal perception of loudness at different frequencies.

(d) Intensity loudness power law
Perceived loudness is approximately the cube root of the intensity.

(e) Autoregressive modeling
Inverse Discrete Fourier Transform (IDFT) is carried out to obtain the autoregressive coefficients and all-pole modeling is then performed.

(f) Cepstral analysis
PLP cepstral coefficients are calculated from the auto-regression (AR) coefficients as the process in LPCC calculation.

3. Mel-frequency cepstral coefficients
MFCC, proposed by Davis and Mermelstein in Reference 27, considers the sensitivity of human perception with respect to frequencies and has become the most popular features used in Automatic Speech Recognition (ASR). The calculation of MFCC involves computing the cosine transform of the real logarithm of the short-time power spectrum on a Mel-warped frequency scale, which can be broken down into the following steps as shown in Figure 9.3.

(a) DFT is applied in each speech frame as

$$X[k] = \sum_{n=0}^{N-1} x[n]\ e^{-j2\pi nk/N}, \quad 0 \leq k \leq N-1 \tag{9.5}$$

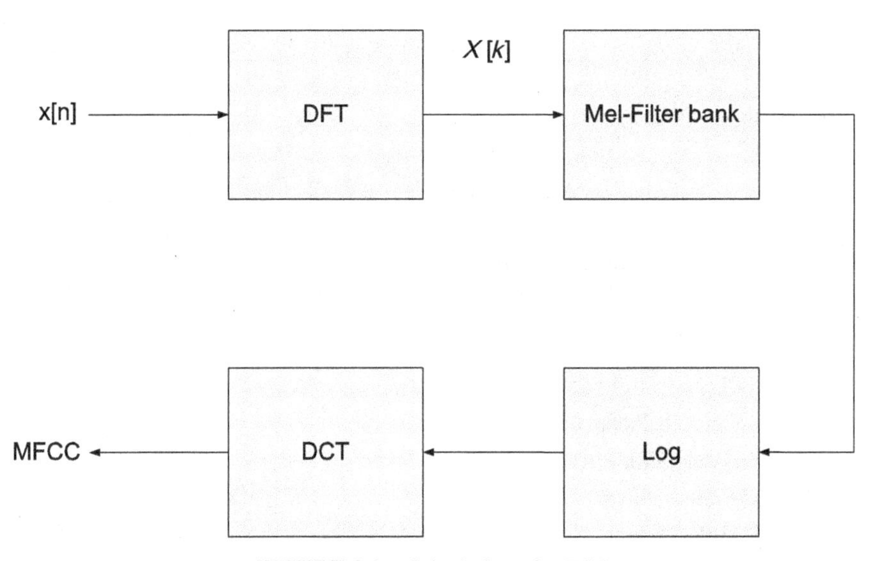

FIGURE 9.3 Calculation of MFCC.

(b) Mel-scale filter bank
The Fourier spectrum is non-uniformly quantized to conduct Mel-filter bank analysis. The window functions that are first uniformly spaced on the Mel-scale and then transformed back to the Hertz scale are multiplied with the Fourier power spectrum and accumulated to achieve the Mel-spectrum filter-bank coefficients. A Mel-filter bank has filters linearly spaced at low frequencies and approximately logarithmically spaced at high frequencies, which can capture the phonetically important characteristics of the speech signal while suppressing insignificant spectral variation in the higher frequency bands [27].

(c) The Mel-spectrum filter-bank coefficients is calculated as

$$F[m] = \log\left(\sum_{k=0}^{N-1} |X[k]|^2 H_m[k]\right) \quad 0 \le m \le M \tag{9.6}$$

(d) The Discrete Cosine Transform (DCT) of the log filter bank energies is calculated to find the MFCC as

$$c[n] = \sum_{m=0}^{M} F[m] \cos\left(\pi n (m-1)/2\mathrm{M}\right) \quad 0 \le n \le M \tag{9.7}$$

where $c[n]$ is the nth coefficient.

Besides the LPCCs, PLP cepstral coefficients, and MFCCs, Delta and Acceleration (Delta Delta) of the raw features are also included and given as follows:

Delta Δx_i:

$$\Delta x_i = \frac{1}{2}(x_{i+1} - x_{i-1}) \tag{9.8}$$

Acceleration (Delta Delta) $\Delta\Delta x_i$:

$$\Delta\Delta x_i = \frac{1}{2}(\Delta x_{i+1} - \Delta x_{i-1}) \tag{9.9}$$

where x_i is the ith value in the feature vector.

The list below shows the feature set to be used for speech emotion recognition.

1. PLP—54 features
   (a) 18 PLP cepstral coefficients
   (b) 18 Delta PLP cepstral coefficients
   (c) 18 Delta Delta PLP cepstral coefficients.
2. MFCC—39 features
   (a) 12 MFCC features
   (b) 12 Delta MFCC features

   (c) 12 Delta Delta MFCC features
   (d) 1 (log) frame energy
   (e) 1 Delta (log) frame energy
   (f) 1 Delta Delta (log) frame energy.
3. LPCC—39 features
   (a) 13 LPCC features
   (b) 13 Delta LPCC features
   (c) 13 Delta Delta LPCC features

By fusing the PLP, MFCC, and LPCC features, a vector with dimension of R^D is achieved, where D represents the total number of the features extracted for each frame, and we have $D = 132$.

To reduce acoustic variation in different speakers and different utterances in phonologically identical utterances and consequently increase recognition accuracy, speaker- and utterance-level normalization are performed by subtracting the mean and dividing by the standard deviation of the features.

9.3 PROPOSED MAP-BASED FUSION METHOD

Most speech emotion analysis methods reported in the literature involves the use of a single classifier [23]. One approach to improve recognition performance is to use multiple classifiers and make the final decision by combining the recognition results from all the base classifiers. The key issue in decision-level fusion is on how to optimally combine the decisions from the base classifiers. This is still an important and open problem in the areas such as signal processing, data mining, and machine learning.

In Reference 25, a fusion algorithm that combines a PNN classifier and a UBM-GMM classifier has been proposed for speech emotion classification. In this section, we formulate a general fusion method based on the MAP criterion that can effectively combine the strengths of several (two or more) classification methods at the decision-level fusion. Figure 9.4 shows the structure of the proposed scheme with M ($M \geq 1$)

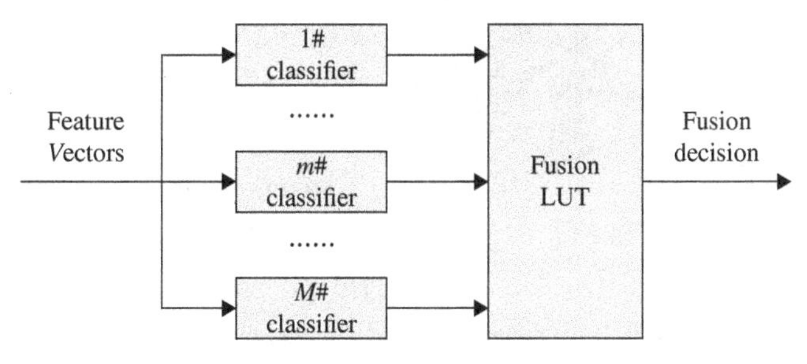

FIGURE 9.4 Structure of proposed hybrid scheme. LUT, Look Up Table.

base classifiers. The outputs of the base classifiers are processed by the proposed MAP fusion method before the final decision is made. In this way, it effectively combines the strengths of base classifiers. From a statistical point of view, the method of MAP yields fusion decisions with good statistical properties. Unlike the other fusion methods such as the weighted linear combination method [28], its implementation is simple without the need to tune any prespecified parameters.

9.3.1 Base Classifiers

In this chapter, the PNN, GMM, and k-NN are used as base classifiers due to their compatibility and good performance. The UBM is incorporated into the GMM to form the UBM-GMM classifier. (This is to improve the recognition performance of the GMM classifier.)

The PNN [16] is a Bayesian statistical classifier that uses Parzen estimator to approximate class-dependent Probability Density Functions (PDFs). It works as a multilayered feed-forward network. There are usually four network layers as shown in Figure 9.5, which are input layer, pattern layer, summation layer, and output layer. It has been employed in many applications, due to its properties such as fast training process, convergence guarantee (if the size of the representative training set is large enough), inherently parallel structure, and adding or removing training samples without retraining process.

The input of the PNN is a multidimensional feature vector. In the pattern layer, supervised training is conducted with a given training set that must be representative of the actual population. The distribution function that estimates the likelihood of an

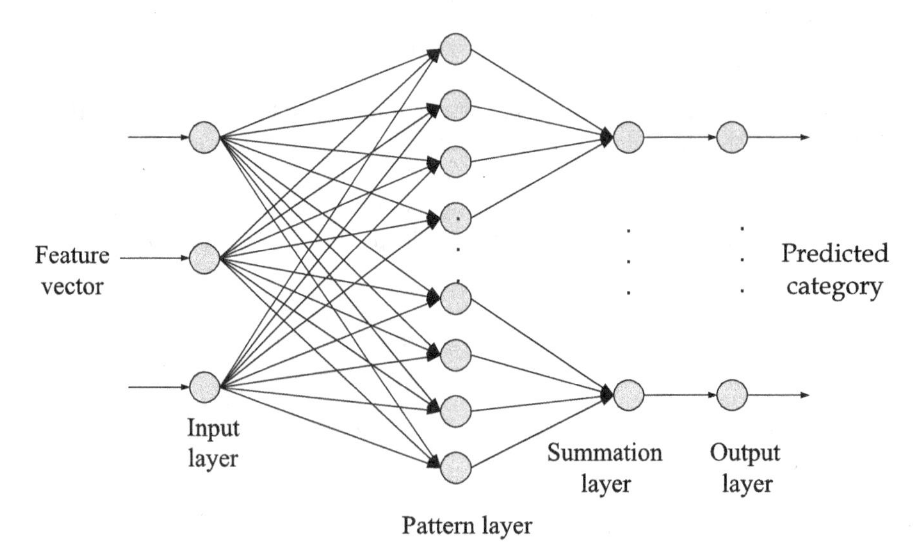

FIGURE 9.5 Structure of a probabilistic neural network.

input feature vector belonging to a learned category is developed. Each unit in the pattern layer represents an exemplar in the training set and commonly a Gaussian function is used as the activated function. Via training process, the outputs from the units that belong to one emotion category are combined in the summation layer. Each unit in the summation layer is associated with one emotion category. The output layer works in a competitive way, where the category that the input vector belongs to is finally predicted.

The GMM is basically a single-state Hidden Markov Model (HMM) [29] with a Gaussian mixture observation density. It assumes that the observed variables are generated via a PDF that is the weighted linear combination of a set of Gaussian PDF. This has been shown to be an effective PDF in text-independent applications, for example, speaker recognition [17], where no *prior* knowledge is available on what the speaker will say.

In order to make the model more reliable and handle mismatches more effectively, the UBM is incorporated into the GMM, and the resultant model is denoted as UBM-GMM [17]. In the UBM-GMM, the hypothesis that an input belongs to a category and the hypothesis that it does not belong to this category are both tested. An emotion model and a background model are trained for each category. The former is trained using the training samples belonging to these emotions, and the latter is trained using the other samples. Since the UBM is trained with a considerable amount of data, it is quite well-defined. To predict the category that an input vector belongs to, both models are used to generate the PDF alone. The predicted category is then determined based on their likelihood ratio.

The k-NN algorithm is a supervised learning algorithm, where a sample is classified based on the majority of k-nearest neighbor category. Here, k is a positive integer, typically small. If $k = 1$, then the object is simply assigned to the class of its nearest neighbor.

It should be noted that, the proposed MAP-based fusion method is a generic scheme and can hence be built from using other base classifiers too.

9.3.2 MAP-Based Fusion

Suppose there are M base classifiers that give M local decisions, $\{c_1, c_2, \ldots, c_M\}$, where c_m for $m = 1, \ldots, M$, is the class decision from the mth base classifier. In emotion recognition, c_m represents the emotional state estimated by the mth base classifier. The problem in decision fusion is, then, on how to optimally generate the global class given $\{c_1, c_2 \ldots, c_M\}$.

Let c be a class label, and c_r be the actual class of a speech sample in the database. The prior probability of class c_r is denoted as $f_c(c_r)$ and is estimated simply as the empirical frequencies of the dataset

$$f_c(c_r) = p(c_r) = \frac{N_{c_r}}{N} \tag{9.10}$$

where N_{c_r} is the number of samples in class c_r and N is the number of total samples. That is,

$$\sum_{c_r \in C} f_c(c_r) = 1 \tag{9.11}$$

where C is the vector of class labels under reorganization. Let C_L be the vector of local decisions, that is, $C_L = \{c_1, c_2, \ldots, c_M\}$, then the likelihood of c_r given C_L, can be expressed as

$$L_{c|C_L}(c_r) = p(c_1, c_2, \ldots, c_M | c_r) \tag{9.12}$$

The posterior probability distribution of c_r given C_L is the conditional probability that the true class is c_r given that the local decision vector is $\{c_1, c_2 \ldots, c_M\}$. That is,

$$f_{c|C_L}(c_r) = p(c_r | c_1, c_2, \ldots, c_M) \tag{9.13}$$

Based on the MAP criterion, we determine the final decision class that maximizes the posterior probability in (9.13) as

$$\hat{c}_r = \arg\max_{c_r \in C} f_{c|C_L}(c_r) \tag{9.14}$$

In the case when

$$f_{c|C_L}(c_r) = 0, \text{for } c_r \in C$$

or when there is more than one class that satisfies (9.14) with equal posterior probabilities, the final decision class is the one that maximizes $p_m(c_m|c_m)$ for $m = 1, 2, \ldots M$ or

$$\hat{c}_r = \arg\max_{m=1}^{M} p_m(c_m | c_m) \tag{9.15}$$

where $p_m(c_m|c_m)$ is the conditional probability that the estimated class is c_m given that the actual class is c_m for the mth base classifier.

9.3.3 Addressing Small Training Dataset Problem—Calculation of $f_{c|C_L}(c_r)$

In Reference 25, $p(c_r|c_1, c_2, \ldots, c_M)$ in (9.13) is approximated by simply counting the corresponding events as

$$p(c_r | c_1, c_2, \ldots, c_M) \cong N_{C_L} / N_{c_r} \tag{9.16}$$

where N_{C_L} represents the number of samples (utterances in speech emotion classification) whose actual class is c_r, given that the local decision vector is $\{c_1, c_2, \ldots, c_M\}$.

When this is approximated by (9.16), the performance can degrade significantly when the training dataset is too small. This is explained as below. Let L be the total number of classes (emotional states) to be classified. With M base classifiers, the number of possible combinations of c_r and $\{c_1, c_2, \ldots, c_M\}$ is L^{M+1}. In order to ensure good recognition performance, a large training dataset is necessary to train all possible combinations.

To address this problem, we calculate the posterior probability of c_r in (9.13) with Bayes' theorem by multiplying the prior probability distribution in (9.10) by the likelihood function in (9.12), and then dividing it by the normalizing constant,

$$f_{c|C_L}(c_r) = \frac{f_c(c_r)L_{c|C_L}(c_r)}{\sum\limits_{c_r \in C} f_c(c_r)L_{c|C_L}(c_r)} \tag{9.17}$$

Without losing the generality, we assume mutual stochastic independence of local decisions from base classifiers given a true class. Under this assumption, the likelihood function in (9.12) can be rewritten as

$$L_{c|C_L}(c_r) = p(c_1|c_r)p(c_2|c_r)\ldots p(c_M|c_r) \tag{9.18}$$

Substituting (9.18) into (9.17), the posterior probability $f_{c|C_L}(c_r)$ can be calculated as

$$\begin{aligned} f_{c|C_L}(c_r) &= \frac{p(c_r)p(c_1|c_r)p(c_2|c_r)\ldots p(c_M|c_r)}{\sum\limits_{c_r \in C} p(c_r)(c_1|c_r)p(c_2|c_r)\ldots p(c_M|c_r)} \\ &= \frac{p(c_r)\prod\limits_{m=1}^{M} p(c_m|c_r)}{\sum\limits_{c_r \in C} p(c_r)\prod\limits_{m=1}^{M} p(c_m|c_r)} \end{aligned} \tag{9.19}$$

where $p(c_m|c_r)$ is the conditional probability that estimated class from the mth base classifier is c_m, given that the actual class is c_r. Substituting (9.19) into (9.14), the final decision class is found by solving the following maximizing problem:

$$\hat{c}_r = \arg\max_{c_r \in C} \left(\frac{p(c_r)\prod\limits_{m=1}^{M} p(c_m|c_r)}{\sum\limits_{c_r \in C} p(c_r)\prod\limits_{m=1}^{M} p(c_m|c_r)} \right) \tag{9.20}$$

9.3.4 Training and Testing Procedure

The proposed fusion scheme consists of two steps in the training stage. Each step uses a different part of the speech data in the training set. In the test stage, each sample is first preceded by all classifiers to achieve the combination of local decisions, $\{c_1, c_2 ..., c_M\}$. Then the final decision class is determined by maximizing the posterior probability. This is elaborated below.

9.3.4.1 Training In the first step of training, each base classifier is trained individually using data samples in training dataset. In the second step, another part of training data are used to test these trained base classifiers. The posterior probability is calculated for each sample. The confusion matrix for each of base classifier is recorded too. With total L emotional states to be classified, the L-by-L confusion matrix for each of the classifier takes the following form:

$$P_m = \begin{pmatrix} p_{m(11)} & \cdots & p_{m(1L)} \\ \vdots & \ddots & \vdots \\ p_{m(L1)} & \cdots & p_{m(LL)} \end{pmatrix}, \text{ for } m = 1, 2, \ldots, M \tag{9.21}$$

In the matrix of mth classifier,

$$p_{m(ij)} = p_m(c_j|c_i) = prob(c_j|c_i) \tag{9.22}$$

is the probability that the estimated class is c_j, given that actual class is c_i. For an effective classifier, the values of the diagonal entries are expected to be much higher than those of the non-diagonal entries.

According to (9.14) or (9.15), a Fusion Look-Up Table (LUT), denoted as, F, is built based on the posterior probabilities and confusion matrices. The dimension of F is $L^M \times (M + 1)$. A typical row of F takes the form

$$[c_1\ c_2\ \ldots\ c_M\ \ c_{r_f}]$$

where $\{c_1, c_2, ..., c_M\}$ are the combination of local decisions, and c_{r_f} is the class that maximizes (9.14) or maximizes (9.15) in the given order. When building a row in F, the first M indices are $c_1, c_2, ...,$ and c_M. The values of

$$f_{c|C_L}(c_l),\ c_l \in C \text{ for } l = 1, \ldots, L$$

are compared to determine the last index c_{r_f} that maximizes $f_{c|C_L}(c_{r_f})$ as given in (9.14). If a unique class that meets (9.14) cannot be found, we compare the values of $p_m(c_m|c_m)$ for $m = 1, 2, ..., M$ in the M confusion matrices and c_{r_f} is then taken to be the class corresponding to the highest $p_m(c_m|c_m)$ as shown in (9.15).

9.3.4.2 ***Testing*** During the testing stage, all of the trained classifiers are employed to test the data samples separately. For each of the data sample, a combination of local decisions, $\{c_1, c_2 ..., c_M\}$, is achieved. The final decision class, $\hat{c}_r$, of a data sample is determined by the fusion LUT obtained in the training stage. We find $\hat{c}_r$ which maximizes the posterior probability shown in (9.14) or maximizes the class-conditional probability shown in (9.15). With the fusion LUT, the global decision is achieved simply by looking up c_{r_f} in the table, where the first M indices are $c_1, c_2, ...,$ and c_M.

The training–testing process can be summarized into the following seven steps.

1. Train each of the base classifiers independently using the training data.
2. Use the trained classifiers to recognize the emotions of another speech data training set.
3. Calculate the posterior probability for each sample.
4. Calculate the confusion matrices for all of the base classifiers.
5. Build the fusion LUT, F, according to the process described earlier.
6. Apply the base classifiers to the test data separately, and obtain the estimated emotional states, $\{c_1, c_2, ..., c_M\}$, respectively.
7. The final decision of the emotional state, $\hat{c}_r$, is taken to be c_r that in the row where the first M indices are $c_1, c_2, ...,$ and c_M.

The training is carried out in the first five steps including the calculation of confusion matrices and fusion LUT. For each test sample, the decision from each base classifier is drawn in Step 6 and then the final decision is made in the last step. Through the above procedure, the proposed fusion method combines the strengths of base classifiers and yields fusion decisions with good statistical properties. Another advantage of the fusion method is its simplicity; it does not need to adjust nor specify any parameter values in fusion process.

9.4 EXPERIMENT

9.4.1 Database

The emotional speech database used in this study is extracted from the Linguistic Data Consortium (LDC) Emotional Prosody Speech corpus (catalog number LDC 2002S28) [30]. It was recorded by the Department of Neurology, University of Pennsylvania Medical School. This database comprises approximately 2300 utterances spoken by three male and four female actors. The speech contents are neutral phrases like dates and numbers, for example, "September fourth" or "eight hundred one," which are expressed in 14 emotional states as well as neutral state.

9.4.2 Experiment Description

The proposed fusion method was used to recognize the emotional states of the spoken utterances in the database. The 15 emotion categories are anxiety, boredom, cold anger, hot anger, contempt, despair, disgust, elation, happiness, interest, panic, pride, sadness, shame, and neutral. A PNN, a GMM, and a k-NN classifier were used as base classifiers in the system. We also tested the same data samples using the PNN method alone, the UBM-GMM method alone, and the k-NN method alone for the sake of comparison.

In the experiment, half of the data were employed to train the PNN classifier, the UBM-GMM classifier, and the k-NN classifier, a quarter of the data were used to calculate the fusion LUT and the confusion matrices, and the rest were used for testing purpose. Note that when using the PNN alone, the UBM-GMM alone, or the k-NN alone, we added those data used to calculate the fusion LUT into the training data to train the PNN, UBM-GMM, or k-NN classifiers, respectively, in order to make a fair comparison to use the same dataset in training and testing stages.

9.4.3 Results and Discussion

To consider the different characteristics of speech among the speakers, the speech data were trained in speaker dependent-training mode. An individual training process was carried out for each speaker.

The numerical results obtained by the four classification methods (i.e., the proposed scheme, the PNN method alone, the UBM-GMM method alone, and the k-NN method alone) are shown in Table 9.1. The average accuracies obtained by the four methods are shown in Figure 9.6. Compared with the average accuracies of 68.60%,

TABLE 9.1 Recognition accuracies (%) of the proposed scheme, PNN, UBM-GMM, and k-NN in speaker-dependent training mode

	Proposed	PNN	GMM	k-NN
Anxiety	81	79	77	76
Boredom	79	71	79	65
Cold angry	70	64	69	57
Contempt	82	73	80	76
Despair	74	65	79	61
Disgust	81	78	89	69
Elation	76	59	81	72
Hot angry	86	75	85	77
Happiness	72	61	76	55
Interest	77	64	70	67
Neutral	81	80	54	84
Panic	79	62	75	71
Pride	64	72	53	61
Sadness	74	74	63	63
Shame	65	52	61	58
Average	**76.07**	**68.60**	**72.73**	**67.47**

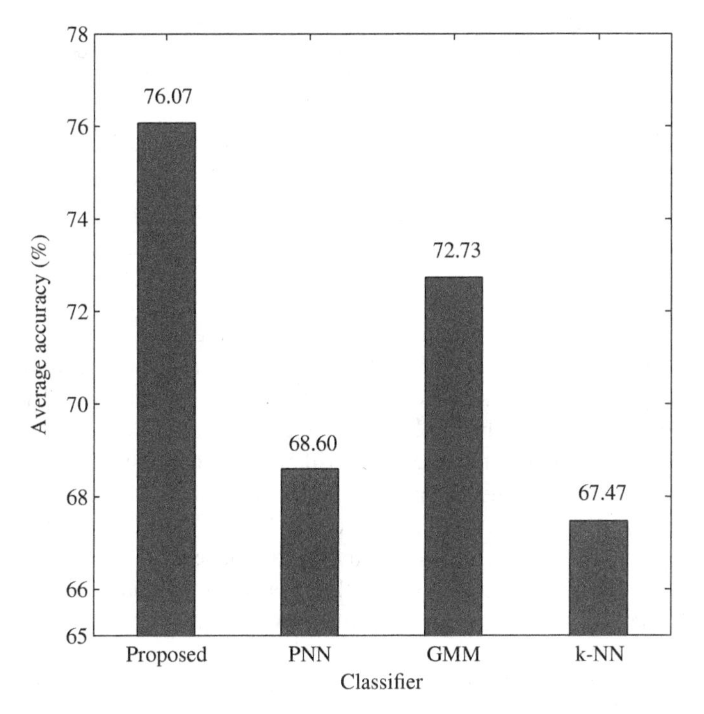

FIGURE 9.6 Average accuracies (%) of the proposed scheme, PNN, UBM-GMM, and k-NN in speaker-dependent training mode.

72.73%, and 67.47% achieved by the PNN, UBM-GMM, and k-NN alone, respectively, the proposed scheme is able to achieve a higher average accuracy of 76.07%. In the literature, usually only two to six different emotional states are classified. Considering the difficulties encountered due to the facts that the number of emotional states is as large as 15, the results are rather satisfying.

It can be seen from Table 9.1 that the proposed hybrid scheme outperforms the PNN, the UBM-GMM, and the k-NN methods for most of 15 emotional states tested. Although the recognition accuracies for a few of emotional states are slightly lower than those from the base classifiers alone, the average accuracies are improved. Using the proposed MAP fusion method, therefore, is better than using any base classifier alone.

The average accuracy of 76.07% given in Table 9.1 was achieved when we used $1/4$ of the total data to calculate the fusion LUT and the confusion matrices. In order to test the performance of the proposed MAP method with a small training dataset, we only included $1/8$ of the total dataset in the second step of the training stage. An average accuracy of 75.32% was achieved, which was only slightly decreased. When only $1/16$ of the total data was used, the average accuracy decreased to 73.41%, as shown in Table 9.2. This indicates that the MAP method can work well with a small training dataset.

TABLE 9.2 Recognition accuracies (%) with different sizes of the training data used to calculate the fusion LUT and the confusion matrices of the dataset.

Accuracy	Data size
76.07	1/4
75.32	1/8
73.41	1/16

Besides the speaker-dependent training mode shown earlier, the speech data are also trained in gender-dependent mode. It takes gender difference into account, where the classifiers are trained individually for males and females. The average accuracy achieved by the hybrid scheme is 73.29%. From the numerical results in the experiments, we can see that the speaker-dependent-based emotion recognition is more reliable than the gender-dependent mode. It is often employed in practical applications, where the speaker information can be known "*a priori*" via, for example, user registration and training. Although the gender-dependent mode is inferior to the speak-dependent mode, it eliminates the need of individual training for each speaker. It is useful in some practical applications when only the gender information is available.

9.5 CONCLUSION

In this chapter, a general MAP-based fusion method is formulated. The proposed method effectively combines the strengths of several classification methods at the decision-level fusion. Experiments have been conducted for the 15 emotional states recorded in the LDC database. Three base classifiers, the PNN, the UBM-GMM, and the k-NN methods, have been used to illustrate how the proposed method can be constructed. It is shown from the numerical results that higher accuracy can be achieved compared with those obtained using base classifiers alone. We also test the performance of the proposed MAP method with small training dataset. The results reveal that it can work well with comparable accuracy.

REFERENCES

[1] D. Ververidis and C. Kotropoulos, "Emotional speech recognition: resources, features, and methods," *Speech Commun.*, 48(9): 1163–1181, 2006.

[2] R. Cowie, "Emotion recognition in human–computer interaction," *IEEE Signal Processing Magazine*, 18(1): 32–80, 2001.

[3] L. Cen, W. Ser, Z. L. Yu, and W. Cen, "Automatic recognition of emotional states from human speeches," in *Pattern Recognition, Recent Advances*, IN-TECH, February 2010, pp. 431–449.

[4] L. Cen, M. H. Dong, H. Z. Li, Z. L. Yu, and P. Chang, "Machine learning methods in the application of speech emotion recognition," in *Application of Machine Learning*, IN-TECH, February 2010, pp. 1–19.

[5] C. Lee and S. Narayanan, "Toward detecting emotions in spoken dialogs," *IEEE Transactions on Speech and Audio Processing*, 13(2): 293–303, 2005.

[6] L. Cen, W. Ser, and Z. L. Yu, "Speech emotion recognition using canonical correlation analysis and probabilistic neural network," in Proceedings of the Seventh International Conference on Machine Learning and Application (ICMLA), San Diego, CA, December 2008.

[7] T. Nguyen and I. Bass, "Investigation of combining SVM and decision tree for emotion classification," in Proceedings of the Seventh IEEE International Symposium on Multimedia, 2005, pp. 540–544.

[8] J. R. Davitz (ed.), *The Communication of Emotional Meaning*, McGraw-Hill, New York, 1964.

[9] G. L. Huttar, "Relations between prosodic variables and emotions in normal American English utterances," *J. Speech Hear. Res.*, 11: 481–487, 1968.

[10] I. Fonagy, "A new method of investigating the perception of prosodic features," *Lang. Speech*, 21: 34–49, 1978.

[11] Z. Havrdova and M. Moravek, "Changes of the voice expression during suggestively influenced states of experiencing," *Act. Nerv. Super.*, 21: 33–35, 1979.

[12] S. McGilloway, R. Cowie, and E. Douglas-Cowie, "Prosodic signs of emotion in speech: preliminary results from a new technique for automatic statistical analysis," in Proceedings of the International Congress of Phonetic Sciences, vol. 1, Stockholm, Sweden, 1995, pp. 250–253.

[13] F. Dellaert, T. Polzin, and A. Waibel, "Recognizing emotion in speech," in Proceedings of the Fourth International Conference on Spoken Language Processing, vol. 3, October 1996, pp. 1970–1973.

[14] V. A. Petrushin, "Emotion recognition in speech signal: experimental study, development, and application," in Proceedings of the Sixth International Conference on Spoken Language Processing, Beijing, China, 2000.

[15] N. Amir, "Classifying emotions in speech: a comparison of methods," in Proceedings of the Eurospeech, 2001.

[16] D. F. Specht, "Probabilistic neural networks for classification, mapping or associative memory," in Proceedings of the IEEE International Conference on Neural Network, vol. 1, July 1988, pp. 525–532.

[17] D. A. Reynolds, T. F. Quatieri, and R. B. Dunn, "Speaker verification using adapted Gaussian mixture model," *Digit. Signal Process.*, 10(1): 19–41, 2000.

[18] J. Zhou, G. Y. Wang, Y. Yang, and P. J. Chen, "Speech emotion recognition based on rough set and SVM," in Proceedings of the Fifth IEEE International Conference on Cognitive Informatics, vol. 1, Beijing, China, July 2006, pp. 53–61.

[19] J. Nicholson, K. Takahashi, and R. Nakatsu, "Emotion recognition in speech using neural networks," in Proceedings of Sixth International Conference on Neural Information Processing, vol. 2, Stockholm, Sweden, 1999, pp. 495–501.

[20] V. A. Petrushin, "Emotion in speech: recognition and application to call centers," in Proceedings of Artificial Neural Networks in Engineering, November 1999, pp. 7–10.

[21] F. Yu, E. Chang, Y. Q. Xu, and H. Y. Shum, "Emotion detection from speech to enrich multimedia content," in Proceedings of the Second IEEE Pacific-Rim Conference on Multimedia, Beijing, China, October 2001.

[22] K. Scherer, "A cross-cultural investigation of emotion inferences from voice and speech: implications for speech technology," in Proceedings of the International Conference on Spoken Language Processing (ICSLP), Beijing, China, October 2000, pp. 379–382.

[23] D. Morrison, R. Wang, and L. C. De Silva, "Ensemble methods for spoken emotion recognition in call-centres," *Speech Commun.*, 49(2): 98–112, 2007.

[24] D. H. Wolpert, "Stacked generalization," *Neural Netw.*, 5(2): 241–259, 1992.

[25] W. Ser, L. Cen, and Z. L. Yu, "A hybrid PNN-GMM classification scheme for speech emotion recognition," in Proceedings of the 19th International Conference on Pattern Recognition (ICPR), Tampa, FL, December 2008.

[26] H. Hermansky, "Perceptual linear predictive (PLP) analysis of speech," *J. Acoust. Soc. Am.*, 87(4): 1738–1752, 1990.

[27] S. B. Davis and P. Mermelstein, "Comparison of parametric representations for monosyllabic word recognition in continuously spoken sentences," *IEEE Transactions on Acoustics, Speech and Signal Processing*, 28(4): 357–365, 1980.

[28] C. C. Vogt and G. W. Cottrell, "Fusion via a linear combination of scores," *Inf. Retr.*, 1(3): 151–173, 1999.

[29] L. R. Rabiner, "A tutorial on hidden Markov models and selected applications in speech recognition," *Proceedings of IEEE*, 77(2): 257–286, February 1989.

[30] *Emotional Prosody Speech Corpus, Linguistic Data Consortium*, University of Pennsylvania, PA.

AUTHOR BIOGRAPHIES

Ling Cen received the B.Eng. degree from the University of Science and Technology, China, in 1997, the M.Eng. degree from Chinese Academy of Sciences in 2001, and the Ph.D. degree in Electrical and Computer Engineering from the National University of Singapore (NUS) in 2006. She was with the General Electronic Technology Institute, China, as a project engineer from 1997 to 1998. In 2005, she joined the Centre for Signal Processing, Nanyang Technological University (NTU), as a Research Associate, and went on to become a research fellow. She is currently working as a scientist in the Human Language Technology Department, Institute for Infocomm Research (I^2R), Singapore. Her research interests include digital filter design, array signal processing, speech synthesis, voice conversion, pattern recognition, and evolutionary computation.

Zhu Liang Yu received the B.S. and M.S. degrees, both in Electronic Engineering, from the Nanjing University of Aeronautics and Astronautics, China, in 1995 and 1998, respectively, and the Ph.D. degree from Nanyang Technological University, Singapore, in 2006. He worked as a software engineer at the Shanghai BELL Company, Ltd., from 1998 to 2000. In 2000, he joined the Centre for Signal Processing, Nanyang Technological University, as a research engineer, then became a Research Fellow. In 2008, he joined the College of Automation Science and Engineering, South China University of Technology, as an associate professor and was promoted to a full professor in 2010. His research interests include array signal processing, acoustic signal processing, and adaptive signal processing.

Wee Ser received his B.Sc. (Hons.) and Ph.D. degrees, both in Electrical and Electronic Engineering, from Loughborough University, UK, in 1978 and 1982, respectively. He joined the Defence Science Organisation (DSO), Singapore, in 1982 and became the Head of the Communications Research Division in 1993. In 1997, he joined NTU as an Associate Professor and was later appointed Director of the Centre for Signal Processing at NTU.

He was a recipient of the Colombo Plan scholarship and the PSC postgraduate scholarship. He was awarded the IEE Prize during his studies in the United Kingdom. While in DSO, he was a recipient of the prestigious Defence Technology (Individual) Prize in 1991 and the DSO Excellent Award in 1992. He is the Associate Editor for the *IEEE Communications Letters* and the *Journal of Multidimensional Systems and Signal Processing* (Springer). He has served in several international and national advisory and technical committees and as reviewer to several international journals. He has published about 120 papers in refereed journals and international conferences. He holds six patents and has three pending patents. His research interests include microphone array and sensor array signal processing in general, signal classification techniques, and channel estimation and equalization techniques.

10

EMOTION RECOGNITION IN NATURALISTIC SPEECH AND LANGUAGE—A SURVEY

FELIX WENINGER, MARTIN WÖLLMER, AND BJÖRN SCHULLER
Institute for Human–Machine Communication, Technische Universität München, Germany

The recognition of emotion and affect has matured to a major topic in the field of speech and language processing over the last one and a half decades. In this chapter, we aim to provide an overview over recent developments in naturalistic emotion recognition based on acoustic and linguistic cues. We start from a variety of use-cases where emotion recognition can improve quality of service and quality of life and describe existing corpora of emotional speech data relating to such scenarios, the underlying theory of emotion modeling, and the need for an optimal unit of analysis. Besides providing an overview over state-of-the-art and novel approaches for implementation of automatic emotion recognition systems, we focus on the challenges for real-life applications that have become evident: non-prototypicality; lack of solid ground truth and data sparsity; generalization across application scenarios, languages, and cultures; requirements of real-time and incremental processing; robustness with respect to acoustic conditions; and appropriate evaluation measures that reflect real-life settings. We conclude by giving further directions for the field, including novel strategies to augment training data by synthesis and (semi-)unsupervised learning, as well as joint learning of other paralinguistic features by mutual information exploitation.

Emotion Recognition: A Pattern Analysis Approach, First Edition. Edited by Amit Konar and Aruna Chakraborty.
© 2015 John Wiley & Sons, Inc. Published 2015 by John Wiley & Sons, Inc.

10.1 INTRODUCTION

Modern intelligent environments which enable natural human–machine interaction consider the principles of human–human communication as the ideal prototype. While automatic speech recognition (ASR) is already an integral part of most intelligent systems such as virtual agents, in-car interfaces, or mobile phones, state-of-the-art systems do not account for the human ability to permanently observe and react to the affective state of the conversational partner in a socially competent way. This ability is crucial for the communication process, as people constantly adapt their manner of speaking based on the assessment they have of their counterpart's mood and exploit this information to make better sense of the interpretation of the counterpart's intentions. Therefore, it is believed that automatic emotion recognition (AER) is an essential precondition to render voice-based systems more human-like and to increase their acceptance among potential users. For instance, in the Seventh Framework Programme for Information and Communication Technology of the European Commission, efforts are devoted to increasing accessibility and efficiency of spoken dialogue systems by integrating emotional and other paralinguistic cues.

Still, even though researchers report outstanding recognition accuracies when trying to assign an affective state to an emotionally colored speech turn [1], systems that apply automatic emotion recognition are only rarely found in every day life. The main reason for this is that emotion recognition performance is often overestimated: Apart from examples such as call-center data [2], databases for interest recognition [3], or other spontaneous speech evaluations [4], most speech-based AER systems are still trained and tested on corpora that contain segmented speech turns with acted, prototypical emotions that are comparatively easy to assign to a set of predefined emotional categories [5]. Often, only utterances that have been labeled equally by the majority of annotators are used to evaluate AER performance. Yet, these assumptions fail to reflect the conditions a recognition system has to face in real-life usage. Next-generation AER systems must be able to deal with non-prototypical speech data and have to continuously process naturalistic and spontaneous speech as uttered by the user (e. g., as in the Interspeech 2009 Emotion Challenge [6]). More specifically, a real-life emotion recognition engine has to model "everything that comes in," which means it has to use all data as recorded, for example, for a dialog system, media retrieval, or surveillance task by using an *open microphone* setting. According to Reference 7, dealing with non-prototypicality is "one of the last barriers prior to integration of emotion recognition from speech into real-life technology."

In this chapter, we aim to provide an overview over recent developments in naturalistic emotion recognition based on acoustic and linguistic cues. First, in Section 10.2, we describe a variety of use-cases (Section 10.2.1), corpora of emotional speech data relating to different application scenarios (Section 10.2.2), the underlying theory of categorization and continuous modeling of emotions (Section 10.2.3), and finally the search for an optimal unit of analysis (Section 10.2.4). We then discuss the actual implementation of these concepts into automatic engines for emotion recognition in Section 10.3: extraction of low-level acoustic and linguistic features from speech and language (Section 10.3.1), selection of appropriate features and salient training instances (10.3.2), design of the training and classification algorithms (10.3.3), their evaluation (10.3.4) and last but not least, existing technology in form of freely available software that can be used as a starting point for application development and serves to advance joint research efforts (10.3.5). As outlined above, the increased focus on real-life AER has revealed interesting challenges for further research, some of which are fleshed out in Section 10.4: Foremost, we deal with non-prototypicality and the related issues of lack of solid ground truth and data sparsity (10.4.1); yet, we also point to generalization of techniques across application scenarios, languages, and cultures (10.4.2), requirements of real-time and incremental processing (10.4.3), and robustness against noise and reverberation (10.4.4). Finally, Section 10.5 outlines promising further directions for the field, including novel strategies to augment training data by synthesis and (semi-)unsupervised learning, joint learning of other paralinguistic features such as gender or personality (multi-task learning), and multimodal analysis.

10.2 TASKS AND APPLICATIONS

10.2.1 Use-Cases for Automatic Emotion Recognition from Speech and Language

For a start, we would like to demonstrate the vast application potential of voice-based AER in real-life use-cases. Mostly, they are motivated from a user-centric perspective: That is, the goal is to increase quality of service, or even quality of life. In some scenarios, the perspective is more system-centric, that is, a primary objective is to improve performance of other pattern recognition applications. A variety of applications are further discussed in References 8–10.

Voice portals and call centers: An obvious application in voice portals such as complaint hotlines analyzes the caller's voice to handle particularly angry customers by specially trained agents or soothe them by special dialog strategies [11]. Furthermore, quality of service provided by call centers [12] can be monitored and optimized by automatically and continuously assessing the emotional state of the agents and/or the caller: The goal is to reliably detect relevant changes, such as an increasing number of angry callers or an increasing average stress level of the call center agents. Finally, recorded call-center conversations can be indexed by emotional categories for the purpose of training agents to cope with special situations.

Tutoring systems, assistive robotics and virtual agents: In edutainment software and tutoring systems, the knowledge of the user's emotional state—including uncertainty, interest, stress, or even deception—can be used to adapt the system and the teaching pace [13], and in general, such paralinguistic cues are essential for tutors and students to make tuition successful [14]. Another field of applications is (assistive) robotics [15], for example, to maintain elderly people in their natural environment. The analysis of the affective states (emotion, feeling) and personality is still very rudimentary in robotics and often limits itself to tactile interactions. It is nevertheless by an understanding of these factors that we are going to be able to add social competence to humanoid or other highly interactive and communicative robots [16] or virtual agents/avatars. In the European SEMAINE project, a Sensitive Artificial Listener (SAL) was built—a multimodal dialog system that can sustain a conversation with a human using social interaction skills rather than verbal capabilities, which are limited mostly to agreement/disagreement and emotionally relevant keywords [17].

Surveillance: There are many security-related situations (surgeries [18], crisis management, and all the tasks connected to piloting) where detection of stress or other negative emotion may play a life decisive role. Speech can be used as modality of analysis for these states. In addition, counter-terrorism or counter-vandalism surveillance may be aided by analysis of aggressiveness of potential aggressors [19] or fear of potential victims [20].

Conversation analysis, computer mediation, and information transmission: There are further use-cases for computer-aided analysis of human–human conversations, for instance, to analyze and summarize meetings [21]. Also for cochlear-implanted (CI) deaf subjects AER is of interest, as CI processors typically alter the spectral cues which are crucial for the perception of emotion in human voices [22]. In that case, an automated system may help such subjects to better understand social cues—thus, the computer acts as a mediator. As another target group in this context, autistic children may profit from computer-aided analysis of emotional cues as they may have difficulties to understand or show these [23]. Finally, AER can be utilized for information transmission, to enrich dictated text messages or to label calls on the voice mailbox by "emoticons" [24]. As a related note, the MPEG4 standard already includes Ekman's "Big Six" emotion categories, which may be used for coding natural speech at high compression rates in future applications.

Speech recognition and understanding: It seems obvious that "what" is being said is to be interpreted by "how" it is being said—thus, natural language understanding can profit from including emotional information. However, since the recognition of spontaneous, emotional speech is still challenging as of today, the emotional speaker state can be valuable information to better recognize "what" is being said, for example, by acoustic model adaptation [25–27].

Media retrieval: In the field of multimedia retrieval, affective information is of interest for manifold types of media searches: For instance, highlights in sports games can be identified by measuring the level of excitement in the reporter's speech [28]; or, emotional tags could be automatically assigned to on-line videos.

Gaming, fun: Last but not least, AER directly offers great economic potential in the entertainment sector. For example, in virtual worlds (e.g., Second Life) and on-line role playing games, the credibility of characters and hence, the immersion of the user, could be enhanced by animating the face of users' avatars according to automatic emotion detection in the users' voice.

Despite this great application potential, the question of user acceptance must not be overlooked. During the emergence of emotion recognition research, technology was advanced without actual proof whether the user really liked a machine judging his or her feelings—apparently, the community was highly convinced of the need for automatic emotion recognition. This belief can now be confirmed through actual user studies: In a real-world product presentation experiment [3], a system with topic switching according to automatically detected user interest has been rated by users to be nearly equal to a human operator, and far better than a system without topic switching, regarding perceived reaction to interest, and understandability of information. Another example of such a study can be found in Reference 29, where 200 paid test subjects called a voice portal with anger feedback strategies: 20% noticed such a strategy of which 70% of these judged it as helpful.

10.2.2 Databases

It should be evident that the great variety of AER applications requires suited speech and language resources for training automatic recognizers. Emotional databases can be coarsely categorized by type of emotion elicitation: For acted databases, speakers are explicitly told to produce specific emotions with low variation of linguistic content, up to producing segmentally identical utterances. It is hoped that by that, generic (maybe universal) expressions of emotions can be obtained—however, it is now known that deliberately acting emotions is not the same as producing "spontaneous" emotions [30]. Second, some databases feature *induced* emotions in monologic speech by forcing subjects to watch film clips, or to imagine specific emotion-prone situations, and so on, rather than instructing them to deliberately act emotions. This might enable (induce) more realistic productions, but the extent of naturalness is questionable in such procedures. Moreover, the resulting speech is usually not dialogic/interactive as featured in many applications. Finally, and perhaps most importantly in the context of naturalistic emotion recognition, *non-prompted* emotions are usually elicited in scenarios that are interesting for potential applications. As such, these data are most likely *non-generic*, that is, not representing some universal emotions. In particular, archetypal emotions such as disgust or sadness will be rarely encountered in scenarios such as interaction with voice-based systems, but most of the data will feature emotion-related states such as frustration, interest, or boredom.

Here, we will only refer to some representative examples of databases; catalogs can be found, for instance, in References 31 and 32. Widely used—because freely available—acted databases are the Danish Emotional Speech (DES) database [33], and the Berlin Emotional Speech (BES) database [5], as well as the Speech Under Simulated and Actual Stress (SUSAS) database [34]. The DES database contains professionally acted nine Danish sentences, two words, and chunks that are located between two silent segments of two passages of fluent text. Affective states contain angry, happy, neutral, sadness, and surprise. The BES database features professional actors speaking 10 emotionally undefined sentences. A total of 494 phrases are commonly used: anger, boredom, disgust, fear, happiness, neutrality, and sadness. The SUSAS database serves as a first reference for spontaneous recordings. Speech is additionally partly masked by field noise in the chosen speech samples of actual stress. SUSAS content is restricted to 35 English air-commands in the speaker states of high stress, medium stress, neutral, and scream. DES and BES are representative for the "early" databases in the nineties but still serve as exemplars for acted emotional databases, for instance, in other languages. For English, there is the 2002 Emotional Prosody Speech and Transcripts acted database [35].

The eNTERFACE [36] corpus is an example for induced emotions. It consists of recordings of subjects from 14 nations speaking predefined spoken content in English. The subjects listened to six successive short stories eliciting a particular emotion out of anger, disgust, fear, happiness, sadness, and surprise. As a further example, the Airplane Behaviour Corpus (ABC) [37] is based on induced mood by prerecorded announcements of a vacation (return) flight, consisting of 13 and 10 scenes. It contains aggressive, cheerful, intoxicated, nervous, neutral, and tired speech.

Scenarios for recording non-prompted, interactive emotions can be based on *human–human* or *human–machine* interaction. The latter can be simulated by using a human operator to play the role of the machine (Wizard-of-Oz scenario). Human–human interaction can be recorded in the laboratory or in some specific real-life settings where recording is possible or already given. Representatives for the latter are TV recordings (The Vera-Am-Mittag (VAM) Corpus [38]), call-center interactions [39], multiparty interaction (the ICSI meeting corpus [21]), or The Speech in Minimal Invasive Surgery (SIMIS) database [18]. In contrast, the Audiovisual Interest Corpus (AVIC) [3] is an example for recording human–human communication in the laboratory. In its scenario setup, a product presenter leads subjects through a commercial presentation. AVIC is labeled in "level of interest" (loi) from 1 to 3.

Human–machine interaction normally takes place in the laboratory or in carefully controlled settings "in the wild." For example, human–machine call-center interaction [40] is representative for telephone conversations in real-life settings. Typical human–machine interactions in the laboratory are stress detection in a driving simulation [26], tutoring dialogs [13, 41], information systems [42], or human–robot communication [43]; a still emerging field is interaction with virtual agents [17]. The Belfast Sensitive Artificial Listener (SAL) data are part of the HUMAINE database and consist of natural conversations of humans and a virtual emotional agent in a Wizard-of-Oz scenario. The data have been labeled continuously in real time with respect to valence and activation. The SmartKom [44] corpus consists of Wizard-of-Oz dialogs. For evaluations, the dialogs recorded during a public environment technical scenario are used. It is structured into sessions which contain one recording of approximately 4.5 min length with one person, and labeled as anger/irritation, helplessness, joy/gratification, neutral, pondering/reflection, surprise, and unidentifiable episodes.

Generally, it is striking that in these sets data are very sparse: Typically, only one hour of speech from only around 10 subjects is provided. In related fields as ASR, several hundreds of hours of speech and subjects are typically contained. This is one of the major problems in this field at the moment. In addition, one still sees a prevalence of acted rather than natural data. Obviously, this is rather an annotation challenge, as emotional speech data per se would be available.

10.2.3 Modeling and Annotation: Categories versus Dimensions

In the psychological and theoretical discourse, a core issue is whether emotions should be modeled as categories or dimensions, and how many categories or dimensions are needed for an adequate modeling. Recent findings in neuroscience [45] suggest that humans most likely perform both categorical and continuous emotion perception and processing—yet, it is fully open whether processing by machine should be modeled along the lines of human processing, or simply output some correlate to the results of higher cortical processes.

Regardless of theoretical foundations for the one or the other type of modeling, in practice, categories can always be mapped onto dimensions and vice versa—albeit not necessarily lossless. For engineering of AER systems so far, performance has been the decisive criterion. However, as a reduction of complexity normally goes along with improved performance, this led to a concentration on a few acted emotions at the beginning, and resulted in a sort of "reality shock" when naturalistic data including non-prompted, realistic, and sparse events were addressed. Of course, in real-life applications the specific requirements and the availability of next-generation machine-learning techniques for learning continuous representations (i.e., performing regression) or categories (i.e., classification) will be decisive in choosing among alternatives.

As is evident from the annotated data that is available (cf. the previous section), in the case of categories, there is some hierarchy from *main categories*—such as the "big N" emotions, for example, anger, fear, sadness, and joy, or simply positive, neutral, negative—to *sub-categories* modeling different shades of the main categories. All these categories can be thought of as *fixed* or *graded* (e. g., weak, medium, strong), or as *pure* or *mixed* —sometimes even antagonistic, if for instance a mixture of anger/joy/irony is being observed. In the case of *dimensions*, it is, foremost, the question of "How many (and which) dimensions we should assume?": Traditionally, arousal (being active or passive) and valence (negative or positive) are modeled, with or without a third-category power/dominance/control. Of course, dimensional modeling can be discretized to form decision problems, such as to discriminate between positive or negative arousal/valence. Furthermore, using more than one dimension automatically models mixed categories as points in the n-dimensional space. If emotion is conceptualized in a broader meaning, most likely, some other dimensions representing social, interactive behavior will be modeled as well [43,46].

10.2.4 Unit of Analysis

Finding an appropriate unit of analysis, that is, a proper segmentation of the input data, is mandatory for future AER applications: Obviously, "emotionally consistent" units will favor an optimal classification performance. Second, incremental processing, that is, providing an estimate of the emotion before a longer utterance is finished, will often be necessary in real-life conversational systems, to enable reasonably fast system reactions (cf. Section 10.4.3).

Often, the unit of analysis corresponds to a *turn*—in conversational speech, it is marked by turn-taking, and generally, it can be defined by voice activity: Then, a turn simply starts when a speaker starts speaking, and ends when she/he stops speaking. This is an objective measure, easily obtainable for acted, read data and for interactions/conversations containing short dialog moves. Longer turns are often sub-segmented by some objective measures such as pauses longer than, for example, 0.5 or 1 second. The crucial question is whether such intuitive units of analysis correlate with meaningful emotional units: Clearly, if the unit is as long as a sentence or dialog move, the marking of specific and shorter emotional episodes might be smeared, resulting in suboptimal classification performance.

Two strategies, coping with these problems, can be observed. One defines "technical units" such as frames, time slices, or proportions of longer units (for instance, the unit is subdivided into three parts of equal length) [1,47,48]. The other one defines meaningful units with varying length such as syllables, words, phrases, that is, chunks that are linguistically and by that, semantically, well-defined. In Reference 49, the word was chosen as the smallest possible, meaningful emotional unit (named *ememe* along the same lines as *phonemes* and *sememes*) which can constitute larger meaningful emotional units; in Reference 50 it has been shown that stressed syllables alone are on par with words as far as classification performance is concerned. Still, for use in "live" systems, the use of more elaborate segmentation schemes requiring linguistic ground truth is limited by performance of ASR, which itself is challenging on spontaneous, emotional data.

Finally, one may want to decide for the optimal unit in a multimodal context if, for example, also video or physiological information is analyzed that typically investigates different units, but shall be fused in a synergistic manner—in fact, this problem can already arise to a certain extent when we want to fuse acoustic and linguistic information.

10.3 IMPLEMENTATION AND EVALUATION

In Figure 10.1, a unified overview on typical emotion recognition systems is given, with dotted boxes indicating optional steps. A speech database (speech signals $s(n)$ with labels y) is used to train the system to predict emotion in unseen test signals. The

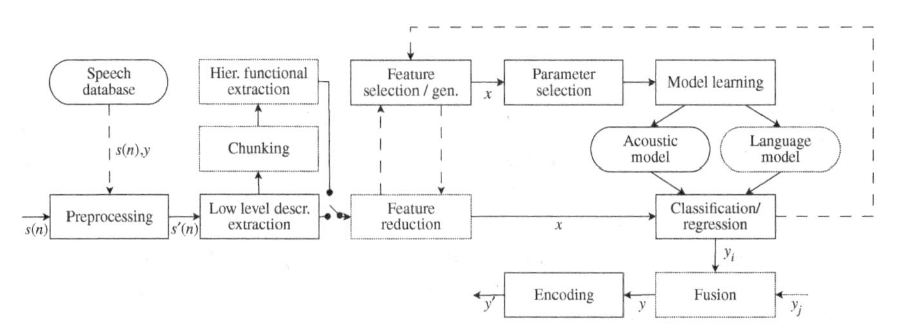

FIGURE 10.1 Unified overview of typical emotion recognition systems. Dotted boxes indicate optional components. Dashed lines indicate steps carried out only during system training or adaptation phases, where $s(n)$, x, and y are the speech signal, feature vector, and target vector, respectively, prime symbol indicates altered versions, and subscripts diverse vectors.

chain of processing usually includes low-level preprocessing (e. g., speech enhancement); generation of Low-Level Descriptors (LLDs) and computation of (hierarchical) statistical functionals for chosen "chunks" of analysis as described below; reduction of the feature space (cf. Section 10.3.2); parameter selection, typically by grid search on a dedicated set of development instances; learning of models (mapping acoustics and/or linguistics to emotion labels, or modeling phonemes and language statistics if ASR is involved in linguistic analysis, cf. Section 10.3.3); application of these models to generate prediction(s) y_i; late fusion of diverse predictions, such as from acoustic and linguistic models, or different classifier/regressor architectures, by majority voting or more powerful techniques such as Bagging, Boosting, or Stacking(C) [51]; and finally encoding of the (fused) prediction y in a standardized format (cf. Section 10.3.5).

10.3.1 Feature Extraction

Subsequent to preprocessing the captured speech signal (e. g., applying speech enhancement, source separation, de-noising, and de-reverberation), feature vectors are extracted at approximately 100 frames per second with typical window sizes of 10–30 ms for acoustics. Windowing functions are usually rectangular for extraction of LLDs in the time domain and smooth (e. g., Hamming or Hann) for extraction in the frequency or time-frequency (TF, e. g., Gaussian or general wavelets) domains. Many systems directly process features on this level, either to provide a frame-by-frame estimate or by dynamical approaches like Hidden Markov Models, that provide some sort of temporal alignment and warping.

Typical *acoustic* LLDs applied for speech-based emotion recognition cover intonation (pitch, etc.), intensity (energy, Teager, etc.), linear prediction cepstral coefficients (LPCCs), perceptual linear prediction (PLP), cepstral coefficients (MFCCs, etc.), formants (amplitude, position, width, etc.), spectrum (Mel frequency bands (MFBs), MPEG-7 audio spectrum projection, roll-off, etc.), TF transformation (wavelets, Gabor, etc.), harmonicity (harmonics-to-noise ratio (HNR), noise-to-harmonics ratio (NHR), etc.), and perturbation (jitter, shimmer, etc.). These are often added by deriving further LLDs based on the raw LLDs (deltas, regression coefficients, correlation coefficients, etc.). Further, diverse filtering (smoothing, normalizing, etc.) may be applied.

Systems that process static feature vectors that are generated for speech chunks usually model statistical functionals of acoustic LLDs. These functionals comprise extremes (minimum, maximum, range, etc.), mean (arithmetic, absolute, etc.), percentiles (quartiles, ranges, etc.), standard deviation, higher moments (skewness, kurtosis, etc.), peaks (number, distances, etc.), segments (number, duration, etc.), regression (coefficients, error, etc.), spectral (discrete cosine transformation coefficients, etc.), and temporal (durations, positions, etc.) as provided, for example, by the openSMILE feature extractor [52].

Commonly used *linguistic* LLDs comprise linguistic strings (phoneme sequences, word sequences, etc.), non-linguistics strings (laughter, sighs, etc.), and disfluencies (pauses, etc.). For processing linguistic LLDs, two methods seem to be predominant, presumably because they are shallow representations of linguistic knowledge and have already been frequently employed in automatic speech processing: (class-based) N-Grams and Bag-of-Words (vector space modeling), cf. Reference 53.

Bag-of-Words is a well-known numerical representation form of texts in automatic document categorization [54]. It has been successfully ported to recognize sentiments or emotion [53] and can equivalently be used for other target problems. Thereby each word in the vocabulary adds a dimension to a linguistic vector representing the term frequency within the actual utterance. Note that easily, very large feature spaces may occur, which usually require intelligent reduction (cf. below). The logarithm of frequency is often used; this value is further better normalized by the length of the utterance and by the overall (log-)frequency within the training corpus.

In addition, exploitation of on-line knowledge sources without domain-specific model training has recently become an interesting alternative or addition [55]—for example, to cope with out-of-vocabulary events. The largely related fields of opinion mining and sentiment analysis in text bear interesting alternatives and variants of methods.

Although we are considering the analysis from spoken text, only a few results for emotion recognition rely on automatic speech recognition (ASR) output [56] rather than on manual annotation of the data. As ASR of affective speech itself is a challenge [27], this step is likely to introduce errors. Note that their extraction usually requires ASR. As the recognition of spontaneous, emotional speech tends to be more challenging than, for example, the recognition of read speech and robust ASR systems, such as tandem [57] or multi-stream architectures, have to be applied for linguistic feature generation.

Apart from low-level feature space reduction techniques as discussed below, more elaborate techniques are available to reduce the number of linguistic features and at the same time data sparsity. Often, *stopping* is used, resembling elimination of irrelevant words. As even for experts it seems hard to judge which words can be of importance in view of the target problem, data-driven approaches as salience or information gain-based reduction are popular. A simple, yet often highly effective way is stopping words that do not exceed a general minimum frequency of occurrence in the training corpus.

Stemming is the clustering of morphological variants of a word (such as "cry," "cries," "cried," "crying," etc.) by its stem into a *lexeme*. Thereby also words that were not seen in the training can be mapped upon their representative morphological variant, for instance, by (Iterated) Lovins or Porter stemmers that are based on suffix lists and rules. Part-of-Speech (POS) tagging is a very compact approach where classes such as nouns, verbs, adjectives, particles, or more detailed subclasses are modeled [58]. Also, *sememes*, that is, semantic units represented by lexemes, can be clustered into higher semantic concepts such as generally positive or negative terms [59]. Finally, note that the aforementioned methods for modeling linguistic information allow to straightforwardly integrate non-linguistic vocalizations such as laughter or sighs, which are often highly meaningful in human–human conversations, into the vocabulary [3].

10.3.2 Feature and Instance Selection

To avoid extremely large feature spaces resulting from a high number of "brute-forced" LLD-functional combination, only the ones well suited for the classification task are kept. Typically, a target function is defined first. In the case of "open loop" selection, typical target functions are of information theoretic in nature such as information gain or statistical nature such as correlation among features and of features with the target of the task at hand. In the case of "closed loop," these are often the learning algorithm's error to be reduced. Usually a search function is needed in addition as exhaustive search in feature space is computationally hardly feasible. Such search may start with an empty set adding features in "forward" direction, with the full set deleting features in "backward" direction or bidirectional starting "somewhere in the middle." As the search is usually based on accepting a suboptimal solution but reducing computation effort, "floating" is often added to overcome nesting effects [60]. That is, in the case of forward search, (limited) backward steps are added to avoid a too "greedy" search. This "Sequential Forward Floating Search" is among the most popular in the field, as one typically searches a small number of final features out of a large set. In addition, generation of further feature variants can be considered within the selection of features, for example, by applying single feature or multiple feature mathematical operations such as logarithm or division which can lead to better representation in the feature space. In some cases, the feature selection and classifier training steps are integrated, for example, in training Deep Neural Networks [32], or by iterative construction (boosting) of ensembles of single-feature regressors as in the Simple Logistic algorithm [61].

Apart from genuine selection of features, the *reduction* of the feature space is often considered to reduce the complexity and the number of free parameters to be learnt for the machine learning algorithms while benefiting from all original feature information. This is achieved by mapping of the input space onto a less-dimensional target one, while keeping as much information as possible. Principal Component Analysis (PCA) [62] and (Heteroscedastic) Linear Discriminant Analysis (LDA) are the most common techniques. While PCA is an unsupervised feature reduction method and thus is often suboptimal for more complex problems, LDA is a supervised feature reduction method that searches for the linear transformation that maximizes the ratio of the determinants of the between-class covariance matrix and the within-class covariance matrix, that is, it is a discriminative method as the name indicates. In fact, none of these methods is optimal: There is no straight forward way of knowing the optimal target space size—typically the variance covered is a decisive measure. Further, a certain degree of normal distribution is expected, and LDA additionally demands linear separability of the input space. PCA and LDA are also not very appropriate for feature mining, as the original features are not retained after the transformation. Finally, Independent Component Analysis (ICA) [63] and Non-negative Matrix Factorization (NMF) [64] can be named. While ICA maximizes statistical independence of features (components), NMF assumes non-negative data and additive features: Thus, NMF is at present mainly employed for large linguistic feature sets, but also increasingly popular for extracting audio features from (magnitude) spectrograms, especially signal mixtures [65].

Generally, it seems important to mention that there is a high danger of over-adaptation to the data that features are selected upon. As a counter-measure, it seems wise to address feature importance across databases [66].

Besides selecting or generating discriminative features, focus of classifier training on "salient" training instances—or *prototypes*—has been considered. A straightforward practical approach is to speak of "prototypical" cases if a multiplicity of human raters agree on how it should be assigned to a category or numeric quantity. Non-prototypical weak and/or mixed emotions [67] can be found when labelers annotate more than one emotion per item, or when we preserve the disagreement of several labelers in some sort of graded/mixed annotation. In Reference 7, non-prototypical emotions are conceptualized under the notion of *hinterland*: In contrast to the big *n* emotions anger, joy, sadness, despair, and so on, emotion-related, affective states such as interest and tiredness, and mixed/weak emotions are conceived as constituting the hinterland of prototypical emotions.

In Reference 68, for instance, the degree of prototypicality of the training material has been varied while keeping everything else constant. The idea behind is, of course that, using only more prototypical cases could yield higher classification performance. This approach is sometimes used in machine learning and is known as data cleaning or pruning: However, suspicious patterns may not always be garbage patterns as "noisy" data too might be needed to make the classifier learn to deal with ambiguities, as are characteristic for spontaneous emotional speech. Similar results obtained on different data sets and with different classifiers are confirmed by Reference 69 where models created for prototypical emotional speech proved to be ineffective for the classification of non-prototypical utterances. Recently, fully automatic methods for instance selection have been introduced to emotion recognition [70].

It is crucial to differentiate the aforementioned selection of training instances from the preselection of "friendly cases" in *testing*: When striving at real-life applications, the goal of such endeavors should always be to improve the expected accuracy in an open-microphone setting. Furthermore, in instance selection, the very same question of generalization as in feature selection remains to be answered.

10.3.3 Classification and Learning

In the training phase, the classifier or regressor model is built based on labeled data and extracted features. Applying appropriate models, the actual target has to be assigned to an unknown test instance. In the case of classification, these are discrete labels such as Ekman's "big six" emotion classes anger, disgust, fear, happiness, sadness, and surprise. In the case of regression, the output is a continuous value like the degree of valence, arousal, or dominance, typically ranging from -1 to $+1$.

Frame-wise classification of emotion presumes a classifier that can access and model long-range context, since emotion mostly affects the long-term *dynamics* of prosodic, spectral, and voice quality speech features. When attempting to predict emotion frame by frame, a large number of preceding speech frames have to be taken into account in order to capture speech characteristics that are influenced by emotion. The *number* of speech frames that should be used to obtain enough context for reliably estimating emotion without affecting the capability of also detecting sudden changes of the speaker's emotional state is hard to determine [47]. While some studies dealt with manual optimization of context length—for example, Reference 71 indicates that one second of context is suitable for valence recognition—classifiers that are able to *learn* the amount of context are promising alternatives to manually defining fixed time windows for emotion recognition. Static techniques such as SVMs do not explicitly model context but rely on either capturing contextual information via statistical functionals of features [3] or aggregating frames using Multi-Instance Learning techniques [72]. Dynamic classifiers like Hidden Markov Models (HMMs) are often used for flexible context modeling and time warping. Yet, HMMs have drawbacks such as the inherent assumption of conditional independence of successive observations, meaning that an observation is statistically independent of past observations, provided that the values of the hidden variables are known. Hidden Conditional Random Fields (HCRFs) are one attempt to overcome this limitation. However, HCRFs also offer no possibility to model a self-learned amount of contextual information. Other classifiers such as neural networks are able to model a certain amount of context by using cyclic connections. These so-called recurrent neural networks can in principle map from the entire *history* of previous inputs to each output. Yet, due to the *vanishing gradient problem*, the range of context that can be modeled via RNNs is limited to about 10 frames. Long Short-Term Memory (LSTM) networks offer an efficient technique to overcome this problem by using *memory blocks* in the hidden layer [73] and are increasingly applied for emotion recognition [74].

However, as emotion is apparently better modeled on a time-scale above frame-level [6], processing static supra-segmental features is a popular alternative to frame-based modeling—note that a combination of static features such as minimum, maximum, onset, offset, duration, and regression implicitly shapes contour dynamics as well. It is straightforward to combine static features with dynamic modeling, for instance, by training context-sensitive classifiers on supra-segmental features [75]; however, usually supra-segmental features are modeled with static classifiers. Among them, SVM can be seen as a sort of state-of-the-art classifier (or regressor, as the related Support Vector Regression allows for handling of continuous problem descriptions). They combine discriminative learning and solving of non-linear problems by a kernel-based transformation of the feature space and are little prone to over-fitting to training data. As an alternative, Random Forests (RFs) can be employed: They consist of an ensemble of trees, each one accounting for random, small subsets of the input features obtained by sampling with replacement [76]. Similar to SVM, they are thus practically insensitive to the "curse of dimensionality" and irrelevant or noisy features, while, at the same time, still providing the benefit of "readable" classifiers rather than "black boxes" such as non-linear SVM or neural networks. Furthermore, model trees [61] are an innovative concept for constructing compact decision trees incorporating regression functions at the leaves. Both RFs and model trees can be used for both classification and regression.

10.3.4 Partitioning and Evaluation

System evaluation is an integral part of iterative system development. Ideally, such evaluation would be carried out in field studies—in practice, standardized data sets are chosen. As pointed out earlier, however, in AER such data sets are often quite sparse. Thus, it is hard to obtain meaningful—that is, statistically significant results, cf. below—when those corpora are further subdivided into training, development, and testing instances. Consequently, many studies still employ J-fold cross-validation (usually, $J = 3$ or $J = 10$) with randomized folds, for example, Reference 66. However, only Leave-One-Subject-Out (cf. Reference 58) or Leave-One-Subject-Group-Out (cf. Reference 6) (cross-) validation would ensure true speaker independence as is needed in many real-life applications such as voice portals.

Once a test (or evaluation) set is defined, the most straightforward—and still widely used—evaluation measure is (weighted) accuracy (WA) or recall: the probability that one test instance is classified correctly. This is equivalent to a weighted sum of the classwise recalls, with a priori class probabilities as weights. In contrast, unweighted accuracy (UA) is obtained when weighting classwise recalls by $1/M$, where M is the number of classes. The difference between WA and UA is visible foremost when the class distribution of test patterns strongly deviates from unit distribution; particularly, in heavily skewed class distributions high WA can be obtained by simply picking the majority class—thus, when evaluating systems on such problems, UA, the official competition measure of the first INTERSPEECH 2009 Emotion Challenge and their follow-ups, seems preferable [6]. In other words, WA evaluates the expected performance of a system while UA measures its ability to discriminate classes. In other studies, the very same metric has been called classwise recognition rate (CL) [42].

Generally, optimizing a classifier to obtain higher recall of one class, or higher number of *true positives* of that class, usually results in false assignments of many patterns from other classes (*false positives*), corresponding to lower *precision*. Precision is defined as the probability that an instance that is assigned to a certain class belongs in fact to that class. F-Measure is the harmonic mean of recall and precision [77].

In the beginning, performance of emotion recognition was usually measured in terms of accuracy obtained on some "big N" categories of emotion—evaluation on the acted BES database being an archetypal example of such methodology. With the recent trend toward real-life settings where the a priori probability of encountering full-blown emotions is fairly low, a paradigm shift can be expected toward evaluating emotion recognition as a *detection* task, such as in Reference 7: In that study, the binary task to discriminate the emotional states "negative valence" or "idle" was turned into a detection of negative valence. Obviously, detection comes with different evaluation measures: detection probability or true positive rate (TPR; in the above example, the probability that negative valence is classified as such), and false alarm or false-positive rate (FPR; in the above, erroneously detecting negative valence). Arguably, false alarm rates are highly relevant in applications where automatic systems react to certain emotional states of the user—for instance, a voice portal erroneously shifting its dialog strategy toward supposedly angry customers might cause users to feel "injustly accused," resulting in low acceptance of the system. Conversely, when preselecting emotional episodes in large amounts of unlabeled speech for subsequent human annotation, one would rather want to optimize the TPR. Usually, TPR and FPR can be adjusted by tuning a detection threshold, corresponding to confidence measure for the "yes" class. Then, an approximation of the receiver-operating characteristic (ROC) can be interpolated from the FPR and TPR obtained for different thresholds. The information contained in the ROC curve (usually, TPR as a function of FPR) can be summarized in a single measure by computing the area under the curve (AUC), or equal error rate (EER), that is, the point of equality of TPR and FPR [78].

As opposed to classification and detection, in case of regression in continuous dimensions of emotion, such as in the INTERSPEECH 2010 Paralinguistic Challenge (Affect Sub-Challenge)—there, the task was to determine the continuous-valued level of customer interest in a product presentation scenario—evaluation measures are mostly based on the mismatch between "ground truth" values and the regressor outputs. This mismatch can be measured in terms of (Pearson or Spearman) correlation coefficient (CC) or by differently weighted versions of the absolute difference, for example, mean absolute error (MAE or MLE) or (root) mean squared error (MSE), across test instances [79].

Finally, as the AER community currently strives for more standardized evaluations [6]—as are standard for a long time in ASR and other related fields—it is of crucial importance to determine whether performance differences encountered in system evaluation on the same data are due to random influences—for example, numerical inconsistencies—or rather caused by the structure of the systems. Significance tests can be utilized to that end. Comparing system accuracies is sometimes performed by the McNemar test [80], related to the well-known Chi-square test. Here, results achieved by both systems are summarized in a contingency table, counting instances that are correctly classified by either of the systems, none of them, or both. The null hypothesis is then that one "loses" as much as one "gains" by choosing system A over system B—that is, the number of newly induced misclassifications by using A is equal to the number of instances that are misclassified by B but not A. In many cases, a simple z-test [80] is used instead—a practical disadvantage of the McNemar test is that it requires the precise predictions of each system on each test instance, which is not given, for example, when comparing results reported in the literature. The z-test is based on the assumption that the accuracy difference between a classifier A and a baseline B with accuracies p_a and p_b is a normally distributed random variable with mean $p_a - p_b$ and variance $2p(1-p)/S$, where $p = (p_a + p_b)/2$ and S is the number of instances of the test set. In a one-tailed test, the null hypothesis is that $p_a \leq p_b$—informally, A is not better than the baseline B. Still, results of this test should only be interpreted as a heuristic measure, since the estimates of p_a and p_b on the test set are not independent [80]. Furthermore, it is not straightforward to measure significance of differences in *unweighted* accuracy. In emotion recognition, caution must be exercised in significane testing since the required assumption of statistical independence among test samples is not necessarily given, especially, when rather short units of analysis such as subsequent words are considered—in ASR, one often assumes that results of a recognition algorithm on parts separated by speech pauses are independent of each other [81]. As a final point of discussion in the context of significance testing, usually repeated measurements are not corrected for: Instead, significance is not used in the inferential meaning but as a sort of descriptive device, as recently suggested, for example, in Reference 82.

10.3.5 Research Toolkits and Open-Source Software

Toolkits and freely available codes have always considerably influenced research—they aid in fast prototyping and provide straightforward reproducibility of experiments reported in the literature. In particular, the field of automatic speech recognition research has been co-shaped by the technology available in the Hidden Markov Model Toolkit (HTK)[1] [83]. The frequent choice of HTK in ASR resulted in an almost exclusive use of MFCC and energy features as implemented in the HTK, if Hidden Markov Models were used for emotion processing. In this respect, two types of available packages influenced the development in the recognition of emotion recognition: first, tools for feature extraction such as PRAAT[2] or the SNACK sound toolkit,[3] and secondly classification software and learning environments such as the WEKA[4] [77] package. Further sparsely encountered toolboxes or alternatives for feature extraction and classification include jAudio,[5] Marsyas,[6] LibXtract,[7] or ESMERALDA.[8] Popular implementations of SVM include SVM-Light[9] or LibSVM[10]. Of course, many individual feature extractions and classification algorithms are found as well.

However, only a few dedicated open source projects are found that aim at a free and comprehensive platform, in particular for emotion recognition from speech. The Munich open-source Emotion and Affect Recognition Toolkit (openEAR) [84] is the first of its kind to provide a free open source toolkit that integrates all three necessary components: feature extraction (by the fast openSMILE [52] back-end), classifiers, and pretrained models. An interesting aspect of the openEAR toolkit is that it provides predefined sets of derived functionals for comparison such as the sets provided for the INTERSPEECH 2009 Emotion Challenge, the INTERSPEECH 2010 Paralinguistic Challenge, INTERSPEECH 2011 Speaker State Challenge, and the first of its kind Audio/Visual Emotion Challenge 2011. This is beneficial, as apart from some publicly available scripts for extractors such as PRAAT, highly individual

[1] http://htk.eng.cam.ac.uk/
[2] http://www.fon.hum.uva.nl/praat/
[3] http://www.speech.kth.se/snack/
[4] http://www.cs.waikato.ac.nz/~ml/weka/
[5] http://sourceforge.net/projects/jaudio
[6] http://marsyas.info/about/projects
[7] http://www.postlude.co.uk/postlude/downloads/LibXtract:_a_lightweight_feature_extraction_library.pdf
[8] http://www.irf.tu-dortmund.de/cms/en/IS/Research/ESMERALDA/index.html
[9] http://svmlight.joachims.org/
[10] http://www.csie.ntu.edu.tw/~cjlin/libsvm/

implementations and solutions are found, also for the functionals. A similar initiative, yet not open source, is the EmoVoice[11] [85] toolkit.

For integration of existing implementations into working systems, interoperability is crucial. Particularly, once a recognition result is produced, it will have to be communicated to the application; for emotional dialog systems, the other direction, that is, the communication of an emotion to synthesize, is relevant as well. A standard was not needed in the early days of emotion recognition from speech because the performances were not sufficiently mature, and no broad application basis existed. However, with a rising need to integrate emotion recognition components into large-scale systems such as voice portal engines, standardized coding of information has to be dealt with. The HUMAINE EARL can be named as one of the first standards tailored to provide a well-defined description of recognized or to be synthesized emotions. Released as a working draft, the W3C EmotionML[12] [86] followed EARL providing more flexibility and a broader coverage including action tendencies, appraisals, meta context or a basis to encode regulation, acting, meta-data, and ontologies. Particularly for the encoding of acoustic and linguistic features, a standard was proposed and used within the CEICES initiative [59]. Further, a number of standards exists which do not exclusively deal with emotion but contain definitions for suited tags, such as EMMA[13] [87]. However, these standards have not been used frequently so far. This will most likely change when more and more applications will be addressed. At the same time, it remains to hope that either their number stays limited, or well-defined translations among standards will be provided, to avoid incompatibilities.

10.4 CHALLENGES

10.4.1 Non-prototypicality, Reliability, and Class Sparsity

In the last years, it has become obvious that one of the greatest challenges for application of emotion recognition in real-life settings is that non-prototypical emotional speech has to be processed. In that context, "non-prototypical" can be considered as not belonging to the well-known, "big N" emotions, cf. References 88 and 89. Nevertheless, the effects of applying AER to non-prototypical data have been studied only recently, cf. References 6, 7, 58, 68, 69, 90. Partly, this might be due to the fact that if we do not select "friendly instances," but deal with all data that is available, for example, in a recorded corpus of emotional speech (open microphone setting), class boundaries might be "noisy" and classification performance rather low. In contrast, performance can often be optimized by selecting the most prototypical cases—at the expense of full realism. For instance, in Reference 7 an upper benchmark is given with respect to performance improvement by selecting exclusively prototypical test patterns: Unweighted accuracy in binary speaker-independent emotion classification of highly spontaneous speech could be boosted to an impressive 85%.

[11] http://www.informatik.uni-augsburg.de/lehrstuehle/hcm/projects/emovoice/
[12] http://www.w3.org/2005/Incubator/emotion/XGR-emotionml/
[13] http://www.w3.org/TR/emma/

A related peculiarity of emotion recognition is uncertainty of ground truth, especially when comparing with ASR, even more so as once one starts dealing with non-prototypicality. Normally, annotations are taken as representing the ground truth. This might not be (fully) true, for several reasons. First, there can be inconsistencies in the annotation process itself. Second, the effort needed often prevents a sufficiently high number of annotators; we might need more if "naive" subjects are employed—for representing the "wisdom of the crowd," or simply due to restricted financial resources—and a bit less if experts are employed. A number close to 10 might be optimal; often, it is much less. Third, annotations rather model perception processes, not production processes, that is, perceived, not felt emotions—which might or might not be problematic, depending on the application. In this respect, physiological signals might come closer to the ground truth, in particular for arousal; however, these are rarely available during recording, and hardly ever available in typical applications of voice-based systems.

Thus, most frequently, an estimate of the ground truth emotion is deduced from multiple human raters. The obvious method is to take a majority vote (for categories) or the mean (for continuous dimensions)—which gives rise to the questions whether such estimates are robust, and whether all of the raters are "reliable": Often, trivial effects such as tiredness or boredom when annotating large amounts of data can deteriorate the quality of human ratings. For identifying such cases, measures of inter-annotator agreement, that is the degree that individual subjects agree on the labeling of an audio signal, have been thoroughly researched in psychology and medicine, but interestingly also in "traditional" electrical engineering in the context of data coding [91]. Usual measures include Cohen's Kappa for nominal ratings, weighted Kappa or Spearman's rank correlation to take into account ordinal scales, Cronbach's Alpha, or Krippendorff's Alpha reliability, which is able to model missing values—an important aspect in practice, as not all instances might be labeled by the same set of raters, or raters might leave out items where they are not able to make a final decision. Measures of agreement also allow for fusing rater decisions by weighting individual labels by their agreement with the majority [92]. Further, sometimes evaluations are restricted to those with high inter-annotator agreement [3]—somewhat corresponding to selection of prototypes as discussed above. Still, in a real-life setting, uncertainty on unseen test data must be detected automatically and be dealt with using suited strategies.

To estimate uncertainty of automatic predictions, naturally, one can think of distances to hyperplanes of SVM or posterior probabilities, and so on, yet evidence from independent sources as often followed in speech recognition, for example, acoustic stability, have not been, to our knowledge, considered in this field so far. One peculiarity of the field appears promising in this respect, though: the fact that usually several labelers annotate the data allows for training of labeler-specific emotion models. As a consequence, one can predict the inter-labeler agreement in addition to the emotion, which may serve as a confidence: In a pioneering study, for example, an error of only one labeler on average out of five is reported on the FAU AIBO corpus [7]. However, the error or confidences when it comes to the temporal alignment, or a change to a spotting paradigm, may be further interesting future directions in this respect. A maybe even more challenging question is how to design systems that can deal with uncertainty. A promising direction for future system design in that respect might be the use of reinforcement learning strategies—there is evidence that reinforcement learning reflects human decision making [93].

A final issue to be discussed in the context of non-prototypicality is the skewness of label distributions that usually results from using non-prompted emotional speech from real-life interactions: Particularly, a prevalence of a "neutral" emotional state is expected, as can be observed in the FAU AIBO corpus of child–robot interaction [6, 7]. From the machine learning point of view, balancing of the training instances with respect to instances per class is often a necessary step before classification [6]. The balancing of the output space can be addressed either by considering proper class weights (e. g., priors) or by resampling, that is, (random) up- or down-sampling. Class priors are implicitly taken into account by discriminative classifiers such as SVM or RNN; hence, for extremely skewed problems, maximum-likelihood class estimation as used, for example, in HMM-based Viterbi decoding might be preferable.

10.4.2 Generalization

It is evident that since emotional corpora are usually tied to specific scenarios, evaluation of automatic analysis, which is an integral part of iterative system development, faces the problem of generalization: It is not clear whether improvements achieved on one database can be exploited in other applications. Thus, there is a need for cross-corpus recognition experiments—that is, training on one or more databases and evaluating on disjoint corpora [94]. The crux, however, is that these data usually come with completely different emotion inventories reaching from Ekman's "big six" to task specific ones. The dimensional space offers us the ability to "translate" these models into, for example, arousal and valence dimensions, and first studies exist on enhancing the robustness of cross-corpus emotion recognition, by combining several emotional speech corpora within the training set and by that reduce the data scarcity problem and extend the variety of acoustic background [94–96]. Still, there is a variety of interesting research issues open to debate: for instance, optimal (early or late) fusion of different training corpora, and normalization approaches to the speaker, corpus, or both to mitigate the divergence between conditions. For instance, it remains to investigate Joint Factor Analysis as often used in speaker recognition for session variability compensation.

Generalization not only across application scenarios but also to different cultural backgrounds of potential system users is highly desirable for real-life systems. In psychology, cross-cultural emotion recognition by humans has been extensively studied, providing evidence for both cultural similarities and differences [97]. While earlier studies mostly focused on facial expression of emotion, recent studies also take into account the auditory or multimodal perception of emotion [98, 99], indicating that emotional cues in speech can be recognized by humans across languages and cultures, particularly in a multimodal setting [98]. In contrast, in the area of automatic affect recognition from speech, it is only in the past decade that the issue of multi-linguality receives increasing attention. At a superficial level, it is clear that spoken language differs across cultures and languages along a multiplicity of dimensions, ranging from acoustic phonetics through grammar, vocabulary, and metaphor to pragmatics and discourse strategies; consequently, in ASR, accuracy typically drops for cross-lingual or multilingual recognition. In AER, while there is evidence that important tasks such as anger detection can be reliably performed in a multi-lingual way—although the features for classification may differ between languages [100]—the task in general has been shown to depend strongly on the language being spoken [48, 101, 102].

10.4.3 Real-Time Processing

Real-time, that is, incremental processing is an issue that is often overlooked in research—which usually evaluates in batch processing—but is crucial when moving technology "out of the lab" toward system integration. Such incremental processing means providing an on-line estimate after the onset, updated continuously until the offset—this is often referred to as "gating" [71]. How fast after the beginning of a speech turn one can provide a reasonable prediction of the likely emotion is for example of importance in dialog systems such as the SEMAINE system [17] where complex multimodal system outputs have to be generated in advance to "fire" the right alternative just when it is needed. For that, an early prediction in the right direction is of course meaningful. A few studies deal with this topic; Reference 47 indicates that one second of speech might be sufficient. An incremental update of the prediction is also possible by classifiers suited for this task that may at the same time learn how much past context information is beneficial [103]. For example, in Reference 104 350 ms is suggested for human-like back-channeling in certain situations. Some toolkits, for example, openEAR, already allow for this type of processing [84].

10.4.4 Acoustic Environments: Noise and Reverberation

Finally, when designing AER systems for usage "in the wild," one has to consider that most research in automatic emotion recognition is dealing with laboratory acoustic environments: Neither the influence of background noise (environmental noise or interfering speakers) or reverberation have been extensively studied. This is in strong contrast to automatic speech recognition, where these robustness issues have been thoroughly researched for years, with a considerable amount of well elaborated techniques available—a survey can be found in Reference 105. A few studies so far dealt with the challenge of noise-robust AER, such as References 106, 107. Besides, at present the tools and particularly evaluation methodologies for noise-robust AER are rather basic: Often, they are constrained to elementary feature enhancement and selection techniques, are characterized by the simplification of additive stationary noise, or are limited to matched condition training [108, 109].

Besides, little is known about the influence of reverberation on emotion analysis. While adaptation techniques to reverberated environments have been discussed in Reference 110, a first attempt toward a systematic study of the influence of non-stationary noise and different microphone conditions on the INTERSPEECH 2009 Emotion Challenge task [6] has been carried out in Reference 111—"systematic" in the sense that typical methodologies from the ASR domain were adopted, such as commonly performed with the "Aurora" tasks. In the context of AER, acoustic feature extraction based on Non-negative Matrix Factorization (NMF) [64], which has lead to considerable success in the ASR domain [112], can lead to considerable improvements in robustness against noise and reverberation [111]. The basic principle of NMF-based audio processing is to find a locally optimal factorization of a spectrogram into two factors, of which the first one represents the spectra of the acoustic events, such as phonemes or noise, occurring in the signal and the second one their activation over time. This factorization can be computed by iteratively minimizing cost functions resembling the perceptual quality of the product of the factors, compared with the original spectrogram. In the future, NMF-based methods for AER could be extended to exemplar-based recognition, reducing unseen test instances to combinations of already labeled data by nearest subspace approximation—a paradigm that has recently led to excellent results in ASR [112] and general audio recognition [65].

Generally, a different, yet promising direction for noise-robustness in real-life AER applications is to augment acoustic by linguistic features derived from ASR output—then, one has access to the full "arsenal" of methodologies for robust ASR. For instance, incorporating features from a multistream tandem speech recognizer in the INTERSPEECH 2009 Emotion Challenge task delivers an accuracy boost even without acoustic feature enhancement [113].

Finally, it is still open to debate whether monaural—or even multichannel, depending on the application—source separation techniques such as NMF or ICA should be exploited for speech enhancement/extraction rather than feature extraction and recognition. These have now evolved to a state of maturity in the last decade, which could be immediately of use for both, enhancement of acoustic and linguistic features for emotion recognition.

10.5 CONCLUSION AND OUTLOOK

Obtaining more realistic data will still be the most important issue in the foreseeable future. It is an ever-lasting belief in pattern recognition that "there is no data like more data." Yet, compared to ASR where researchers have access to thousands of hours of transcribed speech, databases annotated in emotional categories are still sparse—in particular publicly available ones. Furthermore, when striving at realistic data, the sparsity of emotional events is a huge limiting factor. On the other hand, it is commonly believed that the increasing prevalence and robustness of automatic speech recognition in real-life systems is mostly due to increased computational power, allowing model training on huge amounts of data. Thus, the question arises if additional training data for emotion recognition can be automatically generated. A promising result is that synthesized speech can be used for improved cross-corpus model training in emotion recognition [114]—synthesized speech can be produced in high variety to produce very general learning models from data with known ground truth. Similarly, it remains to investigate whether robustness in real-life settings can be addressed by using artificially reverberated or data overlaid with noise—first experiments with realistic and artificial reverberation are quite promising [113]. Semisupervised and unsupervised learning can be a promising approach for the future, if engines are sufficiently robust to label emotional data by themselves: In that case, classification or regression models are built iteratively through retrieval of data from on-line knowledge sources (e. g., YouTube videos), application of (multi-corpus) automatic emotion recognition, and possibly refinement of the annotation by experts. Finally, AER could certainly learn from the ASR field with respect to distributed recognition architectures: Large quantities of emotional data from multiple sources and in multiple languages could be agglomerated on a cloud of servers and used for iterative refinement of publicly available models, instead of building emotion recognition systems from scratch over and over again.

Furthermore, it is striking that at present emotion recognition is mostly considered as an isolated technique next to ASR and automatic retrieval of other paralinguistic states and traits, such as age, gender, personality, sleepiness, or intoxication, although it is known that interdependencies between these phenomena exist; for instance, short-term states such as emotion are dependent on personality traits [115, 116]. Whenever there exist correlations between the characteristics and prediction targets, simultaneously learning multiple classification targets or exploiting contextual knowledge in the form of the automatic estimation of certain speaker or speech characteristics for a related classification task is especially promising. Several studies indicate that considering gender information in an automatic emotion recognition system leads to higher recognition accuracies [117]; however, it is not settled, yet, whether this is simply due to the different pitch registers of male and female voices, or to gender-specific differences in the expression of emotion. Multitask learning has been implemented mostly for recurrent neural networks [118], which may be of particular interest in this field to, for example, assess emotion and personality in one pass, benefiting from mutual dependencies; however, these methods have to be tailored accordingly.

In this chapter, we focused on emotion recognition from speech and language. Frequently, other modalities including video information such as facial expression or, in a few cases, also body postures and gestures is exploited, followed by fewer studies on audiovisual processing, for example, References 3, 119, 120. At the same time, a smaller group of studies dealing with analysis of biosignals, such as heart rate, skin conductivity, or brain-waves are found, for example, References 121–123 also in combination with speech. In such multimodal contexts, typically different units of analysis are investigated, but have to be fused in a synergistic manner. In fact, this problem can already arise to a certain extent when we want to fuse acoustic and linguistic information, or even for processing exclusively acoustic information, to exploit information from various time resolutions. Due to the general trend to multimodal processing and availability of more complex algorithms to this aim such as asynchronous Hidden Markov Models, Dynamic Bayesian Networks, general Graphical Models or Multidimensional Dynamic Time Warp and Meta-Classification [124–127] in combination with higher availability of multimodal data, the future might be expected to be dominated by multimodal approaches, potentially also integrating more contextual information for hopefully higher recognition performance.

ACKNOWLEDGMENT

The research of Felix Weninger and Martin Wöllmer is funded by the Federal Republic of Germany through the German Research Foundation (DFG) under grant nos. SCHU 2508/2-1 and 2508/4-1.

REFERENCES

[1] B. Schuller, M. Wimmer, L. Mösenlechner, C. Kern, D. Arsic, and G. Rigoll, "Brute-forcing hierarchical functionals for paralinguistics: a waste of feature space?" in Proceedings of the ICASSP, Las Vegas, NV, 2008, pp. 4501–4504.

[2] L. Devillers, L. Vidrascu, and L. Lamel, "Challenges in real-life emotion annotation and machine learning based detection," *Neural Netw.*, 18: 407–422, 2005.

[3] B. Schuller, R. Müller, F. Eyben, J. Gast, B. Hörnler, M. Wöllmer, G. Rigoll, A. Höthker, and H. Konosu, Being bored? Recognising natural interest by extensive audiovisual integration for real-life application. *Image Vis. Comput. J.*, (Special Issue on Visual and Multimodal Analysis of Human Spontaneous Behavior), 27: 1760–1774, 2009.

[4] A. Batliner, S. Steidl, and E. Nöth, "Releasing a thoroughly annotated and processed spontaneous emotional database: the FAU Aibo Emotion Corpus," in Proceedings of a Satellite Workshop of LREC 2008 on Corpora for Research on Emotion and Affect (eds L. Devillers, J.-C. Martin, R. Cowie, E. Douglas-Cowie, and A. Batliner), Marrakesh, 2008, pp. 28–31.

[5] F. Burkhardt, A. Paeschke, M. Rolfes, W. Sendlmeier, and B. Weiss, "A database of German emotional speech," in Proceedings of the Interspeech, Lisbon, 2005, pp. 1517–1520.

[6] B. Schuller, A. Batliner, S. Steidl, and D. Seppi, "Recognising realistic emotions and affect in speech: state of the art and lessons learnt from the first challenge," *Speech Commun.*, (Special Issue on Sensing Emotion and Affect—Facing Realism in Speech Processing), 2011.

[7] S. Steidl, B. Schuller, A. Batliner, and D. Seppi, "The hinterland of emotions: facing the open-microphone challenge," in Proceedings of the ACII, Amsterdam, 2009, pp. 690–697.

[8] A. Batliner, F. Burkhardt, M. van Ballegooy, and E. Nöth, "A taxonomy of applications that utilize emotional awareness," in Proceedings of the IS-LTC 2006, Ljubliana, 2006, pp. 246–250.

[9] R. Cowie, E. Douglas-Cowie, N. Tsapatsoulis, G. Votsis, S. Kollias, W. Fellenz, and J. Taylor, "Emotion recognition in human–computer interaction," *IEEE Signal Processing Magazine*, 18(1): 32–80, 2001.

[10] R. Picard, "Affective computing: challenges," *J. Hum.-Comput. Stud.*, 59: 55–64, 2003.

[11] F. Burkhardt, M. van Ballegooy, R. Englert, and R. Huber, "An emotion-aware voice portal," in Proceedings of the Electronic Speech Signal Processing (ESSP), 2005, pp. 123–131.

[12] G. Mishne, D. Carmel, R. Hoory, A. Roytman, and A. Soffer, "Automatic analysis of call-center conversations," in Proceedings of the CIKM'05, Bremen, Germany, 2005, pp. 453–459.

[13] D. Litman and K. Forbes, "Recognizing emotions from student speech in tutoring dialogues," in Proceedings of the ASRU, Virgin Island, 2003, pp. 25–30.

[14] L. Price, J. T. E. Richardson, and A. Jelfs, "Face-to-face versus online tutoring support in distance education," *Stud. High. Educ.*, 32(1): 1–20, 2007.

[15] A. Delaborde and L. Devillers, "Use of non-verbal speech cues in social interaction between human and robot: emotional and interactional markers," in AFFINE'10—Proceedings of the Third ACM Workshop on Affective Interaction in Natural Environments, Co-located with ACM Multimedia 2010, 2010, pp. 75–80.

[16] A. Batliner, S. Steidl, and E. Nöth, "Associating children's non-verbal and verbal behaviour: body movements, emotions, and laughter in a human–robot interaction," in Proceedings of ICASSP, Prague, 2011, pp. 5828–5831.

[17] M. Schröder, R. Cowie, D. Heylen, M. Pantic, C. Pelachaud, and B. Schuller, "Towards responsive sensitive artificial listeners," in Proceedings of the Fourth International Workshop on Human–Computer Conversation, Bellagio, 2008.

[18] B. Schuller, F. Eyben, S. Can, and H. Feussner, "Speech in minimal invasive surgery—towards an affective language resource of real-life medical operations," in Proceedings of the Third ELRA International Workshop on EMOTION (satellite of LREC): Corpora for Research on Emotion and Affect, Valetta, 2010, pp. 5–9.

[19] B. Schuller, M. Wimmer, D. Arsic, T. Moosmayr, and G. Rigoll, "Detection of security related affect and behaviour in passenger transport," in Proceedings of the Interspeech, Brisbane, 2008, pp. 265–268.

[20] C. Clavel, I. Vasilescu, L. Devillers, G. Richard, and T. Ehrette, "Fear-type emotion recognition for future audio-based surveillance systems," *Speech Commun.*, 50(6): 487–503, 2008.

[21] K. Laskowski, "Contrasting emotion-bearing laughter types in multiparticipant vocal activity detection for meetings," in Proceedings of the ICASSP, IEEE, Taipei, Taiwan, 2009, pp. 4765–4768.

[22] Z. Massida, P. Belin, C. James, J. Rouger, B. Fraysse, P. Barone, and O. Deguine, "Voice discrimination in cochlear-implanted deaf subjects," *Hear. Res.*, 275: 120–129, 2011.

[23] J. Demouy, M. Plaza, J. Xavier, F. Ringeval, M. Chetouani, D. Périsse, D. Chauvin, S. Viaux, B. Golse, D. Cohen, and L. Robel, "Differential language markers of pathology in autism, pervasive developmental disorder not otherwise specified and specific language impairment," *Res. Autism Spectr. Dis.*, 5(4): 1402–1412, 2011.

[24] C. Biever, "You have three happy messages," *New Scientist*, 185: 21, 2005.

[25] T. Athanaselis, S. Bakamidis, I. Dologlu, R. Cowie, E. Douglas-Cowie, and C. Cox, "ASR for emotional speech: clarifying the issues and enhancing performance," *Neural Netw.*, 18: 437–444, 2005.

[26] R. Fernandez and R. W. Picard, "Modeling drivers' speech under stress," *Speech Commun.*, 40: 145–159, 2003.

[27] S. Steidl, A. Batliner, D. Seppi, and B. Schuller, "On the impact of children's emotional speech on acoustic and language models," *EURASIP J. Audio Speech Music Process.*, 1–14, 2010.

[28] H. Boril, A. Sangwan, T. Hasan, and J. H. L. Hansen, "Automatic excitement-level detection for sports highlights generation," in Proceedings of the Interspeech 2010, Makuhari, Japan, 2011, pp. 2202–2205.

[29] F. Burkhardt, M. van Ballegooy, K.-P. Engelbrecht, T. Polzehl, and J. Stegmann, "Emotion detection in dialog systems: applications, strategies and challenges," in Proceedings of the ACII, Amsterdam, The Netherlands, 2009, pp. 1–6.

[30] D. Erickson, K. Yoshida, C. Menezes, A. Fujino, T. Mochida, and Y. Shibuya, "Exploratory study of some acoustic and articulatory characteristics of *sad* speech," *Phonetica*, 63: 1–25, 2004.

[31] R. Cowie, E. Douglas-Cowie, and C. Cox, "Beyond emotion archetypes: databases for emotion modelling using neural networks," *Neural Netw.*, 18(3): 388, 2005.

[32] A. Stuhlsatz, C. Meyer, F. Eyben, T. Zielke, G. Meier, and B. Schuller, "Deep neural networks for acoustic emotion recognition: raising the benchmarks," in Proceedings of ICASSP, IEEE, Prague, Czech Republic, 2011, pp. 5688–5691.

[33] I. S. Engberg, A. V. Hansen, O. Andersen, and P. Dalsgaard, "Design, recording and verification of a Danish emotional speech database," in Proceedings of the Eurospeech, Rhodes, Greece, 1997, pp. 1695–1698.

[34] J. H. L. Hansen and S. Bou-Ghazale, "Getting started with susas: a speech under simulated and actual stress database," in Proceedings of EUROSPEECH-97, vol. 4, Rhodes, Greece, 1997, pp. 1743–1746.

[35] J. Hirschberg, J. Liscombe, and J. Venditti, "Experiments in emotional speech," in Proceedings of the ISCA & IEEE Workshop on Spontaneous Speech Processing and Recognition, Tokyo, 2003, pp. 1–7.

[36] O. Martin, I. Kotsia, B. Macq, and I. Pitas, "The enterface'05 audio-visual emotion database," in *IEEE Workshop on Multimedia Database Management*, 2006.

[37] B. Schuller, M. Wimmer, D. Arsic, G. Rigoll, and B. Radig, "Audiovisual behaviour modeling by combined feature spaces," in Proceedings of ICASSP, 2007, pp. 733–736.

[38] M. Grimm, K. Kroschel, and S. Narayanan, "The Vera am Mittag German audio-visual emotional speech database," in Proceedings of the IEEE International Conference on Multimedia and Expo (ICME), Hannover, Germany, 2008, pp. 865–868.

[39] J. Liscombe, G. Riccardi, and D. Hakkani-Tür, "Using context to improve emotion detection in spoken dialog systems," in Proceedings of Interspeech, Lisbon, 2005, pp. 1845–1848.

[40] J. Ang, R. Dhillon, E. Shriberg, and A. Stolcke, "Prosody-based automatic detection of annoyance and frustration in human–computer dialog," in Proceedings of Interspeech, Denver, CO, 2002, pp. 2037–2040.

[41] J. Liscombe, J. Hirschberg, and J. J. Venditti, "Detecting certainness in spoken tutorial dialogues," in Proceedings of INTERSPEECH, Lisbon, Portugal, 2005, pp. 1837–1840.

[42] A. Batliner, V. Zeissler, C. Frank, J. Adelhardt, R. P. Shi, and E. Nöth, "We are not amused—but how do you know? User states in a multi-modal dialogue system," in Proceedings of Interspeech, Geneva, 2003, pp. 733–736.

[43] A. Batliner, S. Steidl, C. Hacker, and E. Nöth, "Private emotions vs. social interaction—a data-driven approach towards analysing emotions in speech," *User Model. User-Adapt. Interact.*, 18: 175–206, 2008.

[44] S. Steininger, F. Schiel, O. Dioubina, and S. Raubold, "Development of user-state conventions for the multimodal corpus in smartkom," in Proceedings of the Workshop on Multimodal Resources and Multimodal Systems Evaluation, Las Palmas, 2002, pp. 33–37.

[45] C. P. Said, C. D. Moore, K. A. Norman, J. V. Haxby, and A. Todorov, "Graded representations of emotional expressions in the left superior temporal sulcus," *Front. Syst. Neurosci.*, 4(6): 2010. doi:10.3389/fnsys.2010.00006

[46] A. Vinciarelli, M. Pantic, H. Bourlard, and A. Pentland, "Social signals, their function, and automatic analysis: a survey," in IMCI '08: Proceedings of the 10th International Conference on Multimodal Interfaces, ACM, New York, 2008, pp. 61–68.

[47] B. Schuller and G. Rigoll, "Timing levels in segment-based speech emotion recognition," in Proceedings of Interspeech, Pittsburgh, PA, 2006, pp. 1818–1821.

[48] M. Shami and W. Verhelst, "Automatic classification of expressiveness in speech: a multi-corpus study," in *Speaker Classification II*, volume 4441 of Lecture Notes in Computer Science/Artificial Intelligence (ed. C. Müller), Springer, Heidelberg, 2007, pp. 43–56.

[49] A. Batliner, D. Seppi, S. Steidl, and B. Schuller, "Segmenting into adequate units for automatic recognition of emotion-related episodes: a speech-based approach," *Adv. Hum.-Comput. Interact.*, vol. 2010, Article ID 782802, 2010, 15 p.

[50] D. Seppi, A. Batliner, S. Steidl, B. Schuller, and E. Nöth, "Word accent and emotion," in Proceedings of Speech Prosody 2010, Chicago, IL, 2010.

[51] B. Schuller, R. Jiménez Villar, G. Rigoll, and M. Lang, "Meta-classifiers in acoustic and linguistic feature fusion-based affect recognition," in Proceedings of ICASSP, Philadelphia, PA, 2005, pp. 325–328.

[52] F. Eyben, M. Wöllmer, and B. Schuller, "openSMILE—the Munich versatile and fast open-source audio feature extractor," in Proceedings of ACM Multimedia, Florence, Italy, 2010, pp. 1459–1462.

[53] B. Schuller, R. Müller, M. Lang, and G. Rigoll, "Speaker independent emotion recognition by early fusion of acoustic and linguistic features within ensemble," in Proceedings of Interspeech, Lisbon, 2005, pp. 805–808.

[54] T. Joachims, "Text categorization with support vector machines: learning with many relevant features," in Proceedings of ECML-98, 10th European Conference on Machine Learning, Chemnitz (eds C. Nédellec and C. Rouveirol), Springer, Heidelberg, 1998, pp. 137–142.

[55] B. Schuller and T. Knaup, "Learning and knowledge-based sentiment analysis in movie review key excerpts," in *Toward Autonomous, Adaptive, and Context-Aware Multimodal Interfaces: Theoretical and Practical Issues*, Volume 6456 of Lecture Notes on Computer Science (LNCS) (eds A. Esposito, A. M. Esposito, R. Martone, V. Müller, and G. Scarpetta), Springer, Heidelberg, 2010, pp. 448–472.

[56] B. Schuller, F. Metze, S. Steidl, A. Batliner, F. Eyben, and T. Polzehl, "Late fusion of individual engines for improved recognition of negative emotions in speech—learning vs. democratic vote," in Proceedings of the 35th IEEE International Conference on Acoustics, Speech and Signal Processing (ICASSP), Dallas, TX, 2010, pp. 5230–5233.

[57] M. Wöllmer, F. Eyben, A. Graves, B. Schuller, and G. Rigoll, "Bidirectional LSTM networks for context-sensitive keyword detection in a cognitive virtual agent framework," *Cogn. Comput.*, 2(3): 180–190, 2010.

[58] S. Steidl, *Automatic Classification of Emotion-Related User States in Spontaneous Children's Speech*, Logos Verlag, Berlin, 2009 (PhD thesis, FAU Erlangen-Nuremberg).

[59] A. Batliner, S. Steidl, B. Schuller, D. Seppi, T. Vogt, J. Wagner, L. Devillers, L. Vidrascu, V. Aharonson, and N. Amir, "Whodunnit—searching for the most important feature

types signalling emotional user states in speech," *Comput. Speech Lang.*, 25: 4–28, 2011.

[60] P. Pudil, J. Novovicova, and J. Kittler, "Floating search methods in feature selection," *Pattern Recognit. Lett.*, 15: 1119–1125, 1994.

[61] N. Landwehr, M. Hall, and E. Frank, "Logistic model trees," *Mach. Learn.*, 161–205, 2005.

[62] I. T. Jolliffe, *Principal Component Analysis*, Springer, Berlin, 2002.

[63] A. Hyvärinen, J. Karhunen, and E. Oja, *Independent Component Analysis*, John Wiley & Sons, New York, 2001.

[64] A. Cichocki, R. Zdunek, A. H. Phan, and S.-I. Amari, *Nonnegative Matrix and Tensor Factorizations*, John Wiley & Sons, 2009.

[65] P. Smaragdis, "Approximate nearest-subspace representations for sound mixtures," in Proceedings of ICASSP, Special Session on "Innovative Representations of Audio," Prague, Czech Republic, 2011, pp. 5892–5895.

[66] F. Eyben, A. Batliner, B. Schuller, D. Seppi, and S. Steidl, "Cross-corpus classification of realistic emotions—some pilot experiments," in Proceedings of the Third International Workshop on EMOTION (Satellite of LREC): Corpora for Research on Emotion and Affect, Valetta, 2010, pp. 77–82.

[67] J.-C. Martin, R. Niewiadomski, L. Devillers, S. Buisine, and C. Pelachaud, "Multimodal complex emotions: gesture expressivity and blended facial expressions," *Int. J. Hum. Robot.*, 3(3): 1–23, 2006.

[68] D. Seppi, A. Batliner, B. Schuller, S. Steidl, T. Vogt, J. Wagner, L. Devillers, L. Vidrascu, N. Amir, and V. Aharonson, "Patterns, prototypes, performance: classifying emotional user states," in Proceedings of Interspeech, Brisbane, 2008, pp. 601–604.

[69] E. Mower, A. Metallinou, C.-C. Lee, A. Kazemzadeh, C. Busso, S. Lee, and S. Narayanan, "Interpreting Ambiguous Emotional Expressions," in Proceedings of ACII, Amsterdam, 2009, pp. 662–669.

[70] C. E. Erdem, E. Bozkurt, E. Erzin, and A. T. Erdem, "RANSAC-based training data selection for emotion recognition from spontaneous speech," in AFFINE'10—Proceedings of the 3rd ACM Workshop on Affective Interaction in Natural Environments, Co-located with ACM Multimedia 2010, 2010, pp. 9–14.

[71] B. Schuller and L. Devillers, "Incremental acoustic valence recognition: an inter-corpus perspective on features, matching, and performance in a gating paradigm," in Proceedings of INTERSPEECH, Makuhari, Japan, 2010, pp. 801–804.

[72] B. Schuller and G. Rigoll, "Recognising interest in conversational speech—comparing bag of frames and supra-segmental features," in Proceedings of Interspeech, Brighton, 2009, pp. 1999–2002.

[73] S. Hochreiter and J. Schmidhuber, "Long short-term memory," *Neural Comput.*, 9(8): 1735–1780, 1997.

[74] M. Wöllmer, B. Schuller, F. Eyben, and G. Rigoll, "Combining long short-term memory and dynamic Bayesian networks for incremental emotion-sensitive artificial listening," *IEEE J. Sel. Top. Signal Process.*, 4(5): 867–881, 2010.

[75] M. Wöllmer, F. Weninger, F. Eyben, and B. Schuller, "Computational assessment of interest in speech—facing the real-life challenge," *Künstl. Intell.*, (Special Issue on Emotion and Computing), 2011.

[76] T. K. Ho, "The random subspace method for constructing decision forests," *IEEE Transactions on Pattern Analysis and Machine Intelligence*, 20: 832–844, 1998.

[77] I. H. Witten and E. Frank, *Data Mining: Practical Machine Learning Tools and Techniques*, 2nd edn, Morgan Kaufmann, San Francisco, CA, 2005.

[78] T. Fawcett, "An introduction to ROC analysis," *Pattern Recognit. Lett.*, 27(8): 861–874, 2006.

[79] B. Schuller, S. Steidl, A. Batliner, F. Burkhardt, L. Devillers, C. Müller, and S. Narayanan, "The INTERSPEECH 2010 paralinguistic challenge—age, gender, and affect," in Proceedings of Interspeech, Makuhari, Japan, 2010, pp. 2794–2797.

[80] T. G. Dietterich, "Approximate statistical tests for comparing supervised classification learning algorithms," *Neural Comput.*, 10: 1895–1923, 1998.

[81] L. Gillick and S. J. Cox, "Some statistical issues in the comparison of speech recognition algorithms," in Proceedings of ICASSP, IEEE, 1989, pp. 532–535.

[82] R. S. Nickerson, "Null hypothesis significance testing: a review of an old and continuing controversy," *Psychol. Methods*, 5: 241–301, 2000.

[83] S. Young, G. Evermann, M. Gales, T. Hain, D. Kershaw, X. Liu, G. Moore, J. Odell, D. Ollason, D. Povey, V. Valtchev, and P. Woodland, *The HTK Book*, Cambridge University Engineering Department, 2006. For HTK Version 3.4.

[84] F. Eyben, M. Wöllmer, and B. Schuller, "openEAR—introducing the Munich open-source emotion and affect recognition toolkit," in Proceedings of ACII, Amsterdam, 2009, pp. 576–581.

[85] T. Vogt, E. André, and N. Bee, Emovoice—a framework for online recognition of emotions from voice," in Proceedings of an IEEE Tutorial and Research Workshop on Perception and Interactive Technologies for Speech-Based Systems (PIT 2008), volume 5078 of Lecture Notes in Computer Science, Springer, Kloster Irsee, 2008, pp. 188–199.

[86] M. Schröder, L. Devillers, K. Karpouzis, J.-C. Martin, C. Pelachaud, C. Peter, H. Pirker, B. Schuller, J. Tao, and I. Wilson, "What should a generic emotion markup language be able to represent?" in *Affective Computing and Intelligent Interaction* (eds A. Paiva, R. Prada, and R. W. Picard), Springer, Berlin, 2007, pp. 440–451.

[87] P. Baggia, D. C. Burnett, J. Carter, D. A. Dahl, G. McCobb, and D. Raggett, *EMMA: Extensible MultiModal Annotation Markup Language*, 2007.

[88] B. Fehr and J. A. Russel, "Concept of emotion viewed from a prototype perspective," *J. Exp. Psychol.: Gen.*, 113: 464–486, 1984.

[89] P. R. Shaver, S. Wu, and J. C. Schwartz, Cross-cultural similarities and differences in emotion and its representation: a prototype approach, *Emotion*, 175–212, 1992.

[90] A. Batliner, S. Steidl, C. Hacker, E. Nöth, and H. Niemann, "Tales of tuning—prototyping for automatic classification of emotional user states," in Proceedings of Interspeech, Lisbon, 2005, pp. 489–492.

[91] K. Krippendorff, *Content Analysis, An Introduction to Its Methodology*, 2nd edn, Sage Publications, Thousand Oaks, CA, 2004.

[92] M. Grimm and K. Kroschel, "Evaluation of natural emotions using self assessment manikins," in Proceedings of ASRU, IEEE, 2005, pp. 381–385.

[93] M. J. Frank and E. D. Claus, "Anatomy of a decision: striato-orbitofrontal interactions in reinforcement learning, decision making, and reversal," *Psychol. Rev.*, 113(2): 300–326, 2006.

[94] B. Schuller, B. Vlasenko, F. Eyben, M. Wöllmer, A. Stuhlsatz, A. Wendemuth, and G. Rigoll, "Cross-corpus acoustic emotion recognition: variances and strategies," *IEEE Transactions on Affective Computing*, 1(2): 119–131, 2010.

[95] I. Lefter, L. J. M. Rothkrantz, P. Wiggers, and D. A. van Leeuwen, "Emotion recognition from speech by combining databases and fusion of classifiers," in Proceedings of Text and Speech and Dialogue, Berlin, Germany, 2010.

[96] B. Schuller, Z. Zhang, F. Weninger, and G. Rigoll, "Using multiple databases for training in emotion recognition: to unite or to vote?" in Proceedings of INTERSPEECH, Florence, Italy, 2011.

[97] H. A. Elfenbein, M. K. Mandal, N. Ambady, S. Harizuka, and S. Kumar, "On the universality and cultural specificity of emotion recognition: a meta-analysis," *Psychol. Bull.*, 128(2): 236–242, 2002.

[98] A. Abelin, "Cross-cultural multimodal interpretation of emotional expressions—an experimental study of Spanish and Swedish," in Proceedings of Speech Prosody, ISCA, 2004.

[99] K. R. Scherer, R. Banse, and H. G. Wallbott, "Emotion inferences from vocal expression correlate across languages and cultures," *J. Cross-Cult. Psychol.*, 32(1): 76–92, 2001.

[100] T. Polzehl, A. Schmitt, and F. Metze, "Approaching multi-lingual emotion recognition from speech—on language dependency of acoustic/prosodic features for anger detection," in Proceedings of Speech Prosody, ISCA, 2010.

[101] A. Chen, "Perception of paralinguistic intonational meaning in a second language," *Lang. Learn.*, 59(2): 367–409, 2009.

[102] A. Esposito and M. T. Riviello, "The cross-modal and cross-cultural processing of affective information," *Front. Intell. Appl.*, 226: 301–310, 2010.

[103] M. Wöllmer, F. Eyben, S. Reiter, B. Schuller, C. Cox, E. Douglas-Cowie, and R. Cowie, "Abandoning emotion classes—towards continuous emotion recognition with modelling of long-range dependencies," in Proceedings of Interspeech, Brisbane, 2008, pp. 597–600.

[104] N. Ward and W. Tsukahara, "Prosodic features which cue backchannel responses in English and Japanese," *J. Pragmat.*, 32: 1177–1207, 2000.

[105] B. Schuller, M. Wöllmer, T. Moosmayr, and G. Rigoll, "Recognition of noisy speech: a comparative survey of robust model architectures and feature enhancement," *EURASIP J. Audio Speech Music Process.*, 2009. Article ID 942617, 17 p.

[106] B. Schuller, D. Arsić, F. Wallhoff, and G. Rigoll, "Emotion recognition in the noise applying large acoustic feature sets," in Proceedings of Speech Prosody 2006, Dresden, 2006.

[107] A. Tawari and M. Trivedi, "Speech emotion analysis in noisy real world environment," in Proceedings of ICPR, Istanbul, Turkey, August 2010, pp. 4605–4608.

[108] M. Grimm, K. Kroschel, H. Harris, C. Nass, B. Schuller, G. Rigoll, and T. Moosmayr, "On the necessity and feasibility of detecting a driver's emotional state while driving," in *Affective Computing and Intelligent Interaction* (eds A. Paiva, R. Prada, and R. W. Picard), Springer, Berlin, 2007, pp. 126–138.

[109] M. Lugger, B. Yang, and W. Wokurek, "Robust estimation of voice quality parameters under real world disturbances," in Proceedings of ICASSP, Toulouse, 2006, pp. 1097–1100.

[110] B. Schuller, "Affective speaker state analysis in the presence of reverberation," *Int. J. Speech Technol.*, 1–11, 2011.

[111] F. Weninger, B. Schuller, A. Batliner, S. Steidl, and D. Seppi, "Recognition of nonprototypical emotions in reverberated and noisy speech by nonnegative matrix factorization," *EURASIP J. Adv. Signal Process.*, (Special Issue on Emotion and Mental State Recognition from Speech), 2011. Article ID 838790, 16 p.

[112] J. F. Gemmeke and T. Virtanen, "Noise robust exemplar-based connected digit recognition," in Proceedings of ICASSP, Dallas, TX, 2010, pp. 4546–4549.

[113] M. Wöllmer, F. Weninger, S. Steidl, A. Batliner, and B. Schuller, "Speech-based nonprototypical affect recognition for child-robot interaction in reverberated environments," in Proceedings of INTERSPEECH, Florence, Italy, 2011.

[114] B. Schuller and F. Burkhardt, "Learning with synthesized speech for automatic emotion recognition," in Proceedings of the 35th IEEE International Conference on Acoustics, Speech and Signal Processing (ICASSP), Dallas, TX, 2010, pp. 5150–5153.

[115] R. Reisenzein and H. Weber, "Personality and emotion," in *The Cambridge Handbook of Personality Psychology* (eds P. J. Corr and G. Matthews), Cambridge University Press, Cambridge, 2009, pp. 54–71.

[116] W. Revelle and K. R. Scherer, "Personality and emotion," in *Oxford Companion to the Affective Sciences*, Oxford University Press, Oxford, 2009.

[117] T. Vogt and E. André, "Improving automatic emotion recognition from speech via gender differentiation," in Proceedings of Language Resources and Evaluation Conference (LREC 2006), 2006.

[118] J. Stadermann, W. Koska, and G. Rigoll, "Multi-task learning strategies for a recurrent neural net in a hybrid tied-posteriors acoustic mode," in Proceedings of Interspeech 2005, Lisbon, Portugal, 2005, ISCA, pp. 2993–2996.

[119] M. Pantic and L. J. M. Rothkrantz, "Toward an affect-sensitive multimodal human-computer interaction," *Proceedings of IEEE*, 91(9): 1370–1390, 2003.

[120] Z. Zeng, M. Pantic, G. I. Roisman, and T. S. Huang, "A survey of affect recognition methods: audio, visual, and spontaneous expressions," *IEEE Transactions on Pattern Analysis and Machine Intelligence*, 31(1): 39–58, 2009.

[121] K. H. Kim, S. W. Bang, and S. R. Kim, "Emotion recognition system using short-term monitoring of physiological signals," *Med. Biol. Eng. Comput.*, 42(3): 419–427, 2004.

[122] F. Nasoz, K. Alvarez, C. L. Lisetti, and N. Finkelstein, "Emotion recognition from physiological signals using wireless sensors for presence technologies," *Cogn. Technol. Work*, 6(1): 4–14, 2004.

[123] R. W. Picard, E. Vyzas, and J. Healey, "Toward machine emotional intelligence: analysis of affective physiological state," *IEEE Transactions on Pattern Analysis and Machine Intelligence*, 23(10): 1175–1191, 2001.

[124] S. Bengio, "An asynchronous hidden Markov model for audio-visual speech recognition," *Advances in NIPS 15*, 2003.

[125] H. Gunes and M. Piccardi, "Affect recognition from face and body: early fusion vs. late fusion," *IEEE International Conference on Systems, Man and Cybernetics*, 4: 3437–3443, 2005.

[126] W. Lizhong, S. Oviatt, and P. R. Cohen, "Multimodal integration—a statistical view," *IEEE Transactions on Multimedia*, 1: 334–341, 1999.

[127] M. Wöllmer, M. Al-Hames, F. Eyben, B. Schuller, and G. Rigoll, "A multidimensional dynamic time warping algorithm for efficient multimodal fusion of asynchronous data streams," *Neurocomputing*, 73: 366–380, 2009.

AUTHOR BIOGRAPHIES

Felix Weninger received his Master's degree (2009) in Computer Science from Technische Universität München (TUM), one of the first three German Excellence Universities. He is currently pursuing his Ph.D. degree as a researcher in the Intelligent Audio Analysis Group at TUM's Institute for Human–Machine Communication. His research focuses on robust techniques for real-life speech and audio recognition tasks, especially the retrieval of paralinguistic information for human–machine interaction and multimedia analysis. Mr. Weninger is a reviewer for the IEEE Transactions on Affective Computing as well as the Speech Communication journal.

Martin Wöllmer works as a researcher in the Intelligent Audio Analysis Group at TUM's Institute for Human–Machine Communication. He obtained his Diploma in Electrical Engineering and Information Technology from TUM where his current research and teaching activity include the subject areas of pattern recognition and speech processing. His focus lies on automatic recognition of emotionally colored and noisy speech and affective computing. His reviewing engagement includes the *IEEE Transactions on Audio, Speech and Language Processing* and the *IEEE Transactions on Affective Computing*. Publications of his in various journals and conference proceedings cover novel and robust modeling architectures for speech and emotion recognition.

Björn Schuller received his Diploma in 1999 and his Doctoral degree in 2006, both in Electrical Engineering and Information Technology from TUM, where he is tenured as a senior researcher and lecturer in pattern recognition and speech processing and is the Head of the Intelligent Audio Analysis Group. From 2009 to 2010 he was with the CNRS-LIMSI in Orsay, France and a visiting scientist in the Imperial College London's Department of Computing in London, UK. Dr. Schuller is a member of the ACM, HUMAINE Association, IEEE, and ISCA and (co-)authored more than 200 peer-reviewed publications leading to more than 1800 citations—his current H-index equals 21.

11

EEG-BASED EMOTION RECOGNITION USING ADVANCED SIGNAL PROCESSING TECHNIQUES

Panagiotis C. Petrantonakis and Leontios J. Hadjileontiadis

School of Electrical and Computer Engineering, Aristotle University of Thessaloniki, Thessaloniki, Greece

The direct association of the EEG signal to the activation of the brain, related to the emotion elicitation and expression, makes it a suitable field to structure effective EEG-based Emotion Recognition (EEG-ER) systems. EEG-ER is a relatively new research area within the Affective Computing (AC) field; hence, a number of issues should be taken under consideration for further justification, during the design of an EEG-ER system. This chapter, by employing advanced signal processing techniques, sheds light upon two main issues of the EEG-ER field, that is, the aspect of quantitatively estimating the degree of emotion elicitation in subjects under suitable stimuli (affective images) and the overall enhancement of the EEG-ER systems performance by introducing new feature vectors. In particular, the use of brain functional concepts, that is, the activation of the Mirror Neuron System and the asymmetry in the brain activity under emotion stimulation, to evaluate the emotion-related content of an EEG signal is described. This includes sophisticated signal processing tools with robust mathematical background, drawn from the field of Multidimensional Directed Information Analysis, resulting in an "emotion-based segmentation" of the related EEG data. Moreover, novel feature extraction algorithms, structured in the time (Higher Order Crossings) and/or time-frequency domains (Empirical Mode Decomposition), are presented, along with their application to experimental

Emotion Recognition: A Pattern Analysis Approach, First Edition. Edited by Amit Konar and Aruna Chakraborty.
© 2015 John Wiley & Sons, Inc. Published 2015 by John Wiley & Sons, Inc.

EEG data, collected from subjects emotionally stimulated by two elicitation protocols (Facial Expressions and Affective Pictures). The chapter concludes by pointing out the potential of the proposed approaches toward efficient EEG-ER systems and discusses some future directions of research for the EEG-ER task.

11.1 INTRODUCTION

Emotion-related research originates back to the 19th century, when William James with his landmark paper [1], "What is an Emotion?", proposed theories of emotion that still affect the related scientific areas. Until recently, emotion research was conducted mostly from psychologists and philosophers who tried to establish the fundamental principles that underpin the experience of emotion. In 1990s, Picard's book "Affective Computing" [2], stressed out the need of imbuing machines with the ability of detecting, recognizing, and processing the human emotion, creating a surge of interest among computer scientists and engineers who were seeking for more pragmatic human–machine interaction.

Toward that more reliable interaction, Picard defined three major applications of affecting computing (AC): (i) systems that detect and recognize emotions, (ii) systems that express emotions (e.g., avatars, agents), and (iii) systems that feel emotions. Among these three system approaches, a great volume of research has been dedicated to the Emotion Recognition (ER) one, where signal processing and machine learning methodologies have been developed using various sources of emotional expressions (e.g., face, voice, physiological signals) [3–5].

A relatively new field in the ER area is the Electroencephalogram (EEG)-based ER (EEG-ER) approach. In contrast with the other sources of emotional expression, brain activity contributes to all three processes proposed by psychologists as important for emotion perception [6], that is: (i) the identification of the emotional significance of a stimulus, (ii) the production of an affective state in response to (i), and (iii) the regulation of the affective state. Moreover, EEG has efficient time resolution, which is a very crucial factor for reliable AC applications. Finally, new technologies have contributed to the development of new EEG recording devices [7] that continuously reduce EEG's intrusiveness and make it a suitable source of real-life AC applications.

EEG-ER is usually realized in three major steps, that is: (i) the emotion elicitation step, where specifically designed experiments are conducted and different emotions are artificially evoked to subjects by pictures [8], videos [9], and/or sounds [10], (ii) the captured data preprocessing step, that is, recorded signals are subjected to frequency band selection, noise canceling, and artifact removal, and (iii) the classification step, where the feature extraction techniques and classification methods are utilized to classify the recorded signals to different emotional groups. One of the longstanding psychological debates that refers to the first aforementioned step has been between the modeling of emotions with discrete categories of emotions (e.g., happiness) and adopting the dimensional models, usually the 2D Valence/Arousal Model (VAM), where valence stands for ones judgment about a situation as positive or negative and arousal spans from calmness to excitement, expressing the degree of ones excitation.

Depending on the model adopted from researchers, different techniques were utilized to evoke the emotions referring to different categories, either in a discrete of a fuzzier manner. Various signal processing and machine learning techniques (see Section 11.3) have been proposed for the two remaining steps, but still there is space for improvement.

In this chapter, advanced signal processing methods are presented for the classification of both discrete emotions and affective states lying in the VAM. The described methods are designed to accomplish two major endeavors. First, the construction of features that, from their definition, are highly related to the expression of the emotional information in the EEG signals, regarding their frequency characteristics. Second, the development of novel signal processing methods to isolate and extract only the valuable emotion-related information from the raw EEG signals, that is, to implement, somehow, emotional filtering processes.

11.2 BRAIN ACTIVITY AND EMOTIONS

The fact that emotions are differentiated by unique patterns of physiological signals, both from the autonomous nervous system (ANS) (e.g., heart rate, galvanic skin response (GSR)) and the central nervous system (CNS) (e.g., brain signals), has extensively confirmed from various studies. Particularly, based on theoretical writings of Darwin [11] and James [1], different emotions are accompanied by discrete patterns of physiological activity. To this end, besides the relative research for the ANS signals, such as References 12 and 13, great volume of research has been conducted for the ability of CNS signals to express unique patterns for different emotions. In particular, relative psychophysiology literature has revealed the most prominent expression of emotion in brain signals, that is, the asymmetry between the left and right brain hemispheres. Davidson *et al.* [14] developed a model that related this asymmetric behavior with emotions, with the latter analyzed with the VAM. According to that model, emotions are: (i) organized around approach-withdrawal tendencies and (ii) differentially lateralized in the frontal region of the brain. The left frontal area is involved in the experience of positive emotions, such as joy or happiness (the experience of positive affect facilitates and maintains approach behaviors), whereas the right frontal region is involved in the experience of negative emotions, such as fear or disgust (the experience of negative affect facilitates and maintains withdrawal behaviors). Furthermore, Davidson *et al.* [15] tried to differentiate the emotions of happiness and disgust with EEG signals captured from the left and right frontal, central, anterior temporal, and parietal regions (F3, F4, C3, C4, T3, T4, P3, P4 positions according to the 10-20 system [16]). The results revealed a more right-sided activation, as far as the power of *alpha* (8–12 Hz) band of the EEG signal is concerned, for the disgust condition for both the frontal and anterior temporal regions. Thus, the results enhanced the applicability of the aforementioned model and confirmed the evidenced extensive anatomical reciprocity of both regions with limbic circuits that have been directly implicated in the control of emotion [17]. Later, Davidson [18] examined the asymmetry concept in regard with the prefrontal cortex

(PFC) based on data from the neuroscience literature on heterogeneity of different sectors of the PFC. He envisaged the PFC's role in affective expression in the brain and considered the use of more EEG signal bands other than *alpha*, that is, *theta* (4–7 Hz), *beta* (13–30 Hz), and *gamma* (31–100 Hz). More recently, a variety of psychophysiological studies of emotion [19] have confirmed and adopted the model proposed by Davidson, expanding it by arguing that frontal EEG asymmetry may serve as both a moderator and a mediator of emotion- and motivation-related constructs [20]. Here, the authors observed that a decrease in power in the alpha band is associated with emotional states. Finally, Aftanas *et al.* [21] used affective pictures and Event-Related Desynchronization/Synchronization (ERD/ERS) analysis to study cortical activation during emotion processing. ERD refers to the desynchronization of a brain rhythm during a stimulus, which leads to a power decrease, whereas ERS to the synchronization of the rhythm and power increases [22]. In accordance with the asymmetry literature, they reported relatively greater right hemisphere ERS for negative emotional states and greater left hemisphere ERS for positive ones.

All above studies reveal the potential of EEG signals to serve as valuable means of differentiating emotions. Based on the frequency characteristics of the emotional expression in EEG signals and the asymmetry phenomenon during their experience, novel EEG-ER methods could be formed, as those presented in this chapter.

11.3 EEG-ER SYSTEMS: AN OVERVIEW

Relatively little research has been conducted in the EEG-ER area compared to emotion recognition from face, voice, and ANS-based signals, mostly in the last decade. In particular, the majority of these studies are user-dependent, confined to feature extraction methods relevant with *alpha* and *beta* bands and their characteristics (e.g., power, frequency peaks) and evoke emotions to subjects with pictures [8], sounds [10], or videos [9] that are assumed to elicit certain affective states within the 2D valence/arousal space (VAS). More specifically, Choppin [23] analyzed EEG signals and used neural networks to classify them in six emotions based on emotional valence and arousal with a 64% success rate. Takahashi [9], by using statistical feature vectors, which were initially used for emotion recognition from physiological signals [5], conducted a user-independent emotion recognition study from either physiological (pulse and GSR) and EEG (Fp1 and Fp2 channels according to 10-20 system [16]) signals or their fusion. With data from 12 subjects, a 41.68% success rate accomplished only from the EEG signals and 41.72% with the fusion of the physiological and EEG signals, for the discrimination of five distinct emotions, that is, *joy*, *anger*, *sadness*, *fear*, and *relax*. Other studies, conducted experiments for emotion elicitation with the participation of four [24], five [10, 25], and 10 [26] subjects and tried to classify EEG and/or physiological signals in emotion or affective states (classes within specific range in the arousal and/or valence axis) with moderate results, that is, success classification rates less than 60% for discrimination between three to five classes. Another feature extraction method for EEG-ER was proposed by Murugappan *et al.* [27], who resulted in a satisfactory clustering within the emotions of *disgust*, *happiness*, and *fear* with fuzzy *c*-means [28] and fuzzy *k*-means [29] used as the classification methods. The feature vector extraction method was based on wavelet analysis using the *4th*-order Daubechies' wavelet [30] and referred to the energy and entropy characteristics of the signal. For the emotion elicitation experiment, clips with specified emotional content were used to induct emotions from six subjects, whose EEG activity was recorded by a 64-channel electroencephalograph. Another prominent feature extraction technique consists of the coefficients of cross-correlation between different EEG sites [31] and has been used to various EEG-ER studies [32, 33], revealing promising results. Finally, other recent works in the field [34, 35] achieve similar performance as those mentioned so far in terms of emotion classification efficiency.

11.4 EMOTION ELICITATION

11.4.1 Discrete Emotions

The most prominent way of evoking emotions to subjects, participating in relative experiments, is to use pictures, videos, or sounds that were accompanied with specific valence and arousal norm values. In order to elicit discrete emotions, for example, the six basic emotions of *happiness*, *surprise*, *anger*, *fear*, *disgust*, and *sadness* [36], the above-mentioned media with norm values that lie in the VAS in an area where the corresponding discrete emotion would also lie, are used. For instance, for the elicitation of anger, which is supposed to be accompanied with high arousal values and relatively negative valence, pictures with corresponding valence and arousal characteristics are used. Nevertheless, taking into account aspects like personality, personal experiences, and, particularly, subject's mood at the time that the experiment is conducted, a significant influence in the way someone emotionally appraises the emotional stimuli is noticed. As a result, it is questioned how effectively the emotion is actually evoked and, consequently, how representative the EEG activity is, with regard to a particular discrete emotion.

To overcome this ambiguity, a novel emotion elicitation process for discrete emotion was proposed in Reference 8 based on the mirror neuron system (MNS) [37]. Neurophysiologic experiments have demonstrated that when individuals observe an action done by another individual seem to have the same or akin brain activity as if they did the corresponding action themselves. This activity is ascribed to the MNS [37], which is also connected with the ability of imitation that, among others, relates to recognition of emotions by others' face expressions and gestures [38]. Furthermore, various studies, for example, References 39, 40, have also highlighted the linkage between the brain activity and the observation of emotional facial expressions and confirmed the role of the MNS to produce brain activity as a result of a mimic process. To this end, the use of images that picture emotional facial expression could be used for the development of a novel emotion elicitation technique where subjects, by watching pictures with facial expressions of the six basic emotions, would express same or akin brain activity as if they indeed experiencing the emotion.

11.4.2 Affective States

Following the widespread practice of evoking affective states lying in the VAS, another dataset was also constructed using pictures depicting situations that are supposed to elicit predefined affective states. International Affective Picture System (IAPS) [41] is a widely used dataset, containing such pictures, which come with their individual values of valence and arousal. Consequently, the projection of the IAPS pictures to a subject with the simultaneous recording of EEG signals formulated an emotion elicitation process originating from the VAS theory.

11.4.3 Datasets

For the construction of datasets of EEG signals recorded during the aforementioned elicitation processes, specifically designed experiments were conducted. In these experiments, 16 healthy, right-handed subjects participated. Totally 60 pictures of the Pictures of Facial Affect (POFA) database [42] were projected (10 for each discrete emotion) resulting in 60 corresponding EEG recordings for each subject. For the construction of the IAPS-based dataset, 40 pictures of the IAPS database were projected (10 for each one state lying in the corresponding quadrant in the VAS) resulting in 40 corresponding EEG recordings for each subject. Moreover, for each emotion elicitation trial, a relaxation trial was preceded and a self-assessment process followed. EEGs were recorded from Fp1 (monopole channel), Fp2 (monopole channel), and F3/F4 (dipole channel) resulting in three-channel recordings. Figure 11.1 (left) depicts brain sites of the EEG recordings. The projection protocol followed in both approaches is shown in Figure 11.1 (right) and incorporates four sequential steps for each emotion elicitation trial: (i) black screen for 5 seconds, (ii) count down one-second frames from five to one, (iii) an one-second projection of a cross to attract the sight of the subject, and (iv) the projection of the corresponding picture for 5 seconds. Elaborate description of the experiment's protocol and datasets' construction can be found in References 8 and 43.

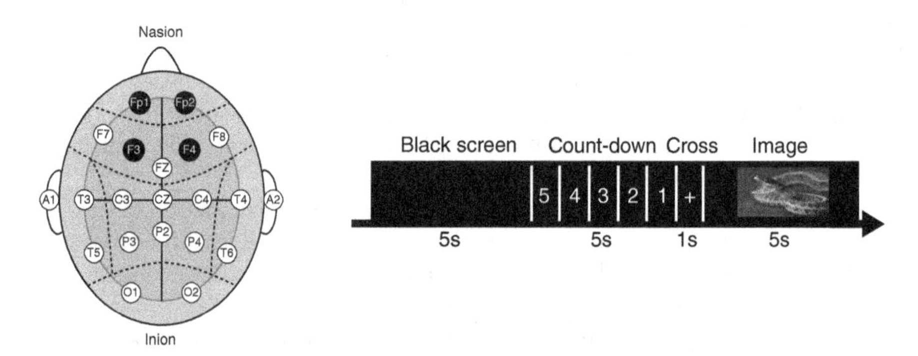

FIGURE 11.1 10/20 positions of the EEG recordings (left) and experimental protocol for pictures' projection (right).

11.5 ADVANCED SIGNAL PROCESSING IN EEG-ER

Although time and spectral characteristics of the EEG signals proved to be quite efficient in the interpretation of the related emotional information, they cannot always provide the full image of the power of EEG to provide efficient features that could further enhance the performance of an EEG-ER system. To this end, some new analysis domains have been proposed in order to enhance some EEG properties, seen from different angles, capturing that way, more efficiently their emotion-related behavior. The presentation of such methodologies aims at not only revealing the importance of EEG as indicators of emotional activity, but also shedding light upon the inherent characteristics of the advanced signal processing tcehniques that manage to adapt to the specific properties of EEG, providing a new standpoint in the EEG-ER realization. These new representations and features of EEG are examined in details in the subsequent sections.

11.5.1 Discrete Emotions

For the classification of the six discrete emotions, that is, *happiness*, *surprise*, *anger*, *fear*, *disgust*, and *sadness*, a novel feature extraction technique is presented, based on Higher Order Crossings (HOC) analysis. The latter originates from the zero-crossings counting after sequential filtering and that way, the frequency characteristics of EEG signals are exploited for the feature vector construction in a time-based implementation. Afterward, an augmentation of the HOC-based feature is attempted by developing a filtering procedure, namely Hybrid Adaptive Filtering, by which the emotion-related information of EEGs is isolated, leading to more representative signals as far as the emotion expression in brain activity is concerned.

11.5.1.1 Higher Order Crossings Analysis The direct relation of emotional expression in brain with the power spectrum of EEG signals in certain brain locations and specific frequency bands (mostly in *alpha* and *beta* bands) has been clearly demonstrated in Section 11.2. Particularly, this differentiation is expressed by the asymmetry concept as far as ERD/ERS neuronal characteristics are concerned. These facts facilitate the development of a feature vector-extraction technique that considers the spectrum-related attitude of the signals and it is highly dependent on the dominancy (power) of certain frequency in a specific sub-band of the whole frequency spectrum. HOC-based analysis [8] provides such perspective by rigorously analyzing the signal in the time domain and without employing spectral transforms.

Consider a finite zero-mean series Z_n, $n = 1, \dots, N$. Its oscillating behavior about the zero level can be expressed by the zero-crossings count. Applying a filter to a time series, generally changes its oscillation attitude and, consequently, its zero-crossings count. Under this perspective, an iterative procedure, regarding the sequential application of a filter to a time series and the counting of the corresponding zero-crossings, can be assumed, that is, repeatedly filter and count (each time the filtering order k increases counts are referred to as HOC [44]. When a specific sequence of filters is applied to a time series, the corresponding sequence of zero-crossings count is obtained, resulting in the so-called HOC sequence. Using different classes of filters, different corresponding types of HOC sequences can be constructed, according to the desired spectral and discrimination analysis. For an extended coverage of the HOC sequence extraction, the reader is encouraged to consult Reference 8.

The resulting feature vector (FV) from the above procedure is constructed as follows:

$$FV_{\text{HOC}} = [D_1, D_2, D_L], 1 < L \leq J \quad (11.1)$$

where J denotes the maximum order of the estimated HOC and L the HOC order up to they were used to form the FV_{HOC}. The L selection is generally related to the classification performance of the corresponding feature vector [8]. The extraction of the feature vector according to the HOC analysis was applied to each one of the three EEG channels (see Section 11.4.3) and for the classification process, the features from each EEG channel were used, either separately or fused. In the fusion case, the feature vector-level fusion procedure was followed. For comparison reasons, two other feature vectors, which have been used in the literature for classification of discrete emotions, were used. Specifically, a statistical based feature vector proposed by Takahashi [9] and a wavelet-based one presented in Reference 27.

As there is no single best classification algorithm, that is one-size-fits-all, the choice of the most efficient classifier is strongly dependent on the examined problem and the relevant dataset to be classified [45]. Thus, four different classification algorithms were employed, that is, Quadratic Discriminant Analysis (QDA) [46], k-Nearest Neighbor (k-NN) [47], Mahalanobis Distance (MD) [48], and Support Vector Machines (SVMs) [49].

Table 11.1 shows the overall performance of the proposed methodology for the most difficult case of the classification of the six basic emotions. Maximum classification rate, that is, 62.30%, for single channel case, was accomplished for channel F3/F4 using the QDA classifier. On the other hand, the globally maximum rate, that is, 83.33%, was achieved by the feature vector fusion of all three EEG channels and after the classification using the SVM algorithm with a fifth-order polynomial kernel. The other baseline approaches of statistical and wavelet-based features exhibited significantly worse classification rates, approximately half of that in the HOC approach. It must be stressed out that different classification setups that examined the discrimination of all combinations of the six emotions per five to two classes resulted in respective rates that followed upward path, reaching even 100% of correct classification [8].

TABLE 11.1 Overall classification rates (%) of HOC approach

Method	Single channel	Fusion
HOC	62.30	83.33
Statistical FV	37.50	44.90
Wavelet FV	34.60	32.70

11.5.1.2 Hybrid Adaptive Filtering In order to focus on the EEG signal characteristics that are mostly related with the emotion information (see Section 11.2), a Hybrid Adaptive Filtering (HAF) approach was developed. Thus, the EEG signals subjected to such a filtering will result in more representative ones as far as emotion expression is concerned and the subsequent extracted features would result in a more robust classification procedure. As hybrid declares, HAF combines two processing tools, that is, the Empirical Mode Decomposition (EMD) algorithm [50] and a simple Genetic Algorithm (GA)-based optimization concept [51]. Particularly, Flaudrin *et al.* [52] qualified EMD as an adaptive filter bank due to its functionality to decompose a signal into intrinsic mode functions (IMFs) and residuals, which are intuitively interpreted as a "spectral" representation. Moreover, they further argued that a consequent result of the above is that selection of modes corresponds to an automatic and adaptive (signal-dependent) time-variant filtering.

In order to exploit the aforementioned capability of the EMD algorithm, a new GA-based approach was developed for the optimized selection of IMFs that correspond to a specific feature of a signal. This specific feature, which is highly related to the emotional expression in EEG signals, is represented in HAF by the Fitness Function (FF) of the GA. The GA used here was based on the realization strategy proposed in Reference 51, including a series of basic steps, that is, initialization (initial population definition), pair selection, cross-over, mutation, elitist strategy (random removal of one string from the current population and replacement with the string that obtained the maximum FF value in the previous generation), and termination. Figure 11.2 shows the block diagram of the HAF algorithm. The initial EEG signal is decomposed into the corresponding IMFs, according to the EMD sifting process [50]. Subsequently, the GA is applied to the extracted IMFs and those who are selected are used either to reconstruct (R) the new signal (R-case), that is, the filtered EEG signal, or provide separate signals (no reconstruction, NR-case) that represent specific modes of oscillation coexisting in the initial EEG signal.

As is expected, based on the above HAF description, one of the most important modules of the filtering procedure is the FF that is selected for the GA, as this is the major criterion according to which the filtering is conducted. In this work, two FFs were used, that is, energy-based (EFF) and a fractal dimension-based one (FDFF). An elaborated description follows.

The aim of using EFF was to conduct a filtering procedure that would lead to the selection of IMFs which embed the majority of the energy of the signal. As described in Section 11.2, emotion differentiation through brain activity is highly related to the power spectrum of specific sub-bands (e.g., *alpha* and *beta* bands, which were also used in the presented analysis) of the EEG signal. Thus, an IMF selection procedure using an energy-based criterion, directly related to the power spectrum of the EEG signal, would result in a filtered signal with boosted information related to emotion expression in EEG signals. The formula used for the EFF is expressed by

$$f(S) = \frac{\sum_{(S|s_r=1)} E\{c_r(n)^2\}}{\sum_{i=1}^{I} E\{c_i(n)^2\}}, n = 1, \dots, N \tag{11.2}$$

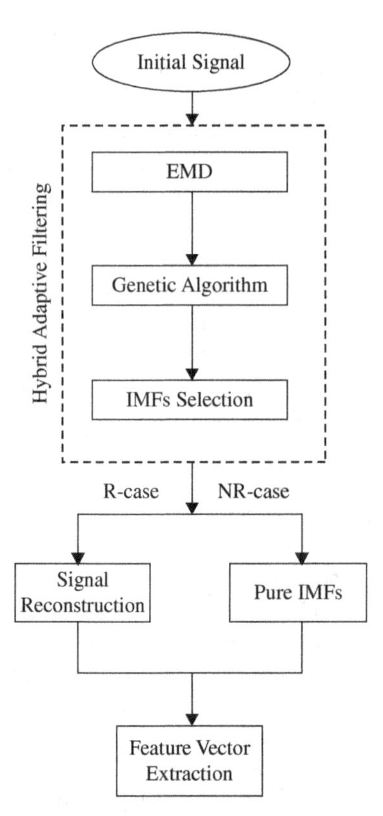

FIGURE 11.2 Block diagram of the HAF algorithm.

where S is the string of 1s and 0s, $s_r = 1$ is the set of the elements of S with value 1, $c(n)$ represents an IMF, and I is the number of IMFs (I is also the number of elements of each string S in the GA realization). According to (11.2), it is obvious that during the GA selection phase, the strings constituted of ones (starting from the initial population) that correspond to more energy-loaded IMFs are more possible to give offspring to the next generation. As a result, a bunch of IMFs that are more likely to embrace the majority of the initial signal's energy is finally selected; in this way, an energy-based filtering is accomplished. In order to exclude any tendency from HAF to select the majority or all of the IMFs the generation number, which was also used as a stopping condition for the GA, was set to be 10 and mutation probabilities were defined as $P_m(1 \to 0) = 0.01$ and $P_m(0 \to 1) = 0.001$. A thorough description of the HAF realization can be found in Reference 53.

The second FF used for the HAF realization was based on the Fractal Dimension (FD) estimation of the IMFs. In general, the FD can be considered as a relative measure of the number of basic building blocks that form a pattern [54] and, as a result, it could be used as a measure of the signal complexity. Calculating the FD of waveforms has the advantage of fast computation. It consists of estimating the dimension of a time-varying signal (waveform) directly in the time domain, which allows significant savings in program run-time. The aim of the resulting FDFF was to capture the variations in the complexity of the EEG signal. As it has been proved in Reference 55, the dynamical complexity of cortical networks, measured by means of the FD, reflects the degree of synergism between neurons and relates to the concepts of ERD/ERS characteristics. On the one hand, synchronization, corresponding to high neural synergism and low complexity, could reflect a resting state of cortical networks. On the other hand, desynchronization, corresponding to low neural synergism and high complexity, could correspond to active information processing in the cortex. To this end, by taking into account the capability of FD to express the complexity of the EEG signal and consequently to monitor the ERD/ERS phenomenon, an FD-based filtering procedure would act as a boosting procedure of the information of the initial EEG signal related to the emotion expression in it. To this end, an FDFF was used in order to select the optimum IMFs from an FD-based perspective.

Here, the a-Petrosian method [56] was used for the estimation of the FD. This method has a preprocessing step, where a new, binary, sequence is generated from the initial signal. According to a-Petrosian method, the binary sequence is generated by assigning one or zero when the waveform value is greater or lower than the mean value of the signal, respectively. The FD is calculated as

$$\mathrm{FD} = \frac{\log_{10} N}{\log_{10} N + \log_{10} \frac{N}{N+0.4N_{\Delta}}} \tag{11.3}$$

where N is the length of the binary sequence (number of points), and N_{Δ} is the number of dissimilar pairs in the binary sequence generated. The a-Petrosian has a relatively low computational burden and its computation is based on a binary sequence that is the same as the one extracted for the computation of the HOC sequence. The formula used for the FDFF is expressed by

$$f(S) = \sum_{S|s_r=1} FD\{c_r(n)\}, n = 1, \ldots, N \tag{11.4}$$

After the filtering process with the HAF algorithm, the same feature vectors with the simple HOC approach were used, that is, HOC-based, statistical-based, and wavelet-based feature vectors, applied on the filtered EEG signals. The classification efficiency of the proposed approach was examined both for the R- and NR-cases and for single and combined channel cases. Table 11.2 shows the overall performance of the proposed methodology for the most difficult case of classification of the six basic emotions. For the FDFF case, the SVM algorithm provides with the best classification performance for both the single and combined channel cases, that is, 77.66% and 85.17%, respectively; for the EFF case, the corresponding classification rates were 67.89% and 77.28%, respectively, both accomplished with the QDA classification algorithm. It is obvious that the filtering procedure introduced by HAF provides with a more robust classification, leading to the assumption that it effectively isolated the

TABLE 11.2 Overall classification rates (%) of the HAF–HOC approach

Method	Single channel	Fusion
HAF–HOC (FDFF)	77.66	85.17
HAF–HOC (EFF)	67.89	77.28
HOC	62.30	83.33
Statistical FV	37.50	44.90
Wavelet FV	34.60	32.70

emotion-related information incorporated in the EEG signals. Moreover, the FDFF appear to be more appropriate for such a filtering, probably due to its direct association with the ERD/ERS characteristics of the emotional expression in EEG. It must be noted that the discrimination of all combinations of six basic emotions per five to two emotion classes was also evaluated, resulting in akin performance as in the HOC approach described in the previous subsection.

Apart from the increase in the classification performance, an additional improvement of the combination of HAF and HOC methods (HAF–HOC approach), over the simple HOC one, was the reduction in the feature vector length. As it was explained in Section 11.5.1.1, the length of the feature vector depends on the selection of the L order, with $L \leq J$. For the case of HAF–HOC approach, J was set equal to 30 [53], whereas for the case of HOC, the value of $J = 50$ was used [8]. Thus, a faster realization of the EEG-ER system is accomplished in the former case. This reduction in the upper bound of the k -order range was adopted by examining the way the HOC sequence monitors the dominant frequencies existing in the analyzed EEG spectra [44]. As the final HAF-filtered EEG signals were representing more nuanced emotion-related information, their spectra were also more restricted, as particular IMFs were removed before their reconstruction, and subsequently the sequential filtering for the HOC feature vector extraction was terminated earlier, that is, in a lower k HOC order. Thus, the HAF–HOC approach provided with a more emotion-oriented processing of the EEG signals, resulting in a more effective and less time-consuming realization of the EEG-ER task.

11.5.2 Affective States

Until now, the effective recognition of the six basic emotions, that is, *happiness*, *surprise*, *anger*, *fear*, *disgust*, and *sadness*, was described. In order to evaluate the efficacy of the proposed methodologies in the recognition of different affective states, defined in the VAS, the same classification setups were followed. Nevertheless, the classification rates obtained from such discrimination were much poorer than the ones obtained from the discrimination of the discrete emotions, of about 30–40%. This result led to the assumption that the emotion elicitation procedure was not as effective in the VAS-based pictures as was in the MNS-based ones. It should be noted again that the same subjects were used in both aforementioned experiments. Consequently, the need to find an index arises, according to which, an emotion-related EEG dataset would be effectively discharged from instances where the emotion elicitation procedure failed. In the following subsection, the development of this new index is described, which provides with the ability to evaluate an EEG dataset in regard to the emotion elicitation efficacy, in a trial base, in a time base, and in a time-frequency base. The aforementioned index is based on the asymmetry concept, described in Section 11.2, and defines a prominent way of emotional expression in the brain. The mathematical tool for its realization is based on the Multidimensional Directed Information (MDI) analysis [57], briefly presented in the following subsection.

11.5.2.1 Multidimensional Directed Information Analysis In an attempt to quantify the shared information between the right and the left hemispheres and ultimately define an index that indicates the existence of the asymmetry between them, a robust mathematical tool was employed, that is, the MDI analysis. Shared information is frequently defined as correlations among multiple EEG recording channels (multiple time series) simultaneously observed from a subject. If a relation of temporal ordering is noted, as the correlation relation among these time series (EEG channels), some are interpreted as causes and others as results, suggesting a cause–effect relation among the time series (causality analysis). In this work, the MDI analysis was employed as a means to identify the causality between any two series (EEG channels) considering all acquired series. One of the main advantages of MDI is that the amount of information propagation is presented as an absolute value in bits and not as a correlation, as in other related methodologies [58].

Consider the simple case of three stationary time series X, Y, and Z of length N, divided into n epochs of $E = N/n$ length; each epoch of $E = P + 1 + M$ length is written as a sequence of two sections of P and M lengths before and after the x_k, y_k, and z_k sampled values of time series X, Y, and Z at time k, respectively, that is,

$$X = x_{k-P}...x_{k-1}x_k x_{k+1}...x_{k+M} = X^P x_k X^M \tag{11.5}$$

$$Y = y_{k-P}...y_{k-1}y_k y_{k+1}...y_{k+M} = Y^P y_k Y^M \tag{11.6}$$

$$Z = z_{k-P}...z_{k-1}z_k z_{k+1}...z_{k+M} = Z^P z_k Z^M \tag{11.7}$$

where $X^P = x_{k-P}...x_{k-1}$; $X^M = x_{k+1}...x_{k+M}$; $Y^P = y_{k-P}...y_{k-1}$; $Y^M = y_{k+1}...y_{k+M}$; $Z^P = z_{k-P}...z_{k-1}$; $Z^M = z_{k+1}...z_{k+M}$.

According to the MDI analysis [57], information that is first generated in X at time k and propagated with a time delay of m to Y taking into consideration information that is propagated to both of them from Z, can be calculated from [57]:

$$I(x_k \rightarrow y_{k+m} | X^P Y^P Z^P y_k z_k) = \frac{1}{2} \log \frac{|R(X^P Y^P Z^P x_k y_k z_k)| \cdot |R(X^P Y^P Z^P y_k z_k y_{k+m})|}{|R(X^P Y^P Z^P y_k z_k)| \cdot |R(X^P Y^P Z^P x_k y_k z_k y_{k+m})|} \tag{11.8}$$

where $R(z_1...z_n)$ is the covariance matrix of the stochastic variables $z_1, \ldots, z_n$. Using (11.8), the total amount of information, namely S, that is first generated in X and propagated to Y taking into account the existence of Z, across the time delay range is

$$S^{XY} : I(x_k \rightarrow Y^M | X^P Y^P Z^P y_k z_k)$$
$$= \sum_{m=1}^{M} \frac{1}{2} \log \frac{|R(X^P Y^P Z^P x_k y_k z_k)| \cdot |R(X^P Y^P Z^P y_k z_k y_{k+m})|}{|R(X^P Y^P Z^P y_k z_k)| \cdot |R(X^P Y^P Z^P x_k y_k z_k y_{k+m})|} \quad (11.9)$$

It must be stressed out that if time series X and Y contain a common component from Z, that is, there is information flow from Z to both X and Y but not between X and Y, in conventional directed information analysis, that is, Z is excluded from (11.8), an information flow would wrongly be identified, as if there exists a flow between X and Y. This ambiguity is clearly circumvented by adopting the MDI approach.

11.5.2.2 Asymmetry Index Frontal brain EEG asymmetry is observed during the experience of negative and positive emotions and is expressed with an increased right frontal and prefrontal hemisphere activity and an enhanced left-hemisphere activity, respectively. Assume an EEG *channel*1 (Fp1) is recorded from the left hemisphere, EEG *channel*2 (Fp1) from the other hemisphere, and *channel*3 (F3/F4) from both hemispheres as a dipole channel (see Section 11.4.3). Let these EEG channels to represent the signals X, Y, and Z, introduced by the MDI analysis, respectively. If we now consider the asymmetry concept, a measure to evaluate this asymmetry information in signals X and Y taking into account the information propagated by signal Z to both of them (and as a result isolating the information shared between the X and Y pair more effectively) would introduce an index of how effectively an emotion has been elicited. Toward this, it is assumed that the total amount of information S (see (11.9)) hidden in the EEG signals and shared between the left and right hemispheres (signals X and Y, respectively) would become maximum when the subject is calm (information symmetry), whereas S would become minimum when the subjects are emotionally aroused (information asymmetry). Thus, two values are estimated according to the MDI analysis, that is, the S_r and S_p values. S_r refers to bidirectional information sharing between X and Y, taking into account Z, when the subject does not feel any emotion; hence, he/she is relaxed, that is,

$$S_r = S_r^{XY} + S_r^{YX} \quad (11.10)$$

whereas S_p is the same sharing information during the period where he/she is supposed to feel an emotion, that is,

$$S_p = S_p^{XY} + S_p^{YX} \quad (11.11)$$

According to the asymmetry concept and the definition of the S_r and S_p values, S_p will be presumably smaller than S_r during an effective emotion elicitation trial. Finally, in order to directly define a measure for evaluating the emotional experience, the Asymmetry Index (AsI) is defined [43], as the distance of the (S_p, S_r) point, corresponding to a specific emotion elicitation trial, from the line $S_r = S_p$, that is:

$$AsI = (S_r - S_p) \times \frac{\sqrt{2}}{2} \tag{11.12}$$

AsI forms a robust measure to evaluate the EEG recordings in trial, time, and time-frequency-based manner, as far as the incorporated emotion-related information is concerned. This is described in the succeeding subsections.

11.5.2.3 Trial-Based Approach The emotion elicitation experiment, as described in Section 11.4.2, was based on the projection of IAPS pictures, which are assumed to evoke certain emotions. The experiment was realized with projection of multiple pictures, defining a multiple-trial emotion elicitation process. As previously discussed, different subjects may be emotionally affected by different pictures. Thus, the AsI value of each emotion elicitation trial of each one of the subjects was estimated. Afterward, the emotion elicitation trials that correspond to either significant emotional responses (AsI values are above a specified threshold [43]) or insignificant ones (AsI values are below a specified threshold [43]). The former trials gathered the Big AsI Group (*BAG*), whereas the latter ones the Small AsI Group (*SAG*). According to the AsI definition, *BAG* should exhibit better discrimination between different affective states rather than the *SAG* or any other randomly extracted group, of the same size, from the initial dataset.

For the classification setup, the HOC-based feature vector was employed. Moreover, a cross correlation-based feature vector, proposed in Reference 31 and extensively used in the literature for the discrimination of affective states, was also used for further evaluation of the AsI approach. The classification procedure was implemented under the following six three-class classification scenarios:

1. S1: class 1:LA, class 2:HA, and class 3: the respective Relax signals.
2. S2: class 1:LV, class 2:HV, and class 3: the respective Relax signals.
3. S3: class 1:LALV, class 2:HALV, and class 3: the respective Relax signals.
4. S4: class 1:LAHV, class 2:HAHV, and class 3: the respective Relax signals.
5. S5: class 1:LALV, class 2:LAHV, and class 3: the respective Relax signals.
6. S6: class 1:HALV, class 2:HAHV, and class 3: the respective Relax signals.

The L, H, A, and V abbreviations stand for Low, High, Arousal, and Valence, respectively. These classification scenarios were chosen in order to both emphasize on the discrimination of the adverse expressions of the valence and arousal coordinates in the VAS and the relax state and to create different classification setups for testing the consistency of the AsI to perform efficiently in every one of them.

Figure 11.3 shows the results for the *BAG*, *SAG*, and the median classification rates from 50 different, with the same size, randomly extracted groups of signals from the initial dataset for both feature vector techniques. In this figure, the initial assumption

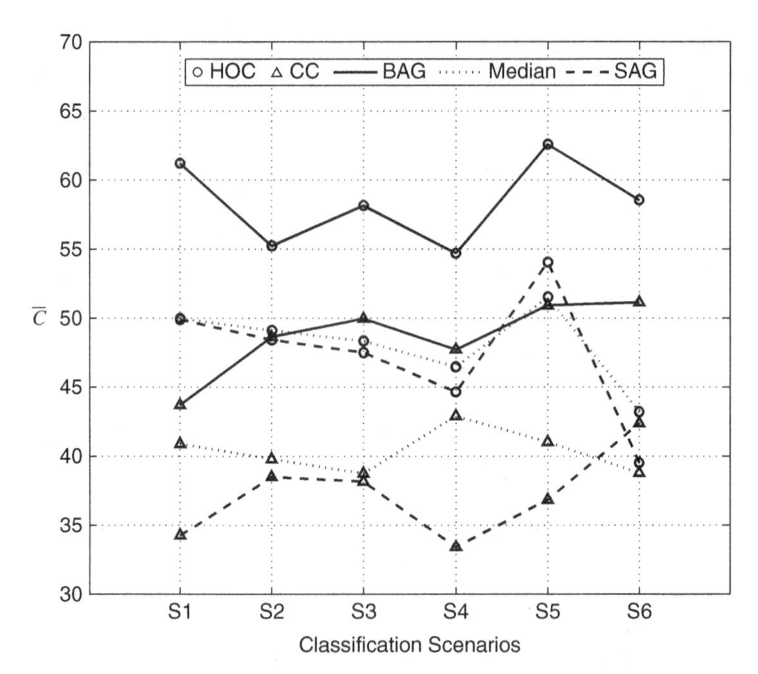

FIGURE 11.3 Classification rates for HOC and CC methods for *BAG* and *SAG* groups, and median rates derived from all 50 randomly created groups with equal signal number with *BAG* group.

that emotion elicitation trials with big *AsI* values lead to better discrimination among different affective states is confirmed and, consequently, they could be supposed to incorporate more nuanced emotion-related information. This is accomplished for both feature extraction techniques, which further strengthens the aforementioned conclusion. Moreover, it is also obvious that BAG rating is much higher than the median rating of the 50 abstract groups. Finally, it should be stressed out that the HOC feature extraction approach overcomes the CC-based one by exhibiting classification rates around 60%, about 10% above of the latter one.

According to the presented findings, it is obvious that the new *AsI* measure effectively indicates the efficacy of the emotion elicitation trial and could be further used for the evaluation of EEG dataset constructed by emotion elicitation procedures. It must also be stressed out that in accordance with the trial-based evaluation, a subject-based evaluation can also be achieved. As it was proved in Reference 43, subjects with large mean *AsI* values tend to exhibit better classification performance in an EEG-ER task than others.

11.5.2.4 Time-Based Approach In the previous subsection, the identification of emotion elicitation trials where an effective elicitation took place was described. Thus, a first step toward a more effective emotion recognition system was achieved, by trying to answer the question of *how effectively has the emotion been evoked* *during an emotion elicitation trial?* Nevertheless, the indexing of the EEG signals as emotional information "carriers" or not was conducted by making the assumption that the emotional information is uniformly distributed in EEGs captured during specific emotion elicitation trial. Thus, another query will be answered in this section, that is, *how is the emotion-related information distributed in the time domain?* In order to localize, in the time domain, the emotion-related information in the EEG signals, an expansion of the AsI calculation will be adopted, by estimating it, instead of using the whole duration of the acquired EEG signal, within a time-window of it, which slides along the trial-related signal and reveals its segments that correspond to big-windowed AsI ($wAsI$) values, that is, over a specified threshold [59].

For each window $q = 1, \ldots, Q$, a $wAsI^q$ value was extracted, according to the concept of the original AsI and was assigned into the middle of the window, that is,

$$wAsI^q = \left(S_r^q - S_p^q\right) \times \frac{\sqrt{2}}{2} \tag{11.13}$$

The resulted sequence of $wAsI$s of an emotion elicitation trial is depicted in Figure 11.4. In order to extract the segment of the signal that is more likely to express emotional EEG activity, the EEG segments of all the three EEG channels of the corresponding trial were selected, by multiplying the initial signals with the unit-amplitude pulse (see Figure 11.4-gray pulse), namely $wAsI - Mask$. The selected segments constitute new $\hat{X}$, $\hat{Y}$, and $\hat{Z}$ signals that are supposed to correspond to "emotionally" filtered signals.

After the extraction of the emotion-related segments of the EEG signals, they were subjected to the adopted feature extraction methods, that is, HOC and CC. The classification scenarios followed were the same as previously. Figures 11.5(a) and 11.5(b) show the classification results for the *BAG* and the whole dataset, respectively. In Figure 11.5(a), the superiority, in regard with the classification performance, of the $wAsI$-filtered signals is obvious, in contrast to the initial EEGs. The filtering introduced by the $wAsI$ approach has obviously augmented the classification performance for all six classification scenarios and for both feature extraction techniques,

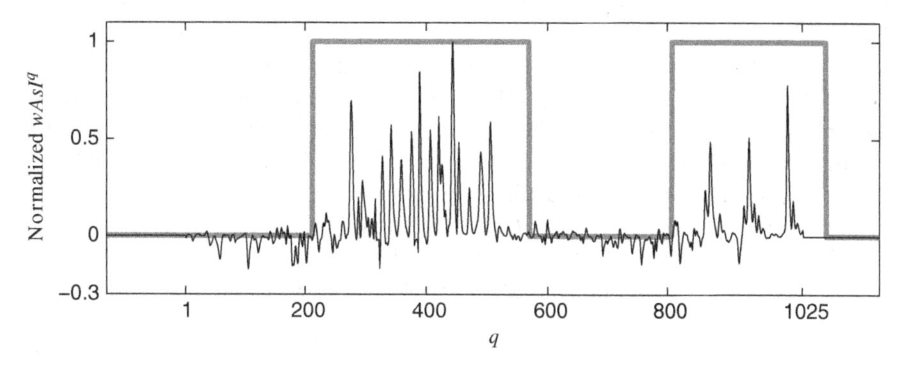

FIGURE 11.4 An example of the $wAsI$ sequence along with the $wAsI - Mask$ (gray line).

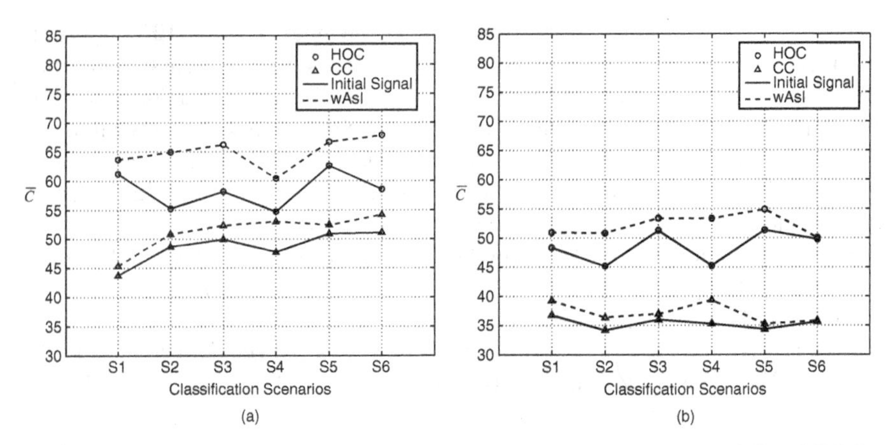

FIGURE 11.5 Classification rates for both feature extraction techniques (HOC and CC) for initial signals and signals subjected to the *wAsI* algorithm. (a) Signals within the *BAG* and (b) whole dataset signals.

exhibiting a mean performance across the classification scenarios of about 65%. The same situation is observed for the whole dataset before and after the *wAsI* filtering is implemented, as depicted in Figure 11.5(b).

All above observations lead to the assumption that the *wAsI* filtering has achieved an efficient isolation of the emotion-related information that consequently led to a more robust EEG-ER system. Finally, it is worth mentioning that the distribution of the $wAsI - Masks$, as proved in Reference 59, was significantly related to the valence and arousal dimension of each affective state; a fact that further justifies the emotion-related character of the filtering conducted of the *wAsI* approach.

11.5.2.5 Time-Frequency-Based Approach Considering the frequency-based characteristics of the emotional expression in the brain (see Section 11.2), the previous trial- and time-based approaches are further expanded to a time-frequency-based one. In order to accomplish this, EMD analysis [50] is utilized, as a means of the time-frequency representation of the EEG signal through the extracted IMFs. Having this, the *wAsI* approach is implemented in the extracted IMFs. The subsequent filtered IMFs ($\widehat{IMFs}$) are used to reconstruct the final filtered, trial-referring, EEG signal by summing them up. The final filtered EEG signal is used to extract the features for classification. The schematic representation of the $EMD - wAsI$ segmentation approach is depicted in Figure 11.6. Figures 11.7(a) and 11.7(b) show the classification results for the *BAG* and the whole dataset, respectively. In Figure 11.7(a) the superiority of the $EMD - wAsI$ approach is observed from both the initial unfiltered EEG signal and the signal which was subjected to the *wAsI* filtering. The same situation is profound in Figure 11.7(b) where the whole dataset results are depicted.

All above results demonstrate the efficacy of the $EMD - wAsI$ algorithm to extract the emotion-related information from the EEG signal and furthermore outperform

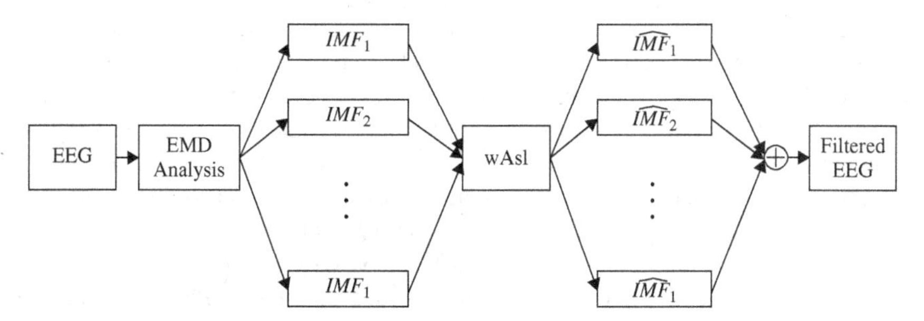

FIGURE 11.6 Block diagram of the *EMD* – *wAsI* algorithm.

over the simple *wAsI* approach due to the implementation of the EEG segmentation, not only in the time level but also in the frequency level, exploiting further the highly related characteristics with the emotional expression in the brain. Finally, it should be noted that, as shown in Reference 60, after the implementation of the *EMD* – *wAsI* algorithm to the EEG dataset the subject-dependent classification rates did not reveal any swings across the subjects. As commented in Section 11.5.2.3, subjects with high mean *AsI* tend to exhibit better classification rates. Nevertheless, after the emotion-oriented filtering, the *EMD* – *wAsI* algorithm implements, the features that are subsequently extracted express more robustly the emotional information in all subjects, whatever the mean *AsI* value was for their initial EEGs.

11.6 CONCLUDING REMARKS AND FUTURE DIRECTIONS

In this chapter, novel signal-processing methodologies for the EEG-ER problem have been presented. The two recognition cases of discrete emotions and fuzzy

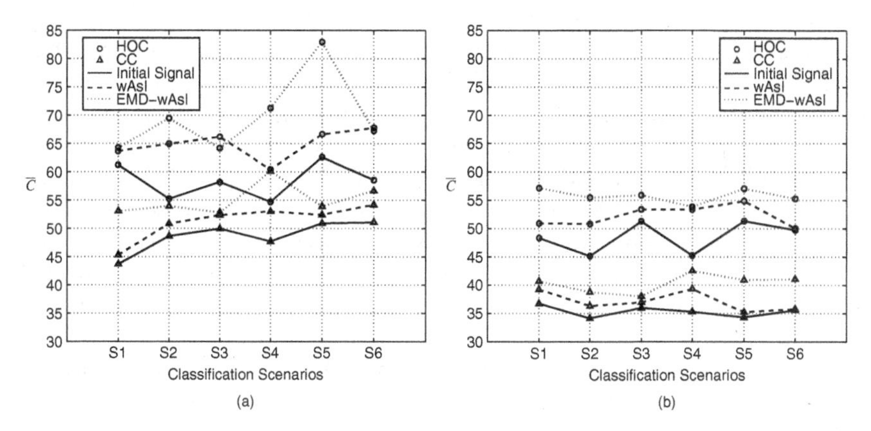

FIGURE 11.7 Classification rates for both feature extraction techniques (HOC and CC) for initial signals and signals subjected to the *wAsI* and *EMD* – *wAsI* algorithms. (a) Signals within the *BAG* and (b) whole dataset signals.

affective states were examined using different emotion elicitation techniques and adaptive processing for the needs of each case. The main approaches for the EEG-ER problem solution presented are based on two main concepts, that is, the fundamental characteristics of the emotional expression in the brain activity and the subject-dependent factors that are inherent in the emotion-oriented arousal and response. From the overall evaluation of the aforementioned methods and approaches, a number of conclusions and future research endeavors can be extracted.

First of all, an important result is that MNS-based emotion elicitation seems to produce more effectively the emotional response than the IAPS-based one. As mentioned in Section 11.5.2, a significant reduction in the classification rates was observed while attempting to classify affective states instead of discrete emotion using the same classification techniques. This fact leads to the assumption that the MNS-based approach indeed elicits the same or akin brain activity, through a mimic procedure, as if the subjects were actually captured by a corresponding emotion. Despite that superiority of MNS-based method over the IAPS-based one, it is argued that Ekman's discrete emotions in an AC application-oriented environment would be of little relevance of the application concept [61]. For instance, in a learning environment, the Ekman's six basic emotions have little relevance to the actual learning process [62] and other learning-related emotions, such as frustration or confusion, should be of much more usability. Thus, research toward application-centered emotions would be of much importance in the EEG-ER area.

Second, taking as a starting point the aforementioned example of a learning AC environment, new concepts, regarding the environments where the EEG-ER task is realized, should be adopted. For instance in the majority of the EEG-ER studies, the elicitation of emotions is realized through strict, trial-centered experiment protocols and, as a result, the context-oriented nature of emotion experience [63] is totally skipped. Thus, EEG-ER research should be directed toward simulations of real-life emotional stimulations in order to develop more pragmatic and adaptive to the environment's context EEG-ER systems.

Furthermore, the aforementioned trial-based emotion elicitation have led to the elimination of the detection phase from the EEG-ER task, that is, researchers have confined their research only to the recognition of the corresponding emotion as they actually know the time location of its elicitation and probably its generation. Nevertheless, in a more realistic emotion elicitation process the detection of such an emotion generation should be conducted before the classification to an emotion category. Toward such an endeavor, many aspects should also be taken under consideration, for example, the mood of the subjects, as they would affect their emotional inertia. A first attempt toward such endeavor is reported in Reference 64 with promising results. Nevertheless, it is still an uncharted territory and much of research is needed.

Although the signal-processing methods described in this chapter do not account for all EEG-ER issues, they create a bed-set for more pragmatic exploitation of the EEG potential and contribute toward the enhancement of their value to serve as a means for effective realization of EEG-ER systems. The ample space for new discoveries and optimization is the driving force that keeps EEG-based emotion recognition research alive; yet, the support and collaboration from complementary disciplines, that is, neurologists, psychologists, and engineers, are the catalyst for the success of such an endeavor.

REFERENCES

[1] W. James, "What is an emotion?" *Mind*, 9: 188–205, 1884.

[2] R. W. Picard, *Affective Computing*, The MIT Press, 1997.

[3] J. J. Lien, T. Kanade, J. F. Cohn, and C. Li, "Automated facial expression recognition based on FACS action units," in Proceedings of the Third IEEE Conference on Automatic Face and Gesture Recognition, Nara, Japan, April 1998, pp. 390–395.

[4] B. Schuller, S. Reiter, R. Mueller, M. Al-Hames, and G. Rigoll, "Speaker independent speech emotion recognition by ensemble classification," in Proceedings of the Sixth International Conference on Multimedia and Expo, Amsterdam, The Netherlands, 2005, pp. 864–867.

[5] R. W. Picard, E. Vyzas, and J. Healey, "Toward machine emotional intelligence: analysis of affective physiological state," *Trans. Pattern Anal. Mach. Intell.*, 23(10): 1175–1191, 2001.

[6] T. Dalgleidh, B. D. Dunn, and D. Mobbs, "Affective neuroscience: past, present, and future," *Emot. Rev.*, 1(4): 355–368, 2009.

[7] G. Gargiulo, P. Bifulco, R. A. Calvo, M. Cesarelli, C. Jin, and A. van Schaik, "A mobile EEG system with dry electrodes," in Proceedings of the Biomedical Circuits and Systems Conference, Baltimore, MD, November 2008, pp. 273–276.

[8] P. C. Petrantonakis and L. J. Hadjileontiadis, "Emotion recognition from EEG using higher order crossings," *Trans. Inf. Technol. Biomed.*, 14(2): 186–197, 2010.

[9] K. Takahashi, "Remarks on emotion recognition from bio-potential signals," in Proceedings of 2nd International Conference on Autonomous Robots and Agents, New Zealand, 2004, pp. 186–191.

[10] D. O. Bos, "EEG-based emotion recognition," Technical report, University of Twente, 2006.

[11] C. Darwin, *The Expression of the Emotions in Man and Animals*, Oxford University Press, Inc., 2002.

[12] P. Ekman, *Expression and the Nature of Emotion*, Chapter 15, Erlbaum, 1984, pp. 319–343.

[13] P. Ekman, R. W. Levenson, and W. V. Friesen, "Emotions differ in autonomic nervous system activity," *Science*, 221: 1208–1210, 1983.

[14] R. J. Davidson, G. E. Schwartz, C. Saron, J. Bennett, and D. J. Goleman, "Frontal versus parietal EEG asymmetry during positive and negative affect," *Psychophysiology*, 16: 202–203, 1979.

[15] R. J. Davidson, P. Ekman, C. D. Saron, J. A. Senulis, and W. V. Friesen, "Approach-withdrawal and cerebral asymmetry: emotional expression and brain physiology," *J. Personal. Soc. Psychol.*, 58: 330–341, 1990.

[16] H. Jasper, "The ten-twenty electrode system of the international federation," *Electroencephalogr. Clin. Neurophysiol.*, 39: 371–375, 1958.

[17] W. J. H. Nauta, "The problem of the frontal lobe: a reinterpretation," *J. Psychiatr. Res.*, 8: 167–187, 1971.

[18] R. J. Davidson, "What does the prefrontal cortex "do" in affect: perspectives on frontal EEG asymmetry research," *Biol. Psychol.*, 67: 219–233, 2004.

[19] J. A. Coan and J. J. B. Allen, "Frontal EEG asymmetry as a moderator and mediator of emotion," *Biol. Psychol.*, 67: 7–49, 2004.

[20] D. Hagemann, E. Naumann, A. Lurken, G. Becker, S. Maier, and D. Bartussek, "EEG asymmetry, dispositional mood and personality," *Personal. Individ. Differ.*, 27: 541–568, 1999.

[21] L. I. Aftanas, A. A. Varlamov, S. V. Pavlov, V. P. Makhnev, and N. V. Reva, "Affective picture processing: event-related synchronization within individually defined human theta band is modulated by valence dimension," *Neurosci. Lett.*, 303: 115–118, 2001.

[22] G. Pfurtscheller and F. H. L. da Silva, "Event-related EEG/MEG synchronization and desynchronization: basic principles," *Clin. Neuropsychol.*, 110: 1842–1857, 1999.

[23] A. Choppin, "EEG-based human interface for disabled individuals: emotion expression with neural networks," Master's thesis, Tokyo Institute of Technology, 2000.

[24] G. Chanel, J. Kronegg, D. Grandjean, and T. Pun, "Emotion assessment: arousal evaluation using EEG and peripheral physiological signals," Technical report, University of Geneva, 2005.

[25] Z. Khalili and M. Moradi, "Emotion detection using brain and peripheral signals," in Proceedings of Biomedical Engineering Conference, 2008, pp. 1–4.

[26] R. Horlings, D. Datcu, and I. J. M. Othkrantz, "Emotion recognition using brain activity," in Proceedings of International Conference on Computer Systems and Technologies, 2008, pp. 1–6.

[27] M. Murugappan, M. Rizon, R. Nagarajan, S. Yaacob, I. Zunaidi, and D. Hazry, "Lifting scheme for human emotion recognition using EEG," in Proceedings of the International Symposium on Information Technology, 2008, pp. 1–7.

[28] K. G. Srinivasa, K. R. Venugopal, and L. M. Patnaik, "Feature extraction using fuzzy c-means clustering for data mining systems," *Int. J. Comput. Sci. Netw. Secur.*, 6: 230–236, 2006.

[29] J. J. De Gruijter and A. B. McBratney, *A Modified Fuzzy k-Means for Predictive Classification*, Elsevier Science, 1988, pp. 97–104.

[30] I. Daubechies, "Orthonormal bases of compactly supported wavelets," *Comm. Pure Appl. Math.*, 41: 909–996, 1988.

[31] T. Musha, Y. Terasaki, H. A. Haque, and G. A. Ivamitsky, "Feature extraction from EEGs associated with emotions," *Artif. Life Robot.*, 1(1): 15–19, 1997.

[32] K. Schaaff and T. Schultz, "Towards emotion recognition from electroencephalographic signals," in Proceedings of International Conference on Affective Computing and Intelligent Interaction, Amsterdam, The Netherlands, September 2009, pp. 175–180.

[33] K. Schaaff and T. Schultz, "Towards an EEG-based emotion recognizer for humanoid robots," in Proceedings of 18th IEEE International Symposium on Robot and Human Interactive Communication, 2009, pp. 792–796.

[34] A. Heraz and C. Frasson, "Predicting the three major dimensions of the learner's emotions from brainwaves," *Int. J. Comput. Sci.*, 2(3): 187–193, 2008.

[35] M. Murugappan, M. Rizon, R. Nagarajan, and S. Yaacob, "Inferring of human emotional states using multichannel EEG," *Eur. J. Sci. Res.*, 48(2): 281–299, 2010.

[36] P. Ekman *et al.*, "Universals and cultural differences in the judgments of facial expressions of emotion," *J. Personal. Soc. Psychol.*, 53(4): 712–717, 1987.

[37] G. Rizzolatti and L. Craighero, "The mirror-neuron system," *Ann. Rev. Neurosci.*, 27: 169–192, 2004.

[38] E. Oztop, M. Kawato, and M. Arbib, "Mirror neurons and imitation: a computationally guided review," *Neural Netw.*, 19: 254–271, 2006.

[39] T. W. Lee, R. J. Dolan, and H. D. Critchley, "Controlling emotion expression: behavioral and neural correlates of nonimitative emotional responses," *Cereb. Cortex*, 8: 104–113, 2008.

[40] L. Carr, M. Iacoboni, M. C. Dubeau, J. C. Mazziotta, and G. L. Lenzi, "Neural mechanisms of empathy in humans: a relay from neural systems for imitation to limbic areas," *Proceedings of the National Academy of Sciences*, 100: 5497–5502, 2003.

[41] P. J. Lang, M. M. Bradley, and B. N. Cuthbert, "International affective picture system (IAPS): affective ratings of pictures and instruction manual," Technical report, University of Florida, 2011.

[42] P. Ekman and W. V. Friesen, "Pictures of facial affect," Technical report, University of California Medical Center, 1976.

[43] P. C. Petrantonakis and L. J. Hadjileontiadis, "A novel emotion elicitation index using frontal brain asymmetry for enhanced EEG-based emotion recognition," *IEEE Transactions on Information Technology in Biomedicine*, 2011.

[44] B. Kedem, *Time Series Analysis by Higher Order Crossings*, IEEE Press, 1994.

[45] R. D. King, C. Feng, and A. Shutherland, "Statlog: comparison of classification algorithms on large real-world problems," *Appl. Artif. Intell.*, 9: 259–287, 1995.

[46] W. J. Krzanowski, *Principles of Multivariate Analysis*, Oxford University Press, 1988.

[47] T. Mitchell, *Machine Learning*, McGraw-Hill, New York, 1997.

[48] P. C. Mahalanobis, "On the generalized distance in statistics," *Proceedings of the National Academy of Sciences India*, 2: 49–55, 1936.

[49] N. Cristianini and J. Shawe-Taylor, *An Introduction to Support Vector Machines and Other Kernel-Based Learning Methods*, Cambridge University Press, Cambridge, UK, 2000.

[50] N. Huang, Z. Shen, S. Long, M. Wu, H. H. Shih, N. C. Zheng, N. C. Yen, C. Tung, and H. Liu, "The empirical mode decomposition and Hilbert spectrum for nonlinear and nonstationary time series analysis," *Proceedings of the Royal Society of London A*, 454: 903–995, 1998.

[51] D. E. Goldberg, *Genetic Algorithms in Search, Optimization & Machine Learning*, Addisson-Wesley, 1989.

[52] P. Flandrin, G. Rilling, and P. Goncalves, "Empirical mode decomposition as a filter bank," *IEEE Signal Processing Letters*, 11(2): 112–114, 2004.

[53] P. C. Petrantonakis and L. J. Hadjileontiadis, "Emotion recognition from brain signals using hybrid adaptive filtering and higher order crossings analysis," *IEEE Transaction on Affective Computing*, 1(2): 81–97, 2010.

[54] M. Katz, "Fractals and the analysis of waveforms," *Comput. Biol. Med.*, 18: 145–156, 1988.

[55] A. Accardo, M. Affinito, M. Carrozzi, and F. Bouquet, "Use of fractal dimension for the analysis of electroencephalographic time series," *Biol. Cybern.*, 77: 339–350, 1997.

[56] A. Petrosian, "Kolmogorov complexity of finite sequences and recognition of different preictal EEG patterns," in Proceedings of IEEE Symposium on Computer-Based Medical Systems, 1995, pp. 212–217.

[57] O. Sakata, T. Shiina, and Y. Saito, "Multidimensional directed information and its application," *Electron. Comput. Jpn.* (Part III: Fundamental Electronic Science), 85(4): 45–55, 2002.

[58] G. Wang and M. Takigawa, "Directed coherence as a measure of interhemispheric correlation of EEG," *Int. J. Psychophysiol.*, 13: 119–128, 1992.

[59] P. C. Petrantonakis and L. J. Hadjileontiadis, "Adaptive EEG segmentation for efficient emotion recognition," *Biomed. Signal Process. Control*, 2011.

[60] P. C. Petrantonakis and L. J. Hadjileontiadis, "Adaptive emotional information retrieval from EEG signals in the time-frequency domain," *IEEE Transactions on Signal Processing*, 2011.

[61] R. A. Calvo and S. D'Mello, "Affect detection: an interdisciplinary review of models, methods, and their applications," *IEEE Transaction on Affective Computing*, 1(1): 18–37, 2010.

[62] S. D'Mello, R. Picard, and A. Graesser, "Towards an affect-sensitive autotutor," *IEEE Intell. Syst.*, 22(4): 53–61, 2007.

[63] J. A. Russell, "Core affect and the psychological construction of emotion," *Psychol. Rev.*, 110: 145–172, 2003.

[64] P. C. Petrantonakis and L. J. Hadjileontiadis, "Towards effective EEG-based detection of affective transitions," *IEEE Transactions on Affective Computing*, 2011.

AUTHOR BIOGRAPHIES

Panagiotis C. Petrantonakis was born in Ierapetra, Crete, Greece, in 1984. He received his Diploma in Electrical and Computer Engineering in 2007 from the Aristotle University of Thessaloniki (AUTH), Thessaloniki, Greece. Currently, he is a Ph.D. researcher at AUTH, affiliated with the Signal Processing and Biomedical Technology Unit of the Telecommunications Laboratory. His current research interests lie in the area of advanced signal processing techniques, nonlinear transforms, and affective computing. He is a member of the Technical Chamber of Greece and a student member of the IEEE.

Leontios J. Hadjileontiadis was born in Kastoria, Greece, in 1966. He received his Diploma in Electrical Engineering in 1989 and the Ph.D. degree in Electrical and Computer Engineering in 1997, both from the Aristotle University of Thessaloniki, Thessaloniki, Greece. In December 1999, he joined the Department of Electrical and Computer Engineering, Aristotle University of Thessaloniki, as a faculty member, where he is currently an associate professor, working on lung sounds, heart sounds, bowel sounds, ECG data compression, seismic data analysis, and crack detection in the Signal Processing and

Biomedical Technology Unit of the Telecommunications Laboratory. His research interests are in higher-order statistics, alpha-stable distributions, higher-order zero crossings, wavelets, polyspectra, fractals, neuro-fuzzy modeling for medical, mobile, and digital signal processing applications.

Dr. Hadjileontiadis is a member of the Technical Chamber of Greece, IEEE, Higher-Order Statistics Society, International Lung Sounds Association, and American College of Chest Physicians. He was the recipient of the second award at the Best Paper Competition of the Ninth Panhellenic Medical Conference on Thorax Diseases97, Thessaloniki. He was also an open finalist at the Student Paper Competition (Whitaker Foundation) of the IEEE EMBS'97, Chicago, IL, a finalist at the Student Paper Competition (in memory of Dick Poortvliet) of the MEDICON'98, Lemesos, Cyprus, and the recipient of the Young Scientist Award of the Twenty-Fourth International Lung Sounds Conference99, Marburg, Germany. At the Imagine Cup Competition (Microsoft), Sao Paulo, Brazil (2004)/Yokohama, Japan (2005)/Seoul, South Korea (2007), with projects involving technology-based solutions for people with disabilities, he organized and served as mentor to three five-student teams that ranked third, second, and seventh worldwide, respectively. Dr. Hadjileontiadis also holds a Ph.D. degree in music composition (University of York, UK, 2004) and he is currently a professor in composition at the State Conservatory of Thessaloniki, Greece.

12

FREQUENCY BAND LOCALIZATION ON MULTIPLE PHYSIOLOGICAL SIGNALS FOR HUMAN EMOTION CLASSIFICATION USING DWT

M. MURUGAPPAN

School of Mechatronic Engineering, Universiti Malaysia Perlis (UniMAP), Campus Ulu Pauh, Arau, Malaysia

Human emotion classification using physiological signals is one of the active research areas on developing intelligent human–machine interface (HMI) systems. This work is proposed to analyze the different frequency band information from two physiological signals (electromyogram (EMG) and electrocardiogram (ECG)) for classifying four different emotions such as disgust, happy, fear, and neutral. Physiological signals are collected using 6 electrodes system from 20 subjects in the age group of 21–26 years with a mean age of 24.5 years. Discrete wavelet transform (DWT) is used for decomposition of the physiological signals into different frequency bands. In EMG signals, we are interested to investigate the four different frequency ranges such as 4–16 Hz, 16–31 Hz, 31–63 Hz, and 4–63 Hz for classifying emotions. Similarly, ECG signals are analyzed over the three different frequency bands namely very low frequency component (VLF) (0.015–0.04 Hz), low frequency component (LF) (0.04–0.14 Hz), and high frequency component (HF) (0.14–0.5 Hz). Butterworth band-pass filter is used for removing the effects of noise and other interferences from the signals. The average spectral power is derived from the preprocessed signals using four different wavelet functions (*db5*, *db7*, *sym8*, and *coif5*). A simple nonlinear classifier (K Nearest Neighbor (KNN)) and linear classifier (Linear Discriminant Analysis (LDA)) are used to map the statistical features into corresponding emotions. As a result of this

Emotion Recognition: A Pattern Analysis Approach, First Edition. Edited by Amit Konar and Aruna Chakraborty.
© 2015 John Wiley & Sons, Inc. Published 2015 by John Wiley & Sons, Inc.

study, EMG signals give the maximum average classification rate on four emotions (disgust—96.25 %, happy—98.75 %, fear—89.03 %, and neutral—89.59 %) using KNN over ECG signals. The spectral power derived over 31–63 Hz of EMG signals and a sum of LF and HF frequency band in ECG signals gives the highest emotion classification rate over other frequency band combinations. We have also presented the average classification accuracy of emotions on two physiological signals over different frequency bands using DWT for justifying the performance of the proposed emotion recognition system.

12.1 INTRODUCTION

In recent years, the researchers are focusing on developing intelligent man–machine system by means of estimating the user affective and emotional states [1]. Human emotional expressions are straightforward to detect and understand by humans, as these are reflected in both voice and body languages [1, 2]. There are several researchers who addressed the subject of emotions and their role in human–computer interaction (HCI) applications [1]. The specific areas of interest span recognition and synthesis of emotion in face and body, speech specifies, and the influence of emotion on information processing and decision making, and interaction metaphors [1]. Emotion directly affects our decision making, perception, cognition, creativity, attention, reasoning, and memory [2]. Most of the researchers focused on assessing emotions through several modalities such as facial expressions, speech, and gestures [3–5]. Though these methods give higher emotion classification rate, the emotions induced and assessed through these approaches are purposefully expressed and it can be more easily concealed by other subjects. Indeed, the performance of this system is severely affected with other factors such as lighting conditions, experimental settings, environmental noises, etc. [6]. In recent years, physiological signal-based emotion assessment is highly preferred for developing more reliable emotion recognition systems. Because, the changes in physiological signals during different emotional states highly reflect the internal body changes through autonomic nervous system (ANS) and central nervous system (CNS) activities [6, 7]. Hence, classifying emotions by retrieving information from physiological signals helps in understand the real and inherent emotional state changes of human. Thereby, it leads to developing intelligent HCI system to interact with computers in a more reliable and efficient manner [6].

Some of the following physiological signals are highly used for classifying the human emotional state changes in a more reliable manner: electroencephalogram (EEG), electrocardiogram (ECG), electromyogram (EMG), electrooculogram (EoG), skin conductive resistance (SCR), skin temperature (ST), and respiration rate (RR) [7–10]. Among these physiological signals, ECG and EMG play a vital role in developing portable, non-intrusive, reliable, and computationally efficient emotion recognition systems [8]. Hence, we have considered these two physiological signals for classifying emotions. More details on the automatic emotion recognition using physiological signals as well as more complete list of reference could be found in Reference 11. At present, researchers are trying to develop a real-time emotion recognition using physiological signals. However, most of the research efforts are focused on analyzing the physiological signals in "off-line" for classifying the emotions.

From the literature, the researchers have considered different types of emotion inducement methods to evoke different emotions. In general, visual stimuli (images/pictures), audiovisual stimuli (film clips/video clips), audio stimuli (sounds/songs), emotional recall paradigm, etc., are used for inducing different emotions [7, 10, 11]. However, no common ground truth has been obtained on defining an efficient emotion inducement through above methods. But most of the researchers have considered audiovisual stimuli for inducing the emotions. In general, physiological signal-based emotion recognition is challenging due to the following aspects. First, physiological signals are highly non-stationary and nonlinear in nature. Thus, characterizing the physiological signals for each emotion is more difficult in real-time applications, Second, physiological signals are highly affected due to the existence of noise and other external interferences. Third, the existence of multiple emotions on single stimuli is more apparent and the emotional experience of one person is different from others for the same stimuli. Hence, developing efficient research methodology for handling physiological signals is highly inevitable for developing intelligent emotion recognition systems.

The rest of this chapter is organized as follows. In Section 12.2, we summarize the related works on classifying the emotions using physiological signals. The research methodology by elucidating the data acquisition process, preprocessing, feature extraction using DWT, and classification of emotions are given in Section 12.3. Section 12.4 presents the overview of the results and discussion of this work, and conclusions are given in Section 12.5.

12.2 RELATED WORK

Determining the rhythmic activity of physiological signals for different emotional stimulus is a promising area of research. It provides an in-depth picture of how the physiological signal characteristics vary for different emotions. Most of the previous studies focus on assessing the two-dimensional human emotion (valence-arousal) through different frequency ranges of physiological signals [12]. Only few works have been proposed on discrete human emotion classification [13, 14]. Furthermore, there is no defined range of frequency suggested for improving the emotion classification rate. Hence, it is necessary to determine the range of frequency over which we achieve the maximum classification accuracy on assessing discrete emotions.

Wagner *et al.* considered five physiological signals: skin conductivity level, frontal electromyography, ST, breathing frequency, and pulse frequency to classify the human emotional states [15]. Takahashi used EEG, EoG, EMG, pulse oxymeter, and skin conductance signals to estimate five different emotional states (joy, anger, sadness, happiness, and relax) [16]. Table 12.1 shows the list of previous works on emotion recognition using physiological signals in specifically EMG and ECG signals. However, a detailed study on previous works on EEG signal-based emotion recognition can be found in Reference 21.

TABLE 12.1 List of previous works on emotion recognition using ECG and EMG signals

Ref.	No. of subjects	Signals	No. of electrodes	Stimuli	Emotions		Classifier	Max % CR
					Nos.	Types		
[13]	2	EMG	4	Audio	4	Joy, anger, sadness, and pleasure	Least square SVM	80
[17]	150	ECG	3	Audiovisual	2	Joy and sadness	KNN	88.88
[18]	391	ECG	3	Audiovisual	2	Joy and sadness	Fisher -KNN	85.78
[19]	40	HRV, SC	2	Visual	2	Valance– Arousal	–	–
[20]	10	GSR, BVP	2	Visual	4	Sad, bliss, joy, and fear	–	–
[14]	2	SC, ECG, EMG, RR	4	Audio	4	Joy, anger, sadness, and pleasure	–	92.05
[12]	391	ECG	3	Audiovisual	2	Joy and sadness	HPSO	88.43

From Table 12.1, most of the researchers used audiovisual stimuli for inducing two emotions (joy and sadness) and achieved the maximum emotion classification rate of 88.88% [17]. DWT-based feature extraction is adopted mostly by the researchers for extracting the statistical features from the physiological signals [12, 14, 17–20]. Among the different physiological signals, the fusion of four physiological signals gives the maximum emotion classification rate of 92.05% on classifying four emotions (joy, anger, sad, and pleasure). However, the researchers have considered only two subjects for classifying the emotions. In Reference 18, the researchers managed to classify the two emotions from 391 subjects and achieved a maximum average classification rate of 88.88% using KNN. Most of the researchers considered simple classifiers such as KNN, Fisher KNN for classifying the emotions using DWT features.

Most of the researchers analyzed the physiological signals over the entire frequency range for classifying the emotions [12–14, 17–20]. None of the researchers indicated a specific range of physiological signal frequency for efficiently classifying the emotions. Selection of specific range of frequency in physiological signals will highly reduce the computation time and memory requirement while going for real-time system development with enhanced performance rate. In this work, we have carried out a set of experiments to analyze four different bands of frequency from EMG signals and three frequency bands from ECG signals for classifying four emotions. We have computed the average classification accuracy for average spectral power feature over different frequency bands from ECG and EMG signals. The result of classification accuracy is reported in Section 12.4.

12.3 RESEARCH METHODOLOGY

12.3.1 Physiological Signals Acquisition

This section provides an experimental procedure for physiological signal acquisition. It begins with the principles of emotion inducement by different types of stimuli. It then briefly narrates about the experimental subjects and setup.

12.3.1.1 Emotion Inducement In general, the researchers used several modalities to induce the emotions on a subject. During the emotion recognition experiments, the subjects are exposed to any of the following one or multiple stimuli to evoke their emotions: (i) visual stimuli: showing some emotional pictures or images to the subjects [19, 20]; (ii) audiovisual stimuli: showing emotional video or film clips to the subjects [12, 17, 18]; (iii) emotional recall stimuli: some of the researchers ask the subjects to recall the past emotional events during their life time; (iv) audio stimuli: playing some emotional music before the subjects [13, 14]. The type of emotional stimuli used for inducing the emotions differs from one research work to the other and no ground truth has been laid on specifically choosing certain stimuli for emotion classification. Most of the researchers considered audiovisual stimuli for evoking discrete emotions. In our earlier work, we have considered both visual and

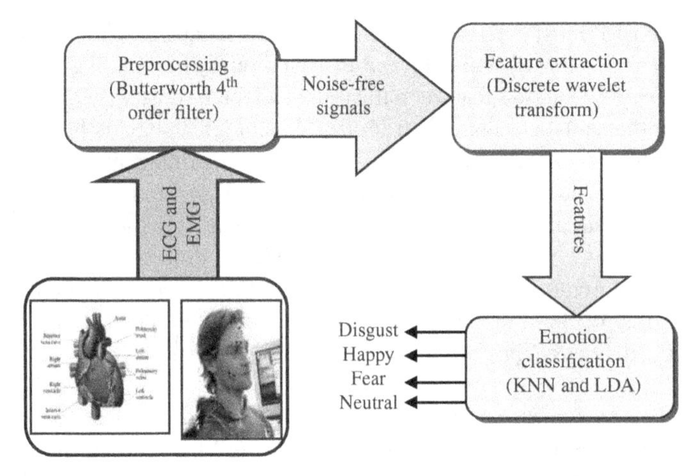

FIGURE 12.1 Overview of emotion recognition system procedure using physiological signals.

audiovisual stimuli for evoking discrete emotions, and the results confirm that audiovisual stimulus performs superior in evoking emotions than visual stimulus [22]. Hence, we have designed an audiovisual stimuli-based data acquisition protocol for eliciting the discrete emotions in this work. The structural overview of emotion recognition system using physiological signals (EMG, ECG) and research methodology of this present work is shown in Figures 12.1 and 12.2, respectively. The audiovisual stimulus protocol for Trail 1 of our experiment is shown in Figure 12.3.

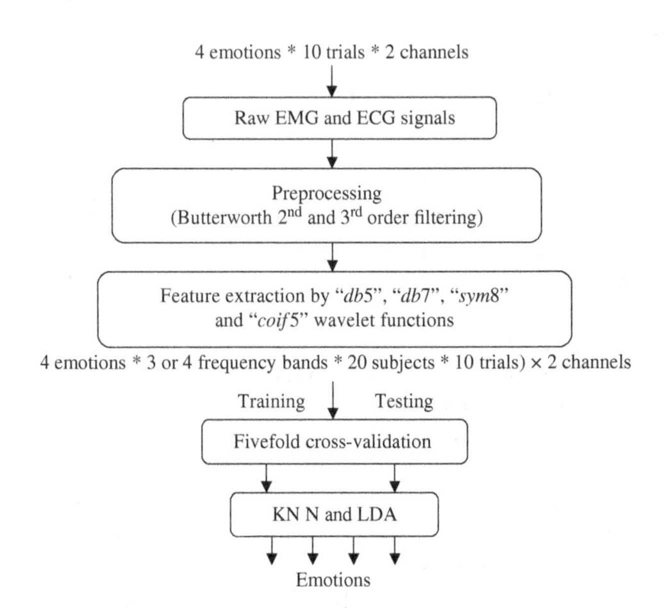

FIGURE 12.2 Research methodology on classifying human emotions.

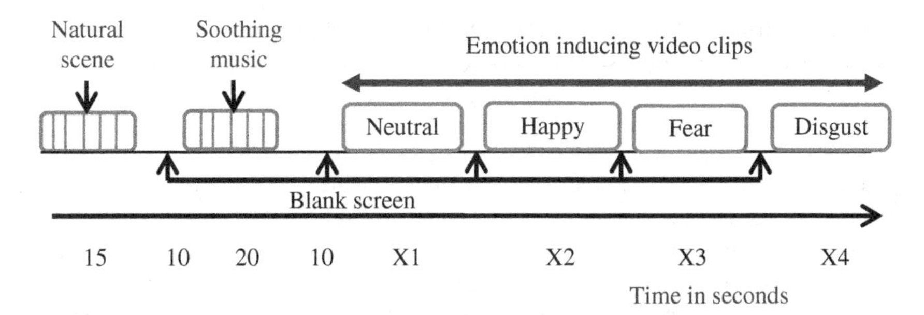

FIGURE 12.3 Audiovisual stimulus-based protocol of trial 1 for inducing emotions.

All the emotional video clips were shown in a random manner over 10 trials (Table 12.2). In Figure 12.3, X1–X4 denote time periods of emotional video clips. Because, all the video clips are having different time duration and it vary from one another. Indeed, most of the emotional video clips are short in time duration and have more dynamic emotional content. Between each emotional stimuli (video clips), a blank screen is shown for 10-second duration to bring the subject to their normal state and to experience a calm mind. A pilot panel study is conducted on 25 students in the age group of 21–30 years to select the 10 most dominating emotional video clips for each emotion from a list of 25 video clips per emotion. A sum of 100 video clips for four emotions is collected for evoking four emotions from various sources such as internet, films, and international standard emotional clips database*. The subjects who have undergone this pilot panel study will not be considered as a subject for data collection experiment. In this work, we have considered 10 trials for inducing each emotion and each trial has four short video clips for each subject. Two images of nature scenes such as hills, skies, ocean, and mountains are displayed between the video clips for 15-second duration. This is necessary to overcome the effect of current emotional state of subjects before continuing the next stimuli presentation [5].

TABLE 12.2 Order of emotional stimuli used over 10 trials in emotion recognition experiment

Trials	Order of emotions			
1	Neutral	Happy	Fear	Disgust
2	Disgust	Neutral	Happy	Fear
3	Fear	Disgust	Neutral	Happy
4	Happy	Fear	Disgust	Neutral
5	Fear	Disgust	Neutral	Happy
6	Disgust	Neutral	Happy	Fear
7	Neutral	Happy	Fear	Disgust
8	Happy	Fear	Disgust	Neutral
9	Fear	Disgust	Neutral	Happy
10	Disgust	Neutral	Happy	Fear

Between each trial, under self-assessment section, the subjects are informed to answer the emotions they experienced [8]. This self-assessment form is used to justify the emotion experienced by the subject in each video clips.

12.3.1.2 Subjects Ten females and ten males in the age group of 21–26 years are employed as subjects in our experiment. All the subjects are healthy volunteers from our university and without any history of medical illness such as mental stress, emotional disorders, etc. Once the consent forms are filled up, the subjects are given a simple introduction about the research work and the various stages of experiment.

12.3.1.3 Experimental Setup Two physiological signals namely ECG and EMG were simultaneously acquired during the experiment. PHYWE Cobra3 is used for acquiring EMG signal with a sampling frequency of 500 Hz and online band-pass filtered between 0.05 and 70 Hz. Three electrodes (two active and one reference) are used to acquire the EMG signals from the subject. Two active electrodes are placed on the face, specifically the zygomatic muscle. One reference electrode is placed on the wrist. Similarly, ECG signals are acquired using AD Instruments with a sampling frequency of 1000 Hz. ECG signals are acquired using three electrodes, and two active electrodes are placed on the left and right wrists, and reference electrode is placed on the right ankle. The computerized data acquisition protocol is shown to the subject through liquid crystal display (LCD) projector screen.

12.3.2 Preprocessing and Normalization

The ECG and EMG signal traces give more useful information for various clinical and non-clinical investigations. These signals are widely used by several researchers for recognizing the human emotions [12–14, 17–20]. However, the signals which are acquired in laboratory settings are highly contaminated by various types of noise and artifacts due to power line frequency, mismatching of electrode impedance, improper electrode interface, subject muscular movements, etc. Some of the common artifacts in ECG signal are power line interference, electrode contact noise, motion artifacts, baseline drift, instrumentation noise generated by electronic devices, electrosurgical noise, etc. [23, 24]. In order to efficiently remove the noises and obtain the utile information from the ECG data, the ECG signals should be pre-processed by using digital filters. Low pass filters are used for avoiding the anti-aliasing effect and to reduce the effect of movement or high frequency interferences. High pass filters are used for removing baseline wander and notch filter for removing power line frequency interference [25, 26]. However, an ideal filter should not affect or remove the significant information from the original ECG and EMG data. The complete removal of artifacts will also remove some of the useful information from EMG and ECG signals. In this work, EMG and ECG signals are filtered by using second- and third-order Butterworth Filter with a cutoff frequency of 0.05–100 Hz, respectively [25–28].

12.3.3 Feature Extraction

Feature extraction is a key step of pattern recognition and statistical features from the preprocessed signals derived and mapped into corresponding emotions. Several feature extraction methods are proposed by different researchers on extracting the statistical features from physiological signals [16]. Two physiological signals (EMG and ECG) used in this experiment have highly nonlinear and non-stationary characteristics. For non-stationary signals, non-parametric methods of feature extraction based on multi-resolution analysis of DWT are widely used in many literatures [13, 17–19]. The joint time–frequency resolution obtained by DWT makes it as a good candidate for the extraction of details as well as approximations of the signal. However, this future cannot be obtained either by Fast Fourier Transform (FFT) or by Short Time Fourier Transform (STFT) [10]. In DWT, the non-stationary nature of ECG and EMG signals expands them onto basis functions by expansion, contraction, and shifting of a single prototype function ($\Psi_{a,b}$, the mother wavelet).

The mother wavelet function $\Psi_{a,b}$ (t) is given as

$$\psi_{a,b}(t) = \frac{1}{\sqrt{a}}\psi\left(\frac{t-b}{a}\right) \tag{12.1}$$

where $a, b \in R$, $a > 0$, and R is the wavelet space. Parameters a and b are the scaling factor and shifting factor, respectively. The only limitation for choosing a prototype function as mother wavelet is to satisfy the admissibility condition (Equation 12.2):

$$C_\psi = \int_{-\infty}^{\infty} \frac{|\Psi(\omega)|^2}{\omega} d\omega < \infty \tag{12.2}$$

where ψ (ω) is the Fourier transform of $\psi_{a,b}$ (t).

The time–frequency representation is performed by repeatedly filtering the signal with a pair of filters called high pass filter (H(n)) and low pass filter (L(n)), that cut the frequency domain in the middle of input signal frequency (Figure 12.4). Specifically,

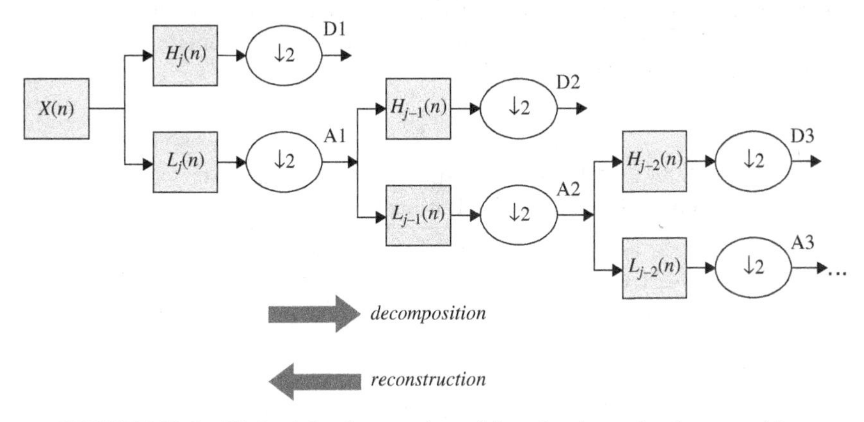

FIGURE 12.4 Filt bank implementation of three-level wavelet decomposition.

TABLE 12.3 Decomposition of ECG signals into different frequency bands using DWT

Frequency range (Hz)	Decomposition level	Frequency bands
0.015–0.04	D15	VLF band
0.04–0.06	D14	LF band
0.07–0.12	D13	LF band
0.13–0.24	D12	HF band
0.25–0.49	D11	HF band

the DWT decomposes the signal into approximation coefficients (CA) and detailed coefficients (CD). The approximation coefficient is subsequently divided into new approximation and detailed coefficients. This process is carried out iteratively producing a set of approximation coefficients and detailed coefficients at different levels or scales [10]. In this work, we have performed the six-level decomposition of EMG signal to retrieve the wavelet coefficients on four different frequency ranges such as 4–16 Hz, 16–31 Hz, 31–63 Hz, and 4–63 Hz. Similarly, ECG signals are decomposed into 14 levels in order to derive the very low frequency component (VLF) (0.015–0.04 Hz), low frequency component (LF) (0.04–0.15 Hz), and high frequency component (HF) (0.15–0.4 Hz). The statistical feature of average spectral power is derived from the above frequency bands in both ECG and EMG signals for classifying the emotions. The DWT-based signal decomposition is carried out using four different wavelet functions (*db5*, *db7*, *sym8*, *coif5*) for emotion classification. These wavelet functions are chosen due to their near optimal time–frequency localization properties. Moreover, the waveforms of these wavelet functions are similar to the waveforms to be detected in the ECG and EMG signals over different emotions. Therefore, extraction of EMG and ECG signals features are more likely to be successful [29]. Tables 12.3 and 12.4 show the wavelet decomposition level and corresponding frequency ranges in ECG and EMG signals using DWT. Thereby, the spectral features calculated from this frequency bands give more meaningful information about the specified frequency ranges. The mathematical formula for computing the spectral power is given in Table 12.5.

TABLE 12.4 Decomposition of EMG signals into different frequency bands using DWT

Frequency range (Hz)	Decomposition level
3.91–7.80	D6
7.81–15.625	D5
15.626–31.25	D4
31.26–62.5	D3

TABLE 12.5 Mathematical computation of average frequency band power using DWT for emotion recognition

Feature	Formula to compute	Description
Spectral power	$P_j = \frac{1}{N}\sum_{k=1}^{N}(d_j(k)^2)$	Computes the squares of the amplitude of wavelet coefficients

j = level of wavelet decomposition; k = No. of wavelet coefficients varies from 1 to N

12.3.4 Emotion Classification

From the literature works in Section 12.2, it is apparent that the previous researchers are using different types of classifiers such as Linear Discriminant Analysis (LDA), K-Nearest Neighbor (KNN), Particle Swarm Optimization (PSO), Support Vector Machine (SVM), etc. to classify the emotions using physiological signals.

From Table 12.1, the maximum average classification rate of 92.05 % is achieved on classifying four emotions using four physiological signals. In Reference 17, the researchers have classified the emotions into two types (joy and sadness) by achieving 88.88% using KNN classifier. However, least square SVM is used to achieve a maximum average classification rate of 80 % on classifying four emotions. In this work, we have adopted LDA and KNN to classify the emotions from the physiological signals (EMG and ECG) for selecting the most appropriate frequency range. Classification accuracy, representing the percentage of accurately classified instances, is adopted to quantify the performance of LDA and KNN.

12.3.4.1 Linear Discriminant Analysis Among the above two classifiers, LDA provides extremely fast evaluations of unknown inputs performed by calculating the distances between a new sample and mean of training data samples. Here, each class weighed by the covariance matrices does not require any additional parameters to be given by the researchers for classifying emotions. An LDA tries to find an optimal hyperplane to separate four classes (disgust, happy, fear, and neutral emotions).

12.3.4.2 K Nearest Neighbor KNN is also a simple and intuitive method of classifier used by many researchers typically for classifying the signals and images. This classifier makes a decision on comparing a new labeled sample (testing data) with the baseline data (training data). In general, for a given unlabeled time series X, the KNN rule finds the K "closest" (neighborhood) labeled time series in the training data set and assigns X to the class that appears most frequently in the neighborhood of k time series. There are two main schemes or decision rules in KNN algorithm, the similarity voting scheme and majority voting scheme. In our work, we used the majority voting for classifying the unlabeled data. It means that, a class (category) gets one vote, for each instance, of that class in a set of K neighborhood samples. Then, the new data sample is classified to the class with the highest amount of votes. This majority voting is more commonly used because it is less sensitive to outliers. However, in KNN, we need to specify the value of Kclosest neighbor for emotion classification. In this experiment, we try different K values ranging from 1 to 10 and suitable value of K is derived based on giving the maximum average emotion classification rate.

12.4 EXPERIMENTAL RESULTS AND DISCUSSIONS

The main motivation of this work is to localize the frequency band information from ECG and EMG signals for enhancing the emotion classification rate. We have analyzed the performance of classifying the emotions from EMG and ECG signals over four and three frequency bands, respectively. Physiological signals acquired from 20 subjects over 4 emotions on 10 trials are preprocessed using Butterworth filters to remove noises, artifacts, and power line interferences. DWT is used to decompose the preprocessed signals into several levels and localize the wavelet coefficients for different frequency ranges from EMG and ECG signals. From the DWT coefficients, we have derived the average spectral power feature for classifying emotions. The derived features are used to form a feature vector and given as input to the classifiers. Among the entire feature vector set, 70% (560 samples out of 800) of feature samples are used for training the classifier and it is tested with remaining 30% (240 samples out of 800) of samples. The classification ability of a statistical feature set can be measured through classification accuracy by taking the average of mean classification rate over 10 times. In this analysis, EMG signal from 4 to 16 Hz frequency band and VLF range (0.0015–0.04 Hz) in ECG signal gives very poor classification rate on four emotions. Hence, the results of the above frequency ranges are discarded from this work. Tables 12.6, 12.7, 12.8, and 12.9 present the results of average classification accuracy obtained from KNN and LDA on classifying emotions using EMG and ECG signals over different frequency bands. Concluding remarks of Tables 12.6, 12.7, 12.8, and 12.9 are given in Tables 12.10 and 12.11.

TABLE 12.6 Average classification rate of emotions from EMG signals using KNN over four different wavelet functions

Wavelet	Frequency range (Hz)	Disgust	Happy	Fear	Neutral	Mean classification classification rate
db5	16–31	77.78	80.28	70.14	73.06	75.32
	31–63	89.165	**98.75**	**89.03**	89.585	**91.63**
	4–63	84.725	98.61	86.665	86.53	89.13
db7	16–31	71.53	69.17	75.695	56.805	68.30
	31–63	90.695	87.775	88.89	83.47	87.71
	4–63	86.805	86.805	83.61	81.67	84.72
sym8	16–31	74.28	77.78	68.09	72.81	73.24
	31–63	85.165	97.53	86.915	87.385	89.25
	4–63	81.225	95.61	84.565	85.03	86.61
coif5	16–31	71.665	73.33	71.39	76.11	73.12
	31–63	**96.25**	87.5	87.64	89.445	90.21
	4–63	90.695	86.11	82.78	**91.67**	87.81

TABLE 12.7 Average classification rate of emotions from EMG signals using LDA over four different wavelet functions

Wavelet	Frequency range (Hz)	Disgust	Happy	Fear	Neutral	Mean classification classification rate
db5	16–31	76.39	69.585	70.835	65.83	70.66
	31–63	78.61	80.275	**81.25**	**83.33**	80.87
	4–63	81.53	81.665	77.365	79.305	79.97
db7	16–31	70.555	71.67	68.19	57.915	67.08
	31–63	80	82.915	80.415	73.19	79.13
	4–63	81.805	82.78	80.28	75.42	80.07
sym8	16–31	78.335	83.475	68.475	64.305	73.65
	31–63	81.67	**95**	78.61	82.64	**84.48**
	4–63	81.665	94.865	78.61	80.83	83.99
coif5	16–31	68.47	71.945	66.945	68.195	68.89
	31–63	**84.025**	84.305	79.305	80.415	82.01
	4–63	83.61	78.335	78.335	80.835	80.28

TABLE 12.8 Average classification rate of emotions from ECG signals using KNN over four different wavelet functions

Wavelet	Frequency range	Disgust	Happy	Fear	Neutral	Mean classification classification rate
db5	0.04–0.15 Hz	71.11	71.945	68.055	77.11	72.06
	0.15–0.49 Hz	57.5	62.5	62.78	62.775	61.39
	LF/HF	71.39	75	62.22	70.555	69.79
	HF/LF	71.53	75.14	62.92	68.885	69.62
	LF+HF	86.53	31.805	73.61	65.555	64.38
db7	0.04–0.15 Hz	73.61	77.22	72.085	75.975	74.72
	0.15–0.49 Hz	56.385	59.03	58.47	57.78	57.92
	LF/HF	70	66.805	69.865	73.89	70.14
	HF/LF	72.085	66.805	67.5	70	69.10
	LF+HF	82.5	74.445	67.5	72.365	74.20
sym8	0.04–0.15 Hz	74.72	75.835	72.365	77.085	75.00
	0.15–0.49 Hz	66.805	67.775	62.36	66.945	65.97
	LF/HF	73.335	71.525	71.665	74.86	72.85
	HF/LF	68.75	72.225	74.165	75	72.54
	LF+HF	90.835	77.225	73.61	**79.025**	80.17
coif5	**0.04–0.15 Hz**	75.97	**79.445**	**78.47**	75.695	77.40
	0.15–0.49 Hz	61.11	59.725	66.25	59.865	61.74
	LF/HF	71.39	77.08	77.225	78.335	76.01
	HF/LF	71.525	76.945	74.445	78.89	75.45
	LF+HF	**91.945**	78.055	75.835	75.28	**80.28**

TABLE 12.9 Average classification rate of emotions from ECG signals using LDA over four different wavelet functions

Wavelet	Frequency range	Disgust	Happy	Fear	Neutral	Mean classification rate
db5	0.04–0.15 Hz	65.56	71.665	73.33	67.5	69.51
	0.15–0.49 Hz	55.695	60	53.335	62.36	57.85
	LF/HF	61.805	69.305	58.06	68.19	64.34
	HF/LF	73.89	70.97	66.665	**78.615**	72.54
	LF+HF	63.75	73.89	70.835	66.665	68.79
db7	0.04–0.15 Hz	69.725	70.555	72.635	68.47	70.35
	0.15–0.49 Hz	55.695	56.385	56.945	57.5	56.63
	LF/HF	66.25	66.665	62.915	65	65.21
	HF/LF	74.03	69.025	70.695	73.61	71.84
	LF+HF	63.055	70.97	72.5	69.305	68.96
sym8	0.04–0.15 Hz	71.11	72.64	72.5	72.22	72.12
	0.15–0.49 Hz	56.945	58.055	57.915	64.03	59.24
	LF/HF	70.42	70.695	66.115	71.665	69.72
	HF/LF	75.275	72.085	71.665	71.53	72.64
	LF+HF	69.86	**74.72**	69.86	68.475	70.73
coif5	0.04–0.15 Hz	71.665	71.665	73.195	71.39	71.98
	0.15–0.49 Hz	56.39	57.78	56.53	61.11	57.95
	LF/HF	69.03	69.03	67.64	72.64	69.59
	HF/LF	**77.92**	71.53	**74.58**	72.36	**74.10**
	LF+HF	65.695	72.64	71.805	72.22	70.59

From Tables 12.6 and 12.7, we found that, average spectral power feature derived in the frequency range of 31–63 Hz gives higher classification rate on three emotions (disgust, happy, and fear) over other frequency band combinations. Indeed, *coif5* wavelet function extracts more meaningful information from EMG signals on distinguishing disgust and neutral emotions with an average classification rate of 96.25% and 91.67%, respectively. However, *db5* wavelet function derives the changes in two emotional states from EMG signals over other wavelet functions. The maximum mean emotion classification rate of 98.75% is achieved on happy emotion and 89.03% on

TABLE 12.10 Comparison between KNN and LDA on classifying emotions using DWT and EMG signals

	KNN			LDA		
Emotions	Frequency range (Hz)	Wavelet function	Average accuracy (%)	Frequency range (Hz)	Wavelet function	Average accuracy (%)
Disgust	31–63	*coif5*	96.25	31–63	*coif5*	84.03
Happy	31–63	*db5*	98.75	31–63	*sym8*	95
Fear	31–63	*db5*	89.03	31–63	*db5*	81.25
Neutral	4–63	*coif5*	91.67	31–63	*db5*	83.33

TABLE 12.11 Comparison between KNN and LDA on classifying emotions using DWT and ECG signals

	KNN			LDA		
Emotions	Frequency range	Wavelet function	Average accuracy (%)	Frequency range	Wavelet function	Average accuracy (%)
Disgust	LF+HF	*coif5*	91.95	HF/LF	*coif5*	77.92
Happy	LF	*coif5*	79.45	LF+HF	*coif5*	74.72
Fear	LF	*coif5*	78.47	HF/LF	*coif5*	74.58
Neutral	LF+HF	*sym8*	79.05	HF/LF	*db5*	78.62

fear emotion over other emotions. In addition, KNN outperforms LDA by giving maximum average classification accuracy on classifying four emotions using EMG signals with a K value of 5. Among the four different wavelet function analysis, *db5* gives the maximum mean emotion classification rate on all four emotions compared to other wavelet functions using KNN and *sym8* wavelet function on LDA.

In ECG signals, combination of LF and HF band information gives the maximum average emotion classification rate on disgust (91.95%) and neutral (79.05%) emotions using KNN. However, LF band plays a vital role on discriminating two emotions such as happy (79.05%) and fear (78.47%) using KNN. Compared to the performance rate of DWT-based emotion classification, KNN performs better over LDA on all four emotions. Among the four different wavelet functions, *coif5* gives the maximum average emotion classification rate on three emotions (happy, fear, and disgust). The characteristic of this wavelet function extracts more meaningful information about the emotional state changes from ECG signals over other wavelet functions. From Tables 12.8 and 12.9, the frequency range of LF+HF gives the maximum mean emotion classification rate of 80.28% on classifying four emotions using KNN and the ratio of LF/HF gives the maximum mean emotion classification rate of 74.10% using LDA. In both classification methods, *coif5* wavelet function performs better over other wavelet functions on discriminating the emotions. Therefore, these results confirm that the characteristic pattern of EMG and ECG signal under different emotional states seems to be similar to the characteristic pattern of *db5* and *coif5* wavelet function, respectively.

Hence, retrieval of information about the emotional state changes from ECG signals in the range of 0.04–0.49 Hz and EMG signals in the range of 31–63 Hz is more significant on classifying the emotions over other frequency ranges. Compared to the two physiological signals (EMG and ECG), EMG gives the higher emotion classification rate on discriminating the emotions over ECG. The effect of inefficient removal of baseline wander and noise from the ECG signals significantly affects the performance rate of emotion classification. Hence, efficient preprocessing on ECG signals is highly essential for classifying the emotions. Comparison of highest emotion classification rate from ECG and EMG signals using KNN and LDA classifier over four different wavelet functions is given in Tables 12.10 and 12.11. All these analysis are performed in off-line using MATLAB 7 software.

12.5 CONCLUSION

Frequency band localization on physiological signals (EMG and ECG) for classifying the discrete emotions is presented in this chapter. Physiological signals are acquired from 20 subjects over 10 trails on 4 emotions. Acquired signals are preprocessed using Butterworth filters, and spectral features are derived using four wavelet functions. DWT is used to decompose the physiological signals into several frequency bands. As a result of this study, the spectral features derived from 31–63 Hz of frequency information from EMG signals and 0.04–0.5 Hz from ECG signals give the highest emotion classification rate over other frequency ranges. Indeed, *coif5* wavelet function performs well on extracting the information about most of the emotional state changes from ECG and EMG signals over other wavelet functions. Therefore, the extracted features successfully capture the emotional changes of the subject through their ECG and EMG signals regardless of the user's cultural background, race, and age. However, EMG signals give the highest emotional classification rate over ECG signals. This study is ongoing to involve different classification algorithms in order to track the emotional status of ANS activation during audiovisual stimuli environment.

12.6 FUTURE WORK

In this work, we mainly focused on analyzing the ECG and EMG signals separately for classifying the emotions. However, we are also interested to fuse the two physiological signals for enhancing the emotion classification rate. This work can be extended to implement the complete analysis on Single Board Computer (SBC TS 7800) for interfacing with the humanoid robot to entertain the completely paralyzed and bedridden people. In addition, we are now focusing on removing the noise and artifacts from ECG signals using empirical mode decomposition (EMD) and Independent Component Analysis (ICA) and extracting ECG morphological features such as QRS complexity for enhancing the emotion classification rate.

ACKNOWLEDGMENTS

I would like to thank my students and colleagues for giving the critical comments on improving the quality of this chapter and for their assistance during the data acquisition. This work is financially supported by two research grants issued by University Malaysia Perlis, Malaysia (Grant Code: 9001-00191) and Ministry of Higher Education, Malaysia (FRGS Grant No: 9003-00214).

REFERENCES

[1] R. W. Picard, E. Vyzas, and J. Healey, "Toward machine emotional intelligence: analysis of affective physiological state," *IEEE Transactions on Pattern Analysis and Machine Intelligence*, 23: 1175–1191, 2001.

[2] HUMAINE Association. Emotion in human–computer interaction, 2007, http://emotion-research.net/workshops_folder/workshop.2007-06-29.5485540239. Last accessed on Sept. 19, 2014.

[3] I. A. Essa and A. P. Pentland, "Coding, analysis, interpretation, and recognition of facial expressions," *IEEE Transactions on Pattern Analysis and Machine Intelligence*, 19: 757–763, 1997.

[4] Y. Wang and L. Guan, "An investigation of speech based human emotion recognition," in 6th IEEE Workshop on Multimedia Signal Processing, 2004, pp. 15–18.

[5] C. Ginevra, D. V. Santiago, and C. Antonio, "Recognizing human emotions from body movement and gesture dynamics," in Lecture Notes in Computer Science, 2007, pp. 71–82.

[6] K. Kim, S. Bang, and S. Kim, "Emotion recognition system using short-term monitoring of physiological signals," *Med. Biol. Eng. Comput.*, 42: 419–427, 2004.

[7] B. Herbelin, P. Benzaki, O. Renault, and D. Thalmann, "Using physiological measures for emotional assessment: a computer-aided tool for cognitive and behavioural therapy," in Proceedings of the 5th International Conference on Disability, Virtual Reality & Associated Technologies, Oxford, UK, 2004.

[8] W. Wan-Hui, Q. Yu-Hui, and L. Guang-Yuan, "Electrocardiography recording, feature extraction and classification for emotion recognition," in 2009 WRI World Congress on Computer Science and Information Engineering, Los Angeles, CA, 2009.

[9] K. Jonghwa and E. Ande, "Emotion recognition based on physiological changes in music listening," *IEEE Transactions on Pattern Analysis and Machine Intelligence*, 30: 2067–2083, 2008.

[10] M. Murugappan, R. Nagarajan, and Y. Sazali, "Classification of human emotion from EEG using discrete wavelet transform," *J. Biomed. Sci. Eng.*, 3: 390–396, 2010.

[11] M. Murugappan, R. Nagarajan, and Y. Sazali, "Inferring of human emotional states using multichannel EEG," *Eur. J. Sci. Res.*, 48: 281–299, 2010.

[12] X. Ya and G. Yuan, "A method of emotion recognition based on ECG signal," in International Conference on Information Technology and Computer Science, 2009, pp. 202–205.

[13] G. Yang and S. Yang, "Study of emotion recognition based on surface electromyography and improved least squares support vector machine," *J. Comput.*, 6: 1707–1714, 2011.

[14] F. Hinoing, J. Wagner, and A. Noth, "Classification of user states with physiological signals: on-line generic features vs specialized feature sets," in 17th European Signal Processing Conference, 2009, pp. 2357–2361.

[15] K. Takahashi, "Remarks on emotion recognition from bio-potential signals," in 2nd International Conference on Autonomous Robots and Agents, 2004, pp. 186–191.

[16] J. Wagner, J. Kim, and E. Andre, "From physiological signal to emotion: implementing and comparing selected method for feature extraction and classification," in IEEE International Conference on Multimedia and Expo, 2005, pp. 940–943.

[17] M. A. Chang and G. Yuan, "Feature extraction, feature selection and classification from electrocardiography to emotions," in International Conference on Computational Intelligence and Natural Computing, 2009, pp. 190–193.

[18] C. Jing, L. Guangyuan, and H. Min, "The research on emotion recognition from ECG signal," in International Conference on Information Technology and Computer Science, 2009, pp. 497–500.

[19] O. Villon and C. Lisetti, "A user modelling approach to build user's psycho-physiological maps of emotions using bio-sensors," in IEEE International Symposium on Robot and Human Interactive Communication, 2006, pp. 269–276.

[20] T. Sharma, S. Bharadwaj, and B. Maringanti, "Emotion recognition using physiological signals," in IEEE Region 10 Conference, 2008, pp. 1–5.

[21] M. Murugappan, R. Nagarajan, and Y. Sazali "Combining spatial filtering and wavelet transform for classifying human emotions using EEG signals," *J. Med. Biol. Eng.*, 31: 45–51, 2011.

[22] M. Murugappan, M. Rizon, R. Nagarajan, and Y. Sazali, "An investigation on visual and audio-visual stimulus based emotion recognition using EEG," *Int. J. Med. Eng. Inform.*, 1: 342–356, 2009.

[23] M. Kaur, B. Singh, and Seema, "Comparisons of different approaches for removal of baseline wander from ECG signal," in IJCA Proceedings on International Conference and Workshop on Emerging Trends in Technology, vol. 5, 2011, pp. 30–34.

[24] M. S. Chavan, R. Agarwala, and M. D. Uplane, "Comparative study of Chebyshev I and Chebyshev II filter used for noise reduction in ECG signal," *Int. J. Circuits Syst. Signal Process.*, 2: 1–17, 2008.

[25] S. Luo and P. Johnston, "A review of electrocardiogram filtering," *J. Electrocardiol.*, 43: 486–496, 2010.

[26] X. Hu, Z. Xiao, and N. Zhang, "Removal of baseline wander from ECG signal based on a statistical weighted moving average filter," *J. Zhejiang Univ. Sci. C*, 12: 397–403, 2011.

[27] S. Wang and S. Jiang, "Removal of power line interference of ECG signal based on independent component analysis," in 2009 First International Workshop on Education Technology and Computer Science (ETCS '09), 2009, pp. 328–330.

[28] B. Cheng, and G. Y. Liu, "Emotion recognition from surface EMG signal using wavelet transform and neural network," in 2nd International Conference on Bioinformatics and Biomedical Engineering (ICBBE 2008), Shanghai, China, 2008.

[29] L. Lan and C. Ji-hua, "Emotion recognition using physiological signals from multiple subjects," in International Conference on Intelligent Information Hiding and Multimedia Signal Processing (IIH-MSP '06), 2006, pp. 355–358.

[30] V. Slavova, H. Sahli, and W. Verhelst, "Multi-modal emotion recognition-more "cognitive" machines," in *New Trends in Intelligent Technologies*, 14th edn, vol. 3, Institute of Information Theories and Applications FOI ITHE, Sofia, Bulgari, 2009.

AUTHOR BIOGRAPHY

M. Murugappan received the B.E. (Electrical and Electronics) (Distinction) from Madras University, India, and M.E. (Applied Electronics) from Anna University, India. He is currently with the School of Mechatronics Engineering, Universiti Malaysia Perlis (UniMAP), Arau, Malaysia as a senior lecturer.

He has been working in the fields of bio-signal processing applications for the past 5 years and has been cited as an expert in Who is Who in the World. He has received several research awards, medals, and certificates on excellent publications and research products. He has several contributions as peer-reviewed journal and conference papers. His current fields of interest are in emotion recognition, stress assessment, hypovigilance detection, and early prediction of cardiovascular diseases due to tobacco smoking. His research interests include bio-medical image and signal processing, pattern recognition and classification.

13

TOWARD AFFECTIVE BRAIN–COMPUTER INTERFACE: FUNDAMENTALS AND ANALYSIS OF EEG-BASED EMOTION CLASSIFICATION

YUAN-PIN LIN AND TZYY-PING JUNG

Brain Research Center, National Chiao Tung University, Hsinchu, Taiwan; and Swartz Center for Computational Neuroscience, University of California, San Diego, CA, USA

YIJUN WANG AND JULIE ONTON

Swartz Center for Computational Neuroscience, University of California, San Diego, CA, USA

Emotion classification from non-invasively measured electroencephalographic (EEG) data has been a growing research topic because of its potential application to affective brain–computer interfaces (ABCI), such as brain-inspired multimedia interaction and clinical assessment. A crucial component in ABCI is to reliably and accurately characterize individuals' brain dynamics into distinct affective states by employing advanced methods of pattern recognition. This chapter explores principles for translating neuroscientific findings into a practical ABCI. It will cover not only an overview of state-of-the-art EEG-based emotion recognition techniques, but also the basic research exploring neurophysiological EEG dynamics associated with affective responses. Although previous studies have demonstrated the use of EEG spectral dynamics for emotion classification, most of them achieved high classification accuracy by using an affective framework involving all available channels as well as

Emotion Recognition: A Pattern Analysis Approach, First Edition. Edited by Amit Konar and Aruna Chakraborty.
© 2015 John Wiley & Sons, Inc. Published 2015 by John Wiley & Sons, Inc.

frequency bands. The issue of feature and electrode reduction/selection has not typically been a primary goal in ABCI research. However, this chapter aims at resolving EEG feature selection and electrode reduction issues by the generalization of subject-independent feature/electrode set extraction techniques that we have proposed in our series of emotion classification studies [1–7]. Furthermore, this study addresses several practical issues and potential challenges for ABCIs as well. We believe a user-friendly EEG cap with a small number of electrodes can efficiently detect affective states, and therefore significantly promote practical ABCI applications in daily life.

13.1 INTRODUCTION

13.1.1 Brain–Computer Interface

Brain–computer interfaces (BCIs) translate human intentions into control signals to establish a direct communication channel between the human brain and output devices [8]. During the past two decades, BCI technology has become a hot research topic in the areas of neuroscience, neural engineering, medicine, and rehabilitation [9, 10]. To detect human brain activities in real time, researchers have used different neuroimaging modalities such as electroencephalogram (EEG), magenetoencephalogram (MEG), electrocorticogram (ECoG), functional magnetic resonance imaging (fMRI), and functional near infrared spectroscopy (fNIRS) to build practical BCI systems [8]. EEG is a measurement of the electrical potential at a particular point on the scalp relative to another point on the head. It is believed that the EEG detected at scalp electrodes is generated by patches of cortical neurons with synchronous activity. A synchronous patch of cortex creates an electrical dipole that alternates in polarity as synaptic depolarization of the apical dendrites alternates with the consequent depolarization of the neuronal cell body. Each end of the dipole (dendrites and cell body) automatically repolarizes while the opposite end receives the depolarizing stimulus. It is generally believed that EEG originates specifically from neocortex because of the laminar, or layered, structure of cortex, which aligns pyramidal neurons parallel to one another and perpendicular to the laminar surface. Synchronous neighboring neurons that are not aligned as they are in cortex would not create a far-field potential because individual neuronal electric fields would cancel each other out and therefore be invisible, electrically speaking, at the scalp. Thus, a limitation of EEG is that it is unlikely to detect subcortical structures, probably primarily because of the architecture of these structures, but also partly because of their physical depth. On the other hand, EEG is regaining popularity because of its excellent temporal resolution that modalities such as fMRI cannot approach. Not only is EEG recorded on a millisecond, as opposed to second time scale, it is also a direct measurement of neuronal activity, rather than a secondary alteration in blood oxygenation level, as with fMRI. The signal that EEG provides is correspondingly much more complex than that of fMRI, a feature that is both a blessing and a curse. While EEG complexity provides many possible features to control a BCI, the connections between EEG phenomena and cognition are still poorly understood. Despite this, due to its advantages of low cost, convenient operation, and non-invasiveness, EEG is the most widely used modality in present-day BCIs. The most widely used EEG features in EEG-based BCIs are visual evoked potential (VEP), sensorimotor mu/beta rhythms, P300 evoked potential, slow cortical potential (SCP), and movement-related cortical potential (MRCP) [8]. With significant progress made over the past two decades, the current EEG-based BCI systems have been able to offer patients with motor disabilities some practical solutions (e.g., cursor movement and word spelling) to communicate with their environment.

Since the 1990s, the main interest of BCI research has been on the clinical use of BCI to improve the quality of life for disabled people. Beyond its successes in clinical research, BCI has also been discovered by many other emerging fields such as motor and neurological rehabilitation, brain-state monitoring, gaming and entertainment, and space applications [8–11]. Recently, with advances in biomedical electronics and computer technology, the BCI community has been making great efforts to develop practical BCI systems for daily use in real life. In the procedure of translating laboratory demonstrations into practical systems, challenges that need to be addressed are related to two major issues: (1) ease-of-use, and (2) robust system performance [12]. A major focus of BCI research today is the development of a more user-friendly BCI, more specifically, new hardware and software platforms for convenient data recording and high-performance data processing.

Beyond verbal or body language, both expressed and perceived emotions are known as a vital and natural communication of human beings in social interactions. Researchers are attempting to learn these emotional cues in order to inform a perceptive human–computer interface. In other words, machines would be augmented with emotional awareness and intelligence to sense and respond to human affective states. With the profound fundamentals established by work on conventional BCIs and the cumulative knowledge in affective neuroscience, the field of affective brain–computer interface (ABCI) is very likely to be the focus of next-generation BCIs. Several interdisciplinary applications can stand on the advances of ABCIs. The first one is to extend BCI's function to help severely disabled persons express their feelings to others. Next, by combining the expertise of multimedia information retrieval, ABCI can also help create an affective-based multimedia system for users to experience multimedia according to their affective feedback. Other potential applications, such as interactive learning, emotion surveillance for clinical patients, emotion-aware robots or machine pets, may be important aspects of our future daily lives.

13.1.2 EEG Dynamics Associated with Emotion

Hans Berger first published a human EEG recording in 1924 [13], but aside from its clinical utility for epilepsy assessment, the technique remained quite cumbersome and primitive for much of the century. Even in the 1970s and 1980s, when computers replaced the paper and pen recordings of previous decades and allowed for digital analysis of EEG data, computer power was rather limited and averaging data across comparable trials was more or less the most sophisticated analysis possible. Today, a smart phone application can process more EEG data than a high-end computer in the 1980s. As a result, we are now able to analyze high dimensional EEG data on a moment-to-moment basis, instead of averaging over trials to conserve computational resources. With this new age of nearly limitless computational power, a fast-growing EEG community, with extensive expertise and/or collaboration with the engineering community, has emerged to tackle the challenge of creating a truly functional BCI system. BCI strives to discover temporal and/or spectral patterns in the EEG data related to certain cognitive or emotional states, the goal being to teach a computer to read the current state of a subject using only his/her brainwaves.

Despite decades of active EEG research, the exact meaning of most EEG variations are still not completely understood. With our current limited understanding of EEG and in light of the physical limitations of detecting subcortical electrical activity in the brain, is it possible and/or advantageous to use EEG as an indicator of cognitive or emotional states? The evidence to date confirms that EEG is indeed a rich source of information pertaining to on-going brain processes, and with the current pace of discovery, a consistent and precise mapping between EEG phenomena and emotional states should be possible.

Traditionally, the most common emotional response studied experimentally is fear, which is usually elicited by viewing frightening pictures or fearful faces. Using fMRI, studies show that these stimuli typically activate the amygdala, deep in the temporal lobes [14]. However, it is not likely that an electrical response in the amygdala would be detected as a change in the scalp EEG pattern. Nevertheless, many other regions of the brain have been associated with different emotional states and are theoretically detectable by EEG, such as subgenual anterior cingulate cortex with depression and rumination [15, 16], and the orbitofrontal and cingulate cortices with reward processing [17, 18]. Furthermore, it is also possible that any region of cortex, even primary sensory cortices, may be directly or indirectly influenced by subcortical structures via fast synaptic and/or slow neuromodulatory factors (i.e., dopamine, serotonin, acetylcholine, etc.) [19, 20] which may modulate cortical processing (and therefore EEG activity) according to the current emotional state of the person.

Several studies have explored the EEG correlates of state or trait emotional experiences. Some have reported differences in low frequency bands (e.g., 4–12 Hz), for example, with anxious rumination [21], natural mood swings [22], affective picture viewing [23], or music-induced emotion [2]. But a larger literature is emerging to suggest that power changes in the gamma frequency range (usually 30–60 Hz, but sometimes up to 150 Hz) may be more accurate in detecting differences in emotional states. For example, intracranial electrodes in the orbital frontal cortex (OFC) of epileptic patients have revealed increased gamma activity in lateral OFC while processing negative facial expressions or negative feedback [24]. Anxiety patients were found to express more posterior gamma activity during a worry induction task, an effect that was less pronounced following psychotherapy [25]. In normal subjects, viewing emotional faces elicited increased gamma power during supra- versus sub-liminal trials, and a right hemisphere-dominant gamma-power increase during emotional as compared to neutral faces [23]. Similar right-dominant gamma was also found during viewing of affective pictures [26], but bilateral temporal gamma power has been correlated with emotional valence during an emotional imagery task [27]. Finally, phobic individuals given a hypnotic suggestion to imagine their phobic object showed increased fronto-central gamma power [28]. Thus, while a comprehensive understanding of the EEG signature of emotion is still largely lacking, these and other reported associations between EEG spectral dynamics and emotional states promote the idea that an affective BCI system is a plausible reality for the near future. The mapping of these complex states using EEG, or any neuroimaging technology, will be difficult and time-consuming if only because of the inherent difficulty in knowing the ground truth of a person's emotional state at a given moment. However, the unprecedented computational power that is currently available along with the rapidly growing interest in brain science provides a fertile environment for the advances previously believed impossible.

13.1.3 Current Research in EEG-Based Emotion Classification

Prior to equipping machines with emotional intelligence, a vital step is learning to actually estimate/classify EEG patterns into desired emotion categories. During the past few years, EEG-based emotion recognition research has seen rapid progress. Table 13.1 provides an overview of relevant emotion classification studies (non-exhaustive) to help readers gain insight into the current state-of-the-art EEG-based emotion classification. Table 13.1 also highlights several noteworthy issues, including emotion elicitation, feature extraction, feature classification, and classification validation, which are all future challenges. First of all, it is crucial to collect EEG data from subjects truly experiencing the intended or reported emotional state, which can be difficult in a laboratory setting. Until now, emotion elicitation is most commonly accomplished through images [29, 30], music [4], music video [31], video clips [32], and emotion recall [33]. The goal of this procedure is to induce specifically desired emotional responses based on either a 2D valence-arousal emotional model [35], where each emotional state can be defined in the 2D space, or a set of six basic emotions found to be universal across human cultures [34], including happiness, surprise, anger, fear, disgust, and sadness. Usually, these visual, audio, or audiovisual stimuli are selected from publicly available websites or a developed affective elicitation database, e.g., the International Affective Picture System (IAPS) [36]. The basic idea is that the selected materials are presumably able to induce similar emotional responses in all subjects participating in a study. While subjects are attending to a chosen sequence of emotional stimuli, a multichannel EEG is acquired simultaneously. Importantly, subject self-reports are generally gathered instead of relying on pre-determined emotional labels because individual differences can substantially affect a subject's emotional experience in response to an emotional cue. It is also worth noting that it is desirable to acquire balanced EEG data samples across emotional classes for each individual.

Emotion is one of the most complex and poorly understood cognitive functions, which may involve various spectral fluctuations in a constellation of different brain regions as mentioned above. Many previous studies attempted to characterize the time-frequency dynamics, obtained by either short-time Fourier transform (STFT) or discrete Wavelet transform (WT) of EEG time series, in distinct frequency bands ($\delta, \theta, \alpha, \beta$, and γ) across multiple channels in different emotional states. So far, most

TABLE 13.1 List of up-to-date EEG-based emotion recognition studies (non-exhaustive)

			EEG frequency band					Feature		
Previous works	Emotion elicitation	No. of electrodes	δ	θ	α	β	γ	Extraction	Classifier	Classification accuracy
Chanel *et al.* (2009) [33]	**No. of subjects:** 10 **No. of emotions:** 3 on the VA space Calm-neutral Positive-excited Negative-excited **Elicitation:** 8-sec emotion recall 100 trials per emotion	64 (whole brain)		√	√	√	√	**Set 1:** STFT features **Set 2:** Mutual information	LDA QDA RVM SVM	**Validation:** Leave-one-trial-out (Subject-dependent) **Accuracy:** 63% (STFT+SVM)
Frantzidis *et al.* (2010) [29]	**No. of subjects:** 28 **No. of emotions:** 4 on the VA space Positive valence-high arousal Positive valence-low arousal Negative valence-high arousal Negative valence-low arousal **Elicitation:** 160 1-s IAPS images 40 trials per emotion	3 (Fz, Cz, Pz)	√	√	√			ERP+WT	MD SVM	**Validation:** 10-fold cross-validation (Subject-independent) **Accuracy:** 81.25% (SVM)
Lin *et al.* (2010) [4]	**No. of subjects:** 26 **No. of emotions:** 4 on the VA space Joy, anger, sadness, pleasure **Elicitation:** 16 30-s music excerpts 4 trials per emotion	30 (whole brain)	√	√	√	√	√	**Set 1:** 12 channel pairs DASM12, RASM12 **Set 2:** 24 channels PSD24 **Set 3:** 30 channels PSD30	MLP SVM	**Validation:** 10-fold cross-validation (Subject-dependent) **Accuracy:** 82.29% (DASM12+SVM)

Petrantonakis *et al.* (2010) [30]	**No. of subjects:** 16 **No. of emotions:** 6 on the VA space Happiness, surprise, anger Fear, disgust, sadness **Elicitation:** 60 5-s facial expression 10 trials per emotionn	4 Fp1, Fp2, F3, F4			√	√		**Set 1:** HAF+HOCs **Set 2:** HOC **Set 3:** Statistics **Set 4:** WT	QDA KNN MD SVM	**Validation:** Leave-subjects-out (Subject-independent) **Accuracy:** 85.17% (HAF-HOCs+SVM)
Murugappan *et al.* (2011) [32]	**No. of subjects:** 20 **No. of emotions:** 5 discrete emotions Happy, surprise, fear Disgust, neutral **Elicitation:** 23 video clips 5 trials for disgust, happy, surprise 4 trials for fear, neutral	62 (whole brain)	√	√	√	√	√	WT (power, STD, variance, entropy) **Set 1:** 62 channels **Set 2:** 24 channels	LDA KNN	**Validation:** 5-fold cross-validation (Subject-independent) **Accuracy:** 83.04% (Entropy+KNN)
Koelstra *et al.* (2011) [31]	**No. of subjects:** 32 **No. of emotions:** 4 on the VA space High valence-high arousal High valence-low arousal Low valence-high arousal Low valence-low arousal) (Dominance, Liking) **Elicitation:** 40 1-min music video 10 trials per emotion	32 (whole brain)		√	√	√	√	Welch method + spectral power asymmetry	Gaussian naïve Bayes	**Validation:** Leave-one-trial-out (Subject-dependent) **Accuracy:** Valence: 62% Arousal: 57.6% Liking: 55.4%

Elicitation: valence-arousal (VA), international affective picture system (IAPS).

Feature type: short-time Fourier transform (STFT), event-related potential (ERP), wavelet transform (WT), differential/rational power asymmetry from 12 channel pairs (DASM12/RASM12), power spectrum density from 24/30 channels (PSD24/30), hybrid adaptive filtering (HAF), higher order crossings (HOCs), standard devidation (STD).

Classifier: linear discriminant analysis (LDA), quadratic discriminant analysis (QDA), relevance vector machine (RVM), support vector machine (SVM), mahalanobis distance (MD), multilayer perceptron (MLP), k-nearest neighbor (KNN) studies focused on developing sophisticated feature-extraction schemes to improve classification performance by using machine-learning algorithms, such as linear discriminant analysis (LDA), quadrant discriminant analysis (QDA), Mahalanobis distance (MD), k-nearest neighbor (KNN), Gaussian naïve Bayes, multilayer perceptron (MLP), relevance vector machine (RVM), support vector machine (SVM), etc. After a comparative analysis, an optimal classification scheme can then be derived for fitting the data of interest. In most cases, using SVM yielded the best performance. However, there are large variations across these studies in terms of the stimuli used to elicit emotion, the number of emotion classes to be classified, and the number of electrodes used for acquiring EEG signals, it is difficult to directly and quantitatively compare these proposed classification frameworks merely relying on resultant classification accuracy. Recently, a dataset, named DEAP, has been released for comparative emotion analysis using physiological signals [31]. Researchers are encouraged to test and report their own classification models in order to directly compare classification accuracy on a common dataset.

Last, depending on the methods used for validating the classification performance, such as leave-*N*-trial-out, *N*-fold cross-validation and leave-*N*-subjects-out, several critical issues need to be further addressed. Most previous BCI studies adopted a leave-*N*-trial-out scheme to test the classification performance of their proposed algorithms. The leave-*N*-trial-out scheme used only *N* trials as test samples to validate a model which trained on the remaining trials. Unlike visual or motor tasks usually used for conventional BCI studies, the number of trials is usually limited in an affective experiment because of habituation or fatigue that is more likely with emotional tasks and may bias the subsequent behavior labels. Besides, inter-trial variability associated with emotional responses is likely to be considerably more prominent than motor or visually induced EEG patterns. In light of the prominent trial-to-trial variability and limited trials for training a model, low performance of single-trial affective classification [31, 33] is usually expected. In addition, as with other physiological classification models, another issue is whether the proposed affective BCI framework can be generalized across subjects, that is, a subject-independent affective BCI. Even though inter-subject variability is expected in emotional perception, presumably due to differences in culture, personal background and taste, general mood, etc., some studies [29,30,32] have reported satisfactory subject-independent classification performance when distinguishing distinct emotion classes, using either an *N*-fold-cross validation scheme [29,32], or even more impressively using a leave-*N*-subject-out strategy [30]. The results suggest that it is feasible to build an affective BCI model that is initially well trained by one group of subjects and then used to estimate the emotional responses of new subjects. However, more work needs to be done to validate the feasibility of the proposed subject-independent affective framework.

13.1.4 Addressed Issues

All the aforementioned studies have demonstrated the use of non-invasively recorded EEG spectral dynamics as a means for developing an EEG-based emotion classification system. Yet, toward an ABCI, one major concern is how to minimize the number of required electrodes to improve ease-of-use, and thus subject acceptance of ABCI. Most previous studies [31–33] focused on achieving high classification accuracy of ABCI. Only a relative few studies [29,30,32] have attempted to construct an affective model using a relative few selected electrodes to increase the feasibility of real-world applications of ABCI. Even in these studies, the selection of electrodes was presumably based on the previous neuroscience findings or functionality of brain regions. No effort has been made to select an optimal set of electrode locations and frequency bands for emotion assessment. This chapter investigates the issues regarding feature and electrode reduction/selection with a special emphasis on a subject-independent feature/electrode set, which chiefly stands on a series of our previous ABCI studies [1–7]. We believe a user-friendly EEG cap using fewer selected electrodes can significantly promote practical ABCI applications in our daily life.

13.2 MATERIALS AND METHODS

13.2.1 EEG Dataset

This study conducts a comparative analysis on the 32-channel EEG data acquired from 26 volunteer students during music-listening experiments reported in Reference 7. Briefly, the 32 scalp electrodes were placed according to a modified international 10–20 system. EEG signals were sampled at 500 Hz and filtered with a 1–100 Hz band-pass filter. The impedances of all electrodes were kept below 10 kΩ referred to linked mastoids. This study used music to induce and examine four basic emotional states based on a 2D valence-arousal emotion model [35], including joy (positive valence and high arousal), anger (negative valence and high arousal), sadness (negative valence and low arousal), and pleasure (positive valence and low arousal) (see Figure 13.1a). In the music-listening experiment, each subject was instructed to sit comfortably, keep eyes closed, and use headphones to carefully listen to music excerpts presented by a PC (see Figure 13.1b). Sixteen 30-second music excerpts extracted from Oscar-winning film soundtracks were selected in advanced based on the consensus of a prior large group study [37]. The goal was to obtain balanced self-reported labels across four emotional classes (joy, anger, sadness, and pleasure). Four of the 16 30-second music excerpts were randomly selected from each class to form a four-trial bout, in which a 15-second silent rest was inserted between excerpts. After each trial, subjects reported their emotional labels to each music excerpt based on what they felt. The dataset of each subject thus comprised 16 30-second 32-channel EEG data and corresponding self-reported labels. Figure 13.1c shows a 5-second portion of 30-channel EEG data during music listening.

13.2.2 EEG Feature Extraction

EEG power spectra at distinct frequency bands, such as delta, theta, alpha, beta, and gamma, are commonly used as indices for assessing the correlates of specific ongoing cognitive processes in EEG research. These distinct power changes are

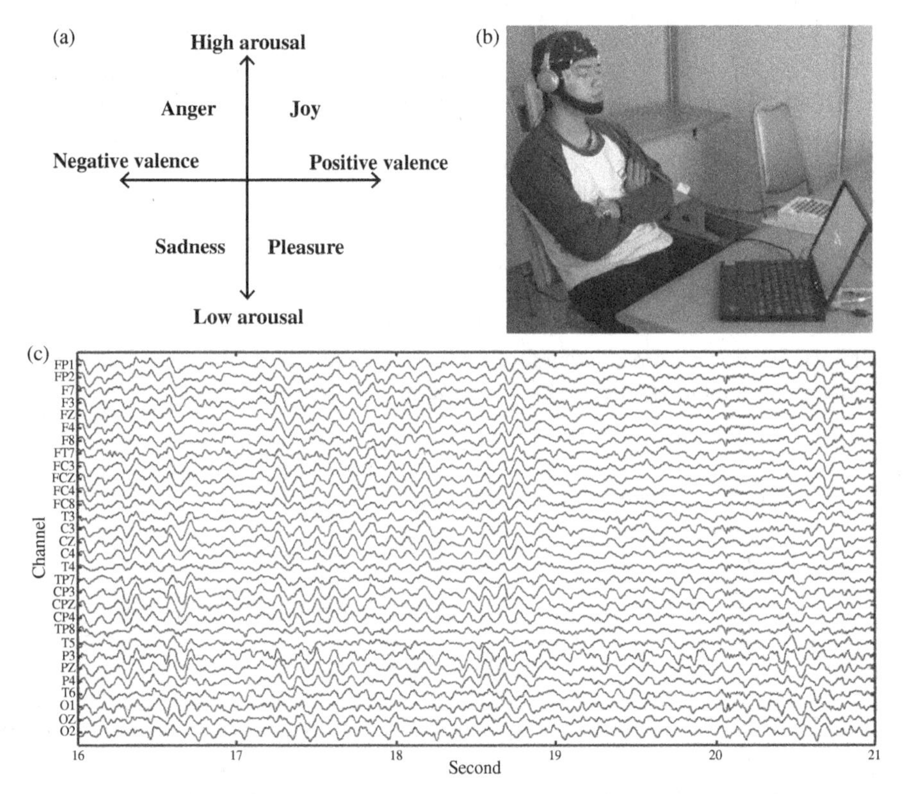

FIGURE 13.1 The adopted model and setup of a musical emotion elicitation experiment. (a) The targeted emotional categories are joy, anger, sadness, and pleasure based on the 2D valence–arousal emotion model. (b) Experiment setup. (c) A sample 5-second portion of the 30-channel EEG time series during the music-listening experiment.

adopted as features in the computer science domain to characterize recorded EEG patterns into pre-defined classes based on prior neuroscience findings. Taking the concept that complex cognitive functions tend to engage several brain oscillations in combination [38], it seems reasonable to explore the association between self-reported emotion labels and all the EEG band power over all available channels. The goal of this investigation was to systematically explore an optimal feature set for obtaining the highest classification performance [6,7]. The first feature type (PSD30 [4]) was composed of five spectral power bands, including δ (1–3 Hz), θ (4–7 Hz), α (8–13 Hz), β (14–30 Hz), and γ (31–50 Hz), across all 30 electrodes (Fp1, Fp2, F7, F3, Fz, F4, F8, FT7, FC3, FCz, FC4, FT8, T7, C3, Cz, C4, T8, TP7, CP3, CPz, CP4, TP8, P7, P3, Pz, P4, P8, O1, Oz, and O2), forming a feature dimension of 150, taking into account information from the whole head. Next, since frontal alpha [39,40] and frontal/parietal theta power [23] asymmetry have been associated with emotional responses, the next set of features characterize the difference or ratio of spectral power between hemispheres, that is, a hemispheric asymmetry index extracted from symmetric electrode pairs [4, 7, 31, 41]. Given a 30-channel EEG dataset, two feature types DASM12 and RASM12 that extracted differential and rational power asymmetry at five frequency bands from 12 electrode pairs (Fp1-Fp2, F7-F8, F3-F4, FT7-FT8, FC3-FC4, T7-T8, P7-P8, C3-C4, TP7-TP8, CP3-CP4, P3-P4, and O1-O2) forming a feature dimension of 60. Last, to compare performance with the power asymmetry index, individual power spectra across five bands extracted from the 12 electrode pairs used for DASM12/RASM12 (24 channels) were also adopted (PSD24 [4, 7]), forming a feature dimension of 120.

13.2.3 EEG Feature Selection

Feature selection is always a critical procedure in solving classification problems. Its intended objective is threefold [42]: improving the classification performance, enhancing the computation efficiency, and better understanding the underlying processes in a given dataset. An additional advantage of employing feature selection is to reduce the complexity of acquisition setups. To promote ABCI for routine use in our daily lives, one needs to reduce the number of EEG sensors. Feature selection methods that explore the discriminating power of each channel spectrum might identify uninformative channels and thus reduce the total number of required electrodes. Several feature selection methods have their own strategies to obtain this objective, including sequential forward selection (SFS), sequential backward selection (SBS), genetic algorithm (GA), F-score index, etc. This study used F-score to select informative EEG features and channels due to its computational simplicity and good performance in our earlier studies [3, 4]. The F-score computes the ratio of between-class and within-class variance and is defined as follows:

$$F(i) = \frac{n_l \sum_{l=1}^{g} (\overline{x}_{l,i} - \overline{x}_i)^2}{\sum_{l=1}^{g} \sum_{k=1}^{n_l} (x_{k,l,i} - \overline{x}_{l,i})^2} \tag{13.1}$$

where $\overline{x}_{l,i}$ and $\overline{x}_i$ indicate the average of the ith feature of class l dataset and the entire dataset ($l = 1 \sim g$, $g = 4$ for four classes in this study), respectively. $\overline{x}_{k,l,i}$ is the ith feature of the kth sample of class l, and n_l is the number of samples of class l. Larger F-score value of a feature *F(i)* implies that it accounts for more discriminative information between classes. It is worth noting that different feature selection schemes would probably result in more or less distinct ranking features. The F-score method used here was intended to easily demonstrate one possible approach to EEG feature selection and electrode reduction. The dissimilarities of using various feature selection methods for summarizing informative EEG features are beyond the scope of this chapter.

13.2.4 EEG Feature Classification

After forming an input feature vector, the next step is to build a computational model, which will be able to discriminate between emotions based on EEG spectra.

A well-established classification model can be obtained through an interactive supervised learning method on sufficient training pairs consisting of EEG band power and desired output classes. Note that the consistency of the training pairs is critical considering the lack of physiological ground truth as well as individual differences in emotional experiences. Most of the satisfactory solutions have used machine-learning models to accomplish high classification rates, as well as be robust against the signal variability over time and across subjects [43]. Neural networks (NN) [44] and SVM [45] are two popular supervised learning classifiers that have been widely used in EEG classification problems. A NN is typically a three-layer structure consisting of input, hidden, and output layers. The number of neurons used for each layer varies depending on the number of feature dimensions (input neurons) and desired prediction classes (output neurons), while the hidden layer, for our purposes, functions best with half the number of neurons as the sum of input and output layers. NNs often adopt a back-projection algorithm to continuously adjust the weight coefficients between layers until the training procedure converges.

On the other hand, an SVM maps input data into a higher dimensional feature space via a transfer function. The iterative learning process of SVM will converge into an optimal hyperplane with maximal margins between classes as a decision boundary for distinguishing input data. The original intention of an SVM was to solve binary classification problems. Currently, two types of methods have been proposed for solving multiclass problems [46]. One basically forms a hierarchical combination of several binary SVMs, such as "one-against-all" [47], "one-against-one" [48] and "model-based" [5, 49], and the other directly solves the optimization problem of multiclass SVM in a single step. Briefly, the difference between hierarchical and multiclass solutions is the number of binary SVMs used for forming their structures. Given samples with k classes, one-against-all method trains k SVM models to separate samples of an ith class from all other samples from other classes $(k - 1)$. In this scheme, a predicted class for a new sample is assigned by the largest value among k decision functions. In the one-against-one scheme, it constructs $k(k - 1)/2$ models to train each pairwise discrimination. The majority vote, called max-wins strategy, is then used to decide the class of a new sample. The model-based scheme was proposed in our previous study [5], intended to mimic the structure of 2D emotional model (valence and arousal) by constructing $(k - 1)$ SVMs. This scheme is able to, for example, separate valence (positive or negative) at the first level, followed by a second level of arousal (high or low) discrimination to make a final prediction on a new sample from k classes. Note that the rule of $(k - 1)$ models constructed for the emotion-model scheme is strictly constrained by the 2D emotion model. A recent comparative study on multiclass SVM schemes using the same emotional dataset showed that the one-against-one method would yield better classification performance than others [5], and the SVM outperformed the NN classifiers [4]. It is worth mentioning that the emotion-model-based scheme provided comparable performance as the one-against-one method but required fewer binary SVMs, using $(k - 1)$ models as opposed to $k(k - 1)/2$ models. This advantage will become even more significant as the number of desired classes increases. Another study [29] also confirmed the feasibility of the emotion-model classification scheme. This study adopted LIBSVM software [50] to build the SVM and employed a radial basis function (RBF) kernel to nonlinearly project data onto a higher dimensional space. A grid-search was then used to seek two optimal parameters of an RBF (C and γ) by selecting the best cross-validation accuracy.

With regard to the validation method, because only 16 30-second EEG trials were available across four emotional responses for each subject, the limited number of samples is insufficient to fully train an SVM model if a trial-based validation method was used. Instead, this study adopted a 10 times 10-fold cross-validation method. STFT, with a 1-second non-overlapping window, was adopted to prevent the potential association between training and test samples during cross-validation. In a 10-fold cross-validation method, all EEG spectral windows were divided into ten subsets. The SVM model was trained with 9 of the 10 subsets, and the remaining subset was kept for testing. This procedure was then repeated 10 times, allowing each spectral sample an equal chance of being the training or test data. Thus, a 10 times of 10-fold cross validation can yield a reliable accuracy. After validation processing, the average subject-dependent classification performance (an average of 100 classification results for each subject) using different feature types as input was evaluated. Note that each feature vector was normalized to a range from 0 to 1 before submitting it to the SVM.

13.3 RESULTS AND DISCUSSION

13.3.1 Superiority of Differential Power Asymmetry

In a recent study, Lin *et al.* [4] systematically tested the discriminative power of a wide range of feature candidates (DASM12, RASM12, PSD24, and PSD30) and the frequency bands of the EEG (delta, theta, alpha, beta, gamma, and ALL-total power) for emotion detection. Table 13.2 summarizes the averaged SVM results using different features. Since SVM outperformed MLP for this dataset in the study [4], only SVM results are listed here. There were several important conclusions from the study [4]. First, the classification accuracy of using all five frequency bands

TABLE 13.2 Comparison of average SVM results (standard deviation) using DASM12, RASM12, PSD24, and PSD30 based on individual and combined frequency bands [4]

	Individual band ($\delta, \theta, \alpha, \beta, \gamma$)		Combined band		
Feature type	Accuracy (%)	No. of features	Accuracy (%)	No. of features	No. of electrodes
DASM12	57.35 (7.37)~69.91 (6.55)	12	82.29 (3.06)	60	24
RASM12	47.61 (7.23)~56.95 (6.79)	12	65.81 (5.09)	60	24
PSD24	51.02 (6.17)~56.80 (6.58)	24	69.54 (5.10)	120	24
PSD30	53.38 (5.79)~59.54 (6.16)	30	71.15 (4.88)	150	30

was better than using only a single frequency band for any feature type. This result was quite reasonable since complex cognitive functions tend to engage several brain oscillations in combination [38]. Next, DASM12 outperformed other feature types by 10~15%. Furthermore, the differential asymmetry of hemispheric EEG power spectra (DASM12) outperformed rational asymmetry (RASM12) and other single electrode-based power spectrum feature extractions (PSD24 and PSD30). Its superiority was especially prominent when comparing DASM12 with PSD24. Although both PSD24 and DASM12 were based on the same set of electrodes (24), that is, feature extractions based on the acquired EEG signals, DASM12 could provide higher accuracy while requiring only half the number of features because of the differential power approach. Amazingly, despite the fact that PSD30 used a full-head coverage with 30 electrodes, DASM12 significantly outperformed PSD30 with fewer electrodes and features. The reason might be due in part to the fact that differential power asymmetry suppressed underlying artifactual sources that contributed equally to hemispheric electrode pairs. In addition, power asymmetry in different frequency bands were already reported to be significantly correlated with emotional states [23, 39, 40] and mental tasks [41]. In short, a one-against-one SVM classifier applied to the 60-dimensional feature type, DASM12 ALL (60: 12 electrode pairs x 5 frequency bands), yielded the best accuracy of 82.29±3.06% in classifying four emotional states [4].

As mentioned above, feature selection might be useful in better understanding the association between feature attributes and specific tasks. Thus, the F-score index was applied to DASM12 to explore which spectral power band would co-vary with the changes in emotional responses. By examining the F-score index of features across 26 subjects, the most informative subject-independent features, Top30 DASM12, could be obtained and used to assess the importance of electrode pairs [4]. In general, spectral dynamics in theta, alpha, beta, and gamma bands were associated with emotional processes specifically in frontal and parietal regions, a finding largely in line with previous studies. Additional features located in other brain regions and frequency bands were also found. The results of the study suggested that the widely distributed power asymmetries were able to meaningfully characterize the information contained in emotional responses.

Although the results were very promising, more validation studies on a large number of datasets, for example, DEAP [31] need to be performed to confirm the superiority and efficacy of differential power asymmetry for emotion classification.

13.3.2 Gender Independence in Differential Power Asymmetry

Flores-Gutierrez *et al.* [51] recently reported that emotional and perceptual processing of music involved different coherent patterns over multiple cortical regions in men and women. For example, unpleasant emotions were accompanied by an increased coherent activity between midline and posterior regions exclusively in the right hemisphere in men but bilateral in women. The study accordingly suggested that men and women should not be analyzed together in neurobiological studies of music-induced emotion. Therefore, this study explores to what extent gender affects the classification accuracy of a Top30 subject-independent DASM12 feature set classification.

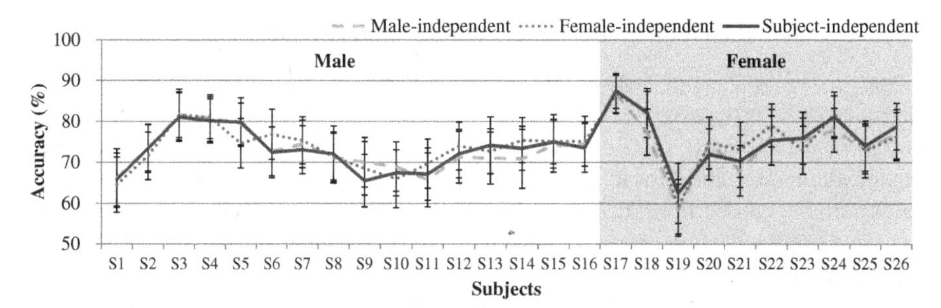

FIGURE 13.2 The comparison of average results using top 30 male-, female-, and subject-independent features.

Since the Top30 DASM12 feature set summarizing the highest F-score features from 26 subjects (16 males and 10 females) yielded an averaged classification accuracy of 74.10±5.85%, only marginally lower than the averaged accuracy of 75.75±5.24% obtained by using Top30 subject-dependent features [4], this study only focuses on the subject-independent DASM12.

Top30 male and female DASM12 feature sets were derived by summarizing the highest F-score features among male and female subjects separately [1]. The resultant topographic feature maps showed that features extracted from the pair F3–F4 at the frontal lobe and the pair P3–P4 at parietal lobe were used more frequently in female subjects, while male subjects showed more features extracted from pair Fp1–Fp2 at the frontal lobe and pair CP1–CP2 over the parietal lobe. Additionally, inspecting the top-ranked gender-specific features revealed that 13 distinct attributes specific to male or female subjects (6 from Top30 male DASAM12 and 7 from Top30 female DASAM12) did not contribute to the Top30 subject-independent (gender-independent) feature set. Yet, both male and female feature set contributed 17 identical features to form over half of the Top30 gender-independent feature set. It was then a natural step to compare the classification performance obtained by gender-specific feature sets with gender-independent feature sets. Figure 13.2 shows the averaged accuracy by applying male-, female-, and subject-independent feature sets to male and female subjects. The averaged classification results of 72.91±4.62%, 73.36±4.69%, and 72.90±4.78% were obtained by using male-, female-, and subject-independent feature sets on male, compared to the averaged results of 74.48±7.34%, 75.93±7.54% and 75.97±6.97% obtained for female subjects. Results showed no significant differences between applying gender-specific and subject-independent feature sets ($p >$ 0.05) on either male or female subjects [1]. In conclusion, although gender-specific features can be further explored, there exist many common features between gender-specific and gender-independent feature sets, and it is possible that adjacent scalp channel pairs that differentiated gender group features extracted similar information to recognize music-induced emotion responses and therefore did not confer significant power to the gender-specific classifications. These results validated the applicability of using a Top30 subject-independent DASM12 feature set for both male and female subjects.

13.3.3 Channel Reduction from Differential Power Asymmetry

Toward ABCIs, an important question is how the accuracy of emotion estimation degrades with progressively fewer electrodes. This issue is of importance to the design of user-friendly EEG caps as well as to overall system costs. As mentioned earlier, redundant and less informative features for characterizing emotion-related EEG patterns can be discarded through a feature selection analysis. Thus, one can use the feature selection results to design a new economical and effective EEG cap for specific BCI tasks.

Adding to the superiority of feature type DASM12 (compared to RASM12, PSD24, PSD30), the number of electrodes required for its feature extraction is reduced to 24 (12 electrode pairs) from a 30-channel EEG cap (see Figure 13.3a). Based on the F-score feature selection analysis of DASM12 summarized across 26 subjects [4], the Top30 subject-independent DASM12 (see Figure 13.3a) gave an average classification accuracy of 74.10±5.85% by using 30 (50%) of all 60 features of DASM12, which greatly enhanced the computational efficiency. However, the number of electrodes used did not decrease (remained at 12 electrode pairs). By inspecting the most used electrode pairs of Top30 DASM12, we found that three electrode pairs were barely used for extracting feature attributes, including FC3–FC4 (δ), P7–P8 (θ), and C3–C4 (δ). Excluding these three features could reduce the number of required electrodes from 24 to 18 (by nine electrode pairs) at the expense of a slight decrease in the classification accuracy from 74.10±5.85% to 72.75±5.69% (see Figure 13.3b).

Interestingly, the electrode arrangement of their Top27 DASM12 electrode pairs seems to conceptually fit a circular head band. A natural next step is to evaluate a new feature type which only used a circular electrode array along the head circumference. Figure 13.3c shows the new feature type, DASM7, consisting of Fp1–Fp2, F7–F8, FT7–FT8, T7–T8, P7–P8, TP7–TP8, and O1–O2. The performance of emotion classification using DASM7 (a 35-dimensional feature vector from 14 electrodes) was 74.86±5.13%, compared to 72.75±5.69% using Top27 DASM12 (a 27-dimensional vector from 18 electrodes). This slight improvement might be attributed to the fact that DASM7 involves seven more features than the Top27 DASM12. As the EEG signals acquired from the temporal lobe (T7 and T8) were often contaminated by muscle

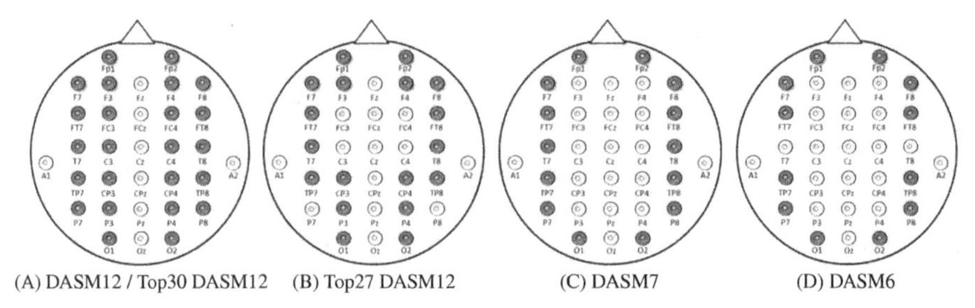

(A) DASM12 / Top30 DASM12 (B) Top27 DASM12 (C) DASM7 (D) DASM6

FIGURE 13.3 Different electrode placements according to the types of feature extraction, including (a) DASM12/Top30 DASM12, (b) Top27 DASM12, (c) DASM7 (seven pairs along head circumference), and (d) DASM6 (six pairs along head circumference).

TABLE 13.3 The comparison of the number of electrodes, the number of features, and averaged SVM results (standard deviation) between using DASM12, DASM7, DASM6, and Top30 and Top27 DASM12 feature sets

				DASM12	
	DASM12	DASM7	DASM6	Top30	Top27
No. of Electrodes	24	14	12	24	18
No. of Features	60	35	30	30	27
Accuracy (%)	82.29 (3.06)	74.86 (5.13)	72.31 (5.32)	74.10 (5.85)	72.75 (5.69)

activities whose spectral range partially coupled with EEG gamma power, we further reduced DASM7 to DASM6 by removing the T7 and T8 channels (see Figure 13.3d). The performance of emotion classification using DASM6 (a 30-dimensional feature vector from 12 electrodes) was 72.31±5.32%, marginally lower that of DASM7.

Table 13.3 summarizes emotion estimation results and the required electrode numbers using different features, including DASM12, DASM7, DASM6, Top30/Top27 DASM12. In short, DASM7 and DASM6 gave acceptable accuracy and are more suitable for routine use in our daily lives.

13.3.4 Generalization of Differential Power Asymmetry

As the emotion estimation results were obtained by a study on 26 subjects listening to 16 selected music excerpts (Ss26), one might raise the question whether the subject-independent feature sets are only applicable for certain music excerpts or subjects. The following study addresses the generalizability of the proposed feature sets for classifying emotion responses induced by new self-chosen music excerpts on additional subjects.

Following the same experiment procedures of the dataset Ss26, new EEG data were acquired from an additional four male subjects (age 23.00±0.00 years), dataset Ss4, when they listened to 16 30-second self-chosen emotional music excerpts according to their own musical preference in daily life [1]. Note that these individually handpicked music excerpts varied largely from subject to subject as well as dramatically differed from those used in dataset Ss26. The music excerpts included Chinese, English, Korean, and Japanese pop music. The purpose of adopting self-preference music was not only to naturally induce their emotional responses by listening to their preferred music excerpts, but to stringently test the applicability of subject-independent feature sets for unknown subjects listening to diverse music excerpts. The dataset Ss4 consisted of 16 30-second EEG segments and the corresponding self-reported emotional labels for each subject.

The same pre-processing and feature extraction procedures used in dataset Ss26 were applied to the new dataset Ss4. Several previously mentioned feature types were compared, including DASM12, RASM12, PSD24, PSD30, DASM7, DASM6, and Top30/Top27 DASM12. Table 13.4 summarizes emotion estimation results using different feature sets. DASM12 outperformed other feature types (RASM12, PSD24,

TABLE 13.4 The comparison of averaged SVM results (standard deviation) by conducting several feature types on dataset Ss26 and Ss4

							DASM12	
Dataset	DASM12	RASM12	PSD24	PSD30	DASM7	DASM6	Top30	Top27
Ss26	82.29	65.81	69.54	71.15	74.86	72.31	74.10	72.75
	(3.06)	(5.09)	(5.10)	(4.88)	(5.13)	(5.32)	(5.85)	(5.69)
Ss4	83.09	64.23	65.61	67.60	72.95	72.76	75.42	73.95
	(0.67)	(1.76)	(3.51)	(4.85)	(2.71)	(2.93)	(3.65)	(3.44)

and PSD30) on the new dataset Ss4, yielding a comparable accuracy of 83.09±0.67% compared to that of Ss26 [4]. This again confirmed the superiority of differential power asymmetry feature extraction. Second, comparable performance on Ss4 was obtained by electrode position-oriented feature types of DASM7 and DASM6 which demonstrated the practicability of wearing a cap or elastic head band embedded with a module of circular electrode array for real-life applications. Last, consistent results of 75.42±3.65% and 73.95±3.44% were obtained by using Top30 and Top27 DASM12 in Ss4, respectively [1], in which Top30 and Top27 DASM12 features were derived from a totally disjointed dataset, Ss26 [4]. The above results apparently validated that the proposed differential power asymmetry can effectively characterize EEG spectral dynamics associated with emotional responses induced by diverse music excerpts and for subjects whose datasets were not previously analyzed.

13.4 CONCLUSION

This chapter focuses on practical issues regarding how a system can accurately estimate emotional responses with a minimal number of electrodes, which is not addressed in the previous studies. In our series of affective studies [1–7], we have explored the feasibility of using several subject-independent feature sets and electrode sets for distinguishing distinct emotional states. Differential power asymmetry derived from symmetric channel pairs across brain hemispheres outperformed other feature types. The generalization of the feature set was further validated on a small group of subjects whose EEG patterns were induced by different types of music excerpts than those used in the original experiment. The montages used for DASM7 and DASM6 are very practical for routine use, improving the applicability of affective BCI in real-world applications.

13.5 ISSUES AND CHALLENGES TOWARD ABCIS

The study reported above was focused on an off-line optimal affective framework. Several critical issues and challenges addressed in the following sections need to be carefully evaluated in the next step toward application of affective BCIs.

13.5.1 Directions for Improving Estimation Performance

This study clearly shows that different emotional states are distinguishable in human EEG signals. It has to be noted that classification performance of an ABCI system might be improved in several directions. First, the features extracted for classification might be enhanced through using advanced signal processing techniques such as the spatial filtering approaches [52] and the data preprocessing methods (e.g., artifact removal [53]). Besides, other EEG features (e.g., EEG coherence [54]), which were not used as feature types in this study, might be extracted and fused with existing features to provide more information for discriminating emotion states. Second, further investigation is required to optimize the classifier(s) used in this study. Third, optimizing user training, which aims to improve a user's ability to produce distinct emotion-related brain patterns, is another way to improve system performance. Details of user-training issues will be described below.

13.5.1.1 Feature Extraction As described in the first section, with the millions of neurons in the brain, the potential number of EEG sources would seem to be immense. However, since a patch of cortex must be relatively large and well synchronized and aligned in parallel to generate an electric field capable of penetrating to the scalp, the number of detectable patches (i.e., EEG "sources") may be somewhat limited. Nevertheless, data collected at a single scalp channel is still a confounded mixture of activity from multiple sources, even if the number of major sources is relatively tractable. These overlapping source projections are one reason that localization of EEG sources is a challenge. To address this problem, independent component analysis (ICA) [55] has been developed and is becoming widely accepted as a means to separate mixed EEG scalp data into maximally independent EEG source activities [56]. ICA decomposition reveals a relatively small number of plausibly brain components (along with artifact and other non-localizable components), which means that their inverse weights (i.e., scalp projections) are nearly identical to a theoretical projection of a single-dipole forward model [57]. Therefore, even though there may be myriad transient patches of synchrony in the brain during a typical EEG recording, ICA allows for a focused analysis on a relatively small number of cortical patches showing strong synchronous activity during the session of interest. But perhaps equally importantly, ICA also separates non-brain artifact components from the data, such as blinks and lateral eye movements, as well as pulse and scalp muscle activities. So even if the final analysis or classification is carried out on mixed-source data from scalp channels, ICA is still a powerful method for removing non-brain artifact activity that is inevitably recorded along with brain signals.

Our previous studies have shown that ICA revealed an independent brain process, which exhibited distinct delta-band and theta-band power changes associated with self-reported emotional states, with an equivalent dipole located in or near the frontocentral region [2]. The spectral profiles of emotion-related independent components might be used as a feature type for characterizing emotional responses. In this way, ICA can be used to obtain spatial filters to improve the signal-to-noise ratio (SNR) of emotion-related brain activities through removing the task-irrelevant background

EEG activities. Therefore, the ICA-based features might achieve better classification performance than the electrode-based features. Furthermore, other spatial filtering techniques such as common spatial pattern (CSP) [58] and beamforming [59] might also be used for improving the SNR of emotion-related brain activities as well.

13.5.1.2 ***Feature Classification*** In general, the choice of a specific classification method plays an important role in EEG-based emotion recognition performance, and improvement might be expected by applying different classification methods. This study employed the non-linear SVM classifier, which has been widely used in machine learning applications. However, in the BCI literature, linear classification methods have already shown good performance in many different BCI paradigms [60]. Therefore, it is worthwhile to compare the performance of different classification methods, thereby selecting a classification method specifically optimal for an ABCI system. In addition, ensemble classification techniques [61] such as AdaBoost and Bagging could potentially improve the classification performance by using classifier combination schemes.

13.5.2 Online System Implementation

This study demonstrates an affective BCI using an off-line data analysis paradigm. The results validated the feasibility of using classification of emotional states to drive a BCI. Due to its requirements of real-time data recording and processing, an online ABCI system might face more challenges in system design and implementation than an off-line demonstration. First, efforts need to be made to develop the hardware and software platforms specific for real-time EEG recording and processing. Second, a subject-specific system configuration always plays an important role in system implementation due to significant individual variability. Third, the long-term non-stationary nature of EEG signals makes user/classifier training a critical procedure in an online BCI operation. These technical problems of an online ABCI might be resolved or alleviated by the following considerations.

13.5.2.1 ***Hardware and Software Design*** The hardware and software design of an online ABCI aims to achieve the following purposes:

(1) Low-cost hardware. In general, a BCI system cannot be popularized if it costs too much. By considering the following issues, system costs might be reduced while system performance can be maintained. First, as general-purpose commercial EEG devices are usually expensive, a customized EEG recording device should be designed to minimally satisfy the requirements of the feature extraction method used. Second, to eliminate the cost of a computer used for data processing, a low-cost data-processing platform such as a digital signal processor (DSP) can be employed to construct a system without dependency on a bulky and expensive computer. Moreover, in real-world applications, some new technologies such as the dry electrode [62] and the wireless/mobile BCI platform [63] could further reduce system cost and improve system practicality of an ABCI system.

(2) Robust performance. Robustness of system performance can be improved by working on hardware and software aspects respectively. Considering data recording in unshielded environments with strong electromagnetic interference, employment of an active electrode for data recording might be better than a passive electrode because it can ensure that the recorded signals are insensitive to interference. During system operation, *ad hoc* functions should be provided in the system to adapt to non-stationarity of the signal caused by changes of electrode impedance or brain states, thus reducing the dependence on technical assistance. For example, the software system should be able to detect bad electrode contacts in real-time and automatically adjust algorithms to use the remaining electrodes with high quality signals.

13.5.2.2 User Training As mentioned before, ease-of-use is one of the main requirements in developing practical BCI systems. One major bottleneck, which limits the applications of current BCI systems, is the procedure of user training. For most systems, user training needs to be performed in two phases. First, training sessions need to be performed by the users to get the ability to control/modulate their brain activities. Second, a calibration procedure is required to adjust the parameters to make the system optimal for intentional online BCI operations. Due to high dimensionality of EEG features, the machine learning classifiers need a large number of training samples to achieve good generalization ability. Therefore, user training is always time- and labor-consuming, and therefore, significantly reduces the practicality of a BCI. To reduce the time required for user training, future work should follow two main directions:

(1) Subject-independent and session-independent approaches. Although individual difference and non-stationarity commonly exist in EEG signals, several studies have recently proposed and demonstrated subject-independent and session-independent methods, including the current study, which aim to facilitate user training and online system implementation [64]. Based on these methods, zero-training classifiers could be achieved using a subject-to-subject or session-to-session scenario, where no training data are required. Furthermore, the adaptive classification approaches [65], which include a short calibration session before online BCI operations, could be integrated to optimize the system performance in a short period of time, making user training much more efficient.

(2) Co-adaptive learning. Currently, most BCI systems use an interleaved paradigm for user training and machine learning. In this strategy, user training was first carried out to collect labeled training data without feedback or with feedbacks determined by using a simple classifier obtained from existing data. After a certain amount of training samples were collected, the machine learning procedure was then performed to train a classifier, which aims to optimize classification performance in the subsequent training sessions and online BCI operations. In practice, because a BCI system consists of a biological system (i.e., the brain) and an artificial intelligence system (i.e., the machine learning system), a co-adaptive learning can significantly improve the user training efficacy [66]. Using a co-adaptive learning paradigm, the brain and the machine can adapt to each other quickly, thus reducing the user training time.

ACKNOWLEDGMENTS

The authors acknowledge the value of collaborations from Jyh-Horng Chen, Jeng-Ren Duann, Chi-Hong Wang, Tien-Lin Wu, Shyh-Kang Jeng to this research. Yuan-Pin Lin especially acknowledges the support by the UST-UCSD International Center of Excellence in Advanced Bio-engineering sponsored by the Taiwan National Science Council I-RiCE Program under Grant Number: NSC-99-2911-I-009-101.

REFERENCES

[1] Y. P. Lin, J. H. Chen, J. R. Duann, C. T. Lin, and T. P. Jung, "Generalizations of the subject-independent feature set for music-induced emotion recognition," in Proceedings of the Annual International Conference on IEEE Engineering in Medicine and Biology Society, 2011.

[2] B. Blankertz, K. R. Muller, D. J. Krusienski, G. Schalk, J. R. Wolpaw, A. Schloge, G. Pfurtscheller, J. R. Millan, M. Schroder, and N. Birbaumer, "The BCI competition III: validating alternative approaches to actual BCI problems," *Electroencephalographic dynamics of musical emotion perception revealed by independent spectral components*, 21(6): 410–415, 2010.

[3] Y. P. Lin, T. P. Jung, and J. H. Chen, "EEG dynamics during music appreciation," in Proceedings of the Conference on IEEE Engineering in Medicine and Biology Society, 2010, pp. 5316–5319.

[4] Y. P. Lin, C. H. Wang, T. P. Jung, T. L. Wu, S. K. Jeng, J. R. Duann, and J. H. Chen, "EEG-based emotion recognition in music listening," *IEEE Transactions on Biomedical Engineering*, 57(7): 1798–1806, 2010.

[5] Y. P. Lin, C. H. Wang, T. L. Wu, S. K. Jeng, and J. H. Chen, "EEG-based emotion recognition in music listening: a comparison of schemes for multiclass support vector machine," in IEEE International Conference on Acoustics, Speech and Signal Processing, 2009, pp. 489–492.

[6] Y. P. Lin, C. H. Wang, T. L. Wu, S. K. Jeng, and J. H. Chen, "Multilayer perceptron for EEG signal classification during listening to emotional music," in IEEE Region 10 Conference, 2007, pp. 1–3.

[7] Y.-P. Lin, C.-H. Wang, T.-L. Wu, S.-K. Jeng, and J.-H. Chen, "Support vector machine for EEG signal classification during listening to emotional music," *IEEE 10th Workshop on Multimedia Signal Process.*, 127–130, 2008.

[8] J. R. Wolpaw, N. Birbaumer, D. J. McFarland, G. Pfurtscheller, and T. M. Vaughan, "Brain–computer interfaces for communication and control," *Clin. Neurophysiol.*, 113(6): 767–791, 2002.

[9] N. Birbaumer, "Brain–computer-interface research: coming of age," *Clin. Neurophysiol.*, 117(3): 479–483, 2006.

[10] M. A. L. Nicolelis and M. A. Lebedev, "Brain–machine interfaces: past, present and future," *Trends Neurosci.*, 28(9): 536–546, 2006.

[11] A. Nijholt and D. Tan, "Brain–computer interfacing for intelligent systems," *IEEE Intell. Syst.*, 23(3): 72–72, 2008.

[12] Y. Wang, X. Gao, B. Hong, and S. Gao, "Practical designs of brain–computer interfaces based on the modulation of EEG rhythms," in *Invasive and Non-Invasive Brain–Computer Interfaces* (eds B. Graimann and G. Pfurtscheller), Springer, Berlin, 2010.

[13] L. F. Haas, "Hans Berger (1873–1941), Richard Caton (1842–1926), and electroencephalography," *J. Neurol. Neurosurg. Psychiatry*, 74(1): 9, 2003.

[14] R. Adolphs, "Fear, faces, and the human amygdala, "*Curr. Opin. Neurobiol.*, 18: 166–172, 2008.

[15] W. C. Drevets, J. Savitz, and M. Trimble, "The subgenual anterior cingulate cortex in mood disorders," *CNS Spectr.*, 13(8): 663–681, 2008.

[16] E. Kross, M. Davidson, J. Weber, and K. Ochsner, "Coping with emotions past: the neural bases of regulating affect associated with negative autobiographical memories," *Biol. Psychiatry*, 65(5): 361–366, 2009.

[17] M. S. Goodkind, M. Sollberger, A. Gyurak, H. J. Rosen, K. P. Rankin, B. Miller, and R. Levenson, "Tracking emotional valence: the role of the orbitofrontal cortex," *Hum. Brain Mapping*, 33(4): 753–762, 2012.

[18] F. L. Stevens, R. A. Hurley, and K. H. Taber, "Anterior cingulate cortex: unique role in cognition and emotion," *J. Neuropsychiatry Clin. Neurosci.*, 23(2): 121–125, 2011.

[19] M. T. Bardo, "Neuropharmacological mechanisms of drug reward: beyond dopamine in the nucleus accumbens," *Crit. Rev. Neurobiol.*, 12(1–2): 37–67, 1998.

[20] T. W. Robbins, "Arousal systems and attentional processes," *Biol. Psychol.*, 45(1–3): 57–71, 1997.

[21] S. B. Andersen, R. A. Moore, L. Venables, and P. J. Corr, "Electrophysiological correlates of anxious rumination," *Int. J. Psychophysiol.*, 71(2): 156–169, 2009.

[22] M. Wyczesany, J. Kaiser, and A. M. L. Coenen, "Subjective mood estimation co-varies with spectral power EEG characteristics," *Acta Neurobiol. Exp.*, 68(2): 180–192, 2008.

[23] L. I. Aftanas, N. V. Reva, A. A. Varlamov, S. V. Pavlov, and V. P. Makhnev, "Analysis of evoked EEG synchronization and desynchronization in conditions of emotional activation in humans: temporal and topographic characteristics," *Neurosci. Behav. Physiol.*, 34(8): 859–867, 2004.

[24] J. L. Jung, D. Bayle, K. Jerbi, J. R. Vidal, M. A. Henaff, T. Ossandon, O. Bertrand, F. Mauguiere, and J. P. Lachaux, "Intracerebral gamma modulations reveal interaction between emotional processing and action outcome evaluation in the human orbitofrontal cortex," *Int. J. Psychophysiol.*, 79(1): 64–72, 2011.

[25] D. J. Oathes, W. J. Ray, A. S. Yamasaki, T. D. Borkovec, L. G. Castonguay, M. G. Newman, and J. Nitschke, "Worry, generalized anxiety disorder, and emotion: evidence from the EEG gamma band," *Biol. Psychol.*, 79(2): 165–170, 2008.

[26] A. Keil, M. M. Muller, T. Gruber, C. Wienbruch, M. Stolarova, and T. Elbert, "Effects of emotional arousal in the cerebral hemispheres: a study of oscillatory brain activity and event-related potentials," *Clin. Neurophysiol.*, 112(11): 2057–2068, 2001.

[27] J. Onton and S. Makeig, "High-frequency broadband modulations of electroencephalographic spectra," *Front. Hum. Neurosci.*, 3, 2009.

[28] A. Gemignani, E. Santarcangelo, L. Sebastiani, C. Marchese, R. Mammoliti, A. Simoni, and B. Ghelarducci, "Changes in autonomic and EEG patterns induced by hypnotic imagination of aversive stimuli in man," *Brain Res. Bull.*, 53(1): 105–111, 2000.

[29] C. A. Frantzidis, C. Bratsas, C. L. Papadelis, E. Konstantinidis, C. Pappas, and P. D. Bamidis, "Toward emotion aware computing: an integrated approach using multichannel

neurophysiological recordings and affective visual stimuli," *IEEE Transactions on Information Technology in Biomedicine*, 14(3): 589–597, 2010.

[30] P. C. Petrantonakis and L. J. Hadjileontiadis, "Emotion recognition from brain signals using hybrid adaptive filtering and higher order crossings analysis," *IEEE Transactions on Affective Computing*, 1(2): 81–97, 2010.

[31] S. Koelstra, C. Muhl, M. Soleymani, J. Lee, A. Yazdani, T. Ebrahimi, T. Pun, A. Nijholt, and I. Patras, "DEAP: a database for emotion analysis using physiological signals," *IEEE Transactions on Affective Computing*, 99(1-1), 2011.

[32] M. Murugappan, R. Nagarajan, and S. Yaacob, "Combining spatial filtering and wavelet transform for classifying human emotions using EEG signals," *J. Med. Biol. Eng.*, 31(1): 45–51, 2011.

[33] G. Chanel, J. J. M. Kierkels, M. Soleymani, and T. Pun, "Short-term emotion assessment in a recall paradigm," *Int. J. Hum.-Comput. Stud.*, 67(8): 607–627, 2009.

[34] P. Ekman, W. V. Friesen, M. Osullivan, A. Chan, I. Diacoyannitarlatzis, K. Heider, R. Krause, W. A. Lecompte, T. Pitcairn, P. E. Riccibitti, K. Scherer, M. Tomita, and A. Tzavaras, "Universals and cultural-differences in the judgments of facial expressions of emotion," *J. Personal. Soc. Psychol.*, 53(4): 712–717, 1987.

[35] J. A. Russell, "A circumplex model of affect," *J. Personal. Soc. Psychol.*, 39(6): 1161–1178, 1980.

[36] P. Lang, M. Bradley, and B. Cuthbert, "International affective picture system (IAPS): affective ratings of pictures and instruction manual," Technical Report A-8, University of Florida, USA, 2008.

[37] T. L. Wu and S. K. Jeng, "Probabilistic estimation of a novel music emotion model," in 14th International Conference on Multimedia Modeling, vol. 4903, 2008, pp. 487–497.

[38] E. Basar, C. Basar-Eroglu, S. Karakas, and M. Schurmann, "Oscillatory brain theory: a new trend in neuroscience—the role of oscillatory processes in sensory and cognitive functions," *IEEE Engineering in Medicine and Biology Magazine*, 618(3): 56–66, 1999.

[39] J. J. B. Allen, J. A. Coan, and M. Nazarian, "Issues and assumptions on the road from raw signals to metrics of frontal EEG asymmetry in emotion," *Biol. Psychol.*, 67(1): 183–218, 2004.

[40] L. A. Schmidt and L. J. Trainor, "Frontal brain electrical activity (EEG) distinguishes valence and intensity of musical emotions," *Cogn. Emot.*, 15(4): 487–500, 2001.

[41] R. Palaniappan, "Utilizing gamma band to improve mental task based brain–computer interface design," *IEEE Transactions on Neural Systems and Rehabilitation Engineering*, 14(3): 299–303, 2006.

[42] G. Isabelle and E. Andre, "An introduction to variable and feature selection," *J. Mach. Learn. Res.*, 3: 1157–1182, 2003.

[43] H. Cecotti and A. Graser, "Convolutional neural networks for P300 detection with application to brain–computer interfaces," *IEEE Transactions on Pattern Analysis and Machine Intelligence*, 33(3): 433–445, 2011.

[44] L. K. Hansen and P. Salamon, "Neural network ensembles," *IEEE Transactions on Pattern Analysis and Machine Intelligence*, 12(10): 993–1001, 1990.

[45] B. E. Boser, I. M. Guyon, and V. N. Vapnik, "A training algorithm for optimal margin classifiers", in Proceedings of the Fifth Annual Workshop on Computational Learning Theory, Pittsburgh, PA, 1992, pp. 144–152. ISBN 0-89791-497-X

[46] C. W. Hsu and C. J. Lin, "A comparison of methods for multiclass support vector machines," *IEEE Transactions on Neural Networks*, 13(2): 415–425, 2002.

[47] L. Bottou, C. Cortes, J. S. Denker, H. Drucker, I. Guyon, L. D. Jackel, Y. LeCun, U. A. Muller, E. Sackinger, P. Simard, and V. Vapnik, "Comparison of classifier methods: a case study in handwritten digit recognition," in Proceedings of the 12th IAPR International. Conference on Pattern Recognition, vol. 2, Conference B: Computer Vision & Image Processing, 1994, pp. 77–82.

[48] U. Kreßel, *Pairwise Classification and Support Vector Machines*, MIT Press, 1999.

[49] J. Kim and E. Andre, "Emotion recognition based on physiological changes in music listening," *IEEE Transactions on Pattern Analysis and Machine Intelligence*, 30(12): 2067–2083, 2008.

[50] C. C. Chang and C. J. Lin, "LIBSVM: a library for support vector machines," *ACM Trans. Intell. Syst. Technol.*, 2(3): 1–27, 2011.

[51] E. O. Flores-Gutierrez, J. L. Diaz, F. A. Barrios, M. A. Guevara, Y. del Rio-Portilla, M. Corsi-Cabrera, and E. O. del Flores-Gutierrez, "Differential alpha coherence hemispheric patterns in men and women during pleasant and unpleasant musical emotions," *Int. J. Psychophysiol.*, 71(1): 43–49, 2009.

[52] D. J. McFarland, L. M. McCane, S. V. David, and J. R. Wolpaw, "Spatial filter selection for EEG-based communication," *Electroencephalogr. Clin. Neurophysiol.*, 103(3): 386–394, 1997.

[53] T. P. Jung, S. Makeig, C. Humphries, T. W. Lee, M. J. McKeown, V. Iragui, and T. J. Sejnowski, "Removing electroencephalographic artifacts by blind source separation," *Psychophysiology*, 37(2): 163–178, 2000.

[54] R. Srinivasan, P. L. Nunez, and R. B. Silberstein, "Spatial filtering and neocortical dynamics: estimates of EEG coherence," *IEEE Transactions on Biomedical Engineering*, 45(7): 814–826, 1998.

[55] A. Hyvarinen and E. Oja, "Independent component analysis: algorithms and applications," *Neural Netw.*, 13(4–5): 411–430, 2010.

[56] S. Makeig, A. J. Bell, T. P. Jung, and T. J. Sejnowski, "Independent component analysis of electroencephalographic data," *Adv. Neural Inf. Process. Syst.*, 8: 145–151, 1996.

[57] R. Oostenveld and T. F. Oostendorp, "Validating the boundary element method for forward and inverse EEG computations in the presence of a hole in the skull," *Hum. Brain Mapp.*, 17(3): 179–192, 2002.

[58] H. Ramoser, J. Muller-Gerking, and G. Pfurtscheller, "Optimal spatial filtering of single trial EEG during imagined hand movement," *IEEE Transactions on Rehabilitation Engineering*, 8(4): 441–446, 2000.

[59] K. Sekihara, S. S. Nagarajan, D. Poeppel, A. Marantz, and Y. Miyashita, "Reconstructing spatio-temporal activities of neural sources using an MEG vector beamformer technique," *IEEE Transactions on Biomedical Engineering*, 48(7): 760–771, 2001.

[60] B. Blankertz, K. R. Muller, D. J. Krusienski, G. Schalk, J. R. Wolpaw, A. Schlogl, G. Pfurtscheller, J. D. R. Millan, M. Schroder, and N. Birbaumer, "The BCI competition III: validating alternative approaches to actual BCI problems," *IEEE Transactions on Neural Systems and Rehabilitation Engineering*, 14(2): 153–159, 2006.

[61] R. O. Duda, P. E. Hart, and D. G. Stork, *Pattern Classification*, Wiley Interscience Press, New York, 2000.

[62] Y. M. Chi, J. Tzyy-Ping, and G. Cauwenberghs, "Dry-contact and noncontact biopotential electrodes: methodological review," *IEEE Rev. Biomed. Eng.*, 3: 106–119, 2010.

[63] Y. T. Wang, Y. J. Wang, and T. P. Jung, "A cell-phone-based brain–computer interface for communication in daily life," *J. Neural Eng.*, 8(2), 2011.

[64] M. Krauledat, M. Tangermann, B. Blankertz, and K. R. Muller, "Towards zero training for brain–computer interfacing," *PLoS One*, 3(8), 2008.

[65] C. Vidaurre, A. Schlogl, R. Cabeza, R. Scherer, and G. Pfurtscheller, "Study of on-line adaptive discriminant analysis for EEG-based brain computer interfaces," *IEEE Transactions on Biomedical Engineering*, 54(3): 550–556, 2007.

[66] C. Vidaurre, C. Sannelli, K. R. Muller, and B. Blankertz, "Machine-learning-based coadaptive calibration for brain–computer interfaces," *Neural Comput.*, 23(3): 791–816, 2011.

[67] M. H. Lee and Q. Zhang, "Analysis of positive and negative emotions in natural scene using brain activity and GIST," *Neurocomputing*, 72(4–6): 1302–1306, 2009.

AUTHOR BIOGRAPHIES

Yuan-Pin Lin received the B.S. degree in Biomedical Engineering from Chung Yuan Christian University, Chung Li, Taiwan, in 2003, and the M.S. and Ph.D. degrees in Electrical Engineering from National Taiwan University, Taipei, Taiwan, in 2005 and 2011, respectively. He is currently a postdoctoral fellow at the Brain Research Center, National Chiao Tung University, Hsinchu, Taiwan. His research interests include areas of human electroencephalogram analysis, affective computing, machine learning, and brain–computer interactions.

Tzyy-Ping Jung (S '91-M '92-SM-06) received the B.S. degree in Electronics Engineering from National Chiao Tung University, Hsinchu, Taiwan, in 1984, and the M.S. and Ph.D. degrees in Electrical Engineering from Ohio State University, Columbus, in 1989 and 1993, respectively. He was a research associate of National Research Council, National Academy of Sciences, USA. He is currently the Codirector of Center for Advanced Neurological Engineering and the Associate Director of the Swartz Center for Computational Neuroscience, University of California, San Diego. He is also an adjunct professor of the Department of Bioengineering, UCSD, and a professor of the Department of Computer Science, National Chiao-Tung University. His research interests include areas of biomedical signal processing, cognitive neuroscience, machine learning, time–frequency analysis of human electroencephalogram, functional neuroimaging, and brain–computer interfaces and interactions.

Yijun Wang received his B.E. and Ph.D. degrees in Biomedical Engineering from Tsinghua University, Beijing, China, in 2001 and 2007, respectively. Currently, he is a postdoctoral researcher at the Swartz Center for Computational Neuroscience, University of California, San Diego. His research interests include brain–computer interface, biomedical signal processing, and machine learning.

Julie Onton received a B.A. degree in Neurobiology from the University of California, Berkeley, in 1996, and a Ph.D. from Tufts University in 2001. She has worked as a postdoctoral fellow and project scientist at the Swartz Center for Computational Neuroscience and the Institute for Neural Computation at the University of California, San Diego. Her research interests include developing advanced EEG analysis techniques based on independent component analysis and other signal processing techniques. She has also investigated frontal midline theta variability during a working memory task and high frequency gamma responses during emotional imagery. She is currently working with the military population to discover EEG biomarkers of traumatic brain injury and posttraumatic stress disorder.

14

BODILY EXPRESSION FOR AUTOMATIC AFFECT RECOGNITION

HATICE GUNES

School of Electronic Engineering and Computer Science, Queen Mary University of London, London, UK

CAIFENG SHAN

Philips Research Eindhoven, High Tech Campus, Eindhoven, The Netherlands

SHIZHI CHEN AND YINGLI TIAN

Department of Electrical Engineering, The City College of New York, NY, USA

This chapter focuses on the why, what, and how of bodily expression analysis for automatic affect recognition. It first asks the question of 'why bodily expression?' and attempts to find answers by reviewing the latest bodily expression perception literature. The chapter then turns its attention to the question of 'what are the bodily expressions recognized automatically?' by providing an overview of the automatic bodily expression recognition literature. The chapter then provides representative answers to how bodily expression analysis can aid affect recognition by describing three case studies: (1) data acquisition and annotation of the first publicly available database of affective face-and-body displays (i.e., the FABO database); (2) a representative approach for affective state recognition from face-and-body display by detecting the space-time interest points in video and using Canonical Correlation Analysis (CCA) for fusion, and (3) a representative approach for explicit detection of the temporal phases (segments) of affective states (start/end of the expression and its subdivision into phases such as neutral, onset, apex, and offset) from bodily expressions. The chapter concludes by summarizing the main challenges faced and discussing how we can advance the state of the art in the field.

Emotion Recognition: A Pattern Analysis Approach, First Edition. Edited by Amit Konar and Aruna Chakraborty.
© 2015 John Wiley & Sons, Inc. Published 2015 by John Wiley & Sons, Inc.

14.1 INTRODUCTION

Humans interact with others and their surrounding environment using their visual, auditory, and tangible sensing. The visual modality is the major input/output channel utilized for next generation human–computer interaction (HCI). Within the visual modality, the body has recently started gaining a particular interest due to the fact that in daily life body movements and gestures are an indispensable means for interaction. Not many of us realize the myriad ways and the extent to which we use our hands in everyday life: when we think, talk, and work. The gaming and entertainment industry is the major driving force behind putting the human body in the core of technology design by creating controller-free human–technology interaction experiences. Consequently, technology today has started to rely on the human body as direct input by reacting to and interacting with its movement [1, 2]. One example of this is the Kinect project [2] that enables users to control and interact with a video game console (the Xbox 360 [3]) through a natural user interface using gestures and spoken commands instead of a game controller.

Bodily cues (postures and gestures) have also started attracting the interest of researchers as a means to communicate emotions and affective states. Psychologists have long explored mechanisms with which humans recognize others' affective states from various cues and modalities, such as voice, face, and body gestures. This exploration has led to identifying the important role played by the modalities' dynamics in the recognition process. Supported by the human physiology, the temporal evolution of a modality appears to be well approximated by a sequence of temporal segments called onset, apex, and offset. Stemming from these findings, computer scientists, over the past 20 years, have proposed various methodologies to automate the affect recognition process. We note, however, two main limitations to date. The first is that much of the past research has focused on affect recognition from voice and face, largely neglecting the affective body display and bodily expressions. Although a fundamental study by Ambady and Rosenthal suggested that the most significant channels for judging behavioral cues of humans appear to be the visual channels of facial expressions and body gestures, affect recognition via body movements and gestures has only recently started attracting the attention of computer science and HCI communities. The second limitation is that automatic affect analyzers have not paid sufficient attention to the dynamics of the (facial and bodily) expressions: the automatic determination of the temporal segments and their role in affect recognition are yet to be adequately explored.

To address these issues, this chapter focuses on the why, what, and how of automatic bodily expression analysis. It first asks the question of "why bodily expression?" and attempts to find answers by reviewing the latest bodily expression perception literature. The chapter then turns its attention to the question of "what are the bodily expressions recognized automatically?" by providing an overview of the automatic bodily expression recognition literature and summarizing the main challenges faced in the field. The chapter then provides representative answers to how bodily expression analysis can aid affect recognition by describing three case studies: (1) data acquisition and annotation of the first publicly available database of affective face-and-body displays (i.e., the FABO database); (2) a representative approach for affective state recognition from face-and-body display by detecting the space-time interest points in video and using Canonical Correlation Analysis (CCA) for fusion, and (3) a representative approach for explicit detection of the temporal phases (segments) of affective states (start/end of the expression and its subdivision into phases such as neutral, onset, apex, and offset) from bodily expressions.

Due to its popularity and extensive exploration, emotion communication through facial expressions will not be covered in this chapter. The interested readers are referred to References 4–11.

14.2 BACKGROUND AND RELATED WORK

Emotion communication through bodily expressions has been a neglected area for much of the emotion research history [12, 13]. This is illustrated by the fact that 95% of the literature on human emotions has been dedicated to using face stimuli, majority of the the remaining 5% on audio-based research, and the remaining small number on whole-body expressions [12]. This is indeed puzzling given the fact that early research on emotion by Darwin [14] and James [15] has paid a considerable attention to emotion-specific body movements and postural configurations. De Gelder argues that the reason why whole-body expressions have been neglected in emotion research is mainly due to the empirical results dating from the first generation of investigations of whole-body stimuli [12]. There are potentially other reasons as to why the body may seem a less reliable source of affective information (i.e., the face bias), its cultural and ideological reasons and heritage, which have been discussed in detail in Reference 12.

Overall, the body and hand gestures are much more varied than facial changes. There is an unlimited vocabulary of body postures and gestures with combinations of movements of various body parts (with multiple degrees of freedom) [13, 16, 17]. Therefore, using bodily expression for emotion communication and perception has a number of advantages:

- Bodily expression provides a means for recognition of affect from a distance. When we are unable to tell the emotional state from the face, we can still clearly read the action from the sight of the body [12]. This has direct implications for designing affective interfaces that will work in realistic settings (e.g., affective tutoring systems, humanoid robotics, affective games).
- Some of the basic mental states are most clearly expressed by the face while others are least ambiguous when expressed by the whole body (e.g., anger and fear) [12]. Perception of facial expression is heavily influenced by bodily expression as in most situations people do not bother to censor their body movements and therefore, the body is at times referred to as the *leaky* source [18]. Consequently, bodily expression, when used as an additional channel for affect communication, can provide a means to resolve ambiguity for affect detection and recognition.

Due to such advantages, automatic recognition of bodily expressions has increasingly started to attract the attention and the interest of the affective computing researchers. In this section, we will first review existing methods that achieve affect recognition and/or temporal segmentation from body display. Second, we will summarize existing systems that combine bodily expression with other cues or modalities in order to achieve multicue and multimodal affect recognition.

14.2.1 Body as an Autonomous Channel for Affect Perception and Analysis

Human recognition of emotions from body movements and postures is still an unresolved area of research in psychology and non-verbal communication. There are numerous works suggesting various opinions in this area. Ekman and Friesen have touched upon the possibility that some bodily (and facial) cues might be able to communicate both the quantity and quality aspects of emotional experience [19]. This leads to two major perspectives regarding the emotion perception and recognition from bodily posture and movement. The first perspective claims that there are body movements and postures that mostly contribute to the understanding of the activity (and intensity) level of the underlying emotions. For instance, Wallbot provided associations between body movements and the arousal dimension of emotion. More specifically, lateralized hand/arm movements, arms stretched out to the front, and opening and closing of the hands were observed during active emotions, such as hot anger, cold anger, and interest [20]. This can somewhat be seen as contributing toward the dimensional approach to emotion perception and recognition from bodily cues. The second perspective considers bodily cues (movements and postures) to be an independent channel of expression able to convey discrete emotions. An example is De Meijer's work that illustrated that observers are able to recognize emotions from body movements alone [21].

In general, recognition of affect from bodily expressions is mainly based on categorical representation of affect. The categories happy, sad, and angry appear to be more distinctive in motion than categories such as pride and disgust. Darwin suggested that in anger, for instance, among other behaviors, the whole body trembles, the head is erect, the chest is well expanded, feet are firmly on the ground, elbows are squared [14, 20]. Wallbot also analyzed emotional displays by actors and concluded that discrete emotional states can be recognized from body movements and postures. For instance, hot anger was encoded by shoulders moving upwards, arms stretched frontally, or lateralized, the execution of various hand movements, as well as high movement activity, dynamism, and expansiveness. Analysis of the arm movements (drinking and knocking) shows that, discrete affective states are aligned with the arousal–pleasure space [22]; and arousal was found to be highly correlated with velocity, acceleration, and jerk of the movement.

To date, the bodily cues that have been more extensively considered for affect recognition are (static) postural configurations of head, arms, and legs [16, 23], dynamic hand/arm movements [20], head movements (e.g., position and rotation) [24], and head gestures (e.g., head nods and shakes) [25, 26].

14.2.1.1 ***Body Posture*** Coulson [16] presented experiments on attributing six universal emotions (anger, disgust, fear, happiness, sadness, and surprise) to static body postures using computer-generated mannequin figures. His experimental results suggested that recognition from body posture is comparable to recognition from voice, and some postures are recognized as well as facial expressions.

When it comes to automatic analysis of affective body postures the main emphasis has been on using the tactile modality (for gross bodily expression analysis) via body-pressure-based affect measurement (e.g., [27]) and on using motion capture technology (e.g., [23]). Mota and Picard [27] studied affective postures in an e-learning scenario, where the posture information was collected through a sensor chair. Kleinsmith *et al.* [23] focused on the dimensional representation of emotions and on acquiring and analyzing affective posture data using motion capture technology [23]. They examined the role of affective dimensions in static postures for automatic recognition and showed that it is possible to automatically recognize the affect dimensions of arousal, valence, potency, and avoidance with acceptable recognition rates (i.e., error rates lower than 21%).

14.2.1.2 ***Body Movement*** Compared to the facial expression literature, attempts for recognizing affective body movements are few and efforts are mostly on the analysis of posed bodily expression data. Burgoon *et al.* discussed the issue of emotion recognition from bodily cues and provided useful references in Reference 28. They claimed that affective states are conveyed by a set of cues and focus on the identification of affective states such as positivity, anger, and tension in videos from body and kinesics cues. Meservy *et al.* [29] focused on extracting body cues for detecting truthful (innocent) and deceptive (guilty) behavior in the context of national security. They achieved a recognition accuracy of 71% for the two-class problem (i.e., guilty/innocent). Bernhardt and Robinson analyzed non-stylized body motions (e.g., walking, running) for affect recognition [30] using kinematic features (e.g., velocity, acceleration, and jerk measured for each joint) and reported that the affective states angry and sad are more recognizable than neutral or happy.

Castellano *et al.* [31] presented an approach for the recognition of acted emotional states based on the analysis of body movement and gesture expressivity. They used the non-propositional movement qualities (e.g. amplitude, speed, and fluidity of movement) to infer emotions (anger 90%, joy 44%, pleasure 62%, sadness 48%). A similar technique was used to extract expressive descriptors of movement (e.g., quantity of motion of the body and velocity of the head movements) in a music performance and to study the dynamic variations of gestures used by a pianist [32]. They found that the timing of expressive motion cues (i.e., the attack and release of the temporal profile of the velocity of the head and the quantity of motion of the upper body) is important in explaining emotional expression in piano performances. Reference 33 presents a framework for analysis of affective behavior starting with a reduced amount of visual information related to human upper-body movements. The work uses the EyesWeb Library (and its extensions) for extracting a number of expressive gesture features (e.g., smoothness of gesture, gesture duration) by tracking of trajectories of head and hands (from a frontal and a lateral view), and the GEMEP corpus (120 posed upper body gestures for 12 emotion classes from 10 subjects) for validation. The authors conclude that for distinguishing bodily expression of different emotions dynamic features related to movement quality (e.g., smoothness of gesture, duration of gesture) are more important than categorical features related to the specific type of gesture.

A number of researchers have also investigated how to map various visual signals onto emotion dimensions. Cowie *et al.* [25] investigated the emotional and communicative significance of head nods and shakes in terms of Arousal and Valence dimensions, together with dimensional representation of *solidarity*, *antagonism*, and *agreement*. Their findings suggest that both head nods and shakes clearly carry information about arousal. However, their significance for evaluating the valence dimensions is less clear (affected by access to words) [25]. In particular, the contribution of the head nods for valence evaluation appears to be more complicated than head shakes (e.g., "I understand what you say, and I care about it, but I don't like it").

14.2.1.3 Gait Gait, in the context of perception and recognition, refers to a person's individual walking style. Therefore, gait is a source of dynamic information by definition. Emotion perception and recognition from gait patterns is also a relatively new area of research [34,35]. Janssen *et al.* [34] focused on emotion recognition from human gait by means of kinetic and kinematic data using artificial neural nets. They conducted two experiments: (1) identifying participants' emotional states (normal, happy, sad, angry) from gait patterns and (2) analyzing effects on gait patterns of listening to different types of music (excitatory, calming, no music) while walking. Their results showed that subject-independent emotion recognition from gait patterns is indeed possible (up to 100% accuracy). Karg *et al.* [35] focused on using both discrete affective states and affective dimensions for emotion modeling from motion capture data. Person-dependent recognition of motion capture data reached 95% accuracy based on the observation of a single stride. This work showed that gait is a useful cue for the recognition of arousal and dominance dimensions.

14.2.1.4 Temporal Dynamics An expression is a dynamic event, which evolves from neutral, onset, apex to offset [36], a structure usually referred to as *temporal dynamics* or *temporal phases*. Evolution of such a temporal event is illustrated, for a typical facial expression, in Figure 14.1. The neutral phase is a plateau where there are no signs of muscular activation and the face is relaxed. The *onset* of the action/movement is when the muscular contraction begins and increases in intensity and the appearance of the face changes. The *apex* is a plateau usually where the intensity reaches a stable level and there are no more changes in facial appearance. The *offset* is the relaxation of the muscular action. A natural facial movement evolves over time in the following order: neutral(N) ⟶ onset(On) ⟶ apex(A) ⟶ offset(Of) ⟶ neutral(N). Other combinations such as multiple-apex facial actions are also possible.

Similarly, the temporal structure of a body gesture consists of (up to) five phases: preparation ⟶ (pre-stroke) hold ⟶ stroke ⟶ (post-stroke) hold ⟶ retraction. The *preparation* moves to the stroke's starting position and the *stroke* is the most

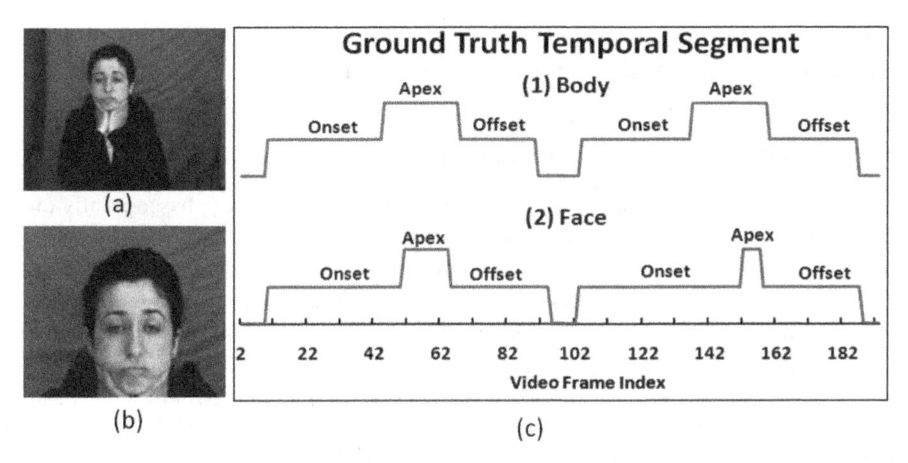

FIGURE 14.1 (a) Sample image of boredom expression to extract body gesture feature (body camera), (b) sample image of boredom expression to extract facial feature (face camera), and (c) the corresponding temporal segments from body gesture and facial features respectively. Taken with permission from the FABO database.

energetic part of the gesture. *Holds* are optional still phases which can occur before and/or after the stroke. The *retraction* returns to a *rest* pose (e.g., arms hanging down, resting in lap, or arms folded). Some gestures (e.g., finger tapping) have multiple strokes that include small beat-like movements that follow the first stroke, but seem to belong to the same gesture [37].

Studies demonstrate that the temporal dynamics play an important role for interpreting emotional displays [38, 39]. It is believed that information about the time course of a facial action may have psychological meaning relevant to the intensity, genuineness, and other aspects of the expresser's state. Among the four temporal phases of neutral, onset, apex, and offset, features during the apex phase result in maximum discriminative power for expression recognition. Gunes and Piccardi showed that, during automatic affect recognition from facial/bodily gestures, decoupling temporal dynamics from spatial extent significantly reduces the dimensionality of the problem compared to dealing with them simultaneously and improves affect recognition accuracy [40]. Thus, successful temporal segmentation can not only help to analyze the dynamics of an (facial/bodily) expression, but also improve the performance of expression recognition. However, in spite of their usefulness, the complex spatial properties and dynamics of face and body gestures also pose a great challenge to affect recognition. Therefore, interest in the temporal dynamics of affective behavior is recent (e.g., [11, 40–42]). The work of Reference 41 temporally segmented facial action units (AUs) using geometric features of 15 facial key points from profile face images. In Reference 37, a method for the detection of the temporal phases in natural gesture was presented. For body movement, a finite-state machine (FSM) was used to spot multiphase gestures against a rest state. In order to detect the gesture phases, candidate rest states were obtained and evaluated. Three variables were used to model the states: distance from rest image, motion magnitude, and duration. Other approaches have exploited dynamics of the gestures without attempting to recognize their temporal phases or segments explicitly (e.g., [31, 43] and [29]).

14.2.2 Body as an Additional Channel for Affect Perception and Analysis

Ambady and Rosenthal reported that human judgment of behaviors based jointly on face and body proved 35% more accurate than those based on the face alone [44]. The face and the body, as part of an integrated whole, both contribute in conveying the emotional state of the individual. A single body gesture can be ambiguous. For instance, the examples shown in the second and fourth rows in Figure 14.2 have similar bodily gestures, but the affective states they express are quite different, as shown by the corresponding facial expressions. In light of such findings, instead of looking at the body as an independent and autonomous channel of emotional expression, researchers have increasingly focused on the relationship between bodily postures and movement with other expressive channels such as voice and face (e.g., [40, 45, 46]).

FIGURE 14.2 Example images from the FABO database recorded by the face (top) and body (bottom) cameras separately. Representative images of non-basic facial expressions (a1–h1) and their corresponding body gestures (a2–h2): (a) neutral, (b) negative surprise, (c) positive surprise, (d) boredom, (e) uncertainty, (f) anxiety, and (g) puzzlement. Taken with permission from the FABO database.

It is important to state that automatic affect recognition does not aim to replace one expression channel (e.g., the facial expressions) as input by another expression channel (e.g., bodily expressions). Instead, the aim is to explore various communicative channels more deeply and more fully in order to obtain a thorough understanding of cross-modal interaction and correlations pertaining to human affective display. An example is the work of Van den Stock *et al.* investigating the influence of whole-body expressions of emotions on the recognition of facial and vocal expressions of emotion [47]. They found that recognition of facial expression was strongly influenced by the bodily expression. This effect was a function of the ambiguity of the facial expression. Overall, during multisensory perception, judgments for one modality seem to be influenced by a second modality, even when the latter modality can provide no information about the judged property itself or increase ambiguity (i.e., cross-modal integration) [48,49]. Meeren *et.al.* [50] showed that the recognition of facial expressions is strongly influenced by the concurrently presented emotional body language and that the affective information from the face and the body start to interact rapidly, and the integration is a mandatory automatic process occurring early in the processing stream. Therefore, fusing facial expression and body gesture in video sequences provides a potential way to accomplish improved affect analysis.

When it comes to using the body as an additional channel for automatic analysis, the idea of combining face and body expressions for affect recognition is relatively new. Balomenos *et al.* [51] combined facial expressions and hand gestures for the recognition of six prototypical emotions. They fused the results from the two subsystems at a decision level using pre-defined weights. An 85% accuracy was achieved for emotion recognition from facial features alone. An overall recognition rate of 94.3% was achieved for emotion recognition from hand gestures. Karpouzis *et al.* [52] fused data from facial, bodily, and vocal cues using a recurrent network to detect emotions. They used data from four subjects and reported the following recognition accuracies for a 4-class problem: 67% (visual), 73% (prosody), 82% (all modalities combined). The fusion was performed on a frame basis, meaning that the visual data values were repeated for every frame of the tune. Neither work has focused on explicit modeling and detection of the (facial/bodily) expression temporal segments. Castellano *et al.* considered the possibility of detecting eight emotions (some basic emotions plus irritation, despair, etc.) by monitoring facial features, speech contours, and gestures [45]. Their findings suggest that incorporating multiple cues and modalities helps with improving the affect recognition accuracy, and the best channel for affect recognition appears to be the gesture channel followed by the audio channel.

Hartmann *et al.* [53] defined a set of expressivity parameters for the generation of expressive gesturing for virtual agents. The studies conducted on perception of expressivity showed that only a subset of parameters and a subset of expressions were recognized well by users. Therefore, further research is needed for the refinement of the proposed parameters (e.g., the interdependence of the expressivity parameters). Valstar *et al.* [54] investigated separating posed from genuine smiles in video sequences using facial, head, and shoulder movement cues, and the temporal correlation between these cues. Their results seem to indicate that using video data from face, head, and shoulders increases the accuracy, and the head is the most reliable source, followed closely by the face. Nicolaou *et al.* capitalize on the fact that the arousal and valence dimensions are correlated, and present an approach that fuses spontaneous facial expression, shoulder gesture, and audio cues for dimensional and continuous prediction of emotions in valence-arousal space [55]. They propose an output-associative fusion framework that incorporates correlations between emotion dimensions. Their findings suggest that incorporating correlations between affect dimensions provides greater accuracy for continuous affect prediction. Audio cues appear to be better for predicting arousal, and visual cues (facial expressions and shoulder movements) appear to perform better for predicting valence.

A number of systems use the tactile modality for gross bodily expression analysis via body-pressure-based affect measurement (measuring participants' back and seat pressure) [56, 57]. Kapoor and Picard focused on the problem of detecting the affective states of high interest, low interest, and refreshing in a child who is solving a puzzle [57]. They combined sensory information from the face video, the posture sensor (a chair sensor) and the game being played in a probabilistic framework. The classification results obtained by Gaussian Processes for individual modalities showed that affective states are best classified by the posture channel (82%), followed by the features from the upper face (67%), the game (57%), and the lower face (53%). Fusion significantly outperformed classification using the individual modalities and resulted in 87% accuracy. D'Mello and Graesser [56] considered a combination of facial features, gross body language, and conversational cues for detecting some of the learning-centered affective states. Classification results supported a *channel∗judgment*-type interaction, where the face was the most diagnostic channel for spontaneous affect judgments (i.e., at any time in the tutorial session), while conversational cues were superior for fixed judgments (i.e., every 20 seconds in the session). The analyzers also indicated that the accuracy of the multichannel model (face, dialog, and posture) was statistically higher than the best single-channel model for the fixed but not spontaneous affect expressions. However, multichannel models reduced the discrepancy (i.e., variance in the precision of the different emotions) of the discriminant models for both judgment types. The results also indicated that the combination of channels yielded enhanced effects for some states but not for others.

14.2.3 Bodily Expression Data and Annotation

Communication of emotions by body gestures is still an unresolved area in psychology. Therefore, the number of databases and corpus that contain expressive bodily gestures and are publicly available for research purposes is scarce, and there exists no annotation scheme commonly used by all researchers in the field.

Data. To the best of our knowledge there exist three publicly available databases that contain expressive bodily postures or gestures. *The UCLIC Database of Affective Postures and Body Movements* [58] contains acted emotion data (angry, fearful, happy, and sad) collected using a VICON motion capture system, and non-acted affective states (frustration, concentration, triumphant, and defeated) in a computer game setting collected using a Gypsy5 (Animazoo UK Ltd.) motion capture system. *The GEMEP Corpus* (The Geneva Multimodal Emotion Portrayals Corpus) [59] contains 120 posed face and upper-body gestures (head and hand gestures), for 12 emotion classes (pride, joy, amusement, interest, pleasure, relief, hot anger, panic fear, despair, irritation, anxiety, and sadness) from 10 subjects recorded by multiple cameras (e.g., frontal and lateral view). The Bimodal Face and Body Gesture Database (the FABO database) comprises of face-and-body expressions [60] and will be reviewed in detail in the next sections.

Annotation. Unlike the facial actions, there is not one common annotation scheme that can be adopted by all the research groups [13] to describe and annotate the body AUs that carry expressive information. Therefore, it is even harder to create a common benchmark database for affective gesture recognition. The most common annotation has been command-purpose annotation, for instance calling the gesture as rotate or click gesture. Another type of annotation is based on the gesture phase, for example, start of gesture stroke-peak of gesture stroke-end of gesture stroke. Rudolf Laban was a pioneer in attempting to analyze and record body movement by developing a systematic annotation scheme called Labanotation [61]. Traditionally Labanotation has been used mostly in dance choreography, physical therapy, and drama for exploring natural and choreographed body movement. Despite the aforementioned effort of Laban in analyzing and annotating body movement, a more detailed annotation scheme, similar to that of Facial Action Coding Scheme (FACS) is needed. A gesture annotation scheme, possibly named as Body Action Unit Coding System (BACS), should include information and description as follows: body part (e.g., left hand), direction (e.g., up/down), speed (e.g., fast/slow), shape (clenching fists), space (flexible/direct), weight (light/strong), time (sustained/quick), and flow (fluent/controlled) as defined by Laban and Ullman [61]. Additionally, temporal segments (neutral-start of gesture stroke-peak of gesture stroke-end of gesture stroke-neutral) of the gestures should be included as part of the annotation scheme. Overall, the most time-costly aspect of current gesture manual annotation is to obtain the onset-apex-offset time markers. This information is crucial for coordinating facial/body activity with simultaneous changes in physiology or speech [62].

14.3 CREATING A DATABASE OF FACIAL AND BODILY EXPRESSIONS: THE FABO DATABASE

The Bimodal Face and Body Gesture Database (the FABO database, henceforth) was created with the aim of using body as an additional channel, together with face, for affect analysis and recognition. The goal was to study how affect can be expressed, and consequently analyzed, when using both the facial and the bodily expression channels simultaneously. Details on the recordings and data annotation are described in the following sections.

Recordings. We recorded the video sequences simultaneously using two fixed cameras with a simple setup and uniform background. One camera was placed to specifically capture the face alone and the second camera was placed in order to capture face-and-body movement from the waist above. Prior to recordings subjects were instructed to take a neutral position, facing the camera and looking straight to it with hands visible and placed on the table. The subjects were asked to perform face and body gestures simultaneously by looking at the facial camera constantly. The recordings were obtained by using a *scenario approach* that was also used in previous emotion research [63]. In this approach, subjects are provided with situation vignettes or short scenarios describing an emotion-eliciting situation. They are instructed to imagine these situations and act out as if they were in such a situation. In our case the subjects were asked what they would do when "it was just announced that they won the biggest prize in lottery" or "the lecture is the most boring one and they can't listen to it anymore," etc. More specifically, although the FABO database was created in laboratory settings, the subjects were not instructed on emotion/case basis as to how to move their facial features and how to exactly display the specific facial expression. In some cases the subjects came up with a variety of combinations of face and body gestures. As a result of the feedback and suggestions obtained from the subjects, the number and combination of face and body gestures performed by each subject varies. A comprehensive list is provided in Table 14.1. The FABO database contains around 1900 gesture sequences from 23 subjects in age from 18 to 50 years. Figure 14.2 shows example images of non-basic facial expressions and their corresponding body gestures for neutral, negative surprise, positive surprise, boredom, uncertainty, anxiety, and puzzlement. Further details on the FABO database recordings can be found in Reference 60.

Annotation. We obtained the annotations for face and body videos separately, by asking human observers to view and label the videos. The purpose of this annotation was to obtain independent interpretations of the displayed face and body expressions and evaluate the performance (i.e., how well the subjects were displaying the affect they intended to communicate using their face and bodily gesture) by a number of human observers from different ethnic and cultural background. To this aim, we developed a survey for face and body videos separately, using the labeling schemes for affective content (e.g., happiness) and signs (e.g., how contracted the body is) by asking six independent human observers. We used two main labeling schemes in line with the psychological literature on descriptors of emotion: (a) verbal categorical labeling (perceptually determined, i.e., happiness) in accordance with Ekman's theory of emotion universality [64] and (b) broad dimensional labeling: arousal/activation (arousal–sleep/activated–deactivated) in accordance with Russell's theory of arousal and valence [65]. The participants were first shown the whole set of facial videos and only after finishing with the face they were shown the corresponding body videos. For each video they were asked to choose one label only, from the list provided: sadness, puzzlement/thinking, uncertainty/"I don't know," boredom, neutral surprise, positive surprise, negative surprise, anxiety, anger, disgust, fear, and happiness. For the temporal segment annotation, one human coder repeatedly viewed each face and body sequence, in slowed and stopped motion, to determine when (in which frame) the neutral–onset–apex–offset–neutral phases start and end [66]. Further details on the FABO data annotation can be found in Reference 49.

TABLE 14.1 List of the affective face and upper-body gestures performed for the recordings of FABO database

Expression	Face gesture	Body gesture
Neutral	Lips closed, eyes open, muscles relaxed	Hands on the table, relaxed
Uncertainty and puzzlement	Lip suck, lid droop, eyes closed, eyes turn right/left/up/down	Head tilt left/right/up/down, shoulder shrug, palms up, palms up + shoulder shrug, right/left hand scratching the head/hair, right/left hand touching the right/left ear, right/left hand touching the nose, right/left hand touching the chin, right/left hand touching the neck, right/left hand touching the forehead, both hands touching the forehead, right/left hand below the chin, elbow on the table, two hands behind the head
Anger	Brows lowered and drawn together; lines appear between brows; lower lid tense/may be raised; upper lid tense/may be lowered due to brows' action; lips are pressed together with corners straight or down or open; nostrils may be dilated	Open/expanded body; hands on hips/waist; closed hands/clenched fists; palm-down gesture; lift the right/left hand up; finger point with right/left hand; shake the finger/hand; crossing the arms
Surprise	Brows raised; skin below brow stretched not wrinkled;horizontal wrinkles across forehead; eyelids opened; jaw drops open or stretching of the mouth	Right/left hand moving toward the head; both hands moving toward the head; moving the right/left hand up; two hands touching the head; two hands touching the face/mouth; both hands over the head; right/left hand touching the face/mouth; self-touch/two hands covering the cheeks; self-touch/two hands covering the mouth; head shake; body shift/backing
Fear	Brows raised and drawn together; forehead wrinkles drawn to the center; upper eyelid is raised and lower eyelid is drawn up; mouth is open; lips are slightly tense or stretched and drawn back	Body contracted; closed body/closed hands/clenched fist; body contracted; arms around the body; self-touch (disbelief)/covering the body parts/arms around the body/shoulders; body shift-backing; hand covering the head; body shift-backing; hand covering the neck; body shift-backing; hands covering the face; both hands over the head; self-touch (disbelief) covering the face with hands

(continued)

TABLE 14.1 (*Continued*)

Expression	Face gesture	Body gesture
Anxiety	Lip suck; lip bite; lid droop; eyes closed; eyes turn right/left/up/down	Hands pressed together in a moving sequence; tapping the tips of the fingers on the table; biting the nails; head tilt left/right/up/down
Happiness	Corners of lips are drawn back and up; mouth may or may not be parted with teeth exposed or not; cheeks are raised; lower eyelid shows wrinkles below it; and may be raised but not tense; wrinkles around the outer corners of the eyes	Body extended; hands clapping; arms lifted up or away from the body with hands made into fists
Disgust	Upper lip is raised; lower lip is raised and pushed up to upper lip or it is lowered; nose is wrinkled; cheeks are raised; brows are lowered	Hands close to the body; body shift-backing; orientation changed/moving to the right or left; backing; hands covering the head; backing; hands covering the neck; backing; right/left hand on the mouth; backing; move right/left hand up
Bored	Lid droop, eyes closed, lip suck, eyes turn right/left/up/down	Body shift; change orientation; move to the right/left; hands behind the head; body shifted; hands below the chin, elbow on the table
Sadness	Inner corners of eyebrows are drawn up; upper lid inner corner is raised; corners of the lips are drawn downwards	Contracted/closed body; dropped shoulders; bowed head; body shift-forward leaning trunk; covering the face with two hands; self-touch (disbelief)/covering the body parts/arms around the body/shoulders; body extended+hands over the head; hands kept lower than their normal position, hands closed, slow motion; two hands touching the head move slowly; one hand touching the neck, hands together closed, head to the right, slow motion.

14.4 AUTOMATIC RECOGNITION OF AFFECT FROM BODILY EXPRESSIONS

14.4.1 Body as an Autonomous Channel for Affect Analysis

In this section, we first investigate affective body gesture analysis in video sequences by approaching the body as an autonomous channel. To this aim, we exploit spatial–temporal features [67], which makes few assumptions about the observed data, such as background, occlusion, and appearance.

14.4.1.1 Spatial–Temporal Features In recent years, spatial–temporal features have been used for event detection and behavior recognition in videos. We extract spatial–temporal features by detecting space-time interest points [67]. We calculate the response function by application of separable linear filters. Assuming a stationary camera or a process that can account for camera motion, the response function has the form

$$R = (I * g * h_{\mathrm{ev}})^2 + (I * g * h_{\mathrm{od}})^2 \tag{14.1}$$

where $I(x, y, t)$ denotes images in the video, $g(x, y; \sigma)$ is the 2D Gaussian smoothing kernel, applied only along the spatial dimensions (x, y), and h_{ev} and h_{od} are a quadrature pair of 1D Gabor filters applied temporally, which are defined as $h_{\mathrm{ev}}(t; \tau, \omega) = -\cos(2\pi t\omega)\mathrm{e}^{-t^2/\tau^2}$ and $h_{\mathrm{od}}(t; \tau, \omega) = -\sin(2\pi t\omega)\mathrm{e}^{-t^2/\tau^2}$. In all cases we use $\omega = 4/\tau$ [67]. The two parameters σ and τ correspond roughly to the spatial and temporal scales of the detector. Each interest point is extracted as a local maxima of the response function. As pointed out in Reference 67, any region with spatially distinguishing characteristics undergoing a complex motion can induce a strong response, while region undergoing pure translational motion, or areas without spatially distinguishing features, will not induce a strong response.

At each detected interest point, a cuboid is extracted which contains the spatio-temporally windowed pixel values. See Figure 14.3 for examples of cuboids extracted. The side length of cuboids is set as approximately six times the scales along each dimension, so containing most of the volume of data that contribute to the response function at each interest point. After extracting the cuboids, the original video is discarded, which is represented as a collection of the cuboids. To compare two cuboids, different descriptors for cuboids have been evaluated in Reference 67, including normalized pixel values, brightness gradient and windowed optical flow, followed by a conversion into a vector by flattening, global histogramming, and local histogramming. As suggested, we adopt the flattened brightness gradient as the cuboid descriptor. To reduce the dimensionality, the descriptor is projected to a lower dimensional PCA space [67]. By clustering a large number of cuboids extracted from the training data using the K-Means algorithm, we derive a library of cuboid prototypes. So each cuboid is assigned a type by mapping it to the closest prototype vector. Following Reference 67, we use the histogram of the cuboid types to describe the video.

14.4.1.2 Classifier We adopt the support vector machine (SVM) classifier to recognize affective body gestures. SVM is an optimal discriminant method based on the Bayesian learning theory. For the cases where it is difficult to estimate the density model in high-dimensional space, the discriminant approach is preferable to the generative approach. SVM performs an implicit mapping of data into a higher dimensional feature space and then finds a linear separating hyperplane with the maximal margin to separate data in this higher dimensional space. SVM allows domain-specific

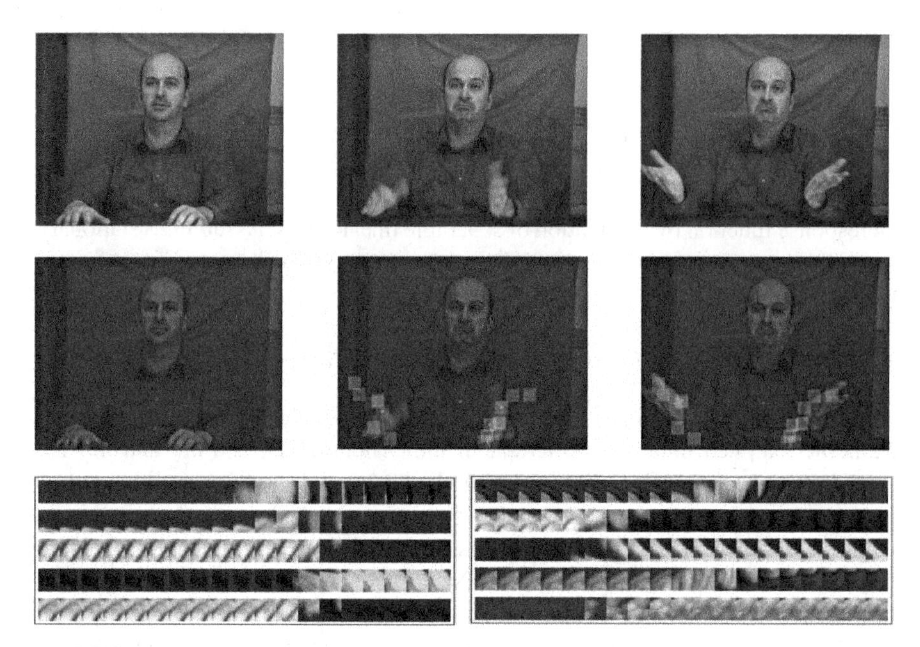

FIGURE 14.3 (Best viewed in color) Examples of spatial–temporal features extracted from videos. The first row is the original input video. Taken with permission from the FABO database. The second row visualizes the cuboids extracted, where each cuboid is labeled with a different color; the third row shows some cuboids, which are flattened with respect to time. Color version of the figure is available in the internet edition.

selection of the kernel function, and the most commonly used kernel functions are the linear, polynomial, and radial basis function (RBF) kernels.

14.4.2 Body as an Additional Channel for Affect Analysis

In this section, we investigate how body contributes to the affect analysis when used as an additional channel. For combining the facial and bodily cues, we exploit CCA, a powerful statistical tool that is well suited for relating two sets of signals, to fuse facial expression and body gesture at the feature level. CCA derives a semantic "affect" space, in which the face and body features are compatible and can be effectively fused.

We propose to fuse the cues from the two channels in a joint feature space, rather than at the decision level. The main difficulties for the feature-level fusion are the features from different modalities may be incompatible, and the relationship between different feature spaces is unknown. Here we fuse face and body cues at the feature level using CCA. Our motivation is that, as face and body cues are two sets of measurements for affective states, conceptually the two modalities are correlated, and their relationship can be established using CCA.

14.4.2.1 Canonical Correlation Analysis CCA [68] is a statistical technique developed for measuring linear relationships between two multidimensional variables. It finds pairs of base vectors (i.e., canonical factors) for two variables such that the correlations between the projections of the variables onto these canonical factors are mutually maximized.

Given two zero-mean random variables $\mathbf{x} \in R^m$ and $\mathbf{y} \in R^n$, CCA finds pairs of directions $\mathbf{w}_x$ and $\mathbf{w}_y$ that maximize the correlation between the projections $x = \mathbf{w}_x^T\mathbf{x}$ and $y = \mathbf{w}_y^T\mathbf{y}$. The projections x and y are called *canonical variates*. More formally, CCA maximizes the function

$$\rho = \frac{E[xy]}{\sqrt{E[x^2]E[y^2]}} = \frac{E[\mathbf{w}_x^T\mathbf{x}\mathbf{y}^T\mathbf{w}_y]}{\sqrt{E[\mathbf{w}_x^T\mathbf{x}\mathbf{x}^T\mathbf{w}_x]E[\mathbf{w}_y^T\mathbf{y}\mathbf{y}^T\mathbf{w}_y]}} = \frac{\mathbf{w}_x^T\mathbf{C}_{xy}\mathbf{w}_y}{\sqrt{\mathbf{w}_x^T\mathbf{C}_{xx}\mathbf{w}_x\mathbf{w}_y^T\mathbf{C}_{yy}\mathbf{w}_y}} \tag{14.2}$$

where $\mathbf{C}_{xx} \in R^{m\times m}$ and $\mathbf{C}_{yy} \in R^{n\times n}$ are the *within-set covariance matrices* of $\mathbf{x}$ and $\mathbf{y}$, respectively, while $\mathbf{C}_{xy} \in R^{m\times n}$ denotes their *between-sets covariance matrix*. A number of at most $k = \min(m, n)$ canonical factor pairs $\langle \mathbf{w}_x^i, \mathbf{w}_y^i \rangle, i = 1, \ldots, k$ can be obtained by successively solving $\arg\max_{\mathbf{w}_x^i,\mathbf{w}_y^i}\{\rho\}$ subject to $\rho(\mathbf{w}_x^j, \mathbf{w}_x^i) = \rho(\mathbf{w}_y^j, \mathbf{w}_y^i) = 0$ for $j = 1, \ldots, i-1$, that is, the next pair of $\langle \mathbf{w}_x, \mathbf{w}_y \rangle$ are orthogonal to the previous ones.

The maximization problem can be solved by setting the derivatives of Equation. 14.2, with respect to $\mathbf{w}_x$ and $\mathbf{w}_y$, equal to zero, resulting in the eigenvalue equations as

$$\begin{cases} \mathbf{C}_{xx}^{-1}\mathbf{C}_{xy}\mathbf{C}_{yy}^{-1}\mathbf{C}_{yx}\mathbf{w}_x = \rho^2\mathbf{w}_x \\ \mathbf{C}_{yy}^{-1}\mathbf{C}_{yx}\mathbf{C}_{xx}^{-1}\mathbf{C}_{xy}\mathbf{w}_y = \rho^2\mathbf{w}_y \end{cases} \tag{14.3}$$

Matrix inversions need to be performed in Equation 14.3, leading to numerical instability if $\mathbf{C}_{xx}$ and $\mathbf{C}_{yy}$ are rank deficient. Alternatively, $\mathbf{w}_x$ and $\mathbf{w}_y$ can be obtained by computing principal angles, as CCA is the statistical interpretation of principal angles between two linear subspaces.

14.4.2.2 Feature Fusion of Facial and Bodily Expression Cues Given $B = \{\mathbf{x}|\mathbf{x} \in R^m\}$ and $F = \{\mathbf{y}|\mathbf{y} \in R^n\}$, where $\mathbf{x}$ and $\mathbf{y}$ are the feature vectors extracted from bodies and faces, respectively, we apply CCA to establish the relationship between $\mathbf{x}$ and $\mathbf{y}$. Suppose $\langle \mathbf{w}_x^i, \mathbf{w}_y^i \rangle, i = 1, \ldots, k$ are the canonical factors pairs obtained, we can use d $(1 \le d \le k)$ factor pairs to represent the correlation information. With $\mathbf{W}_x = [\mathbf{w}_x^1, \ldots, \mathbf{w}_x^d]$ and $\mathbf{W}_y = [\mathbf{w}_y^1, \ldots, \mathbf{w}_y^d]$, we project the original feature vectors as $\mathbf{x}' = \mathbf{W}_x^T\mathbf{x} = [x_1, \ldots, x_d]^T$ and $\mathbf{y}' = \mathbf{W}_y^T\mathbf{y} = [y_1, \ldots, y_d]^T$ in the lower dimensional correlation space, where x_i and y_i are uncorrelated with the previous pairs x_j and $y_j, j = 1, \ldots, i-1$. We then combine the projected feature vectors $\mathbf{x}'$ and $\mathbf{y}'$ to form the new feature vector as

$$\mathbf{z} = \begin{pmatrix} \mathbf{x}' \\ \mathbf{y}' \end{pmatrix} = \begin{pmatrix} \mathbf{W}_x^T\mathbf{x} \\ \mathbf{W}_y^T\mathbf{y} \end{pmatrix} = \begin{pmatrix} \mathbf{W}_x & 0 \\ 0 & \mathbf{W}_y \end{pmatrix}^T \begin{pmatrix} \mathbf{x} \\ \mathbf{y} \end{pmatrix} \tag{14.4}$$

This fused feature vector effectively represents the multimodal information in a joint feature space for affect analysis.

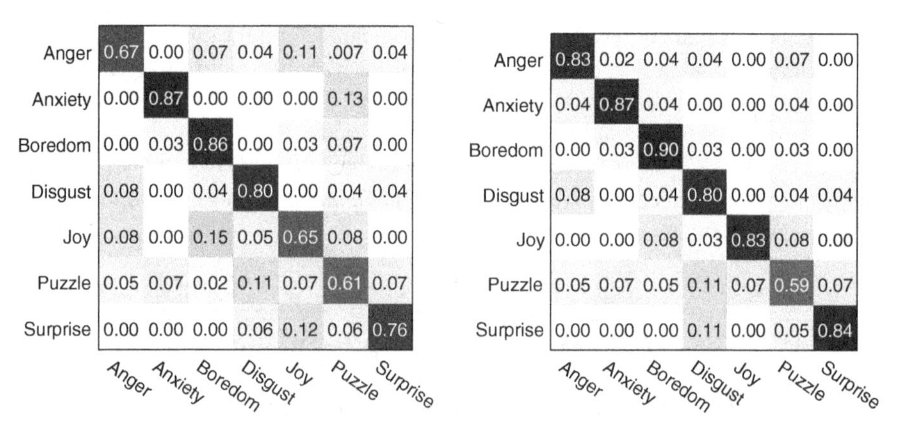

FIGURE 14.4 Confusion matrices for affect recognition from bodily gestures (*left*) and facial expressions (*left*).

14.4.2.3 Experiments and Results In our experiments we used the FABO database [60]. We selected 262 videos of seven emotions (Anger, Anxiety, Boredom, Disgust, Joy, Puzzle, and Surprise) from 23 subjects. To evaluate the algorithms' generalization ability, we adopted a fivefold cross-validation test scheme in all recognition experiments. That is, we divided the data set randomly into five groups with roughly equal number of videos and then used the data from four groups for training, and the left group for testing; the process was repeated five times for each group in turn to be tested. We report the average recognition rates here. In all experiments, we set the soft margin C value of SVMs to infinity so that no training error was allowed. Meanwhile, each training and testing vector was scaled to be between −1 and 1. In our experiments, the RBF kernel always provided the best performance, so we report the performance of the RBF kernel. With regard to the hyper-parameter selection of RBF kernels, as suggested in Reference 69, we carried out grid-search on the kernel parameters in the fivefold cross-validation. The parameter setting producing the best cross-validation accuracy was picked. We used the SVM implementation in the publicly available machine learning library SPIDER[1] in our experiments. To see how the body contributes to the affect analysis when used as an additional channel, we extracted the spatial–temporal features from the face video and the body video and then fused the cues from the two channels at the feature level using CCA.

We first report the classification performance (the confusion matrix) based on bodily cues only in Figure 14.4 (left). The average recognition rate of the SVM classifier using the bodily cues is 72.6%. When we look at the affect recognition using the facial cues only, the recognition rate obtained is 79.2%. Looking at the confusion matrix shown in Figure 14.4, we observe that the emotion classification based on facial expressions is better than that of bodily gesture. This is possibly because there are much variation in affective body gestures.

We then fused facial expression and body gesture at the feature level using CCA. Different numbers of CCA factor pairs can be used to project the original face and

[1]http://kyb.tuebingen.mpg.de/bs/people/spider/index.html

TABLE 14.2 Experimental results of affect recognition by fusing body and face cues

Feature fusion	CCA	Direct	PCA	PCA + LDA
Recognition rate	88.5%	81.9%	82.3%	87.8%

body feature vectors to a lower dimensional CCA feature space, and the recognition performance varies with the dimensionality of the projected CCA features. We report the best result obtained here. We compared the CCA feature fusion with another three feature fusion methods: (1) Direct feature fusion, that is, concatenating the original body and face features to derive a single feature vector. (2) PCA feature fusion: the original body and face features are first projected to the PCA space respectively, and then the PCA features are concatenated to form the single feature vector. In our experiments, all principle components were kept. (3) PCA+LDA feature fusion: for each modality, the derived PCA features are further projected to the discriminant LDA space; the LDA features are then combined to derive the single feature vector. We report the experimental results of different feature fusion schemes in Table 14.2. The confusion matrices of the CCA feature fusion and the direct feature fusion are shown in Figure 14.5. We can see that the presented CCA feature fusion provides best recognition performance. This is because CCA captures the relationship between the feature sets in different modalities, and the fused CCA features effectively represent information from each modality.

14.5 AUTOMATIC RECOGNITION OF BODILY EXPRESSION TEMPORAL DYNAMICS

Works focusing on the detection of the expression temporal segments modeled temporal dynamics of facial or bodily expressions by extracting and tracking geometric or appearance features from a set of fixed interest points (e.g., [40, 70]). However,

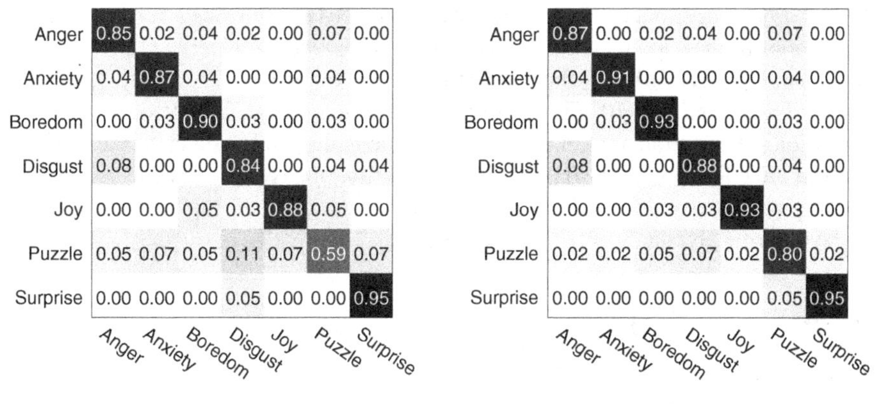

FIGURE 14.5 Confusion matrices of affect recognition by fusing facial expression and body gesture. (*Left*) Direct feature fusion; (*right*) CCA feature fusion. Taken with permission from the FABO database.

such approaches have two limitations. First, the selection of the fixed interest points requires human expertise and mostly needs human intervention. Second, tracking is usually sensitive to occlusions and illumination variations (e.g., the facial point tracking will fail when the hands touch the face). Inaccuracy in tracking will significantly degrade the temporal segmentation performance. To mitigate the aforementioned issues, we propose two types of novel and efficient features in this section, that is, motion area and neutral divergence, to simultaneously segment and recognize temporal phases of an (facial/bodily) expression. The motion area feature is calculated by simple motion history image (MHI) [71, 72], which does not rely on any facial points tracking or body tracking, and the neutral divergence feature is based on the differences between the current frame and the neutral frame.

14.5.1 Feature Extraction

The motion area and the neutral divergence features are extracted from both facial and body gesture information without any motion tracking, so the approach avoids losing informative apex frames due to the unsynchronized face and body gesture temporal phases. Furthermore, both features are efficient to compute.

14.5.1.1 Motion Area We extract the motion area based on the MHI, which is a compact representation of a sequence of motion movement in a video [71, 72]. Pixel intensity of MHI is a function of the motion history at that location, where brighter values correspond to more recent motions. The intensity at pixel (x, y) decays gradually until a specified motion history duration t and the MHI image can be constructed using the equation

$$\begin{aligned} \mathrm{MHI}_\tau(x,y,t) &= D(x,y,t) * \tau + [1 - D(x,y,t)] * U[\mathrm{MHI}_\tau(x,y,t-1) - 1] \\ &\quad * [\mathrm{MHI}_\tau(x,y,t-1) - 1]) \end{aligned} \tag{14.5}$$

where $U[x]$ is a unit step function and t represents the current video frame index. $D(x, y, t)$ is a binary image of pixel intensity difference between the current frame and the previous frame. $D(x, y, t) = 1$ if the intensity difference is greater than a threshold, otherwise, $D(x, y, t) = 0$. τ is the maximum motion duration. In our system, we set $threshold = 25$ and $\tau = 10$. Figure 14.6b shows the generated MHI of a *surprise* expression. The motion area of each video frame is the total number of the motion pixels in the corresponding MHI image. The motion pixels are defined as the pixels with non-zero intensity in the MHI image. The calculation of the motion area $\mathrm{MA}_\tau(t)$ can be described by the following equation:

$$\mathrm{MA}_\tau(t) = \sum_{x=1}^{W} \sum_{y=1}^{H} U[\mathrm{MHI}_\tau(x,y,t) - e]) \tag{14.6}$$

where $0 < e < 1$, $U[x]$ is a unit step function, and W and H are the width and the height of the MHI image.

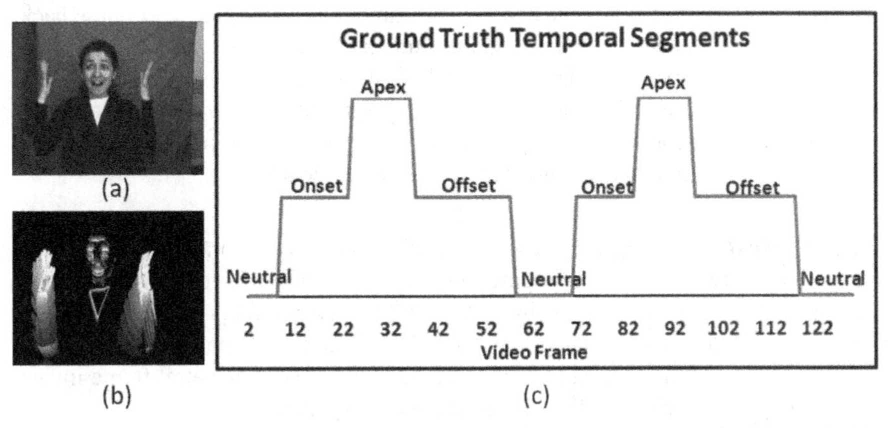

FIGURE 14.6 (a) A *surprise* expression, (b) MHI of the *surprise* expression shown in (a), and (c) ground truth temporal segments of the expression. Part (a) is taken with permission from the FABO database.

Figure 14.7 (left) illustrates how the (normalized) motion area of the *surprise* expression (shown in Figure 14.6) is obtained. The expression starts from the neutral (frames 0 – 10, hands on desk) followed by the onset (frames 11 – 24, hands move up), the apex, the offset and back to the neutral. As shown in Figure 14.7 (left), the motion area $MA_\tau(t)$ is almost 0 at neutral phase and increases and finally reaches the peak at frame 15. As the expression approaches its *apex*, the motion begins to slow down, which causes $MA_\tau(t)$ to decrease between frame 15 to frame 24. The *apex* occurs between frames 25 and 34 in Figure 14.7 (left). During the *apex* phase, the expression reaches its maximum spatial extent and lasts for some time. Hence, there is relatively small (or no motion) during that phase. During the *offset* phase,

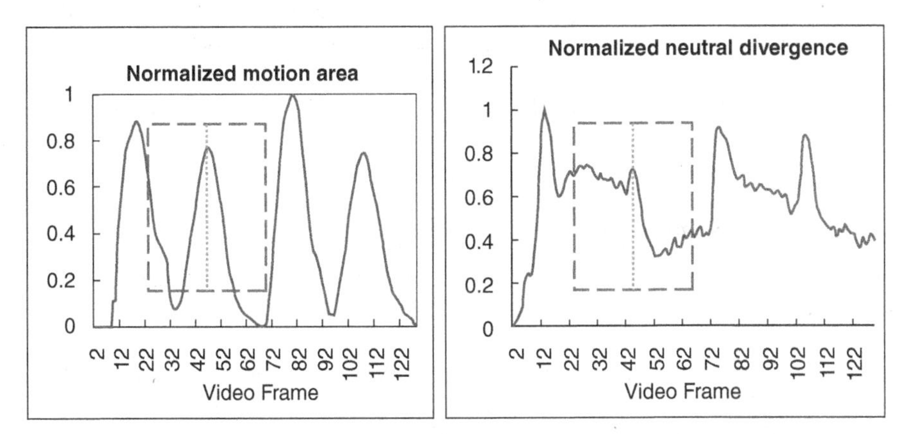

FIGURE 14.7 The motion area feature representation of the current frame is a vector of normalized motion area (*left*), and the neutral divergence feature representation of the current frame is a vector of normalized neutral divergence (*right*).

both the facial expression and the body gesture are moving from the *apex* phase back to the *neutral* phase. This is illustrated in Figure 14.7 (left) between frames 35 and 54. Finally, the expression enters its *neutral* phase between frames 55 and 70 (with very small motion area). The motion area is further normalized to the range of [0, 1] with maximum motion area corresponding to 1. The normalization is done in order to handle variation due to different expressions or subjects.

14.5.1.2 Neutral Divergence The neutral divergence feature measures the degree of difference between the current frame and the neutral frame of an expressive display. Since all videos in the FABO database [60] start from a neutral position, the current frame's neutral divergence ND(t) is calculated by summing up the absolute intensity difference between the current frame image $I(x, y, d, t)$ and the neutral frame image $I(x, y, d, t0)$ over three color channels, as shown in Equation 14.7:

$$\mathrm{ND}(t) = \sum_{d=1}^{3} \sum_{x=1}^{W} \sum_{y=1}^{H} \mathrm{abs}[I(x, y, d, t) - I(x, y, d, t0)] \tag{14.7}$$

where d is the number of color channels of the frame.

Figure 14.7 (right) plots the (normalized) neutral divergences of the *surprise* expression shown in Figure 14.6. Similar to the motion area normalization, the neutral divergence is also normalized to the range of [0, 1]. The neutral divergence is 0 at the *neutral* phase. During the *onset* phase, the neutral divergence increases (as can be observed in Figure 14.7 (right)). During the *apex* phase, the neutral divergence remains relatively stable (with a large neutral divergence value) as there is little movement in the facial expression or the body gesture. However, the *apex* phase is quite different from the *neutral* phase. The neutral divergence decreases at the *offset* phase. When the expression enters its *neutral* phase again, between frames 55 and 70, as shown in Figure 14.7 (right), the neutral divergence does not go back to 0 as would be expected. This indicates that the facial and bodily parts do not return back to their exact starting position. Nevertheless, the difference between the final *neutral* phase and the *apex* phase using the neutral divergence feature is still recognizable (see Figure 14.7 (right)).

14.5.2 Feature Representation and Combination

14.5.2.1 Feature Representation The normalized motion area and the neutral divergence are extracted for every frame in an expression video. To recognize the expression phases of the current frame, we employ a fixed-size temporal window with the center located at the current frame as shown in Figure 14.7a (left). The normalized motion area of every frame within the temporal window is extracted (forming a vector in chronological order). Similar to the motion area feature, as shown in Figure 14.7b (right), the normalized neutral divergence of every frame within the temporal window is also extracted (forming a vector in chronological order). In our experiments, we set the temporal window size to 31, for both the motion area features and the neutral divergence features.

14.5.2.2 Feature Combination The motion area and the neutral divergence features provide complementary information regarding temporal dynamics of an expression. The motion area is able to separate the onset/offset from the apex/neutral phases, since the onset/offset generates large movements. However, the motion area can neither distinguish the *apex* phase from the *neutral* nor the *onset* phase from the *offset.* Nevertheless, the *apex* phase has large intensity deviation from the initial neutral frame. Therefore, the neutral divergence is able to separate the *neutral* phase from the *apex* phase. During the *onset* phase, the neutral divergence is increasing (the opposite occurs during the *offset* phase). Consequently, the neutral divergence is able to separate the *onset* phase from the *offset* phase as well. The combination of both features is obtained by simply concatenating the motion area feature vector with the neutral divergence feature vector.

14.5.2.3 Classifier We employ SVM with an RBF kernel as our multiclass classifier [73]. SVM is used to find a set of hyper-planes which separate each pair of classes with a maximum margin. The temporal segmentation of an expression phase can be considered as a multiclass classification problem. In other words each frame is classified into neutral, onset, apex, and offset temporal phases.

14.5.3 Experiments

14.5.3.1 Experimental Setup We conducted experiments using the FABO database [60]. We chose 288 videos where the ground truth expressions from both the face camera and the body camera were identical. We used 10 expression categories, including both basic expressions (disgust, fear, happiness, surprise, sadness, and anger) and non-basic expressions (anxiety, boredom, puzzlement, and uncertainty). For each video, there are two to four complete expression cycles. Videos of each expression category are randomly separated into three subsets. Then two of these subsets are chosen for training and the remaining subset is kept for testing. Due to the random separation process, the subjects may overlap between the training and the testing sets.

14.5.3.2 Experimental Results We first perform a threefold cross-validation by combining the motion area and the neutral divergence features. Two subsets are used for training, and the remaining subset is used for testing. The procedure is repeated three times, each of the three subsets being used as the testing data exactly once. The accuracy is calculated by averaging the true positive rate of each class (i.e., the neutral, the onset, the apex, and the offset). The average accuracy obtained by the threefold cross validation is 83.1%.

Figure 14.8 shows the temporal segmentation results of the *surprise* expression video shown in Figure 14.6. The ground truth temporal phase of each frame in the expression video is indicated by the solid line, while the corresponding predicted

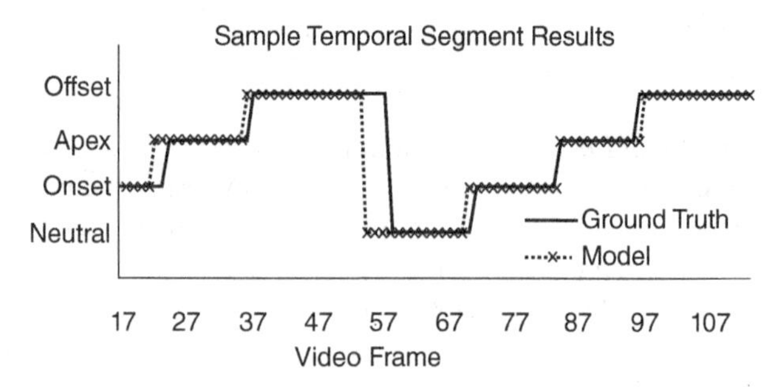

FIGURE 14.8 Temporal segmentation results corresponding to the *surprise* expression video shown in Figure 14.6.

temporal phase is plotted using the dash line with x. The predicted temporal segmentation of the expression video matches the ground truth temporal phase quite well (except at the phase transition frames). For example, frames 22 and 23 are predicted as the apex phase while the ground truth indicates that they are the onset frames right before the apex.

Table 14.3 shows the confusion matrix resulting from the temporal phase segmentation. Each row is the ground truth temporal segment while the columns are the classified temporal segments. Based on the confusion matrix, both the *onset* and the *offset* phase appear to be confused mostly with the *apex* phase. The *apex* phase is temporally adjacent to both the *onset* phase and the *offset* phase. This is mainly due to the fact that the temporal boundary between these phases is not straightforward. However, as shown in the last column of Table 14.3, the overall performance of each temporal phase is fairly stable.

We also conducted an experiment in order to evaluate the effectiveness of the combined feature set (combining the motion area and the neutral divergence). This experiment uses the first subset of expression videos as the testing data and the other two subsets as the training data. Using the motion area alone, the temporal phase detection rate is 68.5%. The neutral divergence feature alone achieves 74.1% detection rate. By combining both the motion area and the neutral divergence, the expression phase segmentation performance has boosted up to 82%. In order to understand why the combined feature set significantly improves the performance, we compare the confusion matrices obtained from the motion area (alone) and the combined feature set. Table 14.4 (top) reports the confusion matrix using the motion area feature alone.

TABLE 14.3 Summary of the threefold cross validation results

True/model	Neutral	Onset	Apex	Offset	Accuracy (%)
Neutral	2631	121	208	161	84.3
Onset	113	2253	324	31	82.8
Apex	187	282	4365	251	85.8
Offset	171	70	227	2539	84.4

TABLE 14.4 Confusion matrices using motion area feature alone (top), and using the combined feature set (bottom)

True/model	Neutral	Onset	Apex	Offset
Neutral	1739	75	757	125
Onset	73	1745	288	429
Apex	691	222	3079	225
Offset	102	472	278	1792
Neutral	2213	107	226	150
Onset	106	2037	246	146
Apex	261	227	3553	176
Offset	150	104	245	2145

Rows indicate the ground truth temporal phases while columns indicate the recognized temporal phases.

From the matrix, we can see that the apex frames are mostly confused with the neutral frames. As an example, there are 757 neutral frames misclassified as the *apex* phase, while there are 691 *apex* frames misclassified as the *neutral* phase. Similarly, the *onset* phase is mostly confused with the *offset* phase. Therefore, we conclude that the motion area can neither distinguish the *apex* phase from the *neutral*, nor the *onset* phase from the *offset*.

As can be observed in Table 14.4 (bottom), combining the motion area and the neutral divergence features reduces the confusion between the *neutral* phase and the *apex* phase, significantly. For instance, there are only 226 neutral frames misclassified as apex, and 261 apex frames misclassified as neutral. Similarly, the confusion between the *onset* phase and the *offset* phase is also reduced. These comparisons confirm the effectiveness of combining both the motion area and the neutral divergence features on the temporal segmentation. The neutral divergence and the motion area provide complementary information for identifying the temporal dynamics of an expression.

14.6 DISCUSSION AND OUTLOOK

Human affect analysis based on bodily expressions is still in its infancy. Therefore, for the interested reader we would like to provide a number of pointers for future research as follows.

Representation-related issues. According to research in psychology, three major approaches to affect modeling can be distinguished [74]: *categorical*, *dimensional*, and *appraisal-based* approach. The categorical approach claims that there exist a small number of emotions that are basic, hard wired in our brain and recognized universally (e.g., [7]). This theory has been the most commonly adopted approach in research on automatic measurement of human affect from bodily expressions. However, a number of researchers claim that a small number of discrete classes may not reflect the complexity of the affective state conveyed [65]. They advocate the use of *dimensional description* of human affect, where affective states are not independent from one another; rather, they are related to one another in a systematic manner (e.g.,

[65, 74–76]). The most widely used dimensional model is a circular configuration called *Circumplex of Affect* introduced by Russell [65]. This model is based on the hypothesis that each basic emotion represents a bipolar entity being a part of the same emotional continuum. The proposed polars are arousal (relaxed vs. aroused) and valence (pleasant vs. unpleasant). Another well-accepted and commonly used dimensional description is the 3D emotional space of pleasure–displeasure, arousal–nonarousal and dominance–submissiveness [75], at times referred to as the *PAD emotion space*. Scherer and colleagues introduced another set of psychological models, referred to as *componential models* of emotion, which are based on the appraisal theory [74, 76, 77]. In the appraisal-based approach, emotions are generated through continuous, recursive subjective evaluation of both our own internal state and the state of the outside world (relevant concerns/needs) [74, 76–78]. Although pioneering efforts have been introduced by Scherer and colleagues (e.g., [79]), how to use the appraisal-based approach for automatic measurement of affect is an open research question as this approach requires complex, multicomponential, and sophisticated measurements of change. Overall, despite the existence of such diverse affect models, there is still not an agreement between researchers on which model should be used for which affect measurement task, and for each modality or cue.

Context. Context usually refers to the knowledge of who the subject is, where she is, what her current task is, and when the observed behavior has been shown. Majority of the works on automated affect analysis from bodily expressions focused on context-free, acted, and emotional expressions (e.g., [40, 45, 80]). More recently, a number of works started exploring automatic analysis of bodily postures in an application-dependent and context-specific manner in non-acted scenarios. Examples include recognizing affect when the user is playing a body-movement-based video game [81] and detecting the level of engagement when the user is interacting with a game companion [82]. Defining and setting up a specific context enables designing automatic systems that are realistic and are sensitive to a specific target user group and target application. Defining a context potentially simplifies the problem of automatic analysis and recognition as the setup chosen may encourage the user to be in a controlled position (e.g., sitting in front of a monitor or standing in a predefined area), wearing specific clothes (e.g., wearing bright-colored t-shirts [82] or a motion capture suit [81]), etc. Overall, however, how to best incorporate and model context for affect recognition from bodily expressions needs to be explored further.

Data acquisition protocol. Defining protocols on how to acquire benchmark data for affective bodily posture and gesture analysis is an ongoing research topic. Currently it is difficult to state whether it is sufficient (or better) to have body-gesture-only databases (e.g., The UCLIC Affective Posture database) or whether it is better to record multiple cues and modalities simultaneously (e.g., recording face and upper body as was done for the FABO database and the GEMEP Corpus). Overall, data acquisition protocols and choices should be contextualized (i.e., by taking into account the application, the user, the task, etc.).

Modeling expression variation. Emotional interpretation of human body motion is based on understanding the action performed. This does not cause major issues when classifying stereotypical bodily expressions (e.g., clenched fists) in terms of emotional content (e.g., anger, sadness). However, when it comes to analyzing natural bodily expressions, the same emotional content (category) may be expressed with similar bodily movements but with some variations or with very different bodily movements. This presents major challenges to the machine learning techniques trained to detect and recognize the movement patterns specific to each emotion category. This, in turn, will hinder the discovery of underlying patterns due to emotional changes. To mitigate this problem, recent works have focused on using explicit models of action patterns to aid emotion classification (e.g., [83]).

Multiple cues/modalities and their dynamics. Although body has been investigated as an additional channel for affect analysis and recognition, it is still not clear what role it should play when combining multiple cues and modalities: Should it be given higher or lower weight? Can it be the primary (or only) cue/modality? In which context? When can gait be used as an additional modality for affect recognition? How does it relate to, or differ from other bodily expression recognition? These questions are likely to stir further investigations. Additionally, when dealing with multiple cues, it is highly likely that the temporal segments of various cues may not be aligned (synchronized) as illustrated in Figure 14.1c where the apex frames for the bodily expression constitute the onset segment for the facial expression. One noteworthy study that investigated fully the automatic coding of human behavior dynamics with respect to both the temporal segments (onset, apex, offset, and neutral) of various visual cues and the temporal correlation between different visual cues (facial, head, and shoulder movements) is that of Valstar *et al.* [54], who investigated separating posed from genuine smiles in video sequences. However, in practice, it is difficult to obtain accurate detection of the facial/bodily key points and track them robustly for temporal segment detection, due to illumination variations and occlusions (see examples in Figure 14.2). Overall, integration, temporal structures, and temporal correlations between different visual cues are virtually unexplored areas of research, ripe for further investigation.

14.7 CONCLUSIONS

This chapter focused on a relatively understudied problem: bodily expression for automatic affect recognition. The chapter explored how bodily expression analysis can aid affect recognition by describing three case studies: (1) data acquisition and annotation of the first publicly available database of affective face-and-body displays (i.e., the FABO database); (2) a representative approach for affective state recognition from face-and-body display by detecting the space-time interest points in video and using CCA for fusion, and (3) a representative approach for explicit detection of the temporal phases (segments) of affective states (start/end of the expression and its subdivision into phases such as neutral, onset, apex, and offset) from bodily expressions. The chapter concluded by summarizing the main challenges faced and discussing how we can advance the state of the art in the field.

Overall, human affect analysis based on bodily expressions is still in its infancy. However, there is a growing research interest driven by various advances and demands (e.g., real-time representation and analysis of naturalistic body motion for affect-sensitive games, interaction with humanoid robots). The current automatic measurement technology has already started moving its focus toward naturalistic settings and less-controlled environments, using various sensing devices, and exploring bodily expression either as an autonomous channel or as an additional channel for affect analysis. The bodily cues (postures and gestures) are much more varied than face gestures. There is an unlimited vocabulary of bodily postures and gestures with combinations of movements of various body parts. Despite the effort of Laban in analyzing and annotating body movement [61], unlike the facial expressions, communication of emotions by bodily movement and expressions is still a relatively unexplored and unresolved area in psychology, and further research is needed in order to obtain a better insight on how they contribute to the perception and recognition of the various affective states. This understanding is expected to pave the way for using the bodily expression to its full potential.

ACKNOWLEDGMENTS

The work of Shizhi Chen and YingLi Tian was partially developed under an appointment to the DHS Summer Research Team Program for Minority Serving Institutions, administered by the Oak Ridge Institute for Science and Education (ORISE) through an interagency agreement between the U.S. Department of Energy (DOE) and U.S. Department of Homeland Security (DHS). ORISE is managed by Oak Ridge Associated Universities (ORAU) under DOE contract number DE-AC05-06OR23100. It has not been formally reviewed by DHS. The views and conclusions contained in this document are those of the authors and should not be interpreted as necessarily representing the official policies, either expressed or implied, of DHS, DOE, or ORAU/ORISE. DHS, DOE, and ORAU/ORISE do not endorse any products or commercial services mentioned in this article.

REFERENCES

[1] EyeToy, http://en.wikipedia.org/wiki/eyetoy (last accessed May 8, 2011).

[2] Kinect, http://en.wikipedia.org/wiki/kinect (last accessed May 8, 2011).

[3] Xbox 360, http://en.wikipedia.org/wiki/xbox_360 (last accessed May 8, 2011).

[4] R. A. Calvo and S. D'Mello, "Affect detection: an interdisciplinary review of models, methods, and their applications," *IEEE Transactions on Affective Computing*, 1(1): 18–37, 2010.

[5] J. F. Cohn, K. Schmidt, R. Gross, and P. Ekman, "Individual differences in facial expression: stability over time, relation to self-reported emotion, and ability to inform person identification," in Proceedings of the IEEE International Conference on Multimodal Interfaces, 2002, pp. 491–496.

[6] P. Ekman and W. V. Friesen, *The Facial Action Coding System: A Technique for Measurement of Facial Movement*, Consulting Psychologists Press, San Francisco, CA, 1978.

[7] P. Ekman and W. V. Friesen, *Unmasking the Face: A Guide to Recognizing Emotions from Facial Clues*, Prentice Hall, New Jersey, 1975.

[8] P. Ekman and E. L. Rosenberg, *What the Face Reveals: Basic and Applied Studies of Spontaneous Expression Using the Facial Action Coding System (FACS)*, Oxford University Press, New York, 1997.

[9] M. Pantic and M. S. Bartlett, "Machine analysis of facial expressions," in *Face Recognition* (eds K. Delac and M. Grgic), I-Tech Education and Publishing, Vienna, Austria, 2007, pp. 377–416.

[10] M. Pantic and L. J. M. Rothkrantz, "Automatic analysis of facial expressions: the state of the art," *IEEE Transactions on Pattern Analysis and Machine Intelligence*, 22(12): 1424–1445, 2000.

[11] Z. Zeng, M. Pantic, G. I. Roisman, and T. S. Huang, "A survey of affect recognition methods: audio, visual, and spontaneous expressions," *IEEE Transactions on Pattern Analysis and Machine Intelligence*, 31: 39–58, 2009.

[12] B. de Gelder, "Why bodies? Twelve reasons for including bodily expressions in affective neuroscience," *Philos. Trans. R. Soc. B: Biol. Sci.*, 364: 3475–3484, 2009.

[13] M. Mortillaro and K. R. Scherer, "Bodily expression of emotion," in *The Oxford Companion to Emotion and the Affective Sciences*, Oxford University Press, 2009, pp. 78–79.

[14] C. Darwin, *The Expression of the Emotions in Man and Animals*, John Murray, London, 1872.

[15] W. James, *Principles of Psychology*, H. Holt & Co., 1890.

[16] M. Coulson, "Attributing emotion to static body postures: recognition accuracy, confusions, and viewpoint dependence," *Nonverbal Behav.*, 28(2): 117–139, 2004.

[17] D. B. Givens, *The Nonverbal Dictionary of Gestures, Signs and Body Language Cues*, Center for Nonverbal Studies Press, Washington, 2010.

[18] P. Ekman, "Darwin, deception, and facial expression," *Ann. N. Y. Acad. Sci.*, 2003.

[19] P. Ekman and W. V. Friesen, "Origin, usage and coding: the basis for five categories of nonverbal behavior," in the Symposium on Communication Theory and Linguistic Models in the Social Sciences, Buenos Aires, Argentina, 1967.

[20] H. G. Wallbott, "Bodily expression of emotion," *Eur. J. Soc. Psychol.*, 28: 879–896, 1998.

[21] M. DeMeijer, "The contribution of general features of body movement to the attribution of emotions," *J. Nonverbal Behav.*, 13(4): 247–268, 1989.

[22] F. E. Pollick, H. Paterson, A. Bruderlin, and A. J. Sanford, "Perceiving affect from arm movement," *Cognition*, 82: 51–61, 2001.

[23] A. Kleinsmith and N. Bianchi-Berthouze, "Recognizing affective dimensions from body posture," in Proceedings of the ACII, 2007, pp. 48–58.

[24] J. F. Cohn, L. I. Reed, T. Moriyama, X. Jing, K. Schmidt, and Z. Ambadar, "Multimodal coordination of facial action, head rotation, and eye motion during spontaneous smiles," in Proceedings of the IEEE International Conference on Automatic Face and Gesture Recognition, 2004, pp. 129–135.

[25] R. Cowie, H. Gunes, G. McKeown, L. Vaclau-Schneider, J. Armstrong, and E. Douglas-Cowie, "The emotional and communicative significance of head nods and shakes in a naturalistic database," in Proceedings of the LREC International Workshop on Emotion, 2010, pp. 42–46.

[26] H. Gunes and M. Pantic, "Dimensional emotion prediction from spontaneous head gestures for interaction with sensitive artificial listeners," in Proceedings of the International Conference on Intelligent Virtual Agents, 2010, pp. 371–377.

[27] S. Mota and R. W. Picard, "Automated posture analysis for detecting learner's interest level," in Proceedings of the IEEE CVPR Workshops, 2003.

[28] J. K. Burgoon, M. L. Jensen, T. O. Meservy, J. Kruse, and J. F. Nunamaker, "Augmenting human identification of emotional states in video," in Proceedings of the International Conference on Intelligent Data Analysis, 2005.

[29] T. O. Meservy, M. L. Jensen, J. Kruse, J. K. Burgoon Jr., J. F. Nunamaker, D. P. Twitchell, G. Tsechpenakis, and D. N. Metaxas, "Deception detection through automatic, unobtrusive analysis of nonverbal behavior," *IEEE Intell. Syst.*, 20(5): 36–43, 2005.

[30] D. Bernhardt and P. Robinson, "Detecting affect from non-stylised body motions," in Proceedings of the ACII, 2007, pp. 59–70.

[31] G. Castellano, S. D. Villalba, and A. Camurri, "Recognising human emotions from body movement and gesture dynamics," in Proceedings of the ACII, 2007, pp. 71–82.

[32] G. Castellano, M. Mortillaro, A. Camurri, G. Volpe, and K. R. Scherer, "Automated analysis of body movement in emotionally expressive piano performances," *Music Percept.*, 26: 103–119, 2008.

[33] D. Glowinski, N. Dael, A. Camurri, G. Volpe, M. Mortillaro, and K. Scherer, "Towards a minimal representation of affective gestures," *IEEE Transactions on Affective Computing*, 2(2): 106–118, 2011.

[34] D. Janssen, W. I. Schöllhorn, J. Lubienetzki, K. Fölling, H. Kokenge, and K. Davids, "Recognition of emotions in gait patterns by means of artificial neural nets," *J. Nonverbal Behav.*, 32(2): 79–92, 2008.

[35] M. Karg, K. Kuhnlenz, and M. Buss, "Recognition of affect based on gait patterns," *IEEE Transactions on Systems, Man and Cybernetics-Part B*, 40: 1050–1061, 2010.

[36] P. Ekman, "About brows: emotional and conversational signals," in *Human Ethology: Claims and Limits of a New Discipline: Contributions to the Colloquium* (eds M. V. Cranach, K. Foppa, W. Lepenies, and D. Ploog), Cambridge University Press, New York, 1979, pp. 169–248.

[37] A. D. Wilson, A. F. Bobick, and J. Cassell, "Temporal classification of natural gesture and application to video coding," in Proceedings of the IEEE Conference on Computer Vision and Pattern Recognition, 1997, pp. 948–954.

[38] J. A. Russell and J. D. Fernndez Dols, *The Psychology of Facial Expression*, Cambridge University Press, Cambridge, 1997.

[39] K. L. Schmidt and J. F. Cohn, "Human facial expressions as adaptations: evolutionary questions in facial expression research," *American Journal of Physical Anthropology*, 116(S33): 3–24, 2001.

[40] H. Gunes and M. Piccardi, "Automatic temporal segment detection and affect recognition from face and body display," *IEEE Transactions on Systems, Man, and Cybernetics-Part B*, 39(1): 64–84, 2009.

[41] M. Pantic and I. Patras, "Dynamics of facial expression: recognition of facial actions and their temporal segments from face profile image sequences," *IEEE Transactions on Systems, Man and Cybernetics-Part B*, 36(2): 433–449, 2006.

[42] M. Valstar and M. Pantic, "Fully automatic facial action unit detection and temporal analysis," in Proceedings of the IEEE CVPR Workshops, 2006, pp. 149–154.

[43] A. Camurri, I. Lager, and G. Volpe, "Recognizing emotion from dance movement: comparison of spectator recognition and automated techniques," *Int. J. Hum.-Comput. Stud.*, 59: 213–225, 2003.

[44] N. Ambady and R. Rosenthal, "Thin slices of expressive behavior as predictors of interpersonal consequences: a meta-analysis," *Psychol. Bull.*, 11(2): 256–274, 1992.

[45] G. Castellano, L. Kessous, and G. Caridakis, "Emotion recognition through multiple modalities: face, body gesture, speech," in *Affect and Emotion in Human–Computer Interaction*, Springer, 2008, pp. 92–103.

[46] K. R. Scherer and H. Ellgring, "Multimodal expression of emotion," *Emotion*, 7: 158–171, 2007.

[47] J. Van den Stock, R. Righart, and B. De Gelder, "Body expressions influence recognition of emotions in the face and voice," *Emotion*, 7(3): 487–494, 2007.

[48] J. Driver and C. Spence, "Multisensory perception: beyond modularity and convergence," *Curr. Biol.*, 10(20): 731–735, 2000.

[49] H. Gunes and M. Piccardi, "Creating and annotating affect databases from face and body display: a contemporary survey," in Proceedings of the IEEE International Conference on Systems, Man and Cybernetics, 2006, pp. 2426–2433.

[50] H. K. Meeren, C. C. Van Heijnsbergen, and B. De Gelder, "Rapid perceptual integration of facial expression and emotional body language," *Proceedings of the National Academy of Sciences of the United States of America*, 102: 16518–16523, 2005.

[51] T. Balomenos, A. Raouzaiou, S. Ioannou, A. Drosopoulos, and K. Karpouzis, "Emotion analysis in man–machine interaction systems," in *Lecture Notes in Computer Science*, vol. 3361, Sprinter, 2005, pp. 318–328.

[52] K. Karpouzis, G. Caridakis, L. Kessous, N. Amir, A. Raouzaiou, L. Malatesta, and S. Kollias, "Modeling naturalistic affective states via facial, vocal and bodily expressions recognition," in *Artificial Intelligence for Human Computing*, Springer, 2007, pp. 91–112.

[53] B. Hartmann, M. Mancini, S. Buisine, and C. Pelachaud, "Design and evaluation of expressive gesture synthesis for embodied conversational agents," in Proceedings of the Third International Joint Conference on Autonomous Agents and Multi-Agent Systems, 2005.

[54] M. F. Valstar, H. Gunes, and M. Pantic, "How to distinguish posed from spontaneous smiles using geometric features," in Proceedings of the ACM International Conference on Multimodal Interfaces, 2007, pp. 38–45.

[55] M. A. Nicolaou, H. Gunes, and M. Pantic, "Continuous prediction of spontaneous affect from multiple cues and modalities in valence-arousal space," *IEEE Transactions on Affective Computing*, 2011.

[56] S. D'Mello and A. Graesser, "Multimodal semi-automated affect detection from conversational cues, gross body language, and facial features," *User Model. User-Adapt. Interact.*, 10: 147–187, 2010.

[57] A. Kapoor and R. W. Picard, "Multimodal affect recognition in learning environments," in Proceedings of the ACM International Conference on Multimedia, 2005, pp. 677–682.

[58] The UCLIC Database of Affective Postures and Body Movements, http://www.ucl.ac.uk/uclic/people/n_berthouze/research (last accessed May 8, 2011).

[59] T. Bänziger and K. R. Scherer, "Using actor portrayals to systematically study multimodal emotion expression: the GEMEP Corpus," in Proceedings of the ACII, 2007, pp. 476–487.

[60] H. Gunes and M. Piccardi, "A bimodal face and body gesture database for automatic analysis of human nonverbal affective behavior," in Proceedings of the International Conference on Pattern Recognition, vol. 1, 2006, pp. 1148–1153.

[61] R. Laban and L. Ullmann, *The Mastery of Movement*, 4th revision edn, Princeton Book Company Publishers, 1988.

[62] J. Allman, J. T. Cacioppo, R. J. Davidson, P. Ekman, W. V. Friesen, C. E. Izard, and M. Phillips. NSF report—facial expression understanding, 2003, http://face–and–emotion.com/dataface/nsfrept/basic_science.html (last accessed Sept. 19, 2014).

[63] H. G. Wallbott and K. R. Scherer, "Cues and channels in emotion recognition," *J. Personal. Soc. Psychol.*, 51: 690–699, 1986.

[64] P. Ekman, *Emotions Revealed*, Weidenfeld and Nicolson, 2003.

[65] J. A. Russell, "A circumplex model of affect," *J. Personal. Soc. Psychol.*, 39: 1161–1178, 1980.

[66] H. Gunes and M. Piccardi, "Observer annotation of affective display and evaluation of expressivity: face vs. face-and-body," in Proceedings of the HCSNet Workshop on the Use of Vision in Human–Computer Interaction, 2006, pp. 35–42.

[67] P. Dollár, V. Rabaud, G. Cottrell, and S. Belongie, "Behavior recognition via sparse spatio-temporal features," in Proceedings of the VS-PETS, 2005, pp. 65–72.

[68] H. Hotelling, "Relations between two sets of variates," *Biometrika*, 8: 321–377, 1936.

[69] C.-W. Hsu, C.-C. Chang, and C.-J. Lin, A practical guide to support vector classification, 2003.

[70] M. Pantic and I. Patras, "Temporal modeling of facial actions from face profile image sequences," in Proceedings of the IEEE International Conference on Multimedia and Expo, vol. 1, 2004, pp. 49–52.

[71] A. Bobick and J. Davis, "The recognition of human movement using temporal templates," *IEEE Transactions on Pattern Analysis and Machine Intelligence*, 23: 257–267, 2001.

[72] J. Davis, "Hierarchical motion history images for recognizing human motion," in Proceedings of the IEEE Workshop on Detection and Recognition of Events in Video, 2001, pp. 39–46.

[73] C. C. Chang and C.-J. Lin, "LIBSVM: a library for support vector machines," *ACM Trans. on Intelligent System and Technology*, 2(3): 27, 2011.

[74] D. Grandjean, D. Sander, and K. R. Scherer, "Conscious emotional experience emerges as a function of multilevel, appraisal-driven response synchronization," *Conscious. Cogn.*, 17(2): 484–495, 2008.

[75] A. Mehrabian, "Pleasure-arousal-dominance: a general framework for describing and measuring individual differences in temperament," *Curr. Psychol.: Dev. Learn. Personal., Soc.*, 14: 261–292, 1996.

[76] K. R. Scherer, A. Schorr, and T. Johnstone, *Appraisal Processes in Emotion: Theory, Methods, Research*, Oxford University Press, Oxford/New York, 2001.

[77] J. R. Fontaine, K. R. Scherer, E. B. Roesch, and P. Ellsworth, "The world of emotion is not two-dimensional," *Psychol. Sci.*, 18: 1050–1057, 2007.

[78] N. H. Frijda, *The Emotions*, Cambridge University Press, 1986.

[79] D. Sander, D. Grandjean, and K. R. Scherer, "A systems approach to appraisal mechanisms in emotion," *Neural Netw.*, 18(4): 317–352, 2005.

[80] P. R. DeSilva, A. Kleinsmith, and N. Bianchi-Berthouze, "Towards unsupervised detection of affective body posture nuances," in Proceedings of the ACII, 2005, pp. 32–39.

[81] A. Kleinsmith, N. Bianchi-Berthouze, and A. Steed, "Automatic recognition of non-acted affective postures," *IEEE Transactions on Systems, Man and Cybernetics-Part B*, 2011.

[82] J. Sanghvi, G. Castellano, I. Leite, A. Pereira, P. W. McOwan, and A. Paiva, "Automatic analysis of affective postures and body motion to detect engagement with a game companion," in Proceedings of the ACM/IEEE International Conference on Human–Robot Interaction, 2011.

[83] D. Bernhardt and P. Robinson, "Detecting emotions from connected action sequences," in Proceedings of the International Conference on Visual Informatics, 2009.

[84] A. Ortony, G. L. Clore, and A. Collins, *The Cognitive Structure of Emotions*, Cambridge University Press, Oxford, 1988.

[85] C. Bartneck, "Integrating the OCC model of emotions in embodied characters," in Proceedings of the Workshop on Virtual Conversational Characters, 2002, pp. 39–48.

[86] H. P. Espinosa, C. A. R. Garcia, and L. V. Pineda, "Features selection for primitives estimation on emotional speech," in Proceedings of the IEEE International Conference on Acoustics Speech and Signal Processing, 2010, pp. 5138–5141.

[87] J. Jia, S. Zhang, F. Meng, Y. Wang, and L. Cai, "Emotional audio-visual speech synthesis based on pad," *IEEE Transactions on Audio, Speech, and Language Processing*, 19(3): 570–582, 2011.

[88] G. McKeown, M. F. Valstar, R. Cowie, and M. Pantic, "The SEMAINE corpus of emotionally coloured character interactions," in Proceedings of the IEEE International Conference on Multimedia and Expo, July 2010, pp. 1079–1084.

[89] M. Pitt and N. Shephard, "Filtering via simulation: auxiliary particle filtering," *J. Am. Stat. Assoc.*, 94: 590–599, 1999.

[90] P. Ravindra De Silva, M. Osano, and A. Marasinghe, "Towards recognizing emotion with affective dimensions through body gestures," in Proceedings of the IEEE International Conference on Automatic Face & Gesture Recognition, 2006.

AUTHOR BIOGRAPHIES

Hatice Gunes is a lecturer at the School of Electronic Engineering and Computer Science, Queen Mary University of London (QMUL), UK. She received her Ph.D. in Computing Sciences from the University of Technology, Sydney (UTS), Australia. Prior to joining QMUL, she was a postdoctoral researcher at Imperial College, London, UK, working on SEMAINE, a European Union (EU-FP7) award-winning project that aimed to build a multimodal dialogue system, which can interact with humans via a virtual character and react appropriately to the user's nonverbal behavior, and MAHNOB that aimed at multimodal analysis of human naturalistic nonverbal behavior. Dr. Gunes (co-)authored over 45 technical papers in the areas of affective computing, visual and multimodal information analysis and processing, human–computer interaction, machine learning, and pattern recognition. She is a guest editor of special issues in *International Journal of Synthetic Emotions* and *Image and Vision Computing Journal*, a member of the Editorial Advisory Board for *International Journal of*

Computer Vision & Signal Processing and the *Affective Computing and Interaction Book* (IGI Global, 2011), a cochair of the EmoSPACE Workshop at IEEE FG 2011, and a reviewer for numerous journals and conferences in her areas of expertise. Dr. Gunes was a recipient of the Outstanding Paper Award at IEEE FG 2011, the Best Demo Award at IEEE ACII 2009, and the Best Student Paper Award at VisHCI 2006. She is a member of IEEE, ACM, and the HUMAINE Association.

Caifeng Shan is a senior scientist with Philips Research, Eindhoven, The Netherlands. He received the Ph.D. degree in Computer Vision from the Queen Mary University of London, UK. His research interests include computer vision, pattern recognition, image/video processing and analysis, machine learning, multimedia, and related applications. He has authored around 40 refereed scientific papers and 5 pending patent applications. He has edited two books *Video Search and Mining* (Springer, 2010) and *Multimedia Interaction and Intelligent User Interfaces: Principles, Methods and Applications* (Springer, 2010). He has been the Guest Editor of *IEEE Transactions on Multimedia* and *IEEE Transactions on Circuits and Systems for Video Technology*. He has chaired several international workshops at flagship conferences such as IEEE, ICCV, and ACM Multimedia. He has served as a program committee member and reviewer for many international conferences and journals. He is also a member of IEEE.

Shizhi Chen is a Ph.D. student in the Department of Electrical Engineering at the City College of New York. His research interests include facial expression recognition, scene understanding, machine learning, and related applications. He received the B.S. degree of Electrical Engineering from SUNY, Binghamton, in 2004, and the M.S. degree of Electrical Engineering and Computer Science from UC, Berkeley, in 2006. From 2006 to 2009, he worked as an engineer in several companies including Altera, Supertex, Inc., and the US Patent and Trademark Office. He is a member of Eta Kappa Nu (electrical engineering honor society), and a member of Tau Beta Pi (engineering honor society). He has also received numerous scholarships and fellowships, including Beat the Odds scholarship, Achievement Rewards for College Scientists (ARCS) Fellowship, and NOAA CREST Fellowship.

YingLi Tian is an associate professor in the Department of Electrical Engineering at the City College of New York since 2008. She received her Ph.D. from the Department of Electronic Engineering at the Chinese University of Hong Kong in 1996 and her B.S. and M.S. from TianJin University, China, in 1987 and 1990. She is experienced in computer vision topics ranging from object recognition, photometric modeling, and shape from shading, to human identification, 3D reconstruction, motion/video analysis, multi-sensor fusion, and facial expression analysis. After she worked in National Laboratory of Pattern Recognition at the Chinese Academy of Sciences, Beijing, China, Dr. Tian joined the Robotics Institute in Carnegie Mellon University as a postdoctoral fellow. She focused on automatic facial expression analysis. From 2001 to 2008, Dr. Tian was a research staff member at IBM T. J. Watson Research Center, Hawthorne, New York. She focused on moving object detection, tracking, and event and activity analysis. She was one of the inventors of the IBM Smart Surveillance Solutions (SSS) and was leading the video analytics team. She received the IBM Invention Achievement Awards every year from 2002 to 2007. She also received the IBM Outstanding Innovation Achievement Award in 2007. As an adjunct professor at Columbia University, she co-taught a course on Automatic Video Surveillance (Spring 2008). Dr. Tian has published more than 90 papers in journals and conferences and has filed more than 30 patents. She is also a senior member of IEEE.

15

BUILDING A ROBUST SYSTEM FOR MULTIMODAL EMOTION RECOGNITION

JOHANNES WAGNER, FLORIAN LINGENFELSER, AND ELISABETH ANDRÉ

Human-Centered Multimedia, Department of Computer Science, University of Augsburg, Augsburg, Germany

This study describes the development of a multimodal, ensemble-based system for emotion recognition covering the major steps in processing: emotion modeling, data segmentation and annotation, feature extraction and selection, classification and multimodal fusion techniques. Special attention is given to a problem often ignored: temporary missing data in one or more observed modalities. In offline evaluation the issue can be easily solved by excluding those parts of the corpus where one or more channels are corrupted or not suitable for evaluation. In online applications, however, we cannot neglect the challenge of missing data and have to find adequate ways to handle it. The presented system solves the problem at the multimodal fusion stage—established and novel emotion-specific ensemble techniques will be explained and enriched with strategies on how to compensate temporarily unavailable modalities. Extensive evaluation, including application of different annotation schemes, is carried out on the CALLAS Expressivity Corpus, featuring facial and vocal modalities. Finally, we present an application, the affective virtual listner Alfred, that was developed based on our online emotion recognition (OER) system.

Emotion Recognition: A Pattern Analysis Approach, First Edition. Edited by Amit Konar and Aruna Chakraborty.
© 2015 John Wiley & Sons, Inc. Published 2015 by John Wiley & Sons, Inc.

15.1 INTRODUCTION

Evoking emotional sensitivity in machines is believed to be a key element toward more human-like computer interaction. Due to the complex nature of human emotions, automatic emotion recognition remains a challenging task since many years. Humans tend express emotions rarely exclusively, but use several channels such as speech and mimics. When trying to recognize these emotional expressions, a machine has the alternatives to focus on one observed modality or to fuse informations from multiple sources in order to draw conclusions from a broader perspective. Studies that have focused on the fusion of multiple modalities, however, often start from too optimistic assumptions, for example, that all data from the different modalities are available at all time. As long as a system is only evaluated on off-line data this assumption can be easily ensured by examining given samples beforehand and excluding parts where one or more channels are corrupted or not suitable for evaluation. In real, application-oriented online systems, however, we cannot neglect the issue of missing data and have to find adequate ways to handle it, so that robustness of recognition performance can be guaranteed.

There are various causes for desired data to be temporarily missing: a sensor device can fail so that an according signal is no longer available or its outcome is corrupted by noise. In another situation a tracked object disappears from the view of the sensor. A system capable of handling missing data must therefore dynamically decide which channels are available and to what extent the present signals can be trusted. For the case that data are partially missing, a couple of treatments have been suggested in literature: multiple imputation predicts missing values using existing values from the previous samples [1]. In data marginalization unreliable features are marginalized to reduce their effect during the decision process [2]. The realization of handling missing data within our aspired multimodal system is engineered within the fusion of available modalities.

Yet, there is another problem often neglected when input from multiple modalities is combined: can we assume that an emotion is always expressed in the various channels in unison? We certainly cannot! We know that humans express emotions in a complementary way. Hence, it is usually argued that a higher recognition accuracy can be expected if all and not only one source is considered. But what are actually the consequences arising for a fusion system? Let us assume we measure emotion from two channels and observe a positive state in the first channel and a neutral in the second. What answer should be given by the system? Probably we would expect it to output positive. But what if it gives neutral? Is this a complete wrong answer or maybe partly right? This becomes even more tricky if we use this sample to train the system. Should we assign a positive or a neutral label? The problem we face here is that as soon as we decide in favor of one channel we automatically introduce errors in the training set of the other channel. This is not just a theoretical issue, but a very realistic one, especially when it comes to naturalistic data as shown by Cowie and colleagues [3]. Their analysis based on two audiovisual emotion databases reveals a large part of data as carrier of contradictory multimodal cues.

While the problem obviously exists, surprisingly little attention has been paid to it when it comes to applications. The study at hand tries to address this shortcoming by examining what influence contradictory multimodal cues actually have on a multimodal recognition system. Therefore, we start from a twofold annotation. First, labels are given to recorded samples based on the audio channel, then on the basis of video respectively.

15.2 RELATED WORK

Recently many studies in multimodal affect recognition have been done by exploiting synergistic combination of different modalities. Most of the previous works focus on fusion of audiovisual information for automatic emotion recognition, for example, combining speech with facial expression. De Silva *et al.* [4] and Chen *et al.* [5] proposed a rule-based decision-level fusion method for a combined analysis of speech and facial expressions. Huang and colleagues [6] used boosting techniques to automatically determine adaptive weights for audio and visual features. In the work of Busso *et al.* [7], an emotion-specific comparison of feature-level and decision-level fusion has been reported by using an audiovisual database containing four emotions, sadness, anger, happiness, and neutral state, deliberately posed by an actress. They observed for their corpus that feature-level fusion was most suitable for differentiating anger and neutral state while decision-level fusion performed better for happiness and sadness. They concluded that the best fusion method depends on the application. Interestingly, in addition to speech and facial expression, the thermal distribution of infra-red images is also integrated to a multimodal recognition system [8] by considering the fact that infra-red images are hardly affected by lighting conditions, which is one of the main problems in facial image analysis.

Humans use several modalities jointly in a complementary manner [5]. For decision-level fusion, however, multiple uni-modal classifiers are trained for each modality separately and those decisions are fused by using specific weighting rules. This means that such kind of fusion method is necessarily based on the assumption of conditional independence between modalities. To address this problem, a number of model-level fusion methods have been proposed, originally in the research field of speaker identification, that are capable to exploit cross-correlations between modalities. Song *et al.* [9], for example, proposed tripled HMM that models correlations between upper face, lower face, and prosodic dynamic behaviors. By relaxing the general requirement of synchronized segmentation for audiovisual streams, Zeng *et al.* [10] proposed a multistream fused HMM which provides the possibility of optimal combination among multiple streams from audio and visual channels. For the estimation of correlation levels between the streams, they used the maximum entropy and the maximum mutual information criterion. Sebe and colleagues [11] suggest the use of dynamic Bayesian Networks to model the interdependencies between audio and video data and handle imperfect data by probabilistic inference. Finally, various types of multimodal correlation models are based on extended artificial neural networks, for example, [12, 13].

Although there are many studies in psychology supporting that the combined face and body approaches are the most informative for the analysis of human expressive behavior [14], there is surprisingly very few effort reported on automatic emotion recognition by combining body gesture with other modalities such as facial expression and speech. For example, bimodal fusion methods at different levels for emotion recognition are presented by Balomenos *et al.* [15] and Gunes and Piccardi [16], using facial expression and body gesture. Kaliouby and Robinson [17] proposed a vision-based computational model to infer acted mental states from head movements and facial expressions. Castellano *et al.* [18] presented a multimodal approach for the recognition of eight emotions that integrate information from facial expressions, body gestures, and speech. They showed a recognition improvement of more than 10% compared to the most successful uni-modal system and the superiority of feature-level fusion to decision-level fusion. All these approaches are based on visual analysis of expressive gestures and dealt with mapping different gesture shapes to relevant emotions. In our experiment, however, we use three-axis accelerometer instead of visual information in order to extract non-proportional movement properties such as relative amplitude, speed, and movement fluidity, under the assumption that distinct emotions are closely associated with different qualities of body movement rather than gesture shapes.

All the studies reviewed above have shown that the performance of automatic emotion recognition systems can be improved by employing multimodal fusion. Some of them highlight the benefits of fusion mechanisms in situations with noisy features or missing values of features, for example, see Reference 11. Nevertheless, surprisingly few fusion approaches explicitly address the problem of non-available information. Most of them are based on the assumption that all data are available at all time. This precondition is not realistic in practical environments. In order to guarantee consistent classification, ways of handling missing sensory input have to be thought of.

15.3 THE CALLAS EXPRESSIVITY CORPUS

The CALLAS Expressivity Corpus [19] was constructed within the European Integrated Project CALLAS and contains affective behavior, incorporating vocal utterances and facial expressions. Recorded probands do not have any acting background, so the featured data can be described as realistic and challenging for classification. This makes the corpus well suited for evaluation of the recognition system to be built. The whole corpus was originally recorded with subjects from Greece, Italy, and Germany, in order to examine cultural differences between emotional expressions of persons from different European countries. As we are not interested in cultural differences in detail, we decided to only use the sub-corpus containing German participants: 21 persons (10 female and 11 male) and almost 5 hours of recorded interaction[1] form the basis for upcoming investigations.

For recording, a set of 120 emotion inducing sentences were successively displayed to the participants and they were asked to perform them. The semantic content of

[1] Since user sessions were continuously captured this includes intermediate parts where users were reading sentences or changing devices.

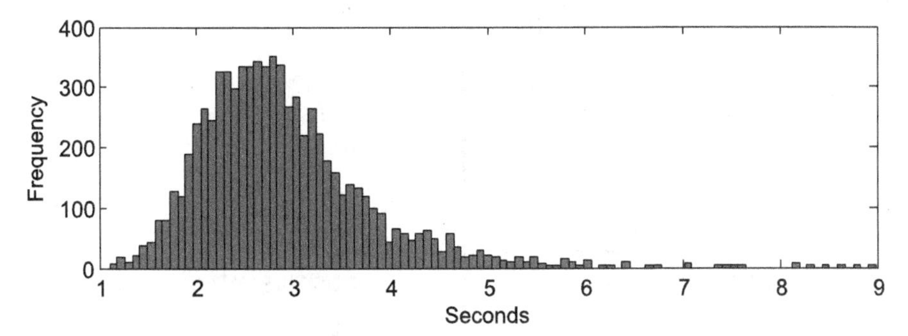

FIGURE 15.1 Histogram of segment lengths.

these sentences, inspired by the Velten mood induction technique [20], ranges from *positive*, over rather *neutral*, to *negative*. After reading a sentence, the projection was blanked out and the subject expressed each sentence in own words and with whatever voice felt fitting. It was left to their discretion to what extent they expressed the emotions. This approach leads to a broader diversity among observed expressions as it would under a more restrictive setup. This freedom of expression has to be expected by an emotion recognition system under realistic settings, for example, in some situations subjects were not using any mimics to accompany their speech and this just reflects what happens in real life. Actions of users were captured with two cameras—one steering at the proband's face and one at the whole body. In the following progress we only analyze videos captured from the face and refer to it as the facial modality. Voice was captured from a microphone hanging above the users head (vocal modality).

15.3.1 Segmentation of Data

Experimental design dictates the voice channel to be the trigger for on- and offset of a recorded sample. Only in very few cases expressive mimics were observed before or after an utterance. Consequently, we decided to segment data according to the speech signal, and beginning and ending of each utterance of a sentence serve as borders. We ended up with 2513 segments,[2] each containing respective snapshots of available audio and video streams. Segments have an average length of 2.9 seconds with a standard deviation of 0.9seconds. Figure 15.1 shows a histogram of the distribution.

15.3.2 Emotion Modeling

Whenever dealing with emotion recognition, a concept of discrete emotion modeling has to be chosen. The term emotion itself is a very abstract human concept, describing a vast amount of possible feelings and expressions. These diverse emotional states are too numerous to use them directly for recognition tasks, so they have to be

[2]Seven segments were skipped either because subjects refused to perform or due to technical problems.

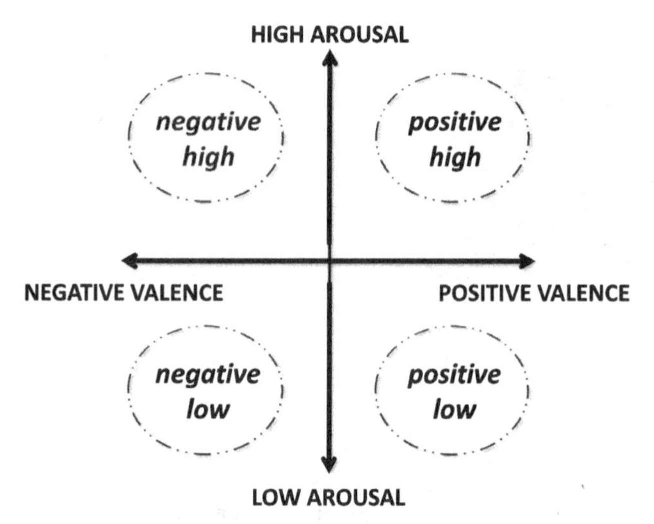

FIGURE 15.2 Arousal–valence-based emotion model.

integrated into quantifiable categories of emotions. A discrete emotion model is necessary to define categories of target emotions, so that the recognition system is able to understand the classification problem it has to solve. Moreover, an emotion model is definitely needed for convergent annotation of data (15.3.3). One possibility to narrow the amount of identifiable feelings and group the wide field of possible individual emotions into a small amount of discrete emotion classes is to have all recorded emotional experiences labeled by external specialists and subsumed under predefined expressions like love, hate, sadness, surprise, etc. However, this approach is a kind of restricting, as many blended feelings and emotions cannot adequately be described by chosen categories. This selection of some particular expressions can not be expected to cover the broad range of emotional states and could suffer from randomness.

A more flexible way of categorizing emotions is to attach the experienced stimuli to continuous scales. Lang [21] proposes arousal and valence as measurements (Figure 15.2). These scales describe multiple aspects of an emotion, the combination of stimuli's alignments on these scales defines single emotions. More precisely the valence scale describes the pleasantness of a given emotion. A positive valence value indicates an enjoyable emotion such as joy or pleasure. Negative values are associated with unpleasant emotions like sadness and fear. This designation is complemented by the arousal scale which measures the agitation level of an emotion. Combination of the two scales forms four emotion-quadrants, representing a four-class classification problem to be dealt with by the emotion recognition system.

15.3.3 Annotation

Though mood-inducing sentences used in the CALLAS corpus are rather categorized along the valence axis (positive, neutral, and negative), it becomes obvious that an arousal categorization is also needed. Especially when looking at negative sentences, there are samples tending to a depressed and sad mood (*negative-low*), while others are expressed in an aroused and angry way (*negative-high*). Nearly all neutral and part of the positive observations share a calm and optimistic sub-tone (*positive-low*) in contrast to fewer positive examples bearing clear hints of joy and laughter (*positive-high*). Based on these impressions we refrained from including neutral moods as an own class and labeled calm and non-negative emotions as *positive-low*. For annotating recorded samples, three experts were asked to independently label observations in terms of high or low arousal as well as positive or negative valence. The term experts is used to denote that the annotators have some sort of knowledge about emotions and their recognition that goes beyond everyday experience.

As we deal with the information gathered from facial and vocal modalities we can choose among three options for presenting recorded samples to the annotators: (1) audio only, (2) video only, or (3) audiovisual. Obviously, we could decide in favor of the latter as it provides the most comprehensive information to the raters. But in this case it remains unclear whether the annotator relies more on what he sees or what he hears in order to assign categories to the samples. This will depend on the annotating person and might even alter between samples. Theoretically this should not affect annotation as long as emotions are equally expressed in both channels. However, a study by Cowie *et. al* [3] raises reasonable doubts that this assumption can hold. By means of two multimodal databases, the EmoTV corpus [22] and the Belfast naturalistic database [23], they showed that in everyday interaction emotions are expressed through different channels that are as likely to conflict as to complement. In this fact they also see the reason for the low agreement that was observed among the raters.

As we may reckon similar effects on the present database we decided to follow a twofold annotation. First, labels are given to recorded samples based on the audio channel, then on basis of video only. Comparing decisions made for samples in both annotations tells us more about how emotions are expressed and allows us to investigate the discrepancies a fusion system has to deal with. Both runs were completed independently of each other with a temporal distance of several weeks. During annotation phase segments were replayed in chronological order to each expert independently. Annotators could loop an utterance as often as needed and even jump forth and back in order to repeat older segments and re-assign labels. Final combination of differing annotations is done via majority decision, as three assessments are given to each orientation of valence or arousal respectively; decisions are definite. For instance, if the first rater assigns label *low* and *positive*, the second *low* and *negative*, and the third *high* and *negative*, the segment is finally labeled as *low and negative*.

To report inter-rater reliability we calculate the kappa value according to Fleiss [24]. Fleiss' Kappa value is a common way to measure the agreement over multiple raters. It is expressed as a number between 0 and 1, where 1 indicates a perfect agreement. Table 15.1 gives Kappa values for both conditions (i.e., when labeled on audio only vs. labeled on video only). In both cases the kappa value for all four classes amounts to 0.52, which expresses a moderate agreement. If we look at the

TABLE 15.1 Agreement between raters under both conditions

	Fleiss' kappa value		
	Valence–arousal	Valence	Arousal
Audio	0.52	0.84	0.38
Video	0.52	0.71	0.48

kappa value for valence and arousal independently, there are two facts to notice. For both conditions the agreement for valence is much higher (0.84 and 0.71, which implies high agreement) as it is for arousal (0.38 and 0.48, which implies only a fair agreement). This drift obviously follows from the fact that expressed sentences were selected to be either negative, neutral, or positive. Thus, it should be easier for the raters to agree on the valence of an utterance than its level of arousal. Apart from this, we observe a higher agreement for valence if annotation is based on audio recordings, while the agreement for arousal is higher if based on video recordings. This is a first sign that emotions in the studied corpus are indeed not always expressed in a coherent way across modalities. We will pick up on this in Section 15.6.4.

15.4 METHODOLOGY

We now introduce possible feature extraction methods for available modalities and describe the classification scheme and feature selection strategies we applied to the CALLAS corpus. We focus on methods that can be computed within in a reasonable amount of time, as the developed system is meant to be capable of online recognition. In particular, feature extraction and classification should be applicable in (near) real time and must not require manual tuning at runtime.

15.4.1 Classification Model

Naive Bayes (NB) is a simple but efficient classification scheme. It is based on the Bayes Theorem which states:

$$P(E_i|f_1, \ldots, f_n) = \frac{P(E_i)\prod_{j=1}^{n} P(f_j|E_i)}{P(f_1, \ldots, f_n)}$$

This means that the probability of the emotion E_i given an observed feature vector $(f_1, \ldots, f_n)$ of dimension n depends on the a priori probability $P(E_i)$ of the emotion, multiplied by the product of the probability of each feature f_i given the emotion, divided by the a priori probability of the feature vector. As classification result, the emotion E_i from a set of N emotions $E_1, \ldots, E_N$ that maximizes the equation is chosen. This is simplifying in so far (and hence the name *Naive* Bayes), as the Bayes Theorem assumes the features to be independent from each other. Parameters for the probability distributions $P(E_i)$ and $P(f_j|E_i)$ are gained from the annotated training material.

Many other classification schemes, like for example, support vector machines (SVM), would of course also be applicable in a multimodal recognition system. The Naive Bayes classification is not guaranteed to produce the best classification results, in fact it is often outperformed by more complex approaches. The reason we nevertheless used it in our presented system lies within its compactness and time-efficiency. Naive Bayes has a linear execution time, therefore it is extremely fast in training and testing, even for high-dimensional feature vectors and large training databases, which enables extensive testing and comparison of conceivable approaches in an appropriate time span.

15.4.2 Feature Extraction

The extraction of descriptive features brings the raw signals into the compact form required by the classification model. Often features are computed after a pre-processing phase during which additional properties of the signals are carved out and unwanted aspects are suppressed. Depending on whether the features are extracted on a small running window of fixed size or for longer chunks of variable length, we denote them as short- or long-term feature. While in off-line analysis the whole signal is available from the beginning and processing can fall back on global statistics, such as global mean and standard deviation, or perform zero-phase filtering by processing the input data in both the forward and reverse directions, such treatments are not possible in online processing. In our experiments signals are processed in small blocks with a fixed window size.[3] This implies that only the information of the current and previously seen blocks is available to our algorithms. Table 15.2 offers a summary of the applied processing methods. A detailed description is given in the following paragraphs.

15.4.3 Speech Features

CALLAS corpus includes mono audio recordings from a single USB microphone (Samson C01U) placed near the subject's head. The audio stream was captured at a sample rate of 16 kHz and quantized by 16 bit PCM. Recording quality and noise level are similar among all sessions.

The list of proposed features suited for emotion recognition from speech is long. A co-operation of different sites under the name CEICES (Combining Efforts for Improving Automatic Classification of Emotional user States) has carried out experiments based on a pool of more than 4000 features including acoustic and linguistic feature types [25]. Results for the individual groups, as well as a combined set, have led to the following assumptions: among acoustic features duration and energy seem

[3] At each processing step a small portion of the signal equal to the window size is processed. Afterwards the window is moved by a certain number of samples, which is defined by the frame shift.

TABLE 15.2 Overview of pre-processing steps and feature extraction methods applied in our system

Modality	Channels	Pre-processing	Short-term feature	Long-term feature	Total
Voice	Mono audio, 16 kHz	Pre-emphasis filter	Pitch, energy, MFCCs, spectral, voice quality	Mean, median, maximum, minimum, variance, median, lower/upper quartile, absolute/ quartile range	1316
Face	RGB video, 720 × 576, 25 fps	Conversion to gray image	Bounding box of face, position of eyes, mouth and nose, opening of mouth, facial expression happy/angry/ sad/surprised	Mean, energy, standard deviation, minimum, maximum, range, position minimum/ maximum, number crossings/peaks, length	264

to be most relevant, while voice quality showed less impact. Yet, no single group outperformed the pool of all acoustic features.

Human language encodes emotional information in two different ways—what is said and how it is said. So a spoken message can be broken down into two parts, a semantic and a paralinguistic one. The importance of the paralinguistic part is pointed out in a study by Mehrabian [26]. It shows that the listener's feelings toward liking or disliking depends 55% on facial expressions and 38% on paralinguistic cues, but only 7% on the spoken word. During the recordings for this study the subjects were asked to repeat a set of pre-selected sentences with a strong connection to the expressed emotion. Therefore only the "how" is of interest for our analysis.

In our experiments we restricted the set of features to those that can be extracted in real time and in a fully automatic manner. For example, no features have been included that require information on the spoken words or grammatical context, as such information is difficult to get without manual annotation.

In our previous studies [27] MFCC and spectral features turned out to be good candidates. In addition we also compute features from pitch, energy, duration, voicing, and voice quality and use feature selection to reduce the full set of 1316 features to the most relevant ones. In a previous study the feature set was evaluated on the Berlin Database of Emotional Speech [28] that is commonly used in off-line research (7 emotion classes, 10 professional actors) and achieved an average recognition accuracy of 80%. On the FAU Aibo Emotion Corpus [29] as part of the INTERSPEECH Emotion Challenge 2009 [30], we were able to slightly exceed the baseline given by the organizers for a five-class problem (anger, emphatic, neutral, positive, and rest) [31]. Both corpora have been intensively studied in the past by many researchers working on emotion recognition from speech and serve as a kind of benchmark in this area.

15.4.4 Facial Features

As mentioned before we use only video recordings of the subject's face. The according camera was placed in a distance of about 2 m and captured frames include a close-up of shoulder and head. The resolution of the video is 720×576 pixels at 25 fps. Videos are stored in uncompressed 24-bit RGB format. Video processing is provided by SHORE, a library for facial emotion detection developed by Fraunhofer IIS[4] [32]. In the first place, SHORE offers a robust tracking of in-plane rotated faces up to 60 degrees. For each face that is found, SHORE reports the bounding box of the face, as well as position of the left/right eye and the nose tip. These features measure head movement. In addition the left and right corners of the mouth and its degree of opening are reported. Most important, SHORE also calculates scores for four facial expressions—namely happy, angry, sad, and surprised. These scores are also extracted for each frame and used in addition to the geometric features. In total, for each segment a series of 24 short-term features is derived by joining features extracted for each frame in the clip. Finally, we extract 11 long-term measurements, leading to an overall feature set with 264 entries.

15.4.5 Feature Selection

Especially on small corpora a large number of features can lead to a problem known as "curse of dimensionality." This term was introduced by Richard Bellman in 1961 as a mathematical problem, and in machine learning it describes the exponentially rising need for number of samples for a sufficient description of a high dimensional feature space. In short this means that the more features are given to a classification model, the more samples are needed to train it. In most cases, however, not all of the features add useful information to the classification problem, while some may carry redundant information. This fact is related to the challenge of finding that subset of features, which tweaks the best recognition performance.

To select a set of most relevant features in a feasible amount of time, we apply a combination of two selection approaches. First, we select the best 150 features according to correlation-based feature subset selection (CFS, [33]). CFS aims at finding a subset of features where the correlation of each feature with the class is maximized, while the correlation of the features among each other is low. This strategy is especially beneficial for the Naive Bayes classifier which performs badly when features are highly correlated since it assumes features to be independent for simplification reasons.

Afterwards, sequential forward selection (SFS) is applied on this subset and stopped after 100 iterations (i.e., after 100 features have been selected). SFS is a simple but popular selection method. Like other wrapper approaches it uses a classifier to measure the performance gain of different feature subsets. SFS starts from an empty subset and adds at each step the feature that brings the highest performance gain. In order to avoid overfitting, cross-validation should be used to evaluate

[4]http://.iis.fraunhofer.de/en/bf/bv/ks/gpe/demo/

the feature sets. Finally, the subset, which by then gives the best performance, is selected.

15.4.6 Recognizing Missing Data

Appropriate handling of missing data is a crucial task when establishing a robust emotion recognition system, and the CALLAS corpus is well suited for exploring this problem. The facial modality is missing at points in time when the video processing component loses track of a recorded person, and therefore no meaningful facial features can be extracted. The vocal modality is used as an indicator for segmentation (see Section 15.3.1) and is therefore always accessible during off-line experiments. Handling of missing data is modeled within the multimodal fusion process. Recognition of missing data has to be executed beforehand. Therefore we keep track of when SHORE looses the bounding box around an observed face—this happening marks recorded data as missing until the face is recognized again.

15.5 MULTISENSOR DATA FUSION

The examined CALLAS dataset features two modalities—video and speech data. As a special challenge, the video channel is not always accessible, so dedicated ways of fusing all available information have to be thought of. In order to discuss preconditions and advantages of decision-level fusion in a multisensor environment, we begin with a description of differences between the possible levels on which fusion can be executed and follow up with exemplary reviews on possible fusion methods and inherited strategies meant to deal with missing data.

15.5.1 Feature-Level Fusion

Feature-level fusion is a common and straightforward way to fuse all recorded observation channels. All desired features are merged into a single high dimensional feature set. One single classifier is then trained for the task of classification. As the fused data contains a bigger amount of information than single modalities, an increase in classification accuracy can theoretically be expected. In practice these classifiers yield reliable classification results. But this very accessible approach to data fusion comes along with a couple of major problems: first drawback is the eventually occurring "curse of dimensionality" on small datasets (see Section 15.4.5). If the available data is not ample, the classification results become non-meaningful. As a second, it has to be mentioned that a growing feature vector may stress computational resources for training and evaluation of the classification model. In some examinations these obstacles may be not of interest due to a fair availability of time and resources; other ones may refuse the feature-level approach solely because of these reasons and consider decision-level approaches to data fusion instead.

Handling missing data. A crucial shortcoming of the naive feature fusion approach lies within the single classifier, trained on the whole feature set. It is by default not capable of handling the problem of missing data. One possibility to handle incoming incomplete data is to use the pre-trained classifiers for decision making whenever data from single source is missing. So if no missing data is detected, the feature fusion classifier is asked to categorize respective sample, when information from one or more modalities is missing; classification models associated with available sources are used to create the needed decision.

15.5.2 Ensemble-Based Systems and Decision-Level Fusion

Contrary to feature level fusion and its reliance on a single classifier that deals with a high dimensional feature vector, decision-level fusion focusses on the usage of small classifiers and their combination. The assembly of these classifiers is called an ensemble. Outcomes of these classifier models are taken into account for the final decision making process. The term decision-level fusion sums up a variety of methods designed in order to merge the decisions of ensemble members into one single ensemble decision. Classification models used in creating the underlying ensemble for all discussed decision making algorithms stem from multimodal emotion observations. Our implemented system forms an ensemble by providing features from each listed channel with a classification model. Neither must they provide perfect performance on some given problem, nor do their outputs need to resemble each other. It is preferable that the chosen classifiers make mistakes, at best on different instances. A base idea of ensemble-based systems is to reduce the total error rate of classification by strategically combining the members of the ensemble and their errors. Therefore the single classifiers need to be diverse from one another.

Following these guidelines, the emerging ensemble-based systems can offer some advantages over the use of a single classifier. In many applications a vast amount of data are gathered and computational efficiency can greatly suffer from training and evaluation of a single classifier with huge datasets. Partitioning of data, training of independent classifiers with different subsets, and combination for a final decision often prove to be more practical, time-saving, and yield at least competitive results in most cases.[5] Given the contrary case that too little data is available, various resampling techniques can be used to form overlapping sub-samples of the dataset. Each of the resulting subsets can be applied for training of classifiers, which then are capable of decision making via combination.

The training and testing of classification models typically takes place on data gained from some kind of laboratory environment. Statements about generalized classification performance experienced in field testing—whenever previously unknown samples appear—are difficult to estimate. The risk of performing below average in

[5]Several smaller classifiers may not save training time when using a classification model of linear time complexity (e.g., Naive Bayes—$O(n)$), as the training will consume as much time as for an overarching classifier. But as time complexity rises (e.g., support vector machines classification scheme (SVM)—$O(n^2)$ or worse), this behavior changes in favor of small classifiers, using only a section of the original, high dimensional feature vector or training data.

the field is much higher for a single classifier than for an assemblage of classifier models. Some by chance poorly trained classifiers within a set are a much less of a menace than a single classifier performing poorly. Concerned with systems for multi-sensor data fusion and real-time applications, these can be implemented in a way that they resist the breakdown of one or more attached sensors. If the classifiers involved in decision making represent the observations of an associated sensory device, the absence of a single contribution to the final decision is unlikely to result in a drastic quality fall-off for overall classification accuracy—especially if the sensory malfunction is recognized and the corresponding classifier's (most likely counter-productive) contribution is accordingly rated.

Diverse decisions of ensemble members have to be merged into a single ensemble decision. For this purpose we can choose from various fusion strategies.[6] Among these we find all forms of algebraic combinations of the classifiers' continuous outputs, ranking methods as well as varying voting schemes, and other ways of class label combination. Some fusion methods apply weights to the ensemble members. Some methods utilize prior knowledge about ensemble members, for example, gained by evaluation of training performance, and single classifiers are associated with a certain weight. This way their importance within the ensemble is reflected. Note the immense importance of not taking any knowledge of data to be classified into account for the calculation of mentioned weights. Otherwise, unrealistic prior knowledge is hypothesized, and regarded experimental results can no longer be rated as significant. Among the vast range of possible combination strategies, we chose some established and well-known fusion schemes along with examples of more elaborated approaches and with these examples we describe the ways of handling temporarily unavailable data sources.

Weighted majority voting. Majority Voting is perhaps the most known and intuitive way to generate an ensemble decision. It simply sums up decisions of T classifiers. The ensemble decision for an observed sample x is chosen to be the class ω_n which received the most votes (decisions) v_n. A definite decision is only guaranteed if an odd number of ensemble members handle a two-class problem. In weighted majority voting, each vote is associated with the pre-calculated weight of the ensemble member. The ensemble decision for an observed sample x is chosen to be the class ω_n which received the most weighted votes v_n. Ties are not likely to happen this way, which makes the weighted variant more suited for practical application:

$$v_n(x) = \sum_{t=1}^{T} w_t d_{t,n}(x)$$

Handling missing data. Ensemble member t containing training data from modalities not featured in an observed sample is not included in the poll.

[6]For the explanation of reviewed algorithms the following annotations are used: The decision of ensemble member t for class n is denoted as $d_{t,n} \in \{0, 1\}$, with $t = 1...T$ and $n = 1...N$ and $d_{t,n} = 1$ if class ω_n is chosen, $d_{t,n} = 0$ otherwise. Respectively the support given to each class n (i.e., the calculated probability for the observed sample to belong to single classes) by classifier t is described as $s_{t,n} \in [1...0]$.

Sum rule. The sum rule can be categorized as algebraic combination of classifier outputs. It simply sums up the support given to each class ω_n in order to generate total support μ_n for each class. The ensemble decision for an observed sample x is chosen to be the class ω_n for which support $\mu_n(x)$ is largest:

$$\mu_n(x) = \sum_{t=1}^{T} s_{t,n}(x).$$

Handling missing data. Ensemble member t containing training data from modalities not featured in an observed sample gives support s_t of value zero to each of the n classes.

Cascading specialists. The cascading specialists method [34] does not focus on merging outputs from all ensemble members, but on selecting experts for each class and bringing them in a reasonable order. Based on evaluation of training data, experts for every class of the classification problem are chosen. Next, classes are rank ordered, from worst classified class across the ensemble's members to the best one. Given these preparations, classification works as follows: First class in the sequence is chosen and the corresponding specialist is asked to classify the sample. If the output matches the currently observed class, this classification is chosen as ensemble decision. If not, the sample is passed on to the next weaker class and corresponding expert whilst repeating the strategy. Whenever the case occurs that none of the experts classifies its connected class, the classifier with the best overall performance on the training data is selected as final instance and is asked to label the sample (Figure 15.3). This strategy aims at a flattening effect among class accuracies that will—at best—improve overall classification performance.

Handling missing data. The concept of choosing experts for certain classes has to be broadened, so that an expert unable to handle the given sample (because of missing data) can be adequately replaced. Instead of selecting one single ensemble member for expert and final classification tasks, ordered lists containing all classifiers—ranked by their qualification for the given task—replace sole classifiers. If in the classification step missing data is detected and the most qualified ensemble member is trained with

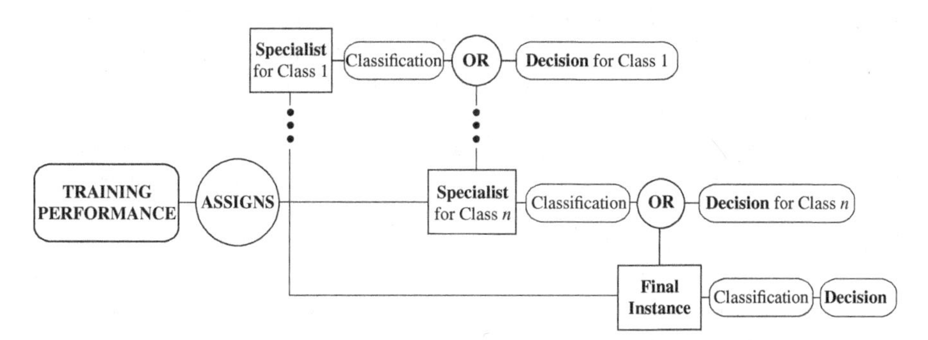

FIGURE 15.3 Cascading specialists scheme.

data of that type, we move down in the prepared list to find the next best classifier that is able to handle the observed sample.

OvR. The One Versus Rest approach trains N classifiers on every available feature-set, each specialized in recognizing one of the$N$ classes. This breakdown on several two-class classification problems is done by relabeling. Afterwards it chooses among the two-class classifiers the most promising one for every class. Given test-sample x, each specialist provides its support for the one associated class, and the class with highest support is chosen as ensemble decision.

Handling missing data. As this approach does utilize specialists for given classes, the same way of keeping track of available classifiers for specialist selection as in the cascading specialists approach can be used.

Emotion-adapted fusion—VAC. The generic approach to classification in a multi-class environment is to train classifiers and corresponding ensembles to categorize among the available classes. However, the structure of the chosen emotion model, consisting of two scales for valence and arousal, is employed to arrive at an ensemble decision [35]. Multiple ensembles are trained to recognize the observed emotion's axial alignments. Resulting outputs are logically combined for final decision. Because of the mapping to orientation in the emotion model, this strategy cannot be generalized for common classification problems. Ensembles for orientation on the valence and arousal axes and as a further addition, the cross-axis can be defined. Just like arousal and valence axes, the cross-axis divides the emotion model into two separate parts, each containing two emotion-quadrants. These parts contain the respective, complementary quadrants and therefore split the model in a diagonal way. According to arousal and valence, a proper ensemble for cross-axis is constructed. A stepwise algorithm is used for combination of the sources of information:[7] In *Step 1* each ensemble distributes its votes among the two quadrants that fit the recognized alignments in the emotion model. This step results in one of the two possible outcomes (Figure 15.4). If the ensembles agree on one emotion-quadrant, it receives three votes and can already be chosen as final decision, otherwise a voting tie occurs. No final decision can be chosen, instead the draw has to be dissolved and the algorithm moves on to *Step 2*. A direct classification ensemble designates exactly one vote to the class it predicts. Two situations can arise through this supplemental vote: if the ensemble chooses an emotion-quadrant that already holds two votes, the tie is resolved and the corresponding emotion is determined to be the final decision. Otherwise the ensemble chooses the emotion-quadrant that has not yet received any votes, the tie is not resolved (Figure 15.5) and *Step 3* to be executed: the emotion class that was originally determined by arousal and valence ensembles is chosen as final decision. In practice this case rarely occurs, but it is definitely needed to guarantee that no sample passes the decision process unclassified.

[7] An additional ensemble for direct classification—as established for generic approaches—is needed for *Step 2* of the combination strategy.

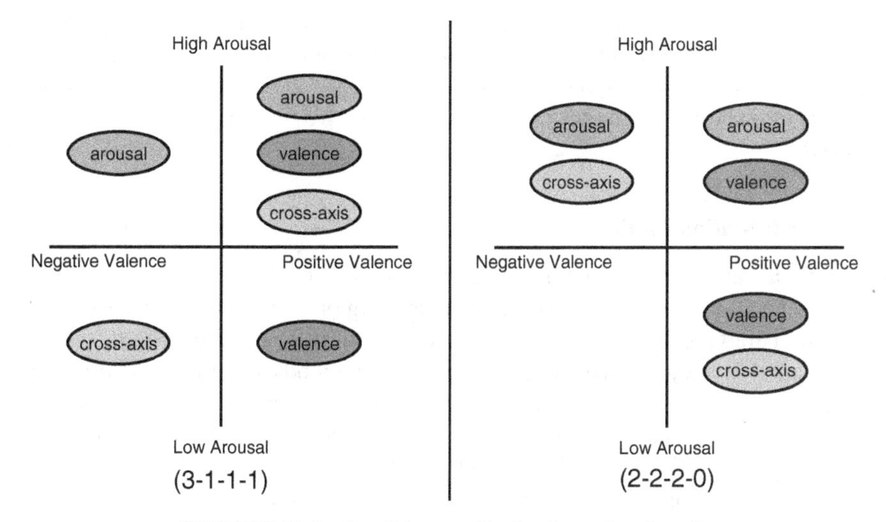

FIGURE 15.4 Possible vote distributions after *Step 1*.

Handling missing data. In our implementation, weighted majority voting is used to generate the ensembles' decisions on axial alignments. Therefore the handling of missing data described for respective fusion scheme is adopted.

15.6 EXPERIMENTS

Presented experiments are done on the described CALLAS Expressivity corpus. Missing data is included in the facial modality, the vocal modality is always accessible, as samples represent one spoken sentence. In detail, audio signals are

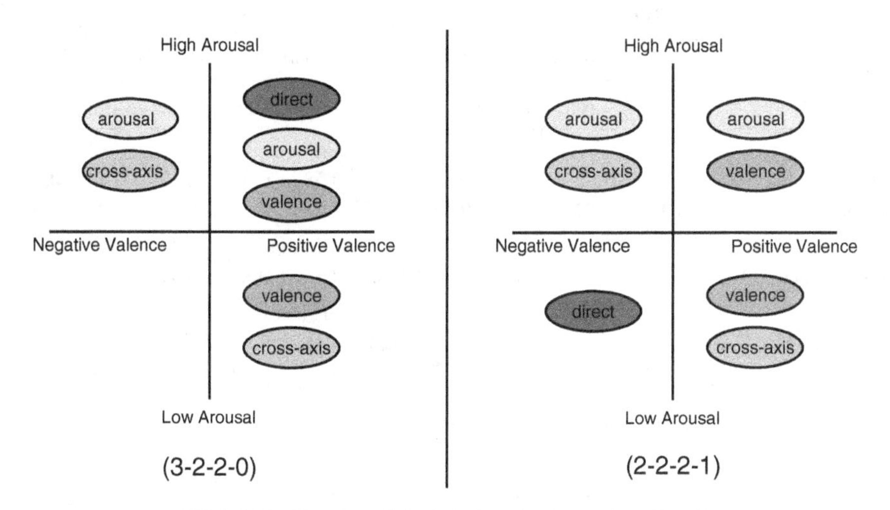

FIGURE 15.5 Possible Vote Distributions after *Step 2*.

constantly present in all 2513 samples; facial features can successfully be extracted throughout 2251 observations (90%). In consequence of the experimental design, the samples are more or less equally distributed among the 21 subjects, however, if we calculate the relative portion of samples per emotion, we find a high variety between the users.

15.6.1 Evaluation Method

Choosing an adequate evaluation method is crucial for meaningful experiments. Among others, possibilities involve random drawing of samples for testing, percentaged subset drawing, *k*-fold splits, or the leave-one-out strategy. As this work focusses strongly on the practical adaptability of presented methods and a good field performance under lifelike or even disadvantageous circumstances, the chosen evaluation method should reflect this intention. We agreed on a very realistic, user-independent approach for evaluation of our experiments: leave-one-speaker-out. As the employed corpus contains 10 female and 11 male participants, we consecutively draw samples belonging to one single subject out of the set. Remaining samples are used for training of classification models which then are tested against the isolated samples. Another important decision concerns the way recognition rates are presented. In consequence of the dominant presence of *positive-low* and a general imbalance in class distribution, we base our studies on the class-wise recognition rate (sometimes referred to as unweighted average recall), which is the mean of the recognition rates observed for each class.

15.6.2 Results

The results obtained are given in Table 15.3.

TABLE 15.3 Recognition results

	Labelled on audio channel					*Labelled on video channel*				
	p-low	p-high	n-low	n-high	**avg**	p-low	p-high	n-low	n-high	**avg**
					Single modalities					
Voice	0.61	0.50	0.49	0.43	**0.51**	0.53	0.43	0.58	0.44	**0.49**
Face	0.42	0.72	0.32	0.53	**0.50**	0.48	0.82	0.39	0.37	**0.51**
					Feature-level fusion					
Voice+Face	0.60	0.64	0.44	0.45	**0.53**	0.54	0.81	0.47	0.35	**0.54**
					Decision-level fusion					
WeightedMajVote	0.59	0.51	0.50	0.42	**0.51**	0.48	0.82	0.39	0.37	**0.51**
SumRule	0.59	0.65	0.43	0.47	**0.54**	0.55	0.77	0.51	0.38	**0.55**
Cascading	0.57	0.55	0.50	0.45	**0.52**	0.58	0.36	0.62	0.49	**0.51**
OvR	0.62	0.65	0.47	0.37	**0.53**	0.57	0.80	0.46	0.32	**0.54**
					Emotion-adapted decision-level fusion					
VAC	0.63	0.67	0.50	0.40	**0.55**	0.61	0.76	0.55	0.30	**0.55**

15.6.3 Discussion

Results shown in Table 15.3 are split into two parts: the left side illustrates the results gained for the audio annotation, the right side for the video annotation. In both cases, single channel performance is shown for each modality. Whenever missing data is found, the respective sample is not included for evaluation, so these results stem from different quantities of samples. As to be expected, the audio channel produces better results than the facial modality, when annotators used this channel for their estimations. This behavior is observable on the right side vice versa. General characteristics of observed channels are however independent from applied annotation: the video channel is in both scenarios well suited for recognizing the *positive-high* emotion category while neglecting other classes (especially the ones with low arousal)—presumably because it is well suited for the detection of smiles and movements of the face associated with laughter. Vocal expressions generate a more balanced distribution among recognition rates for all considered categories.

Next line compares results for feature-level fusion. Interestingly do results on the left side reflect characteristics of the vocal modality, as recognition rates are rather balanced among classes (though the influence of facial cues is observable in higher recognition accuracy for *positive-high*). The right side clearly resembles the results gained on the video channel. This phenomenon is also obvious in the following generic fusion schemes. Weighted Majority Voting's inherent weighting method causes a strong reliance on the dominant modality, resulting in both cases in almost the same accuracies across all classes as the respective annotated channel. The Sum Rule merges probabilistic outputs from both modalities, but as no weighting is included, the mirroring effect of Weighted Majority Voting does not occur. Yet do results more or less correlate to the mentioned characteristics of feature-level fusion. The Cascading Specialist scheme favors the classes that are weakly recognized by the respective dominant modality, in fact best results for these classes across all tested fusion schemes are achieved. But as a trade-off, this approach loses too much accuracy on other classes to generate an overall improved recognition rate. The specialized classifiers, selected for the One versus Rest fusion strategy, again tend to produce recognition results similar to the dominant channel, so no real improvements compared to the rather simple Sum Rule can be achieved. Another approach to enhance recognition rates compared to single channel classification on the dominant modality is to exploit deeper knowledge about the classification problem at hand with emotion-adapted fusion strategies that employ more than a single generic ensemble. The combination of valence, arousal, and cross-axis axial alignments shows stable classification accuracy across both annotations and seems to be the most promising approach in both cases. Nevertheless, the effect of mimicking the annotated channel stays in effect. All in all, no strategy does generate worse results than the best modality—actually they perform well compared to single modality classification.

15.6.4 Contradictory Cues

As mentioned earlier we can observe a variance in rater agreement for certain classes depending on the channel that was used as reference during annotation. The

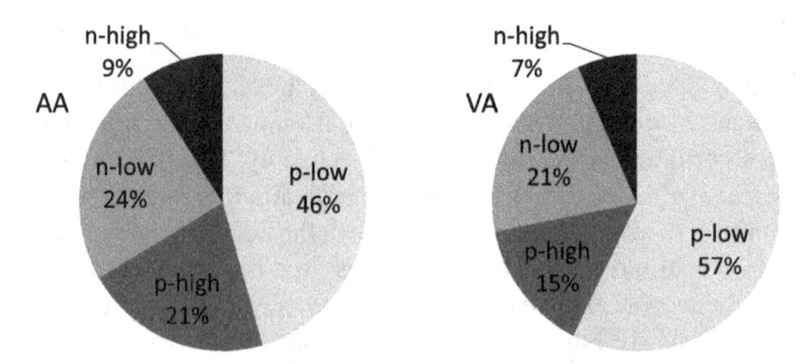

FIGURE 15.6 Differences in the distribution of labels suggest that probands are more expressive through the audio channel. For instance, in the audio-based annotation class p-low occurs with a 10% smaller frequency compared to the video-based annotation. AA = audio-based annotation, VA = video-based annotation.

consequential assumption that emotions are not equally expressed across modalities is further backed up by the distribution of labels as shown in Figure 15.6. Here we see, for instance, that label *positive-low* has been assigned with a 10% higher frequency if annotation was video based. Such differences explain the variations in Table 15.3 for the different annotation conditions. In the end we obviously face the same problem as Cowie and colleagues [3] faced: contradictory multimodal cues.

Now, what are the consequences for a system that tries to fuse information from two channels that do not necessarily agree? First of all this is surely a matter of frequency. For the examined corpus in 71% of the time, a sample receives the same label in both annotations. In other words: almost one-third of the samples seem to carry contradictory cues. It appears only natural that these samples are also preferred candidates for mistakes if fused by a classifier. Indeed is the amount of correctly classified samples significantly higher when both annotations agree. Exact numbers are given in Figure 15.7. Allowing both labels for samples, which were assigned to

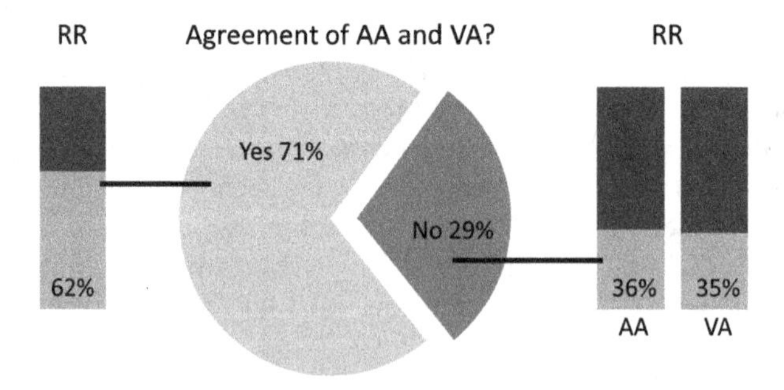

FIGURE 15.7 Amount of correctly classified samples is significantly higher for samples where both annotations agree. AA = audio-based annotation, VA = video-based annotation, RR = recognition rate.

different classes, would actually boost recognition accuracy for the four classes up to 65% (for Sum Rule).

Such reflections reveal the limitations of what we may expect from a fusion system similar to those discussed in this chapter. In the end, it has to face the same uncertainties humans have to cope with in everyday social interaction.

15.7 ONLINE RECOGNITION SYSTEM

In the recent years affective computing has established to a research area on its own, which has led to an increasing number of theories and tools dealing with the automatic recognition of emotions. Some groups have dealt with the recording of affective databases, containing acted (e.g., Emo-DB [28]) and—in the past years more and more frequently—spontaneous emotions (e. g. AIBO corpus [29]), while others have developed appropriate tools for annotation, such as the free video annotation software Anvil [36]. Computational models of emotion have been developed, such as EMA [37] and ALMA [38]. Developers of emotion recognition systems can count on powerful tools, such as Octave or Matlab for signal processing, and Weka, which offers a large collection of machine learning algorithms for data mining tasks. However, so far most effort has been put toward off-line analysis, whereas to date only few applications exist, which are able to react to a user's emotion in real time. The few systems for OER usually concentrate on one modality only, such as speech (EmoVoice [27] and OpenSmile [39]) or mimics (SHORE [32] and [40]), while multimodal approaches, such as Gaze-X [41], are based on the components that deliver affective output, but do not support their development.

15.7.1 Social Signal Interpretation

Building a multimodal online recognition system from scratch involves several tasks and each of them bears its own challenges. These include recording of training data, creating annotations, feature extraction, and model training. Finally, we have to set up a pipeline that detects interesting segments, extracts features, and applies classification in real time. In the following we will introduce the Social Signal Interpretation (SSI) framework [42,43] as a coherent solution for building multimodal online recognition systems. In particular, SSI supports the design and evaluation of machine learning pipelines by offering tailored tools for signal processing, feature extraction, and pattern recognition, as well as tools to apply them off- (training phase) and online (real-time recognition). In fact, the reported processing and classification methods, as well as, all fusion-based approaches used in our experiments have been developed entirely with SSI. Apart from that, SSI has been used to collect and annotate the CALLAS corpus [19], too.

To ease the use of SSI, a graphical user interface called SSI/ModelUI has been developed, which enables even novice users to complete all steps to create own models and set up an online recognition pipeline (see Figure 15.8).

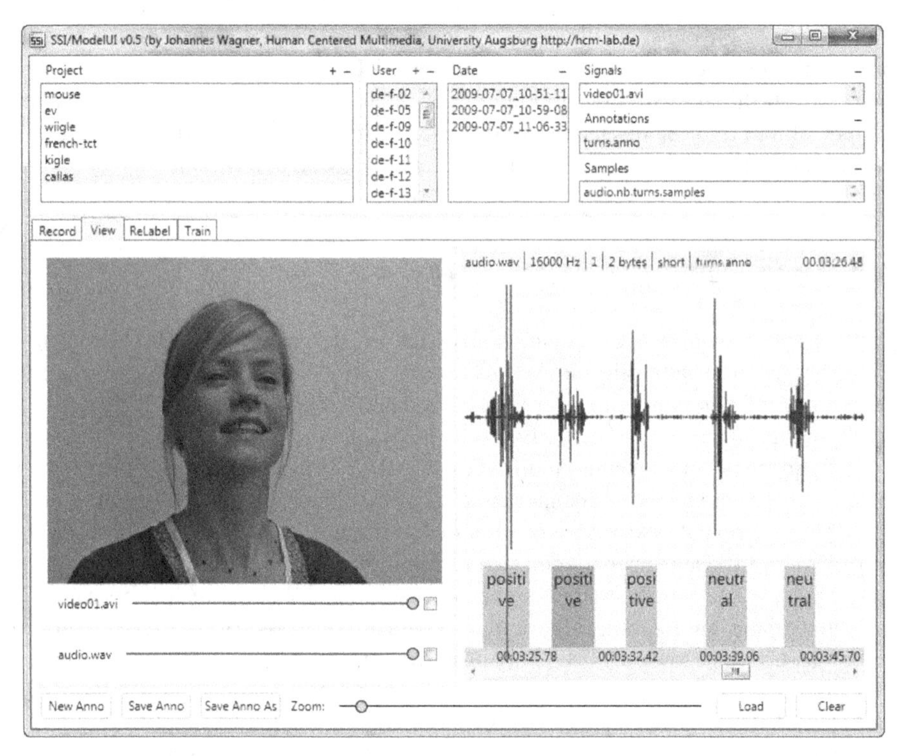

FIGURE 15.8 SSI/ModelUI is a graphical interface for SSI, which allows even novice users to complete all necessary steps to train their own classification model. The screenshot shows a user recording together with its annotation line.

15.7.2 Synchronized Data Recording and Annotation

First of all, a sufficient amount of training data is required. Of course we may consider available corpora, but it is neither likely to find a database that meets our requirements in terms of affective states, available modalities, and required quality, nor is it possible to adopt the pre-trained system to the user. This approach usually results in higher recognition accuracies. Hence, our system should be capable of recording synchronized data from all involved modalities. SSI supports various kinds of sensory devices. These include web/dv cameras and multichannel ASIO audio, as well as the wii remote control and the kinect and various physiological sensors like NeXus, ProComp, AliveHeartMonitor, IOM, or Emotiv. The runtime engine of SSI allows parallel and synchronized recording from an arbitrary number of devices. Common formats such as AVI for video streams and WAV for audio streams, as well as data exchange formats such as CSV (Comma Separated Value) or ARFF (Attribute-Relation File Format) are available for storage of data. During a recording, SSI/ModelUI offers the possibility to present stimuli to the user as a series of html pages. The pages may include textual instructions, for example, the Velten sentences from Section 15.3, but also images or videos.

Once a sufficient amount of training data has been collected, we have to add appropriate descriptions. To speed up annotation, SSI detects events during recording and synchronizes them with the signals. By adding a label that corresponds to the current presented stimuli to the event interval, that is, the intended user state, a pre-annotation is created. Afterwards, pre-annotations and recorded signals can be reviewed and played back in SSI/ModelUI. Now, existing annotations can be adjusted and new annotations can be added. This is important; the reaction of the user often differs from the intended emotion.

15.7.3 Feature Extraction and Model Training

As a next step we can apply feature extraction algorithms to our database and train a classification model. Here, we are restricted to algorithms that can be computed in real time and even though training happens off-line, we must ensure that features are calculated in an "online manner," for example, based on previous data only. The same counts for the applied learning algorithm. SSI offers a large number of feature extraction and pattern recognition methods, which have been implemented to suit the requirement of real-time processing. Apart from the feature extraction methods and fusion techniques explained in Section 15.4, SSI also includes moving average filters, low/band/high pass filters, derivative and integral filters, as well as additional classification models such as K-Nearest Neighbor (KNN), Linear Discriminant Analysis (LDA), Support Vector Machines (SVM) or Hidden Markov Models (HMM). Several feature selection algorithms (e.g., SFS) and over-sampling techniques (e.g., SMOTE) for boosting under represented classes are available, too.

The interface of SSI/ModelUI allows for direct comparison of different configurations of feature sets and classifiers in terms of recognition accuracy. Therefore, the user chooses one or more recording sessions and an appropriate evaluation method (e.g., k-folds or leave-one-user-out). Feature extraction and training runs fully automatically. In order to save computation time, features are extracted only once and stored for re-use. By selecting sessions recorded from different users it becomes possible to compare the performance of user-independent classification with user-dependent classification. For the setting, which gives the most promising recognition rates, a final model is trained and is exported to a file, which stores learned model parameters, as well as applied feature extraction methods and indices of selected features.

15.7.4 Online Classification

For the final online recognition system, a simultaneous execution of all involved sub-parts is required, that is, sensor data must be permanently captured and processed, while at the same time classification has to be invoked on detected segments. The pipeline architecture of SSI provides the necessary tools for this step. Feature extraction is encapsulated in transforming blocks, which receive input from the sensor devices (or according filter blocks if pre-processing is applied). A trigger

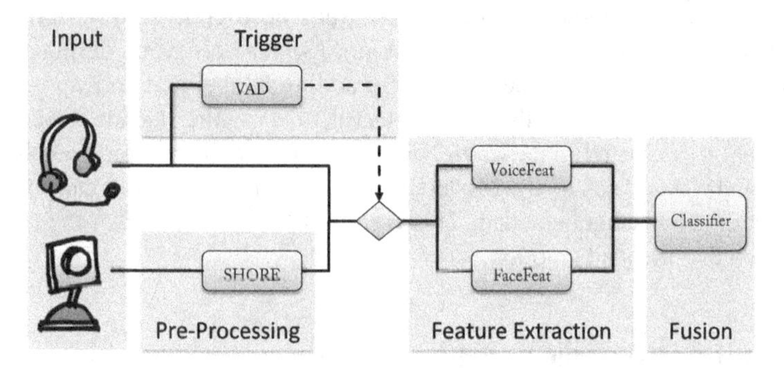

FIGURE 15.9 Flow chart of the final online recognition system.

mechanism—running in a separate block—invokes classification when a new segment is detected, that is, the fusion component collects data from all input blocks and applies classification. To build an online recognition system for the experiment discussed in Section 15.4, we need to connect an audio and video sensor with a voice activity detection and the learned fusion models. A flowchart of the pipeline is depicted in Figure 15.9.

An exemplary application that was developed with SSI is the Affective Listener "Alfred." Alfred is a butler-like virtual character [44] that is aware of the user and reacts to his or her affective expressions (see Figure 15.10). The user interacts with Alfred via acoustics of speech and facial expressions. As a response, Alfred simply mirrors the user's emotional state by appropriate facial expressions. This behavior can be interpreted as a simple form of showing empathy.

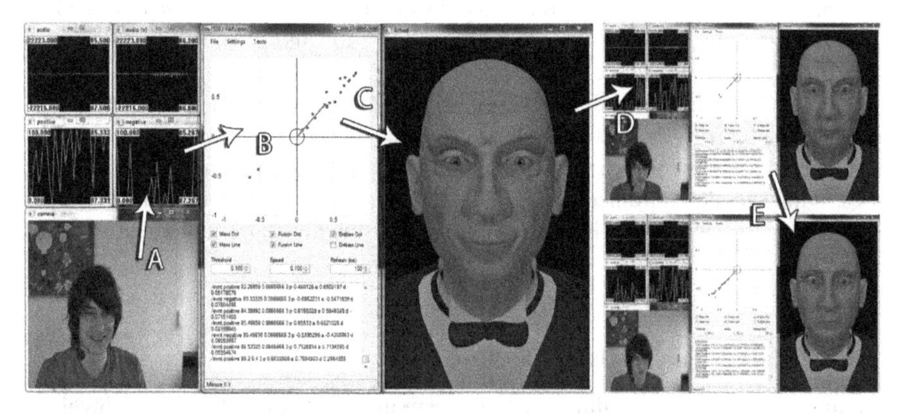

FIGURE 15.10 Affective Listener Alfred: the current user state is perceived using the SSI framework (A); observed cues are mapped onto the VA space (B); VA values are combined to a final decision and transformed to a set of FACS parameters, which are visualized by Alfred (C); a change in the mood of the user creates a new set of VA values, which moves the fused vector to a new position on the VA space (D); this finally leads to a change in the expression of Alfred (E).

To allow smooth changes of the users emotional state we interpose an additional fusion mechanism between recognition system and avatar. Like the fusion methods in Section 15.4 it is also based on valence (V) and arousal (A) axis, but maps discrete categories like "positive-low" or "negative-high" into a continuous space. Details on the fusion algorithm can be found in Reference 45. Note that Gilroy *et.al* generate one vector per modality; we generate one vector for each detected event. This way we prevent sudden leaps in case of a false detection. Since the strength of a vector decreases with time, the influence of older events is lessened until the value falls under a certain threshold and is completely removed.

15.8 CONCLUSION

In this work we described the process of constructing a robust multimodal system for emotion recognition from audio and video data. For off-line evaluation we chose the CALLAS Expressivity corpus, featuring 21 male and female participants and the realistic Leave-One-User-Out evaluation method. We theoretically introduced the necessary background needed to prepare the corpus and an ensemble-based system for emotion recognition. This includes emotion theory, preprocessing, segmentation, and feature extraction from raw audiovisual data, as well as classification and fusion techniques for multimodal information. In order to guarantee robustness of the developed system, the need to recognize and handle temporarily missing sensory input has been pointed out and ways to technically deal with this problem are included within the fusion stage of emotion recognition. We concluded with the description of the SSI framework, a tool that enables even inexperienced users to set up online recognition pipelines, including all the mentioned techniques, by covering all complex tasks with an easy to use, XML–based editor.

For evaluation of the developed ensemble-based system, we first compared the single channel classification performance to exemplary chosen fusion techniques on different levels (feature-level fusion and decision-level fusion) and finally presented a possibility of a fusion technique especially adapted to emotion classification in the valence–arousal domain. Results foremost show that multimodal fusion techniques clearly outperform single channel performance. Characteristics of the single modalities stay however observable within results of applied fusion schemes, like, for example, the good recognition rates for *positive-high* emotions via the facial modality, that mostly carries on throughout fused results. Definite accuracy of multimodal fusion has shown to rely on chosen combination strategy, whereupon an emotion-adapted approach seems to be a stable technique.

We also took a close look on problems concerning the annotation and evaluation of a corpus that features more than one modality. Therefore, comparisons between annotations based solely on audio information as well as on the video channel respectively have been included in the results. As expected, classification performance of single modalities depends on the annotation modality to be used, fused results do rather not. Closer investigation on contradictory cues from different modalities has shown to negatively influence the results of multimodal fusion;

this depicts problems and limitations of ensemble-based techniques for multisensor data fusion.

For future work we also aim at improving the segmentation techniques applied to the CALLAS Expressivity corpus. So far, recorded modalities have been segmented in a very straightforward way. Based on the user's speech, beginning and ending of a recorded sample coincide with the boundaries of the spoken stimuli sentence. Facial and gestural signals are simply observed and segmented over the according time span. This strategy suffers from two major problems: the hopefully expressed emotion could occur within a much shorter period somewhere within the spoken sentence and therefore information before and afterwards is not of great meaning for recognition. Furthermore significant hints from different modalities are not guaranteed to emerge at exactly the same time interval. Classification accuracy could be expected to improve if modalities were segmented individually and the succession and corresponding delays between occurrences of emotional hints in different signals could be investigated more closely. However, this approach gives room for a whole new set of hypotheses and experiments.

ACKNOWLEDGMENT

The work described in this chapter is funded by the EU under research grant CEEDS (FP7-ICT-2009-5) and the IRIS Network of Excellence (Reference: 231824).

REFERENCES

[1] M. Pantic and L. J. M. Rothkrantz, "Toward an affect-sensitive multimodal human–computer interaction," *Proceedings of the IEEE*, 91: 1370–1390, 2003.

[2] C. Demiroglu, D. V. Anderson, and M. A. Clements, "A missing data-based feature fusion strategy for noise-robust automatic speech recognition using noisy sensors," in International Symposium on Circuits and Systems (ISCAS), 2007, pp. 965–968.

[3] E. D. Cowie, L. Devillers, J. C. Martin, R. Cowie, S. Savvidou, S. Abrilian, and C. Cox, "Multimodal databases of everyday emotion: facing up to complexity," in INTERSPEECH, 2005, pp. 813–816.

[4] L. C. De Silva and P. C. Ng, "Bimodal emotion recognition," in 4th IEEE International Conference on Automatic Face and Gesture Recognition (FG), 2000, pp. 332–335.

[5] L. S. Chen, T. S. Huang, T. Miyasato, and R. Nakatsu, "Multimodal human emotion/expression recognition," in Proceedings of the 3rd International Conference on Face & Gesture Recognition, IEEE Computer Society, Series FG, 1998, p. 366.

[6] L. Huang, L. Xin, L. Zhao, and J. Tao, "Combining audio and video by dominance in bimodal emotion recognition," in Proceedings of the 2nd International Conference on Affective Computing and Intelligent Interaction, 2007, pp. 729–730.

[7] C. Busso, Z. Deng, S. Yildirim, M. Bulut, C. M. Lee, A. Kazemzadeh, S. Lee, U. Neumann, and S. Narayanan, "Analysis of emotion recognition using facial expressions, speech and multimodal information," in Proceedings of the 6th International Conference on Multimodal Interfaces (ICMI), 2004, pp. 205–211.

[8] Y. Yoshitomi, S.-I. Kim, T. Kawano, and T. Kitazoe, "Effect of sensor fusion for recognition of emotional states using voice, face image and thermal image of face," in 9th IEEE International Workshop on Robot and Human Interactive Communication, RO-MAN, 2000, pp. 178–183.

[9] M. Song, J. Bu, C. Chen, and N. Li, "Audio-visual based emotion recognition—a new approach," in Proceedings of the 2004 IEEE Computer Society Conference on Computer Vision and Pattern Recognition, 2004, pp. 1020–1025.

[10] Z. Zeng, J. Tu, B. Pianfetti, M. Liu, T. Zhang, Z. Zhang, T. S. Huang, and S. Levinson, "Audio-visual affect recognition through multi-stream fused HMM for HCI," in Proceedings of the 2005 IEEE Computer Society Conference on Computer Vision and Pattern Recognition (CVPR'05), vol. 2, 2002, pp. 967–972.

[11] N. Sebe, T. Gevers, I. Cohen, and T. S. Huang, "Multimodal approaches for emotion recognition: a survey," in Proceedings of SPIE—The International Society for Optical Engineering, vol. 5670, 2005, pp. 56–67.

[12] N. F. Fragopanagos and J. G. Taylor, "Emotion recognition in human–computer interaction," *Neural Netw.*, 18(4): 389–405, 2005.

[13] G. Caridakis, L. Malatesta, L. Kessous, N. Amir, A. Raouzaiou, and K. Karpouzis, "Modeling naturalistic affective states via facial and vocal expressions recognition," in ACM Proceedings of the 8th International Conference on Multimodal Interfaces (ICMI), 2006, pp. 146–154.

[14] A. Nalini and R. Robert, "Thin slices of expressive behavior as predictors of interpersonal consequences: a meta-analysis," *Psychol. Bull.*, 111(2): 256–274, 1992.

[15] T. Balomenos, A. Raouzaiou, S. Ioannou, A. I. Drosopoulos, K. Karpouzis, and S. D. Kollias, "Emotion analysis in man–machine interaction systems," in First International Workshop in Machine Learning for Multimodal Interaction (MLMI), vol. 34, Martigny, Switzerland, June 21–23, 2004, pp. 318–328.

[16] H. Gunes, M. Piccardi, and T. Jan, "Face and body gesture recognition for a vision-based multimodal analyzer," in Tenth Annual ACM Symposium on Principles of Programming Languages, 2004, pp. 88–98.

[17] R. El Kaliouby and P. Robinson, "Generalization of a vision-based computational model of mind-reading," in First International Conference on Affective Computing and Intelligent Interaction, LNCS 3784, Springer, 2005, pp. 582–589.

[18] G. Castellano, L. Kessous, and G. Caridakis, "Multimodal emotion recognition from expressive faces, body gestures and speech," in *Affect and Emotion in Human–Computer Interaction, Lecture Notes in Computer Science*, Spinger, 2007, pp. 375–388.

[19] G. Caridakis, J. Wagner, A. Raouzaiou, Z. Curto, E. André, and K. Karpouzis, "A multimodal corpus for gesture expressivity analysis," in Multimodal Corpora: Advances in Capturing, Coding and Analyzing Multimodality, Malta, May 17–23, 2010.

[20] E. Velten, "A laboratory task for induction of mood states," *Behav. Res. Ther.*, 6(4): 473–482, 1967.

[21] P. J. Lang, M. M. Bradley, and B. N. Cuthbert, "Motivated attention: affect, activation, and action," in *Attention and Orienting: Sensory and Motivational Processes* Psychology Press, 1997, pp. 97–135.

[22] S. Abrilian, L. Devillers, S. Buisine, and J. C. Martin, "EmoTV1: annotation of real-life emotions for the specification of multimodal affective interfaces," in Proceedings of the HCI International, 2005.

[23] E. D. Cowie, N. Campbell, R. Cowie, and P. Roach, "Emotional speech: towards a new generation of databases," *Speech Commun.*, 40(1): 33–60, 2003.

[24] J. L. Fleiss, B. A. Levin, and M. C. Paik, "Statistical methods for rates and proportions," in *Wiley Series in Probability and Mathematical Statistics*, 2013.

[25] A. Batliner, S. Steidl, B. Schuller, D. Seppi, K. Laskowski, T. Vogt, L. Devillers, L. Vidrascu, N. Amir, L. Kessous, and V. Aharonson, "Combining efforts for improving automatic classification of emotional user states," in Proceedings of the IS-LTC, Ljubliana, 2006, pp. 240–245.

[26] A. Mehrabian, "Framework for a comprehensive description and measurement of emotional states," *Genet. Soc. Gen. Psychol. Monogr.*, 121(3): 339–361, 1995.

[27] T. Vogt, E. André, and N. Bee, "EmoVoice—a framework for online recognition of emotions from voice," in Proceedings of Workshop on Perception and Interactive Technologies for Speech-Based Systems, 2008.

[28] F. Burkhardt, A. Paeschke, M. Rolfes, W. F. Sendlmeier, and B. Weiss, "A database of German emotional speech," in INTERSPEECH-Eurospeech, 9th European Conference on Speech Communication and Technology, Lisbon, Portugal, September 4–8, 2005, pp. 1517–1520.

[29] S. Steidl, *Automatic Classification of Emotion-Related User States in Spontaneous Children's Speech*, Logos Verlag, Berlin, 2009.

[30] B. Schuller, S. Steidl, and A. Batliner, "The INTERSPEECH 2009 emotion challenge," in Proceedings of the Interspeech, 2009, pp. 312–315.

[31] T. Vogt and E. André, "Exploring the benefits of discretization of acoustic features for speech emotion recognition," in Proceedings of 10th Conference of the International Speech Communication Association (INTERSPEECH), 2009, pp. 328–331.

[32] C. Küblbeck and A. Ernst, "Face detection and tracking in video sequences using the modified census transformation," *Image Vis. Comput.*, 24: 564–572, 2006.

[33] M. A. Hall, "Correlation-based feature subset selection for machine learning," Ph.D. Thesis, University of Waikato, New Zealand, 1998.

[34] F. Lingenfelser, J. Wagner, T. Vogt, J. Kim, and E. André, "Age and gender classification from speech using decision level fusion and ensemble based techniques," in INTERSPEECH, Vol. 10, pp. 2798–2801, 2010.

[35] J. Kim and F. Lingenfelser, "Ensemble approaches to parametric decision fusion for bimodal emotion recognition," in BIOSIGNALS 2010—Proceedings of the Third International Conference on Bio-inspired Systems and Signal Processing, Valencia, Spain, January 20–23, 2010, pp. 460–463.

[36] M. Kipp, "Anvil—a generic annotation tool for multimodal dialogue," in Proceedings of the 7th European Conference on Speech Communication and Technology (Eurospeech), pp. 1367–1370, 2001.

[37] S. Marsella and J. Gratch, "EMA: a computational model of appraisal dynamics," *Cybern. Syst.*, 601–606, 2006.

[38] P. Gebhard, "ALMA: a layered model of affect," in Proceedings of the Fourth International Joint Conference on Autonomous Agents and Multiagent Systems, 2005, pp. 29–36.

[39] F. Eyben, M. Wöllmer, and B. Schuller, "openEAR—introducing the Munich open-source emotion and affect recognition toolkit," in 3rd International Conference on Affective Computing and Intelligent Interaction and Workshops, 2009, pp. 1–6.

[40] R. El Kaliouby and P. Robinson, "Real-time inference of complex mental states from facial expressions and head gestures," in Computer Vision and Pattern Recognition Workshop, 2004, pp. 1063–6919.

[41] L. Maat and M. Pantic, "Gaze-X: adaptive, affective, multimodal interface for single-user office scenarios," *Artif. Intell. Hum. Comput.*, 4451: 251–271, 2007.

[42] J. Wagner, F. Lingenfelser, and E. André, "The social signal interpretation framework (SSI) for real time signal processing and recognition," in Proceedings of Interspeech, 2011.

[43] J. Wagner, F. Lingenfelser, N. Bee, and E. André, "Social signal interpretation (SSI)—a framework for real-time sensing of affective and social signals," *Künstl. Intell.*, 25: 251–256, 2011.

[44] N. Bee, B. Falk, and E. André, "Simplified facial animation control utilizing novel input devices: a comparative study," in International Conference on Intelligent User Interfaces (IUI), 2009, pp. 197–206.

[45] S. W. Gilroy, M. Cavazza, M. Niiranen, E. André, T. Vogt, J. Urbain, H. Seichter, M. Benayoun, and M. Billinghurst, "PAD-based multimodal affective fusion," in Affective Computing and Intelligent Interaction (ACII), 2009.

[46] A. Batliner, "Releasing a thoroughly annotated and processed spontaneous emotional database: the FAU Aibo Emotion Corpus," Marrakesh, 2008.

[47] A. Battocchi, F. Pianesi, and D. Goren-Bar, "DaFEx: database of facial expressions," in International Conference on Intelligent Technologies for Interactive Entertainment (INTETAIN), 2005, pp. 303–306.

[48] N. Bee, E. André, T. Vogt, and P. Gebhard, "The use of affective and attentive cues in an empathic computer-based companion," in *Close Engagements with Artificial Companions: Key Social, Psychological, Ethical and Design Issues*, 2010, pp. 131–142.

[49] A. Camurri, P. Coletta, G. Varni, and S. Ghisio, "Developing multimodal interactive systems with EyesWeb XMI," Proceedings of the 7th International Conference on New Interfaces for Musical Expression (NIME), 2007, pp. 305–308.

[50] G. Caridakis, A. Raouzaiou, K. Karpouzis, and S. Kollias, "Synthesizing gesture expressivity based on real sequences," in Workshop on Multimodal Corpora: From Multimodal Behaviour Theories to Usable Models, LREC, Genoa, Italy, May 24–26, 2006.

[51] M. Cavazza, D. Pizzi, F. Charles, T. Vogt, Thurid, and E. André, "Emotional input for character-based interactive storytelling," in Proceedings of the 8th International Conference on Autonomous Agents and Multiagent Systems (AAMAS), Budapest, Hungary, 2009, pp. 313–320.

[52] C. Conati, R. Chabbal, and H. Maclaren, "A study on using biometric sensors for detecting user emotions in educational games," in Proceedings of the Workshop on Assessing and Adapting to User Attitude and Affects: Why, When and How? 2003.

[53] R. V. Bezooijen, A. S. Otto, and T. A. Heenan, "Recognition of vocal expressions of emotion: a three-nation study to identify universal characteristics," *Cross-Cult. Psychol.*, 14(4): 387–406, 1983.

[54] P. Ekman, *Universals and Cultural Differences in Facial Expressions of Emotion*, University of Nebraska Press, 1971.

[55] P. Ekman and W. Friesen, *Facial Action Coding System: A Technique for the Measurement of Facial Movement*, Consulting Psychologists Press, 1978.

[56] P. Ekman, "Strong evidence for universals in facial expressions: a reply to Russell's mistaken critique," *Psychol. Bull.*, 268–287, 1994.

[57] P. Ekman, "An argument for basic emotions," *Cogn. Emot.*, 6(3): 169–200, 1992.

[58] H. A. Elfenbein, M. Beaupré, M. Lévesque, and U. Hess, "Toward a dialect theory: cultural differences in the expression and recognition of posed facial expressions," *Emotion*, 7(1): 131–146, 2007.

[59] H. A. Elfenbein and N. Ambady, "Universals and cultural differences in recognizing emotions," *Curr. Dir. Psychol. Sci.*, 12: 159–164, 2003.

[60] F. Eyben, M. Wöllmer, and B. Schuller, "openSMILE: the Munich versatile and fast open-source audio feature extractor," in Proceedings of the International Conference on Multimedia, 2010, pp. 1459–1462.

[61] S. W. Gilroy, M. Cavazza, R. Chaignon, S. M. Mäkelä, M. Niranen, E. André, T. Vogt, J. Urbain, H. Seichter, M. Billinghurst, and M. Benayoun, "An affective model of user experience for interactive art," in Proceedings of the 2008 International Conference on Advances in Computer Entertainment Technology, 2008, pp. 107–110.

[62] S. Gilroy, M. Cavazza, and V. Vervondel, "Evaluating multimodal affective fusion with physiological signals," in Proceedings of the International Conference on Intelligent User Interfaces, 2011.

[63] J. Gratch and S. Marsella, "A domain-independent framework for modeling emotion," *J. Cogn. Syst. Res.*, 5(4): 296–306, 2004.

[64] M. Hall, E. Frank, G. Holmes, B. Pfahringer, P. Reutemann, and I. H. Witten, "The WEKA data mining software: an update," *SIGKDD Explor. Newsl.*, 11: 10–18, 2009.

[65] B. Hartmann, M. Mancini, and C. Pelachaud, "Implementing expressive gesture synthesis for embodied conversational agents," in *Gesture in Human–Computer Interaction and Simulation* (eds S. Gibet, N. Courty, and J.-F. Kamp), Springer, Berlin/Heidelberg, 2006.

[66] F. Hönig, J. Wagner, A. Batliner, and E. Nöth, "Classification of user states with physiological signals: on-line generic features vs. specialized," in Proceedings of the 17th European Signal Processing Conference (EUSIPCO), 2009, pp. 2357–2361.

[67] G. Jacucci, A. Spagnolli, A. Chalambalakis, A. Morrison, L. Liikkanen, S. Roveda, and M. Bertoncini, "Bodily explorations in space: social experience of a multimodal art installation," in Proceedings of the 12th IFIP TC 13 International Conference on Human–Computer Interaction: Part II, 2009, pp. 62–75.

[68] J. Kim, E. André, M. Rehm, T. Vogt, and J. Wagner, "Integrating information from speech and physiological signals to achieve emotional sensitivity," in Interspeech 2005-Eurospeech, September 2005, pp. 809–812.

[69] J. Kim and E. André, "Emotion recognition based on physiological changes in music listening," *IEEE Transactions on Pattern Analysis and Machine Intelligence*, 30: 2067–2083, 2009.

[70] J. Kim, A. Ragnoni, and J. Biancat, "In-vehicle monitoring of affective symptoms for diabetic drivers," in International Conference on Health Informatics (HEALTHINF 2010), 2010, pp. 367–372.

[71] P. Laukka, "Vocal expression of emotion: discrete-emotions and dimensional accounts," PhD thesis, 2004.

[72] J. A. Russell, "A circumplex model of affect," *J. Personal. Soc. Psychol.*, 39(6): 1161–1178, 1980.

[73] M. Pantic, A. Pentland, A. Nijholt, and T. S. Huang, "Human Computing and Machine Understanding of Human Behavior: A Survey," in Artificial Intelligence for Human Computing, Springer, pp. 47–71, 2007.

[74] R. Polikar, "Ensemble based systems in decision making," *IEEE Circuits and Systems Magazine*, 6(3): 21–45, 2006.

[75] P. Boersma and D. Weenink. Praat: doing phonetics by computer, Computer Program (version 5.1.41), 2010, http://www.praat.org. Last accessed on Sept. 19, 2014.

[76] M. Puckette, "Pure data: another integrated computer music environment," in Proceedings of the Second Intercollege Computer Music Concerts, 1996, pp. 37–41.

[77] A. Rabie, B. Wrede, T. Vogt, and M. Hanheide, "Evaluation and discussion of multimodal emotion recognition," in Proceedings of the Second International Conference on Computer and Electrical Engineering (ICCEE), vol. 1, 2009, pp. 598–602.

[78] I. Rish, "An empirical study of the naive Bayes classifier," in Workshop on Empirical Methods in Artificial Intelligence, IJCAI, 2001.

[79] A. K. Seewald and J. Fuernkranz, "An evaluation of grading classifiers," in 4th International Conference on Advances in Intelligent Data Analysis (IDA), vol. 10, Springer, 2001, pp. 271–289.

[80] M. Serrano, L. Nigay, J.-Y. L. Lawson, A. Ramsay, R. Murray-Smith, and S. Denef, "The openinterface framework: a tool for multimodal interaction," in Extended Abstracts on Human Factors in Computing Systems, CHI, 2008, pp. 3501–3506.

[81] M. Shami and W. Verhelst, "Automatic classification of expressiveness in speech: a multi-corpus study," in *Speaker Classification II*, Springer, 2007, pp. 43–56.

[82] J. Urbain, R. Niewiadomski, E. Bevacqua, T. Dutoit, A. Moinet, C. Pelachaud, B. Picart, J. Tilmanne, and J. Wagner, "AVLaughter cycle," *J. Multimodal User Interfaces*, 4: 47–58, 2010.

[83] T. Vogt, E. André, J. Wagner, S. Gilroy, F. Charles, and M. Cavazza, "Real-time vocal emotion recognition in artistic installations and interactive storytelling: experiences and lessons learnt from CALLAS and IRIS," in Proceedings of the International Conference on Affective Computing and Intelligent Interaction (ACII), 2009.

[84] T. Vogt and E. André, "An evaluation of emotion units and feature types for real-time speech emotion recognition," *Künstl. Intell.*, 25: 213–223, 2011.

[85] J. Wagner, J. Kim, and E. André, "From physiological signals to emotions: implementing and comparing selected methods for feature extraction and classification," in IEEE International Conference on Multimedia and Expo (ICME), 2005, pp. 940–943.

[86] J. Wagner, E. André, and F. Jung, "Smart sensor integration: a framework for multimodal emotion recognition in real-time," in Third IEEE Conf. on Affective Computing and Intelligent Interaction (ACII), pp. 1–8, 2009.

[87] J. Wagner, F. Jung, J. Kim, E. André, and T. Vogt, "The smart sensor integration framework and its application in EU projects," in Workshop on Bio-inspired Human–Machine Interfaces and Healthcare Applications, 2010, pp. 13–21.

[88] J. Wagner, F. Lingenfelser, E. André, and J. Kim, "Exploring fusion methods for multimodal emotion recognition with missing data," *IEEE Transactions on Affective Computing*, vol. 99, 2011.

[89] K. M. Ting and I. H. Witten, "Issues in stacked generalization," *J. Artif. Intell. Res.*, 10: 115–124, 1999.

[90] J. O. Wobbrock, A. D. Wilson, and Y. Li, "Gestures without libraries, toolkits or training: a $1 recognizer for user interface prototypes," in Proceedings of the 20th Annual ACM Symposium on User Interface Software and Technology, 2007, pp. 159–168.

AUTHOR BIOGRAPHIES

Johannes Wagner graduated as a Master of Science in Informatics and Multimedia from the University of Augsburg, Germany, in 2007. Afterward he joined the Chair for Human-Centered Multimedia of the same university. Among other projects, he has been working on multimodal signal processing in the framework of CALLAS and is currently developing a general framework for the integration of multiple sensors into multimedia applications called Social Signal Interpretation (SSI).

Florian Lingenfelser received his M.Sc. degree in Informatics and Multimedia from the University of Augsburg, Germany, in 2009. In 2010 he joined the Chair for Human-Centered Multimedia of the same university as Ph.D. student. He is currently contributing to multimodal data fusion within the CEEDS project and developing the Social Signal Interpretation framework (SSI).

Elisabeth André is full professor of computer science at Augsburg University and chair of the Laboratory for Human-Centered Multimedia. Prior to that, she worked as a principal researcher at DFKI GmbH, where she has been leading various academic and industrial projects in the area of intelligent user interfaces. In 2007 Elisabeth André was nominated a fellow of the Alcatel-Lucent Foundation for Communications Research. In 2010, she was elected a member of the prestigious German Academy of Sciences Leopoldina and the Academy of Europe. Her research interests include affective computing, intelligent multimedia interfaces, and embodied agents.

16

SEMANTIC AUDIOVISUAL DATA FUSION FOR AUTOMATIC EMOTION RECOGNITION

DRAGOS DATCU
Faculty of Military Sciences, Netherlands Defence Academy, Den Helder, The Netherlands

LEON J. M. ROTHKRANTZ
Department of Computer Science, Delft University of Technology, Delft, The Netherlands

Emotions play an important role in everyday life of human beings. Our social behavior is essentially based on communication, both in terms of language and non-verbal interaction. Information from the environment determines individuals to mutually interpret other persons' intentions, goals, thoughts, feelings, and emotions and to change the behavior accordingly. Within the last couple of years, automatic multimodal recognition of human emotions has gained a considerable interest from the research community. From the technical point of view, the challenge is, in part, supported also by the successes that have been noticed in the development of methods for automatic recognition of emotion from separate modalities. By taking into account more sources of information, the multimodal approaches allow for more reliable estimation of the human emotions. They increase the confidence of the results and decrease the level of ambiguity with respect to the emotions among the separate communication channels. This chapter provides a thorough description of a bimodal emotion recognition system that uses face and speech analysis. We use hidden Markov models—HMMs to learn and to describe the temporal dynamics of the emotion clues in the visual and acoustic channels. This approach provides a powerful method enabling to fuse the data we extract from separate modalities. We report on

Emotion Recognition: A Pattern Analysis Approach, First Edition. Edited by Amit Konar and Aruna Chakraborty.
© 2015 John Wiley & Sons, Inc. Published 2015 by John Wiley & Sons, Inc.

the results we have achieved so far for the discussed models. The last part of the chapter relates to conclusions and discussions on the possible ways to continue the research on the topic of multimodal emotion recognition.

16.1 INTRODUCTION

Within the last couple of years, automatic multimodal recognition of human emotions has gained a considerable interest from the research community. From the technical point of view, the challenge is, in part, supported also by the successful developments that have been noticed in the development of methods for automatic recognition of emotion from separate modalities. By taking into account more sources of information, the multimodal approaches allow for more reliable estimation of the human emotions. They increase the confidence of the results and decrease the level of ambiguity with respect to the emotion, among the separate communication channels.

This chapter provides a thorough description of a bimodal emotion recognition system that uses face and speech analysis. The multimodal models we build are based on algorithms which are different than the ones used. Basically, we use HMMs to learn and to describe the temporal dynamics of the emotion clues in the visual and acoustic channels. The approach provides a powerful method which allows for fusing the data we extract from separate modalities.

The complexity of emotion recognition using multiple modalities is higher than the complexity of unimodal methods. Some causes for that relate to the asynchronous character of the emotion patterns and the ambiguity and the correlation which possibly occur in the different informational channels. For instance, speaking while expressing emotions implies that the mouth shape corresponds to a combination of the influence of both the pronounced phoneme and the internal emotional state. In this case, the use of the regular algorithms we have used so far for facial expression recognition shows limited performance and reliability. In order to apply fusion, the model differentiates the silence video segments and the segments that show the subject speaking. Figure 16.1 depicts an example of the speech–silence-based segmentation. The models

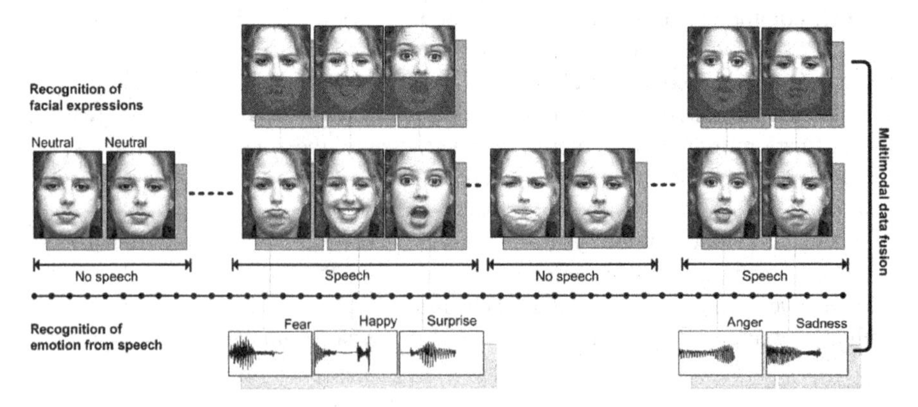

FIGURE 16.1 Fusion model using speech and silence audio segments. Extracted with permission from Enterface 2005 database.

we build in this chapter run emotion analysis on the data segments which regard activity in both visual and audio channels. In the following section, we present the algorithms and the results achieved in some recent and relevant research works in the field of multimodal emotion recognition. Then, we describe our new system. We present the details of all the steps involved in the analysis, from the preparation of the multimodal database and the feature extraction to the classification of six prototypic emotions. Apart from working with unimodal recognizers, we conduct experiments on early fusion and focus more on the decision-level fusion of visual and audio features.

The novelty of our approach consists of the dynamic modeling of emotions using the combination of local binary patterns (LBPs) [1] as visual features and mel-frequency cepstral coefficients (MFCCs) as audio features. In the same time, we propose a new method for visual feature selection based on the multiclass Adaboost.M2 classifier. A cross database method is employed to identify the set of most relevant features from a unimodal database and to proceed with applying it in the context of the multimodal setup.

We report on the results we have achieved so far for the discussed models. The last part of the chapter relates to conclusions and discussions on the possible ways to continue the research on the topic of multimodal emotion recognition.

16.2 RELATED WORK

A noticeable approach for approaching the recognition of emotion represents the multimodal analysis. The multimodal integration of speech and face analysis can be done by taking into account features at different levels of abstraction. Depending on that, the integration takes the form of fusion at the low, intermediate, or high levels. The low-level fusion is also called early fusion or fusion at the signal level and the high-level fusion is also called semantic, late fusion, or fusion at the decision level. Several researchers have recently tackled these types of integration.

Paleari *et al.* [2] presented an extensive study conducted on feature selection for automatic, audiovisual real-time and person-independent emotion recognition. Han *et al.* [3] proposed a method for bimodal recognition of four emotion categories plus the neutral state, based on hierarchical support vector machine (SVM) classifiers. Binary SVM classifiers made use of fusion of low-level features to determine the dominant modality that, in turn, leads to the estimation of emotion labels. The video processing implied the use of skin color segmentation for face detection and optical density and edge detection for face feature localization. The algorithm extracted 12 geometrical features based on the location of specific key points on the face area. In case of speech analysis, 12 feature values were computed using the contours of pitch and the energy from the audio signal. On a database of 140 video instances, the authors reported an improvement of 5% compared to the performance of the facial expression recognition and an improvement of 13%, compared to the result of the emotion recognition from speech.

The paper [4] studied early feature fusion models based on statistically analyzing multivariate time series for combining the processing of video-based and audio-based low-level descriptors (LLDs). Paleari and Huet [5] researched the multimodal recognition of emotions on eNTERFACE 2005 database. They used mel-frequency cepstral coefficients(MFCCs) and linear predictive coding (LPC) for emotion recognition and optical flow for facial expression recognition together with support vector machines and neural network classifiers. The recognition rate of emotion classification was less than 35% for speech-oriented analysis and less than 30% in case of face-oriented analysis. Though, combining the two modalities leads to an improvement of 5% in case of fusion at the decision level and to almost 40% recognition rate in case of early fusion.

Another study research on the eNTERFACE 2005 database was presented by Mansoorizadeh and Charkari [6]. They applied principal components analysis—PCA to reduce the size of the audio and visual feature vectors and binary SVM for the bimodal person-dependent classification of basic emotions. Depending on the type of fusion, the inputs of the SVM models contained either separate or combined audiovisual feature vectors. For speech analysis, the features related to the energy, the pitch contour, the first four formants, their bandwidth, and 12 MFCC components of the audio signal. For face analysis, the features represented geometric features that were computed based on a set of specific key points on the face area. The likelihood results of the binary SVM classifiers were used in a rule-based system to determine the emotion labels of the video instances. The authors reported the 53% the classification rate for emotion recognition from speech, 36.00% for facial expression recognition, 52.00% for feature-level fusion, and 57.00% for decision-level fusion.

The work of Hoch *et al.* [7] presented an algorithm for bimodal emotion recognition in automotive environment. The fusion of results from unimodal acoustic and visual emotion recognizers was realized at abstract decision level. For the analysis, the authors used a database of 840 audiovisual samples that contain recordings from seven different speakers showing three emotions. By using a fusion model based on a weighted linear combination, the performance gain became nearly 4% compared to the results in the case of unimodal emotion recognition.

The work of Song *et al.* [8] presented a emotion recognition method based on active appearance models (AAM) for facial feature tracking. Facial animation parameters (FAPs) were extracted from video data and are used together with low-level audio features as input for an HMM to classify the human emotions.

The paper [9] presented a multimodal fusion framework for emotion recognition that relies on Multimodal Affective User Interface (MAUI) paradigm. The approach was based on the Scherer's theory Component Process Theory (CPT) for the definition of the user model and to simulate the agent emotion generation.

Sebe *et al.* [10] proposed a Bayesian network topology for recognizing emotions from audio and facial expressions. The database they used included recordings of 38 subjects who show 11 classes of affects. According to the authors, the achieved performance results pointed to around 90% for bimodal classification of emotions from speech and facial expressions compared to 56% for the face-only classifier and about 45% for the prosody-only classifier.

Zeng *et al.* [11] conducted a series of experiments related to the multimodal recognition of spontaneous emotions in a realistic setup for Adult Attachment Interview. They used Facial Action Coding System (FACS) [12] to label the emotion samples.

Their bimodal fusion model combined facial texture and prosody in a framework of Adaboost multi-stream hidden Markov model (AdaMHMM).

Joo *et al.* [13] investigated the use of S-type membership functions for creating bimodal fusion models for the recognition of five emotions from speech signal and facial expressions. The achieved recognition rate of the fusion model was 70.4%, whereas the performance of the audio-based analysis was 63% and the performance of the face-based analysis was 53.4%.

Caridakis *et al.* [14] described a multi-cue, dynamic approach in naturalistic video sequences using recurrent neural networks. The approach differs from the existing works at the time, in the way that the expressibility was modeled using a dimensional representation of activation and valence instead of the prototypic emotions. The facial expressions were modeled in terms of geometric features from MPEG-4 FAPs and were computed using the location of 19 key points on the face image. Combining FAPs and audio features related to pitch and rhythm leads to the multimodal recognition rate of 79%, as opposed to facial expression recognition rate of 67% and emotion from speech detection rate of 73%.

The work of Meng *et al.* [15] presented a speech-emotion recognizer that worked in combination with an automatic speech recognition system. The algorithm used HMM as a classifier. The features considered for the experiments consisted of 39 MFCCs plus pitch, intensity, and three formants, including some of their statistical derivates.

An emotion recognition study on a language-independent database has been done in Reference 16. The authors extracted MFCC and formant frequency features from the speech signal and Gabor wavelet features from the face images. The classification of six emotions was done using neural networks and Fisher's linear discriminant analysis (FLDA). The results indicated the higher efficiency of using the audio signal with 66.43% recognition rate over the visual processing with 49.29% recognition rate. The audiovisual fusion had the classification rate of 70%.

Busso *et al.* [17] explored the properties of both unimodal and multimodal systems for emotion recognition in case of four emotion classes. In this study, the multimodal fusion was realized separately at the semantic level and at the feature level. The overall performance of the classifier based on feature-level fusion is 89.1% which was close to the performance of the semantic fusion-based classifier when the product-combining criterion was used.

Go *et al.* [18] used Z-type membership functions to compute the membership degree of each of the six emotions based on the facial expression and the speech data. The facial expression recognition algorithm used multi-resolution analysis based on discrete wavelets. An initial gender classification was done by the pitch of the speech signal criterion. The authors reported final emotion recognition results of 95% in case of male and 98.3% for female subjects. Fellenz *et al.* [19] used a hybrid classification procedure organized in a two-stages architecture to select and fuse the features extracted from face and speech to perform the recognition of emotions. In the first stage, a multilayered perceptron (MLP) was trained with the back propagation of error procedure. The second symbolic stage involved the use of PAC learning paradigm for Boolean functions.

Kanluan *et al.* [20] described emotions using three continuous valued emotion primitives namely valence, activation, and dominance in a 3D emotion space. The values of the emotion primitives were estimated using decision-level fusion and support vector regression on prosodic features, spectral features, and visual features based on 2D discrete cosine transform. More recently, Wang *et al.* [21] approached the problem of multimodal emotion recognition by fusing kernel matrices at feature and score levels, through algebraic operations and using dimensionality reduction methods to map the original disparate features to a nonlinear joint subspace. An HMM was used to handle the temporal structure of the features and the statistical dependence across successive frames.

16.3 DATA SET PREPARATION

The models we are going to build for multimodal emotion recognition are based on the use of HMM classifier. In the current research context, HMM is used as a supervised machine learning technique. Based on that, the HMM training and testing processes rely on the use of fully labeled samples of audiovisual data instances.

At the moment of starting this research, finding a fully annotated database turned to be difficult to fulfill. This was first because of the lack of multimodal databases. Some databases had no emotion labels and were not proper for audiovisual processing. We specifically avoided using multimodal data sets that have recordings with noise and utterance overlapping in the audio signal, or with occlusion and too much rotation of the subjects' face.

The database we have eventually decided to use for our research is eNTERFACE 2005 [22]. This database contains audiovisual recordings of 42 subjects who represent 14 different nationalities. A percentage of 81% are men, while the remaining 19% are women. At the recording time, 31% of the subjects wore glasses and 17% had beard.

The recording procedure first consisted of listening to six successive short stories, each of them eliciting a particular emotion. The emotions relate to the prototypic emotions which are happiness, sadness, surprise, anger, disgust, and fear, as identified by Ekman [12]. Then, the subjects had to read, memorize, and finally utter five different reactions to each story, all by using English language. For each story, the subjects were asked to produce messages that contain only the emotion to be elicited and to show as much expressiveness as possible. The recording setup implied the use of a monochromatic dark gray panel for the image background and constant illumination. The audiovisual data was encoded using Microsoft AVI format. The image frames were stored using the image resolution of 720×576 pixels, at the frame rate of 25 frames per second. The audio samples were stored using uncompressed stereo 16-bit format at the sample rate of 48,000 Hz. Figure 16.2 illustrates an example from eNTERFACE 2005 database.

We have started the data pre-processing step from the set of 1293 samples from eNTERFACE 2005 database. We have used the database and have extracted the whole set of audio instances for building models for emotion extraction from the speech signal. In the context of multimodal processing, we had to first verify the appropriateness of each video sample. As a result, we have removed a subset of

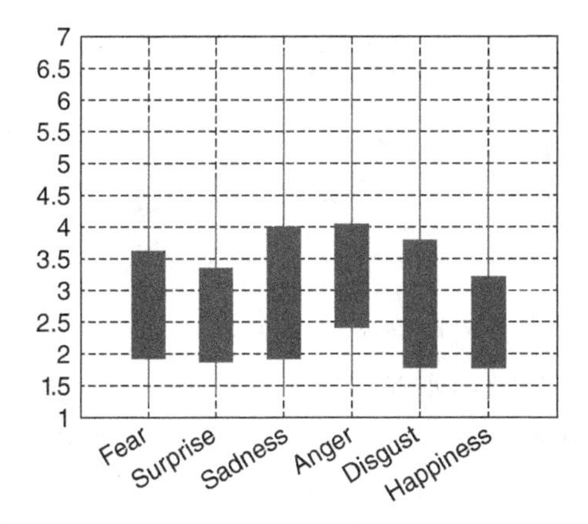

FIGURE 16.2 An example from eNTERFACE 2005 database.

463 instances. From the set of 830 remaining samples, 135 accounted for emotion class fear, 143 for surprise, 137 for sadness, 145 for anger, 141 for disgust, and 129 for happiness. This subset represents a well-balanced multimodal database of simulated emotion recordings from 30 subjects. Figure 16.3 illustrates the duration in seconds of the utterances from the final multimodal database. Like in the case of unimodal emotion recognition, the data pre-processing step involved using specific vision-oriented methods for extracting and normalizing the actual face images from each video frame. At first, we used Viola&Jones face detection algorithm [23] and AAMs [24] to obtain the location and the shape of the faces. Then, we have removed the unnecessary image patches and scaled down the face images to 60 pixels width

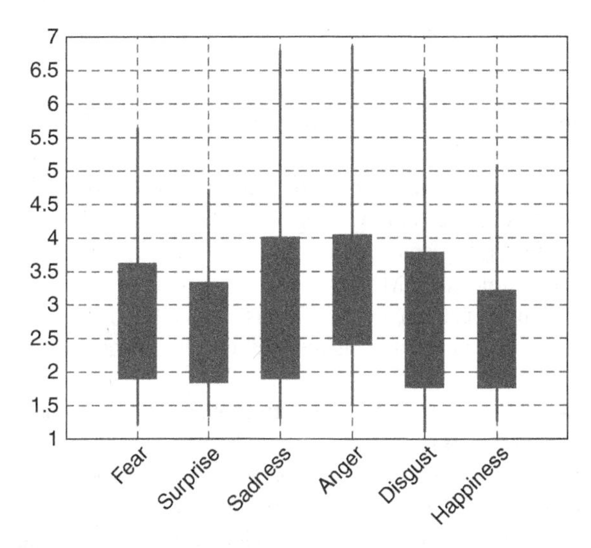

FIGURE 16.3 Utterance duration (in seconds) for each emotion class.

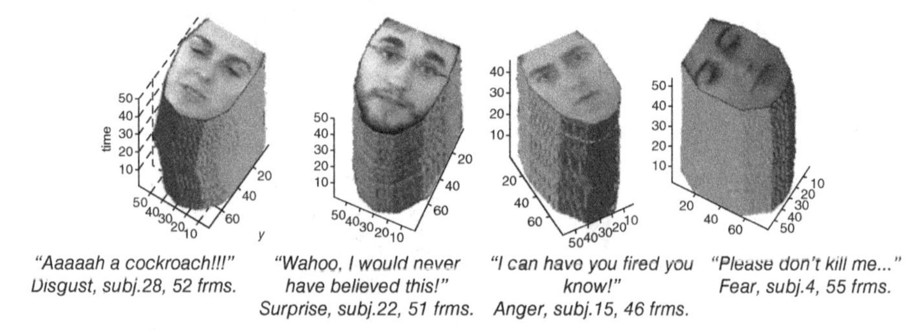

FIGURE 16.4 60×80 video samples containing the face area only.

by 80 pixels height. Here, unnecessary image patches relate to the visible parts of background, subject's hair and cloth.

For aligning the faces, we used the reference key point located at the middle of the line segment delimitated by the inner corners of the eyes. Figure 16.4 illustrates the result of applying the previously described methods on four video samples containing faces.

16.4 ARCHITECTURE

16.4.1 Classification Model

HMM represents a statistical method for modeling data that can be characterized in terms of an underlying process which generates measurable and observable sequences. The causality of the observation sequence $O = (o_1, o_2, \dots, o_T)$ is interpreted through the so-called hidden states $Q = (q_1, q_2, \dots, q_T)$. The observations and the states are part of a state alphabet set $S = (s_1, s_2, \dots, s_N)$ and of an observation alphabet set $V = (v_1, v_2, \dots, v_M)$.

By definition, the HMM model has the form: $\lambda = (A, B, \pi)$. The term A refers to the state transition table that gives the time-independent probability of one state j following state i: $A = [a_{ij}], a_{ij} = P(q_t = s_j | q_{t-1} = s_i)$. The term B is the observation model which gives the time-independent probability of observation k being generated from state j: $B = [b_i(k)], b_i(k) = P(x_t = v_k | q_t = s_i)$. The last element of the HMM model π refers to the initial state probability: $\pi = [\pi_i], \pi_i = P(q_1 = s_i)$. There are two assumptions that hold for HMM. According to the first assumption, the current state depends only on the previous state. This is also called the Markov assumption: $P(q_t | q_1^{t-1}) = P(q_t | q_{t-1})$. The relation is also interpreted as the base for the memory of the HMM model.

The second assumption states that the output observation at time t depends only on the current state and so, it is independent on the previous states and observations: $P(o_t | o_1^{t-1}, q_1^t) = P(o_t | q_t)$. Given the definition of HMM, it is possible to identify the parameters A, B, and λ that best explain the observed data. Further on, it is possible to compute the probability that the model produced the observed data. Eventually, the most probable sequence of steps for some observation data can be determined, given the model.

In a multiclass classification setup, the decision is taken according to the model that gives the best prediction for an observation sequence. The sequence of hidden states which is most likely to have produced an observation sequence can be decoded using Viterbi algorithm. For the given face data set, the goal of training HMM is to determine the parameters $\lambda = (A, B, \pi)$ of the model. One solution for that is to first divide the training observation vectors equally amongst the states of the model and to compute the mean and variance of each state. Then, finding the maximum likelihood state sequence with Viterbi algorithm allows us to reassign the observation vectors to states and to recompute the mean and variance of each state. The process can be repeated until the mean and variance estimates do not change. The re-estimation procedure is done using Baum–Welch re-estimation. For each class of emotions, one HMM model is constructed. Let these models be designated as l_i and $i = 1, \ldots, C$, where C equals the number of emotion categories. A given observation sequence O is evaluated by computing the probabilities of individual models $P(l_i|O)$. Applying the Bayes rule and assuming equal a priori probability of each model $P(l_i)$, the pattern classification can be performed by maximizing the likelihood function $P(O|l_i)$.

All the unimodal and multimodal experiments with HMMs we have conducted in this research and which are presented in the following parts of the chapter are based on the use of HTK toolkit for HMMs [25] developed by Microsoft Corporation and Cambridge University Engineering Department.

16.4.2 Emotion Estimation from Speech

The assessment of the emotion levels from speech can be naturally done by identifying patterns in the audio data and by using them in a classification setup. The features we extract are the energy component and 12 MFCCs together with their delta and the acceleration terms from 25 ms audio frames, with 10 ms frame periodicity from a filter bank of 26 channels. A Hamming window is used on each audio frame during the application of Fourier transform. Using the feature extraction procedure determines the conversion of the original audio sampling rate of 48 kHz to the MFCC frame rate of 100 Hz. Each MFCC frame contains 39 terms, as indicated previously.

The recognition of emotions is realized using the HMM algorithm. Each emotion has associated one distinct HMM and the set of HMMs forms a multiclass classifier. For evaluation, we use threefold cross-validation. The samples from the same subject are part of either the training set or the test set. This restriction assumes that the testing is done on instances of subjects other than those of the subjects included in the training data set. The method is supposed to give a better estimation of the performance of the classifiers. For finding the best HMM model, we conduct experiments in which we investigates the optimal values for the HMM parameters. In this way, we build and test models which use two, three, and four HMM states. The two-state HMMs encode the emotion onset and offset. The three-state HMMs encode the emotion onset, apex, and offset. The models with four states encode the neutral state and emotion onset, apex, and offset states.

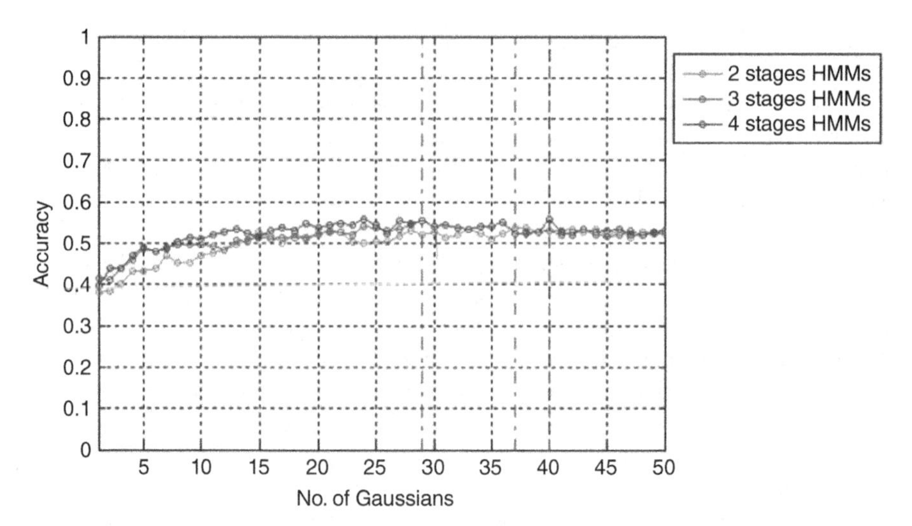

FIGURE 16.5 The accuracy of HMM-based classifiers of emotions in speech signal. The number of states is 2, 3, 4 and the number of Gaussians varies from 1 to 50. Threefold cross-validation method is used for performance estimation.

For each state configuration, we build distinct models of HMMs with Gaussian mixtures with different number of components (1–50 components). The results of testing all the models are illustrated in Figure 16.5. Following the evaluation, it results that the most efficient configuration is to use 4 states and 40 Gaussians per mixture and that the accuracy of this classifier is 55.90%. Table 16.1 presents the confusion matrix of this classification model.

16.4.3 Video Analysis

The goal of the video analysis is to build models to dynamically process the video data and to generate labels according to the six basic emotion classes. The input data is represented by video sequences that can have different number of frames, as

TABLE 16.1 The confusion matrix of the HMM that has 4 states and 40 Gaussian components; the accuracy of the emotion recognition from speech model is 55.90% for six basic emotion categories

	Fear	Surprise	Sadness	Anger	Disgust	Happy
Fear	**91.72**	2.07	0.69	1.38	2.76	1.38
Surprise	24.11	**44.68**	9.22	11.35	4.26	6.38
Sadness	25.19	14.81	**41.48**	6.67	5.19	6.67
Anger	19.38	18.60	3.88	**48.06**	6.98	3.10
Disgust	23.78	9.09	8.39	8.39	**38.46**	11.89
Happy	5.84	5.84	10.22	2.19	6.57	**69.34**

determined by the utterance-based segmentation method. The limits of each video data segment are identified by using the information obtained during the analysis of the audio signal. Based on the set of frames, a feature extraction is applied for preparing the input to the actual classifier. HMM models are then employed to classify the input sequence in terms of the emotion classes.

One problem that has to be taken into account while developing the facial expression recognizers is that both the input set of features and the classifier models should be chosen in such a way so as to be able to handle the time-dependent variability of the face appearance. More specifically, some of the inner dynamics of the face are generated due to the effect of the speech process that is present in the data. Taking into account the aforementioned issues, the focus of the research is to study the selection of most relevant visual features and to use the values of these features as data observations for the HMM-based classifiers.

16.4.3.1 Database Preparation The first problem of feature selection is tackled by conducting a separate research using a second database namely Cohn–Kanade [26]. In this context, the multi-class classification method Adaboost.M2 is used as a feature selection algorithm. The procedure is based on the primary property of Adaboost.M2 to identify the most important features while running the training phase of the classification process. We use the same set of prototypic emotions as for the main study on eNTERFACE 2005 dataset.

The first problem is to make a proper data set of representative face image samples. For this, we use the results of the study on Cohn–Kanade Action Units-based facial expression labeling. The basic set of non-ambiguous facial expression samples from Cohn–Kanade database include 251 instances. Each instance corresponds to the last frame of the video sequence and represents the face at the apex of one facial expression. As it can be seen in Table 16.2, the structure of this set indicates a rather unbalanced data. While emotion class happiness accounts for 99 samples, there are only seven instances for each of the emotion classes surprise and disgust. To obtain a balanced dataset, we adopt the solution of adding new instances for the undersized classes and to remove data instances from the oversized classes.

TABLE 16.2 The structure of the non-ambiguous set of 251 samples in Cohn–Kanade database

	No. non-ambiguous samples							No. mixed emotion samples						
Distance	**0**	**1**	**2**	**3**	**4**	**5**	Total	**0**	**1**	**2**	**3**	**4**	**5**	Total
Fear	2	8	13	6	6	3	**38**	0	0	0	0	0	0	**0**
Surprise	0	6	1	0	0	0	**7**	0	0	0	0	0	0	**0**
Sadness	0	4	23	13	7	0	**47**	0	0	0	0	0	0	**0**
Anger	0	4	12	12	18	7	**53**	0	0	0	0	0	0	**0**
Disgust	0	0	6	1	0	0	**7**	0	0	0	0	0	0	**0**
Happiness	18	55	22	4	0	0	**99**	0	0	0	0	0	0	**0**

In case of emotion class fear, we have added 12 image samples by selecting one frame which precedes the apex in 12 video sequences of emotion fear, from the non-ambiguous set of 251 samples. The Action Unit-based patterns of two selected instances have distance 1, four instances have distance 2, five instances have distance 3, and one has distance 4 to the Action Unit pattern of class fear. For class surprise we add 21 non-ambiguous new face images by selecting three frames preceding the frame showing the facial expression apex in seven different videos. The selected frames are not consecutive and show considerable degree of facial expression onset. Furthermore, 22 new instances are added to the set of surprise class from the sample subset of mixed emotions. The Action Unit-based patterns of 10 of these instances have distance 1 and the rest have distance 2 to the Action Unit pattern of class surprise. The size of emotion class sadness is increased by adding three instances as preceding frames to the apex of three video sequences. They have AU pattern distances 1, 2, and 3, respectively.

The emotion class anger has neither received new samples nor it has undergone sample removal. For emotion disgust, we add 21 non-ambiguous instances from 7 videos and 22 mixed emotion face instances each being related to the apex in the sequence. Similarly to the case of emotion class surprise, the non-ambiguous samples precede the sample associated to the apex frame and are not consecutive. From each video sequence, we have selected three such instances. For the samples of mixed emotions, we selected four instances with Action Unit patterns at distance 1 and 18 samples with Action Unit patterns at distance 2 to the Action Unit pattern of emotion disgust. In the case of the last emotion class, happiness, we have removed 49 image samples from the initial non-ambiguous set. The structure of the resulting balanced database is depicted in Table 16.3.

16.4.3.2 Visual Feature Selection Because the sets of the visual features we derive from the face samples are too large to be used directly as observations in the HMM classification setup, we have to decrease their size. This can be done by transforming the original visual features to other set of more representative features. Such a method is used by Mansoorizadeh and Charkari [6] by applying PCA technique to obtain a reduced set of visual features. Another approach is to select only a limited

TABLE 16.3 The structure of the balanced set of 303 samples selected from Cohn–Kanade database

	No. non-ambiguous samples							No. mixed emotion samples						
Distance	**0**	**1**	**2**	**3**	**4**	**5**	Total	**0**	**1**	**2**	**3**	**4**	**5**	Total
Fear	2	10	17	11	7	3	**50**	0	0	0	0	0	0	**0**
Surprise	0	24	4	0	0	0	**28**	0	10	12	0	0	0	**22**
Sadness	0	5	24	14	7	0	**50**	0	0	0	0	0	0	**0**
Anger	0	4	12	12	18	7	**53**	0	0	0	0	0	0	**0**
Disgust	0	0	24	4	0	0	**28**	0	4	18	0	0	0	**22**
Happiness	17	21	10	2	0	0	**50**	0	0	0	0	0	0	**0**

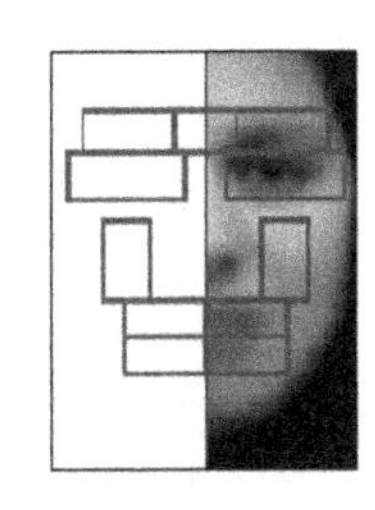

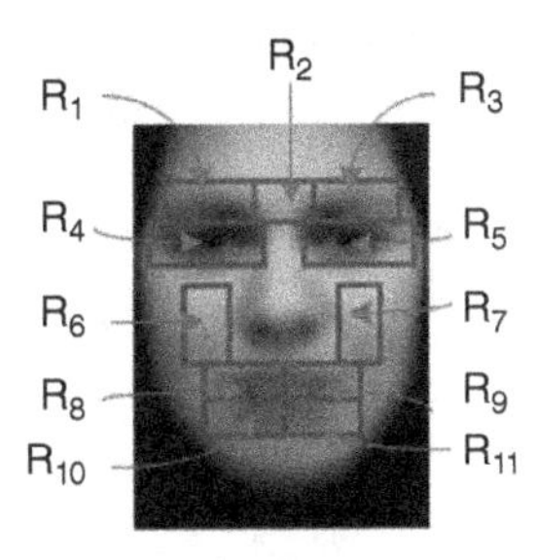

FIGURE 16.6 Average face sample from Cohn–Kanade balanced database. A symmetric facial feature model is used to delimitate rectangular face regions from which specific visual features are extracted. Taken with permission from Cohn–Kanade balanced database.

number of relevant visual features. Wang and Guan [16] used a stepwise method based on Mahalanobis distance. Genetic algorithms can also be employed at this step [27]. Shan *et al.* [28] proposed matrix-based canonical correlation analysis (MCCA) as a method to reduce the feature set by identifying the factor pairs from mouth and eye regions, that best represent the facial expressions. The boosting methods represent a specific class of algorithms that can be successfully used to select representative features.

We do the feature selection by following the same steps we have made for unimodal facial expression recognition. We use LBPs and Adaboost.M2 classifier. The result of this part of research will be later applied for the facial expression recognition in video data of speaking subjects. Previous studies on facial expression recognition in single images, like the one of Dubuisson *et al.* [29], showed that different face regions produce features with different informative power for classification.

In our dynamic recognition setup, we want also to investigate the contribution of speaking mouth region and other face regions to the classification. In addition to using the whole face image, we define two symmetric models of face regions around the face features. Figure 16.6 illustrates the face regions taken into account. Regions R_8, R_9, R_{10}, and R_{11} are located on the mouth area and therefore are considered to be essentially influenced during the production of speech and during expressing emotions. The first face region model consists of using regions R_1 to R_7 and the second model consists of using regions R_1 to R_{11}.

We generated 27.226 LBP features located on the whole face image, 22.276 LBP features based on the second face region model and 14.176 LBP features based on the first face region model. Adaboost.M2 classifier was then used to identify the features that provided the best facial expression recognition results. For evaluation we used 20-folds cross-validation method.

Figures 16.7, 16.9, and 16.11 show the train and test mismatch rates achieved by Adaboost.M2 for each LBP-based data set. The left side of these figures illustrates the test mismatch rate for each emotion category. The minimum test mismatch rate in the case of using the whole face LBP feature set is 30.69%. The result has been achieved for the optimal number of 30 stages of training, with the train mismatch rate of 0.42%.

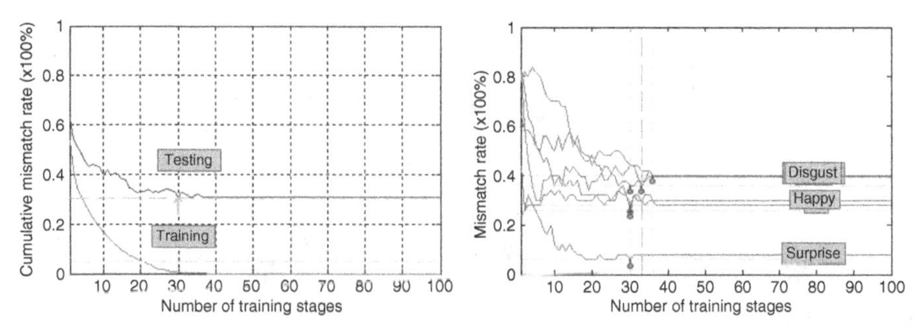

FIGURE 16.7 Train and test mismatch rate of Adaboost.M2 using LBPs from the whole face image.

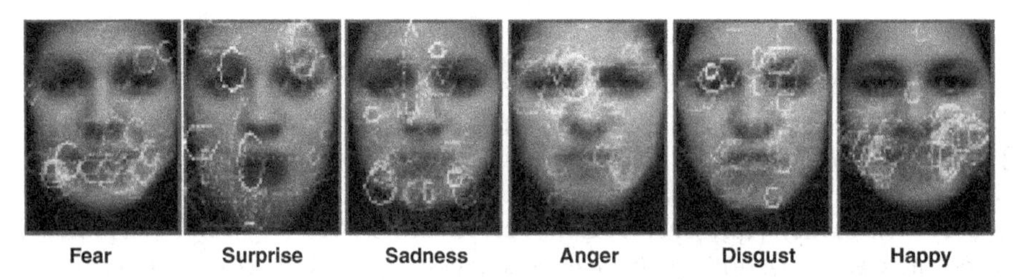

FIGURE 16.8 Projection of the set of 30 LBPs of the whole face model, on average face images showing the six basic emotions.

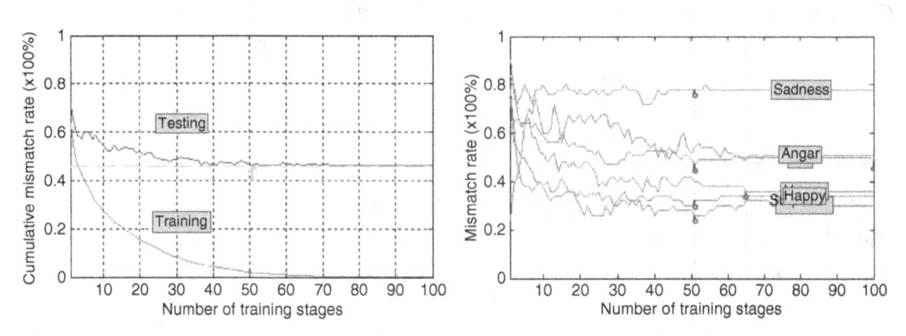

FIGURE 16.9 Train and test mismatch rate of Adaboost.M2 using LBPs from seven face regions.

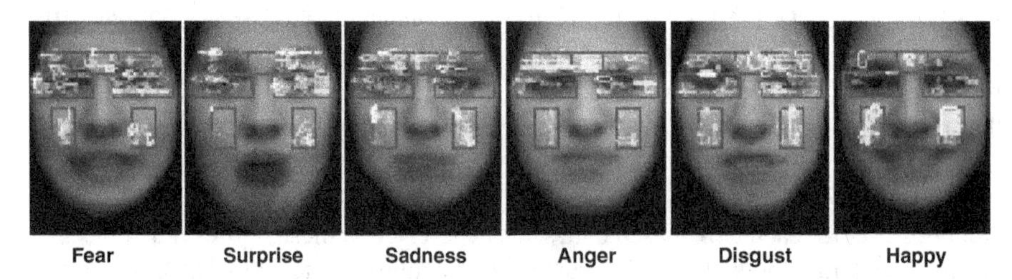

FIGURE 16.10 Projection of the set of 51 LBPs of the model of seven face regions, on average face images showing the six basic emotions.

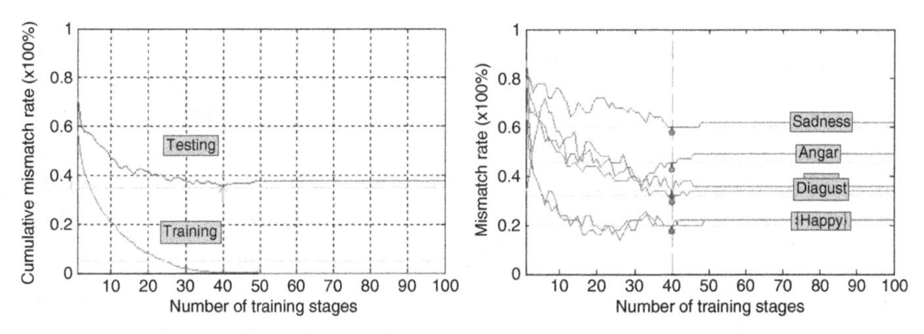

FIGURE 16.11 Train and test mismatch rate of Adaboost.M2 using LBPs from the 11 face regions.

Using LBP features according to the first face model of seven regions generates a classifier with the minimum test mismatch rate of 46.20%, the train mismatch rate of 1.82% for 51 stages of training. Finally, using LBP features according to the second face model of 11 regions generates a classifier with the minimum test mismatch rate of 35.31% and the train mismatch rate of 0.26% for 40 stages of training. Tables 16.4, 16.5, and 16.6 show the confusion matrices of the three LBP-based Adaboost.M2 classifiers.

Based on the previous feature selection procedures, we find a relevant set of features for each facial expression category. Figures 16.8, 16.10, and 16.12 show graphical representations for the projection of the optimal LBP features on average faces for each facial expression category. The graphical representations account for the LBP features which are selected by Adaboost.M2 during training of all the 20 folders of the cross-validation method. For instance, for each facial expression class in Figure 16.8, the projection contains 600 LBP features (30 features × 20 folders).

Tables 16.7 and 16.8 show the proportion of optimal LBP features from each face region, for the model of 7 face regions and for the model of 11 face regions.

16.4.3.3 HMM-Based Facial Expression Recognition Making facial expression recognizers with HMMs implies the identification of the optimal model parameters.

TABLE 16.4 Confusion matrix of the Adaboost.M2 facial expression classifier using LBP features extracted from the whole face image

	Fear	Surprise	Sadness	Anger	Disgust	Happy
Fear	**74.00**	4.00	4.00	2.00	6.00	10.00
Surprise	2.00	**94.00**	2.00	0.00	2.00	0.00
Sadness	10.00	4.00	**56.00**	8.00	14.00	8.00
Anger	5.66	0.00	18.86	**64.15**	7.54	3.77
Disgust	8.00	0.00	24.00	10.00	**56.00**	2.00
Happy	6.00	0.00	6.00	6.00	10.00	**72.00**

TABLE 16.5 Confusion matrix of the Adaboost.M2 facial expression classifier using LBP features extracted from seven face regions

	Fear	Surprise	Sadness	Anger	Disgust	Happy
Fear	**46.00**	4.00	26.00	8.00	2.00	14.00
Surprise	12.00	**74.00**	6.00	2.00	6.00	0.00
Sadness	28.00	12.00	**22.00**	16.00	16.00	6.00
Anger	18.86	1.88	16.98	**52.83**	5.66	3.77
Disgust	8.00	8.00	10.00	14.00	**60.00**	4.00
Happy	6.00	0.00	8.00	4.00	14.00	**68.00**

TABLE 16.6 Confusion matrix of the Adaboost.M2 facial expression classifier using LBP features extracted from 11 face regions

	Fear	Surprise	Sadness	Anger	Disgust	Happy
Fear	**66.00**	4.00	18.00	4.00	0.00	8.00
Surprise	6.00	**80.00**	10.00	0.00	4.00	0.00
Sadness	6.00	12.00	**40.00**	14.00	18.00	10.00
Anger	13.20	0	18.86	**54.71**	7.54	5.66
Disgust	2.00	4.00	20.00	4.00	**68.00**	2.00
Happy	2.00	0.00	8.00	6.00	4.00	**80.00**

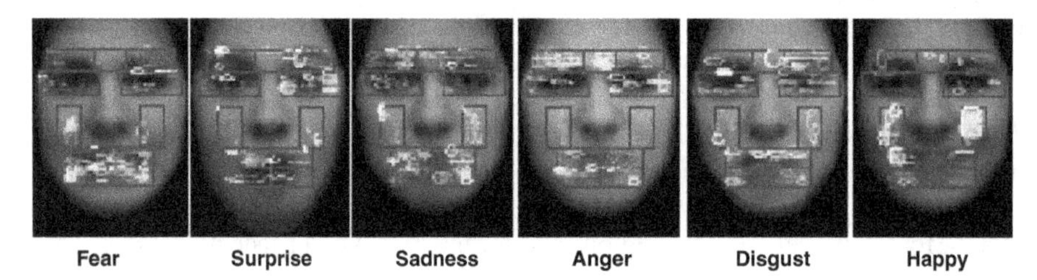

FIGURE 16.12 Projection of the set of 40 LBPs of the model of 11 face regions, on average face images showing the six basic emotions.

TABLE 16.7 Proportion of LBP features in seven regions selected by Adaboost.M2 classifier

(%)	R1	R2	R3	R4	R5	R6	R7
Fear	11.56	10.67	7.24	21.35	29.48	8.01	11.69
Surprise	9.80	1.22	23.95	14.69	35.10	0.95	14.29
Sadness	20.03	8.53	16.02	15.37	6.20	11.11	22.74
Anger	33.12	25.22	8.28	9.03	19.07	2.26	3.01
Disgust	14.32	6.66	22.74	19.35	20.85	7.04	9.05
Happy	4.92	4.67	2.90	3.41	6.06	30.18	47.85

TABLE 16.8 Proportion of LBP features in 11 regions selected by Adaboost.M2 classifier

(%)	R1	R2	R3	R4	R5	R6	R7	R8	R9	R10	R11
Fear	1.2	0.6	4.5	4.6	8.5	6.1	2.1	5.6	18.8	30.6	16.8
Surprise	10.4	1.6	13.6	7.6	27.2	0.5	8.6	11.9	7.8	6.1	4.2
Sadness	10.4	4.3	7.2	8.7	2.8	6.5	5.7	13.6	10.6	8.3	21.4
Anger	23.7	19.4	5.3	6.9	17.1	0.9	0.7	3.0	6.4	10.4	5.7
Disgust	8.1	4.8	19.2	16.8	10.6	5.3	2.8	11.8	11.8	4.6	3.8
Happy	2.1	1.7	2.7	2.3	4.0	21.0	41.3	15.4	6.0	2.4	0.6

Finding the best number of states, the best number of Gaussian mixture components and the best set of local binary features—LBPs, represents a non-trivial task. We start from the results of the Adaboost.M2 classifiers. In the case of facial expression recognition using LBP features extracted from seven face regions, we have found that the optimal number of training stages is 51. At each training stage, Adaboost.M2 selects a subset of six LBP features, one for each facial expression category. As a consequence, there would be 306 features which account for the six facial expressions of the final optimal classifier. Taking into account the fact that for evaluation we have used cross-validation method with 20 folders, it results that the final LBP feature set contains 6120 LBP features. However, the set obtained by concatenating all the subsets of the 20 folders of the cross-validation method does not include only distinct features. In fact, an important part of the subset relates to features that are commonly selected by Adaboost.M2 during training multiple folders. In addition, Adaboost.M2 may select the same feature multiple times during the training at the same cross-validation folder. This is depicted graphically in Figure 16.13b. For example, the set of features collected by taking the first 45 most important LBPs from seven face regions, for all emotion categories, includes 5400 features, though the same set contains only 1599 distinct LBP features.

We define the importance of LBP features based on the number of times an LBP is selected by the optimal Adaboost.M2 classifier during the training. Figure 16.14 illustrates the accumulated percentage of importance for the set of LBPs for the whole face and for the two face region models. In the figure, the LBPs are presented in the descending order. Using the feature importance measure, we make separate data sets by gradually choosing the first most important features for all emotion classes, from the feature sets of the 20 cross-validation folders.

For the evaluation of facial expression recognition, we have generated HMM models for each emotion category. The training data sets have been created by taking into account the emotion label of each video sample. At the testing stage, each video instance is analyzed using six HMM models, one for each emotion. Figure 16.15 shows the performance of different HMM classifiers on the data set of LBPs extracted from the whole face region and on the data sets of LBPs extracted from 7 and 11 face regions. The best facial expression recognition model uses 268 distinct features that corresponds to the selection of 45 features from each facial expression category. The

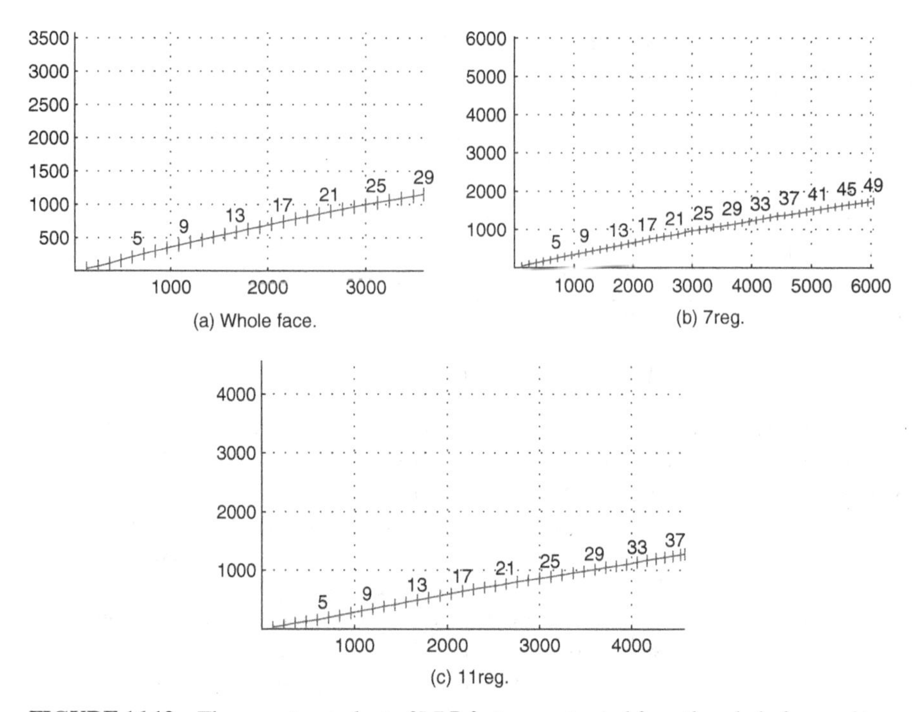

FIGURE 16.13 The concatenated set of LBP features extracted from the whole face and two types of face regions. The x-axis represents the size of the feature set; the y-axis represents the number of distinct LBP features selected by Adaboost.M2.

accuracy of this classifier is 37.71%. Table 16.9 shows the confusion matrix of the HMM classifier.

Apart from using LBP visual features, we also used features derived from optical flow estimation method. For that, we have applied pyramidal Lucas-Kanade optical flow algorithm [30] to obtain the displacement of texture pixels between consecutive video frames from all the video instances from eNTERFACE 2005 database. Based on the dense optical flow estimation, we used the previously described models of 7 and 11 face regions to determine the magnitude and the orientation of the aggregated motion vectors in each face region. This results to a set of 14 features in the case of using the model of 7 face regions and to a set of 22 features in the case of using the model of 11 face regions. The two feature sets represent the time-dependent characteristics of the emotions. Building two-state HMM-based facial expression classifiers on the databases obtained by using the two sets of optical flow features, we obtained 19.51% accuracy for the model of 7 face regions and 19.63% for the model of 11 face regions. The low results are comparable to random guessing.

16.4.4 Fusion Model

Considering the previous results of unimodal emotion estimation, it turns out that the use of audio data leads to better recognition rate (55.9%) when compared to the use

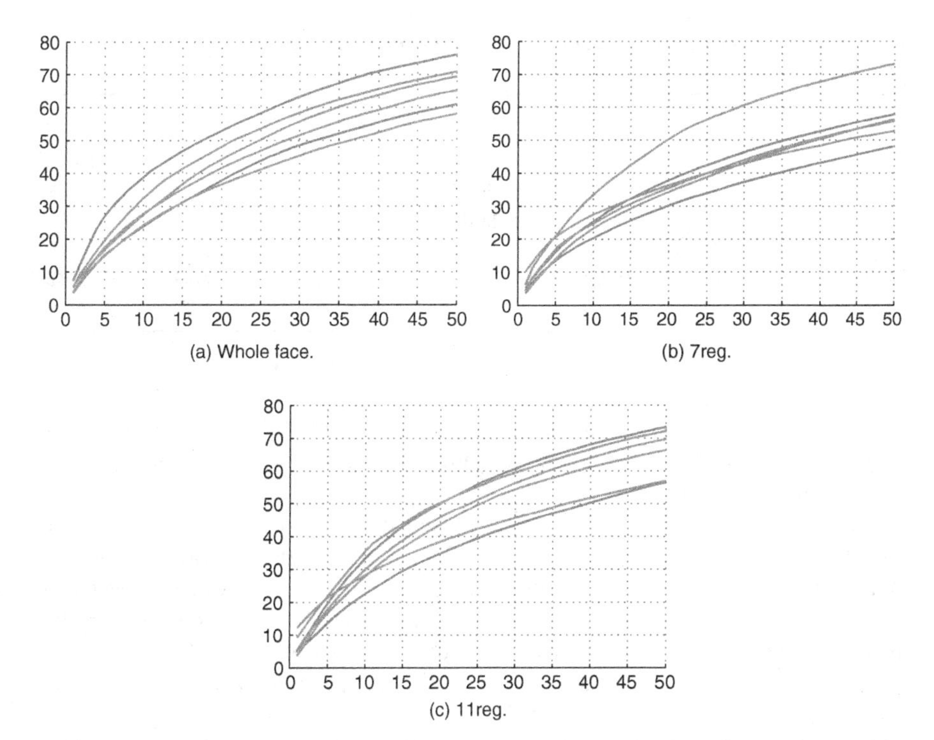

FIGURE 16.14 Importance of LBP features extracted from the three face region models. The features are sorted in the descending order of the selections (%) by Adaboost.M2 classifier, for each basic emotion category.

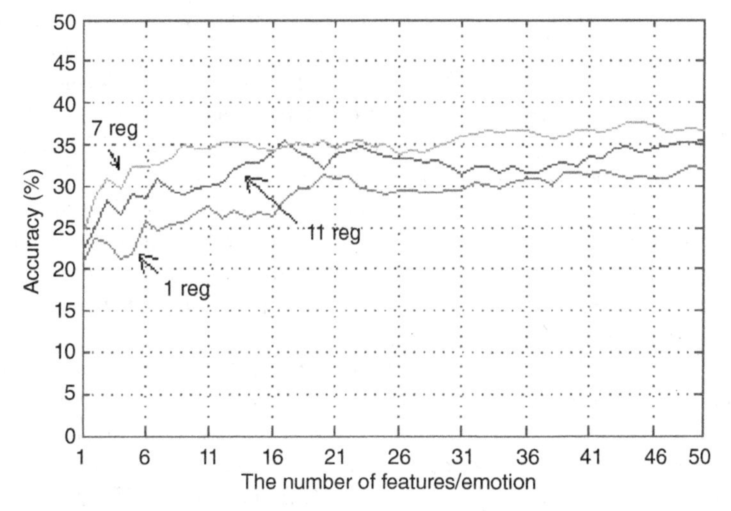

FIGURE 16.15 Facial expression recognition recognition results by using HMM models with different number of LBP features. The best HMM uses 45 LBP features for each facial expression category, from seven different face regions.

TABLE 16.9 Confusion matrix of the best HMM facial expression classifier using LBP features

(%)	Anger	Disgust	Happy	Surprise	Sadness	Fear
Anger	**18.62**	15.86	11.03	24.13	17.24	13.10
Disgust	9.92	**60.28**	10.63	63.82	6.38	6.38
Happy	6.20	17.05	**48.06**	18.60	5.42	4.65
Surprise	10.48	2.79	9.09	**53.14**	16.08	8.39
Sadness	17.51	10.94	10.94	25.54	**19.70**	15.32
Fear	9.62	16.29	6.66	26.66	14.07	**26.66**

of facial expressions-oriented models (37.71%). The next step in the attempt to get higher performance for the emotion recognition is to combine the information from the two unimodal approaches.

Depending on the type of information taken into consideration, we can define separate categories of integration. Using combined sets of audio and video features as input for the classification models is considered to fall in the category of low-level data fusion. This approach is also called early fusion or signal-level fusion. Conversely, the use of final emotion estimates from unimodal face and speech analysis is defined as high level fusion. This alternative is also called late fusion or fusion at the decision level.

Prior to building models which integrate audio and video data, the first problem that regard to the video segmentation must be solved. We identify the beginning and end points of audio–video data chunks based on the turn-based segmentation. The long pauses in conversation are used as indicators for identifying the edges of one segment. Once the audio–video segments are obtained, we then proceed with removing the sub-segments that denote the lack of speech. Based on the resulting data segments, the distinct sets of audio and video features are further extracted following the same procedures as in the case of unimodal emotion recognition.

In case of the audio signal, we extracted the sequences of MFCC frames at the rate of 100 frames per second, each frame being sized to 39 acoustic features. As a result in video processing, we extracted visual feature sets at the rate of 25 feature sets per second. Extracting LBP features from the whole face image leads to sets of 331 features per set. Similarly, using LBP features from 7 face regions generates sets of 307 features per set and using LBP features from 11 face regions generates sets of 335 features per set.

From the point of view of using HMM for modeling the emotions, each feature set represents one observation. Unlike in the case of high level fusion which runs separate classification on each modality, in the case of low level fusion the integration implies the concatenation of the visual and acoustic feature observation vectors. Because of the difference between the 100 Hz rate of MFCC frames and the 25 Hz rate of video frames, a special feature formatting procedure has to be done to first synchronize the unimodal sets of features. This additional step can be done by upscaling the observation rate of the visual feature sets to the observation rate of the audio feature sets. The second solution is to proceed with downscaling the observation rate of the audio feature sets to the observation rate of the visual feature sets. In the current approach we opt for the first solution.

The recognition of emotions based on low level fusions of audiovisual data is done by using the synchronized bimodal observation vectors with HMMs. For each emotion category, we create a separate HMM and combine all the models to obtain a multiclass emotion classifier. For evaluation, we have used threefold cross-validation method with the additional restriction that the train and test data sets do not contain samples on the same subject, for all subjects.

The simplest model consists of HMMs with one Gaussian for each state. Combining the acoustic features and LBP features extracted from 7 face regions leads to the final model accuracy of 38.55%. Using acoustic features and LBP features extracted from 11 face regions leads to the accuracy of 39.15%. Both results are superior to the results of the recognition of emotions using visual data. Still, they are worse than the results from the emotion extraction from speech. Setting the number of HMM Gaussian components to 40 and the number of HMM states to 4 like in the case of the best speech-oriented emotion classifier and combining with LBP features from the seven regions lead to a classifier which shows 22.18% accuracy. The recognition of emotions based on decision-level fusion implies the combination of the final classification results obtained by each modality separately.

For this, we take into consideration four sets of unimodal classification results namely from the speech-oriented analysis and from the separate LBP-oriented analysis which use visual features from the whole face image, from 7 face regions and from 11 face regions. We use these sets together with a weighting function that allows for setting different importance levels for each set of unimodal results. This weight-based semantic fusion approach models the asynchronous character of the emotion in visual and auditory channels accordingly. The best model obtained in this way has the accuracy of 56.27%. Although this result reflects an improvement when compared to the emotion recognition from the unimodal approaches considered, it represents only a slight increase of performance.

16.5 RESULTS

Studying the unimodal recognition of emotions on eNTERFACE 2005 shows that the speech-oriented analysis proves to be more reliable than the facial expression analysis. The best classifier we obtained in case of using HMM models with MFCC features has the accuracy of 55.90%. Conversely, the best HMM-based facial expression recognition model we got uses LBP features and has the accuracy of 37.71%. The difference of 18.19% between the classification rates achieved on separate modalities is close to the same difference between the unimodal performances reported in References 5 and 6. However, we obtained better results than the results from these two research papers, for the emotion analysis on separate modalities. Moreover, as opposed to the work [6] which attempts the person-dependent recognition of emotions, our models are completely independent on the identity of the users. To support this approach, we use n-folds cross-validation method and separate the samples of each subject in the train set from the test set.

The best facial expression recognition classifier we have obtained is based on the use of LBPs. The rather low results of the models based on optical flow estimation can be explained by the limited visual representation of the feature set. Extracting feature observations from consecutive frames of the 25 Hz video sequences does not offer enough information to describe the dynamics of the emotion generation process. A solution is to calculate and to derive features from the face motion flow applied over large integration windows. The fusion of audio and video features leads to results that are at best, close to the best unimodal classification result. In order to improve the fusion results, more investigations are needed.

16.6 CONCLUSION

The current chapter has proposed a method for bimodal emotion recognition using face and speech data. The advantage of such a method is that the resulting models overcome the limited efficiency of single modality emotion analysis. We focus on the person-independent recognition of prototypic emotions from audiovisual sequences.

The novelty of our approach is represented by the use of HMMs for the classification process. Furthermore, we introduced a new technique to select the most relevant visual features by running a separate modeling study on a separate database of facial expressions. The HMM and Adaboost.M2 algorithms we have used for the recognition relate to multiclass classification methods. Finally, we show that the fusion at the semantic level provides the best performance for the multimodal emotion analysis.

REFERENCES

[1] T. Ojala, M. Pietikainen, and T. Maenpaa, "Multiresolution gray-scale and rotation invariant texture classification with local binary patterns," *IEEE Transactions on Pattern Analysis and Machine Intelligence*, 24(7): 971–987, 2002.

[2] M. Paleari, R. Chellali, and B. Huet, "Features for multimodal emotion recognition: an extensive study," in IEEE Conference on Cybernetics and Intelligent Systems, Singapore, June 28–30, 2010.

[3] M. J. Han, J. H. Hsu, K. T. Song, and F. Y. Chang, "A new information fusion method for bimodal robotic emotion recognition," *J. Comput.*, 3(7): 39–47, 2008.

[4] M. Wimmer, B. Schuller, D. Arsic, B. Radig, and G. Rigoll, "Low-level fusion of audio and video feature for multi-modal emotion recognition," in 3rd International Conference on Computer Vision Theory and Applications. VISAPP, vol. 2, 2008, pp. 145–151.

[5] M. Paleari and B. Huet, "Toward emotion indexing of multimedia excerpts," in International Workshop on Content-Based Multimedia Indexing, 2008, pp. 425–432.

[6] M. Mansoorizadeh and N. M. Charkari, "Bimodal person-dependent emotion recognition comparison of feature level and decision level information fusion," in Proceedings of the

1st International Conference on Pervasive Technologies Related to Assistive Environments, Athens, Greece, July 15–19, 2008, pp. 1–4.

[7] S. Hoch, F. Althoff, G. McGlaun, and G. Rigoll, "Bimodal fusion of emotional data in an automotive environment," in Proceedings of IEEE International Conference on Acoustics, Speech, and Signal Processing, vol. 2, 2005.

[8] M. Song, J. Bu, C. Chen, and N. Li, "Audio-visual based emotion recognition. A new approach," in Proceedings of IEEE Computer Society Conference on Computer Vision and Pattern Recognition, vol. 2, 2004, pp. 1020–1025.

[9] M. Paleari and C. L. Lisetti, "Toward multimodal fusion of affective cues," in Proceedings of the 1st ACM International Workshop on Human-Centered Multimedia, 2006, pp. 99–108.

[10] N. Sebe, I. Cohen, T. Gevers, and T. S. Huang, "Emotion recognition based on joint visual and audio cues," in International Conference on Pattern Recognition, 2006, pp. 1136–1139.

[11] Z. Zeng, Y. Hu, G. I. Roisman, Z. Wen, Y. Fu, and T. S. Huang, "Audio-visual spontaneous emotion recognition," in *Artificial Intelligence for Human Computing*, vol. 4451, Lecture Notes in Computer Science, Springer, 2007, pp. 72–90.

[12] P. Ekman and W. V. Friesen, *Facial Action Coding System: Investigator's Guide*, Consulting Psychologists Press, 1978.

[13] J. T. Joo, S. W. Seo, K. E. Ko, and K. B. Sim, "Emotion recognition method based on multimodal sensor fusion algorithm," in Proceedings of the 8th Symposium on Advanced Intelligent Systems, 2007.

[14] G. Caridakis, L. Malatesta, L. Kessous, N. Amir, A. Raouzaiou, and K. Karpouzis, "Modeling naturalistic affective states via facial and vocal expressions recognition," in Proceedings of the 8th International Conference on Multimodal Interfaces, 2006, pp. 146–154.

[15] H. Meng, J. Pittermann, A. Pittermann, and W. Minker, "Combined speech-emotion recognition for spoken human–computer interfaces," in IEEE International Conference on Signal Processing and Communications, 2007, pp. 1179–1182.

[16] Y. Wang and L. Guan, "Recognizing human emotion from audiovisual information," in Proceedings of IEEE International Conference on Acoustics, Speech, and Signal Processing, vol. 2, 2005, pp. 1125–1128.

[17] C. Busso, Z. Deng, S. Yildirim, M. Bulut, C. M. Lee, A. Kazemzadeh, S. Lee, U. Neumann, and S. Narayanan, "Analysis of emotion recognition using facial expressions, speech and multimodal information," in Proceedings of the 6th International Conference on Multimodal Interfaces, 2004, pp. 205–211.

[18] H. J. Go, K. C. Kwak, D. J. Lee, and M. G. Chun, "Emotion recognition from the facial image and speech signal," in SICE 2003 Annual Conference, vol. 3, 2003, pp. 2890–2895.

[19] W. A. Fellenz, J. G. Taylor, R. Cowie, E. Douglas-Cowie, F. Piat, S. Kollias, C. Orovas, and B. Apolloni, "On emotion recognition of faces and of speech using neural networks, fuzzy logic and the ASSESS system," in Proceedings of the IEEE-INNS-ENNS International Joint Conference on Neural Networks, 2000.

[20] I. Kanluan, M. Grimm, and K. Kroschel, "Audio-visual emotion recognition using an emotion space concept," in Proceedings of 16th European Signal Processing Conference (EUSIPCO 2008), Lausanne, Switzerland, August 25–29, 2008.

[21] Y. J. Wang, R. Zhang, L. Guan, and A. N. Venetsanopoulos, "Kernel fusion of audio and visual information for emotion recognition," in International Conference on Image Analysis and Recognition (ICIAR), Burnaby, BC, Canada, June 22–24, 2011, pp. 140–150.

[22] O. Martin, I. Kotsia, B. Macq, and I. Pita, "The eNTERFACE'05 audio-visual emotion database," in Proceedings of the 22nd International Conference on Data Engineering Workshops, 2006.

[23] P. Viola and M. J. Jones, "Robust real-time face detection," *Int. J. Comput. Vis.*, 57(2): 137–154, 2004.

[24] G. J. Edwards, C. J. Taylor, and T. F. Cootes, "Interpreting face images using active appearance models," in Proceedings of the 3rd International Conference on Face & Gesture Recognition, 1998.

[25] S. Young, G. Evermann, M. Gales, T. Hain, D. Kershaw, X. Liu, G. Moore, J. Odell, D. Ollason, D. Povey, V. Valtchev, and P. Woodland, *The HTK Book (For HTK Version 3.4)*, 2006.

[26] T. Kanade, J. F. Cohn, and Y. Tian, "Comprehensive database for facial expression analysis," in Proceedings of the Fourth IEEE International Conference on Automatic Face and Gesture Recognition, 2000, pp. 46–53.

[27] J. Yu and B. Bhanu, "Evolutionary feature synthesis for facial expression recognition," *Pattern Recognit. Lett.* (Special Issue: Evolutionary Computer Vision and Image Understanding), 27(11), 2006.

[28] C. F. Shan, S. G. Gong, and P. W. McOwan, "Capturing correlations among facial parts for facial expression analysis," in BMVC'07, 2007.

[29] S. Dubuisson, F. Davoine, and M. Masson, "A solution for facial expression representation and recognition," *Signal Process.: Image Commun.*, 17(9): 657–673, 2002.

[30] B. D. Lucas and T. Kanade, "An iterative image registration technique with an application to stereo vision (DARPA)," in Proceedings of the 1981 DARPA Image Understanding Workshop, 1981, pp. 121–130.

AUTHOR BIOGRAPHIES

Dragos Datcu received his Ph.D. degree in Computer Science from the Delft University of Technology, The Netherlands, in 2009. He is currently working as a postdoc researcher at the Faculty of Military Sciences, Netherlands Defence Academy, and at the Delft University of Technology. His research interests include image and video analysis, face analysis, multimodal emotion recognition, and automatic surveillance.

Leon J. M. Rothkrantz received his M.Sc. Mathematics from the University of Utrecht and his M.Sc. Psychology from the University of Leiden. He got a Ph.D. degree in Mathematics from the University of Amsterdam. He joined the Computer Science Department of the Delft University of Technology in 1980 and has been an associate professor since 2001. In 2008 he was appointed as a professor at the Netherlands Defence Academy. The research interests of Prof. Rothkrantz are artificial intelligence, multimodal communication, speech technology, and pattern recognition. He is (co-) author of more than 200 scientific publications.

17

A MULTILEVEL FUSION APPROACH FOR AUDIOVISUAL EMOTION RECOGNITION

GIRIJA CHETTY, MICHAEL WAGNER, AND ROLAND GOECKE
Faculty of Information Sciences and Engineering, University of Canberra, Canberra, Australia

In this Chapter, we address two important issues in the facial expression and emotion recognition area. First, the need for quantification of expressions so as to detect subtle and micro expressions is addressed. Next, we discuss the need to utilize multiple channels of information for improvement in recognition performance of emotions. The applicability of our approach is demonstrated on various expressions at varying levels of intensity. Further, we report the results of experiments on analyzing the strengths and weaknesses of unimodal facial expression and acoustic emotion classification approaches, where some pairs of emotions are often misclassified. We propose an audiovisual fusion approach at multiple levels to show that most of these confusions could be resolved.

17.1 INTRODUCTION

An expression change in a face is characterized and quantified through combinations of nonrigid facial deformations corresponding to several regions in the face. In order to model the subtle expression variations, it is necessary to quantify the variations in emotion, as the intensity of an emotion evolves from neutral to high expression. In this Chapter, we address the aspects of facial expression quantification to detect low, medium, and high levels of expressions, and develop an automatic emotion

Emotion Recognition: A Pattern Analysis Approach, First Edition. Edited by Amit Konar and Aruna Chakraborty.
© 2015 John Wiley & Sons, Inc. Published 2015 by John Wiley & Sons, Inc.

classification technique for recognizing six different facial emotions—anger, disgust, fear, happiness, sadness, and surprise. We evaluate two different facial features for this purpose: (i) facial deformation features and (ii) marker-based features for extracting facial expression features. The performance evaluation done on the DaFEx database shows the ability to classify facial expressions at different intensities (low, medium, and high), thereby ensuring a significant improvement in expression classification using facial deformation features as compared to marker-based features.

Further, we propose using multiple information channels to enhance the accuracy and robustness in detecting emotions. The human–computer interaction will be more natural if computers are able to perceive and respond to human nonverbal communication such as emotions. Although several approaches have been proposed to recognize human emotions based on facial expressions or speech, relatively limited work has been done to fuse these two, in order to improve the accuracy and robustness of the emotion recognition system. In this chapter, we address this aspect, by analyzing the strengths and the limitations of systems based only on facial expressions or acoustic information, and propose a fusion of multiple information channels such as audio and video. Further, we examine two commonly used methods of fusion: feature fusion and score fusion and propose a new multilevel fusion approach for enhancing the person-dependent and person-independent classification performance for different emotions. Two different audiovisual emotion data corpora were used for evaluating the proposed fusion approach—DaFEx [1, 2] and eNTERFACE [3] comprising audiovisual emotion data from several actors eliciting six different emotions—anger, disgust, fear, happiness, sadness, and surprise. The results of the experimental study reveal that the system based on fusion of facial expression with acoustic information yields better performance than the system based on just acoustic information or facial expressions for the emotions considered. Results also show an improvement in classification performance of different emotions with a multilevel fusion approach as compared to either feature-level or score-level fusion.

17.2 MOTIVATION AND BACKGROUND

Facial expression is the main means for humans and other species to communicate emotions and intentions [4, 5]. Such expressions provide information not only about the affective state, but also about the cognitive activity, temperament, personality, and psychopathology of an individual [4–6]. Facial expression analysis has been increasingly used in basic research on healthy people and in clinical investigations of neuropsychiatric disorders, including affective disorders and schizophrenia. Since brain disorders likely affect emotional expressions, and their perception [5–7], the problem of expression quantification and analysis is extremely challenging due to individual physical facial differences such as wrinkles, skin texture etc. These applications may require efforts to quantify differences in expressiveness, degree of facial mobility, frequency, and rate of expression, all of which could be associated with brain disorders. A major neuropsychiatric disorder characterized by deficits in emotional expressiveness is schizophrenia, where "flat affect" is a hallmark of the illness [5, 7]. Similar expression quantification methods find immense applications in human–computer interaction applications, biometric security systems and forensic investigations, where subtle micro expressions can reveal the emotional state of a person, or discriminate a true client from impostor, or detect a lie or deceit [7, 8]. Methods of expression rating currently employed are time-consuming and depend on the subjective judgment of raters.

Further, just facial expressions may not accurately reflect the true emotions, as people try to put on a fake expression, and "facial expression" is what you see and "emotion" is what you feel, and is much deeper than the superficial facial expression that is visible. However, humans tend to use multiple channels of information to infer the true emotional state of a person, and currently automatic human computer interaction systems do not take complete advantage of multiple channels of interpersonal human communicative abilities. Being an inherently multimodal phenomenon, emotions involve both verbal and non verbal cues such as facial expressions and tone of the voice, and if used appropriately, can result in computers utilizing multiple emotional inputs for a natural human–computer interaction, and/or better lie and fake expression detection. It is widely accepted from psychological theory that human emotions can be classified into six archetypal emotions: anger, disgust, fear, happiness, sadness, and surprise [9], which are manifested in terms of motion variation in facial and speech articulators. Facial motion and the tone of the speech play a major role in expressing these emotions. The muscles of the face can be changed and the tone and the energy in the production of the speech can be intentionally modified to communicate different feelings. Human beings can recognize these signals even if they are subtly displayed, by fusion of these sensor inputs or by simultaneously processing information acquired by ears and eyes. Motivated by these clues, De Silva *et al.* conducted experiments, in which 18 people were required to recognize emotion using visual and acoustic information separately from an audiovisual database recorded from two subjects [10, 11]. They concluded that some emotions are better identified with audio such as sadness and fear, and others with video, such as anger and happiness. Moreover, Chen *et al.* [12] showed that these two modalities give complementary information, and argued that the performance of the system increased when both modalities were considered together. Although several automatic emotion recognition systems have explored the use of either facial expressions [13–17] or speech [18–23] to detect human emotion states, relatively few efforts have focused on emotion recognition using both modalities [24–27].

Further, the previously proposed methods fused facial expressions and acoustic information either at the decision level, in which the outputs of the unimodal systems are integrated by the use of suitable criteria, or at feature level, in which the data from both modalities are combined before classification. However, there is no clear reasoning behind why one type of fusion is better than another, or whether it is motivated by studies in neuroscience on fusion of emotion related inputs. It is apparent that complex multilevel fusion of emotion inputs are the reason why humans possess superior capabilities of detecting even subtle emotions, and if automatic systems can attempt to emulate these processes, it would result in better emotion recognition technologies. This chapter evaluates these two traditional fusion approaches and investigates a new multilevel fusion approach involving both feature-level and score-level fusion, to enhance the classification performance of six different emotions.

The Chapter is organized as follows. The scheme for expression quantification is described in Section 17.3. The technique for facial feature extraction from the video clips, using the facial deformation approach is described in Section 17.4. The details of the experiment design and the two audiovisual emotion corpora used are described in Section 17.4.1. The proposed multilevel fusion approach is described in Section 17.4.4, and results of the experimental work are given in Section 17.5. Finally, the chapter concludes with a discussion of the results, and a plan for further work in Section 17.6.

17.3 FACIAL EXPRESSION QUANTIFICATION

Automatic expression analysis has attracted attention in the computer vision literature because of its importance in clinical investigations, but the efforts have been focused on mainly expression recognition [11–17, 24]. We use a powerful multilevel method for expression quantification as well as expression classification that is applicable sector-wise and can be extended to automate the recognition of facial expressions. For the expression recognition, an SVM-based classifier is used for classifying the expression as one of the several possible emotions [24–26]. For the expression quantification, the intensity of the emotion is quantified on a sector-wise basis to understand how much each region contributes to an expression as well as to its intensity.

Several techniques based on quantitative morphological analysis of complex structures such as the brain and heart, which transform nonrigidly over time, have been developed in medical imaging [18–20, 24]. These techniques normalize spatially, that is, elastically register deformable, elastic body parts of a subject to those of a template. We use one such technique for modeling facial deformations. Facial deformation, in general, provides a powerful way to obtain a detailed and localized structural characterization of complex objects such as the brain [24–26].

In the expression quantification method proposed here, the facial deformation maps a neutral face, taken as a template, to a face with expression, which is the query-face. It is based on point correspondences determined on distinct region boundaries along with some landmark points, demarcated on the query-face and template face. We compute an elastic transformation, which warps the template to the query-face, mapping the corresponding boundaries to each other and elastically interpolating the enclosed region [8, 9]. Comparison of the properties of these facial deformations helps to quantify recognized emotions. The advantages of the technique used for expression quantification are:

- The method can quantify expression changes between various intensities, for example, low, medium, and high, and across individuals expressing the same or a different emotion, by using high-dimensional facial deformation. The current literature on face expression recognition is able to recognize only high emotion, that is, when all the action units are activated, and not grades of that emotion.

- As the method involves sector-based volumetric difference maps using the facial deformations, which quantify the deformation between two faces, the average of each expression is based on these volumetric difference maps. Such models designed using expressions of normal people allow examining individual differences in people with expression-related disorders. Such analysis, in conjunction with other recognition techniques, can be helpful for comparison of expressions of people with different degrees of neuron-psychiatric disorders. The approach used in the work reported here also allows predicting the expression of a neutral face using the deformation field generated by the facial deformation.

The details of the expression quantification method are presented here. We choose a neutral face (face with no expression) as a template. All the other expressions of the same individual are categorized as query-faces; to be analyzed against this template. A query-face is one that has expression. The expression could vary in intensity from low to medium to high. Our general approach can be outlined in the following steps:

1. Identify sectors or areas on the template that characterize the various facial features. These sectors are demarcated by marking their boundaries (as shown in Figure 17.1). Some landmark points are also marked; curve segments in between are parameterized via constant-speed parameterization.
2. For each area picked on the template, identify the corresponding area on the query-face.
3. Compute the elastic transformation from one face to the other, so that the different areas of the face from the query-face are mapped to their counterparts in the template.

The facial deformation map provides us with the information regarding the deformation produced by the expression change. Also, we define a sectored volumetric

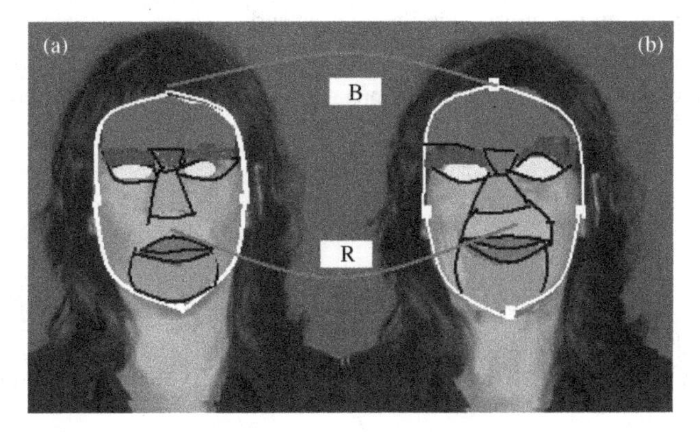

FIGURE 17.1 Regions sectored in the face. B depicts one of the boundaries and R indicates one of the regions. Corresponding regions are painted with the same shade (color in electronic version of the book). (a) The neutral face chosen as the template. (b) The happy face taken as the query-face.

difference function (SVDF), which provides a numeric value for each pixel on the face. This value quantifies the expansion and contraction of the area on a pixel-wise basis. The facial deformation map used in this chapter is adapted from the work in Reference 28.

Let Ω_q and Ω_t denote the query space and template space respectively.

Then,
$\Omega_t = F_t|$ denotes face with neutral expression, and
$\Omega_q = F_q|$ denotes face with expression at various intensities |(low, medium, high)

Now we define each template and query-face as a union of sectors and their boundaries.

$$F_t = \cup_i (B_t^i, R_t^i)$$
$$F_q = \cup_i (B_q^i, R_q^i)$$

where B_t^i and B_q^i are the boundaries of the corresponding regions R_t^i and R_q^i respectively.

Figure 17.1 shows some of the sectors or areas that have been demarcated on the face. Some landmark points (typically two to four) are also selected on the boundaries. The boundary segments in between two consecutive landmarks are parameterized by evenly spaced points (constant-speed parameterization). This parameterization effectively sets up point correspondences between the boundaries of the query-face areas and the boundaries of the corresponding template areas. These point correspondences can be used to define a transformation between the template face and the query-face. In order to determine a shape transformation that accounts for the morphological or elastic changes occurring in the enclosed areas, we compute the elastic transformation S which

- Maps the corresponding points between the boundaries of the template to their counterparts on the query-face boundaries and
- Warps the enclosed regions from the template to the query-face elastically.
  The elastic shape transformation (facial deformation) S is computed as

$$S : \Omega_t \ni F_t \to F_q \in \Omega_q \tag{17.1}$$

- The boundaries are mapped to each other,

$$f : B_t^i \to B_q^i \ldots \forall_i \tag{17.2}$$

  by mapping the corresponding points on the boundary, and,
- The enclosed template regions are elastically warped to the enclosed regions on the subject

$$R_t^i \to R_q^i \tag{17.3}$$

Thus S elastically matches the boundaries and accounts for the elastic (deformable) changes not only on the boundary but also within the enclosed region. The resulting shape transformation (facial deformation) S quantifies the shape properties of the query-face, F_q with respect to the template face, F_t. Therefore, two faces with different emotion expression can be analyzed on the basis of a point-wise comparison of the facial deformations. The shape transformation (facial deformation) produces a map of vectors (one vector at each pixel) called a deformation vector field. These vectors provide the direction of movement of the pixel as the result of the deformation caused by the change in expression. Each vector is denoted by a pair (dx, dy) which denotes the displacement in x and y direction that each pixel on the template undergoes, when it is transformed to the face with expression. On adding the vector (dx, dy) to the position of the template (x, y), we get the deformation $(x + dx, y + dy)$ which the pixel on the template undergoes. The position to which the pixel on the neutral template moves to as a result of the action of this vector denotes the deformed position in the query-face after undergoing an expression change. This is known as the deformation vector field. Additional details of the method can be found in several related works [8, 29–33].

We obtain two quantities from the shape transformation (facial deformation) that we use as expression quantification features for the second level—the expression classification stage.

- The scalar field of values of the sectored volumetric difference function (SVDF), is evaluated at each pixel on the face. We define the $SVDFas : SVDF(s) = det(\Gamma(S(s))) = det|\text{Jacobian}(S)|$. The determinant of the Jacobian of S is evaluated at each point s on the query-face. The SVDF value quantifies the volumetric difference (variability in expansion and contraction) between regions. The map containing the SVDF values for each pixel of the face is called the SVD map of the face. Various inferences may be drawn from the values of the SVD function.

  If $SVDF1(u) \geq SVDF2(u)$ for the same sector/area on two faces which have the same expression at the same intensity, then it quantifies that the area on face 1 has deformed (expanded or contracted) more than the area on face 2, relative to their respective neutral states. It, therefore, quantifies the variability in expressing the same emotion across individuals. If this is on the same face and same region, it indicates and quantifies the change in expression. Figure 17.2c shows the color visualization of an SVD map.
- The vector displacement fields of the deformation. These characterize the direction and degree of movement of each pixel of the face during the course of an expression change.
  1. These vectors can be used to quantify temporal changes in an expression. Figure 17.2d shows the deformation of each pixel of the template face 17.2a as a result of the expression change from Figures 17.2a to 17.2b.
  2. For the second level—the expression classification stage, SVD map of the shape transformation of a query-face with respect to a template, when fused with the vector deformation field (VDF map), provides information about the direction of movement of the regions and forms the feature sets for training the Hidden Markov Model (HMM) for each expression classification.

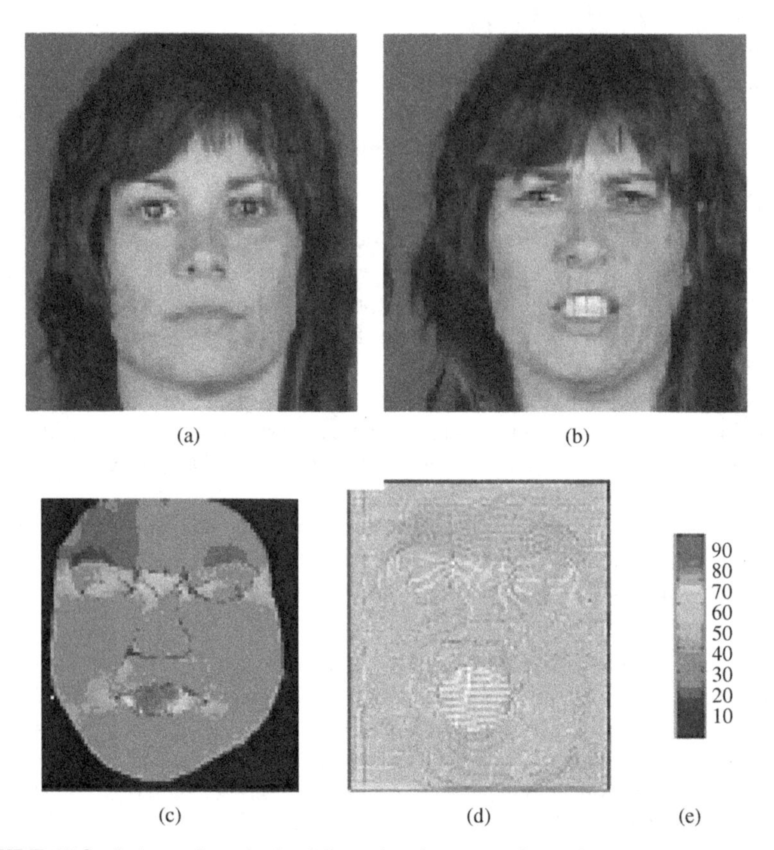

FIGURE 17.2 Information obtained from the shape transformation: (a) template face, (b) query—with expression, (c) intensity normalized SVD map, (d) vector deformation field, and (e) color map for visualization of SVD (color will be available in e-copy of the book).

17.4 EXPERIMENT DESIGN

In this Section, the details of the emotion data corpora used, and details of marker based facial features, and speech-related audio features used are described.

17.4.1 Data Corpora

Two different corpora, DaFEx [1, 2] and eNTERFACE [3], with actors eliciting six different emotions, were used for evaluating the facial expression quantification and emotion recognition based on a multilevel fusion approach.

DaFEx is an Italian audiovisual database of posed human facial expressions collected with the purpose of creating a valid benchmark for the evaluation of synthetic faces and embodied conversational agents. DaFEx can also be used as a general reference for research on emotions and facial expressions. DaFEx is composed of 1008 short videos in which Ekman's prototypic emotions (happiness, sadness, anger, fear, disgust, and surprise) plus the neutral expression are shown. Facial expressions were recorded by eight Italian professional actors (four male and four female) on three intensity levels (low, medium, and high) and in two different conditions. The database also comprises a spoken utterance subset with the actors playing these emotions while uttering a phonetically rich and visemically balanced sentence "*In quella piccola stanza vuota c'era per soltanto una sveglia*", Italian for "In that little empty room there was only an alarm clock."

For the nonutterance subset, the actors played emotions without pronouncing any sentence. In addition, each video started and ended with the actor showing a neutral expression. Both video and audio signals were recorded, with each actor recording a subset of 126 videos, which includes all the emotions considered at three intensity levels and in two different conditions. The recording was done with a digital camera (Canon MV630i) and a directional microphone (Sennheiser MKH 406T). Videos were then compressed with Indeo 5.10 compressor and the audio signal was filtered in order to eliminate external noise. Finally, the videos were made available as .avi files with 360 × 288 pixel images. Figure 17.3a shows some images from this corpus.

The eNTERFACE corpus comprised English audiovisual emotion data collected by a particular project group for a European Similar Network of excellence workshop [3]. This database is composed of over 1300 emotionally tagged videos portraying nonnative English speaker displaying a single emotion while verbalizing a semantically relevant English sentence. The six universal emotions from Ekman and Friesen [18] were portrayed, namely anger, disgust, fear, happiness, sadness, and surprise. Videos have a duration ranging from 1.2 to 6.7 seconds. This database is publicly available on the Internet but carries a few drawbacks mainly due to the low quality of the video compression and actor performances.

17.4.2 Facial Deformation Features

We now describe the method used for extracting shape transformation (facial deformation) features for facial expression quantification purposes. In Figure 17.2, we show the information that is produced by the shape transformation applied to a segmented face shown in Figure 17.2. Figure 17.2a shows the template neutral face and Figure 17.2b shows the corresponding query-face expressing anger. We then compute the shape transformation that elastically warps the regions demarcated on the template to the corresponding sectors or areas identified on the subject (see Figure 17.1 for sectors). A positive SVD value indicates an expansion and a negative SVD value indicates a contraction. These are the values used in the analyses. However, these SVD values are normalized to a specific range for visualization of the expression changes in the form of a color map. In our case, we choose the range to be 0 to 90, as it provides the best demarcation. In doing so, the base value of 0, indicating no change, is shifted to 30. The range for displaying the color map can be changed by the user. Figure 17.2c shows the color coded SVD map of SVD values (shown by dark shade) computed at each pixel of the face, the color map for which (shown by light shade) is in Figure 17.2e. After normalization, an increase in SVDF values from the template to the subject indicates the expansion and a decrease indicates the contraction of the corresponding region in the subject. These maps are computed at varying intensities of the same emotion to study expression changes. In general,

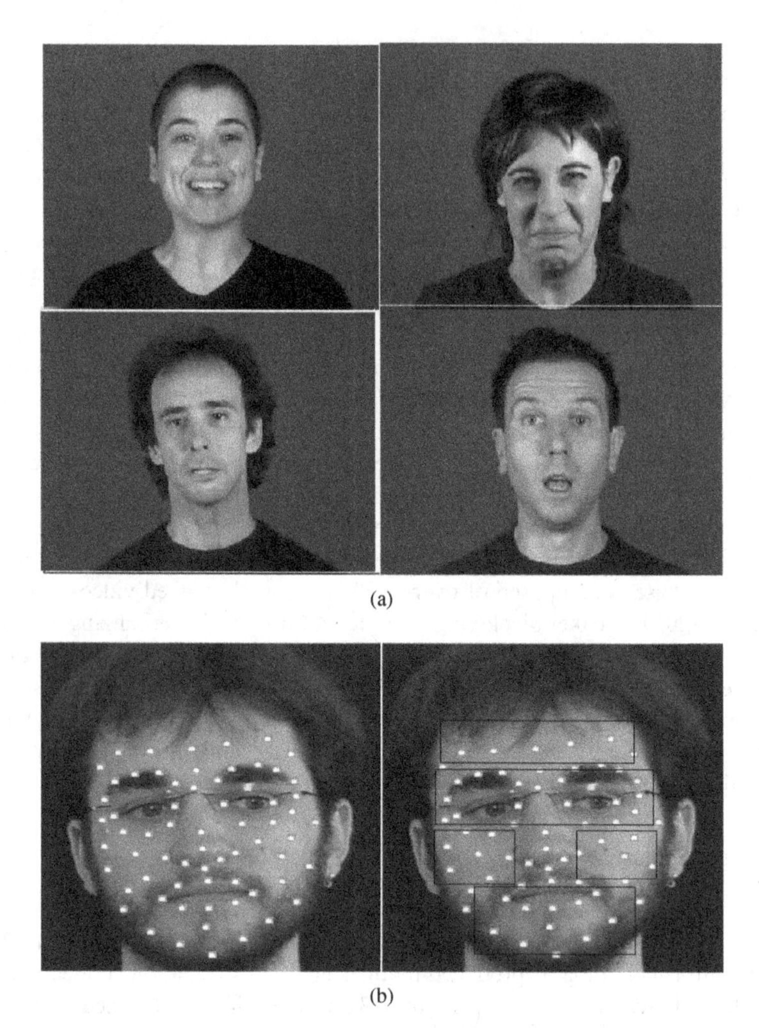

(a)

(b)

FIGURE 17.3 (a) Audiovisual emotion corpus images. Extracted from publicly available free DaFEx database. (b) Visual feature markers and five face regions. Taken with permission from eNTERFACE database.

darker shades (blue in e-edition) indicate contraction and lighter shades (yellow to red in e-edition) depicts increasing expansion. Figure 17.2d depicts the deformation field indicating the pixel movements due to the expression change from Figures 17.2a to 17.2b. In order to create Figure 17.2d, we represented the pixels of the template face as a grid of the same size. Then to each position of the grid, the displacement of the vector field produced as a result of the shape transformation was applied. This produces the deformed grid shown in Figure 17.2d. In this chapter, we will use SVD maps (as shown in Figure 17.2c) for quantification as these provide a numeric value of the changes at each pixel as a result of the expression change.

The SVD maps can also be used to determine movement of facial regions. This is achieved by studying the sector or the region as a combination of expansion and contraction of several regions. In Figure 17.2c, the forehead contracts and the upper lid and the region between the eyes expand. This indicates that the eyebrows have been raised. The mouth and upper lip expand. This in conjunction with the fact that the chin and cheeks expand slightly and lower face expands, indicating a jaw drop. The actual expansion and contraction of these areas and the direction of their movement can be verified with the actual changes in face in Figures 17.2a and 17.2b. We now describe the conventional marker-based approach for extracting facial expressions for comparison.

17.4.3 Marker-Based Audio Visual Features

Though display of emotion is an inherently multimodal process comprising motion variations in facial articulators such as mouth, eye, eyebrows, and forehead, head pose and tilt, coupled with motion in speech articulators manifested as variations in tone, pitch, prosody, and volume of speech, along with variations in bodily gestures. Relatively few efforts have focused on using these multiple information channels to build automatic emotion recognition systems. De Silva *et al.* proposed a rule-based audiovisual fusion approach, in which the outputs of the unimodal classifiers are fused at the decision-level [9, 17]. From audio, they used prosodic features, and from video, they used the maximum distances and velocities between six specific facial points. A similar approach was also presented by Chen *et al.* [12], in which the dominant modality, according to the subjective experiments conducted in [12], was used to resolve discrepancies between the outputs of unimodal systems. In both studies, they concluded that the performance of the system increased when both modalities were fused. Yoshitomi *et al.* proposed a multimodal approach that not only considers speech and visual information but also the thermal distribution acquired by infrared camera [18]. They argue that infrared images are not sensitive to lighting conditions, which is one of the main problems when the facial expressions are acquired with conventional cameras. They used a database recorded from a female speaker that read a single word acted in five emotional states. They integrated these three modalities at decision-level using empirically determined weights. The performance of the system was better when three modalities were used together. In [19] and [20], a bimodal emotion recognition system was proposed to recognize six emotions, in which the audiovisual data was fused at feature level. They used prosodic features from audio, and the position and movement of facial organs from video. The best features from both unimodal systems were used as input in the bimodal classifier. They showed that the performance significantly increased from 69.4% (video system) and 75% (audio system) to 97.2% (bimodal system). However, they use a small database with only six clips per emotion, so the generalizability and robustness of the results should be tested with a larger data set.

For investigating the improvements achieved with multiple channels of information in this study, we used a simple marker-based visual features as described below:

The face was divided into several sectors or regions as in facial expression quantification technique described in Section 17.3. However, the regions used for marker-based facial feature extraction were different from the expression quantification technique. Markers were placed in all regions at points expected to display variations under expression changes. All markers were translated in order to make a nose marker the local coordinate centre of each frame, and then one frame with neutral and close-mouth head pose was picked as the reference frame. The three approximately rigid markers (manually chosen and illustrated as white points in Figure 17.3b define a local coordinate origin for each frame, and each frame was rotated to align it with the reference frame. Each data frame is divided into five regions: forehead, eye, lower mouth, right cheek, and left cheek area (see Figure 17.3b. For each region, the *x*-*y* coordinates of markers concatenated together to form a data vector. Then, principal component analysis (PCA) method was used to reduce the number of features per frame into a 10D vector for each area (covering more than 95% of the variation).

17.4.4 Expression Classification and Multilevel Fusion

For expression classification and quantification experiments, we examined different features and classifier methods. First, we performed recognition experiments with i) audio only features, ii) marker-based facial features only, iii) feature-level fusion of audio and marker-based facial features, and iv) score-level fusion and multilevel (combination of feature level and score level fusion). Six emotions—anger, disgust, fear, happiness, sadness, and surprise were recognized.

For all the experiments, a support vector machine classifier (SVM) with second order polynomial kernel functions was used [34]. Also, for all experiments, the training and testing using the corpora was done using a 10-fold cross-validation method. The schematic for this set of experiments is shown in Figures 17.4, 17.5, and 17.6. Next, we performed facial expression quantification experiments by using a different set of features. We used LDA optimized SVDF and VDF feature vectors and an SVM classifier for evaluating expression quantification as HIGH, LOW, and MEDIUM. The schematic for expression quantification is shown in Figure 17.7.

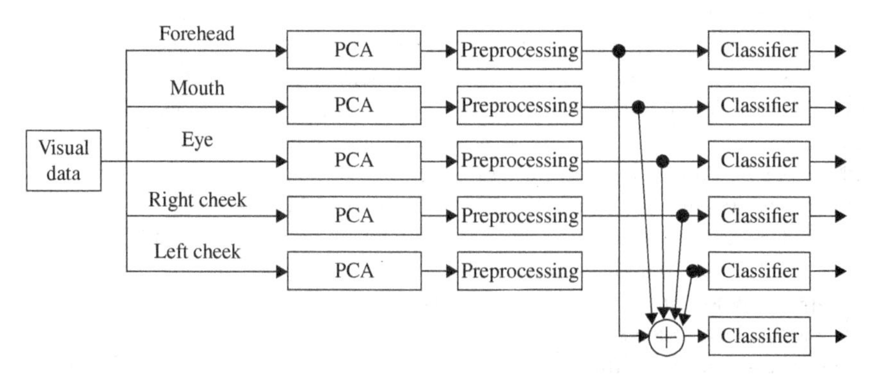

FIGURE 17.4 Facial features from different facial regions for expression classification.

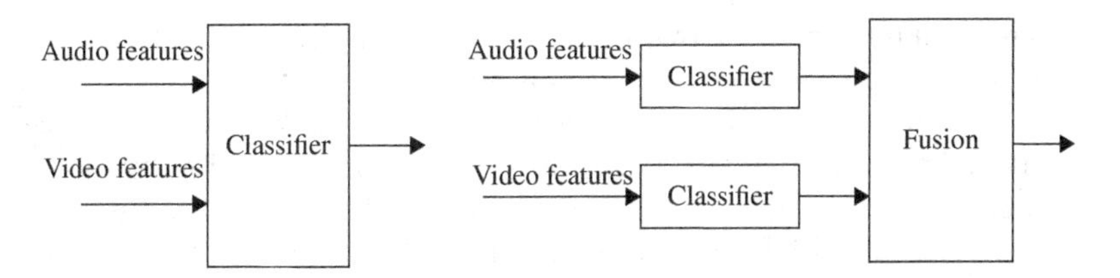

FIGURE 17.5 Feature-level and score-level fusion of audio and marker-based facial features.

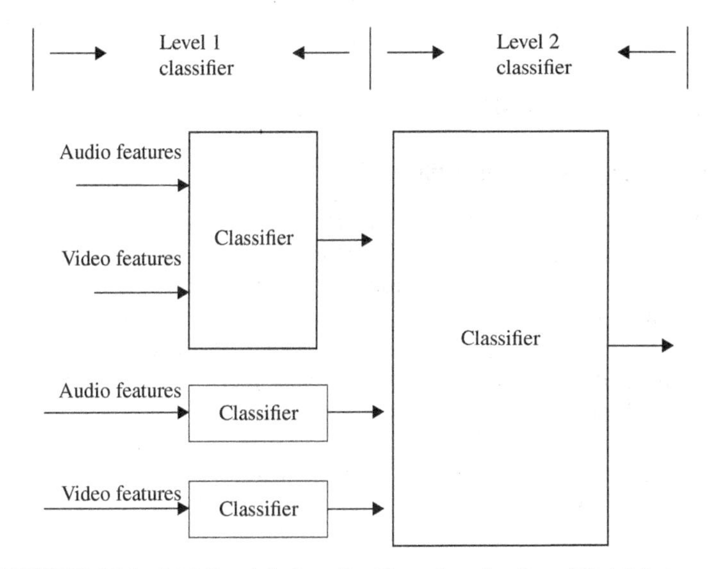

FIGURE 17.6 Multilevel fusion of audio and marker based facial features.

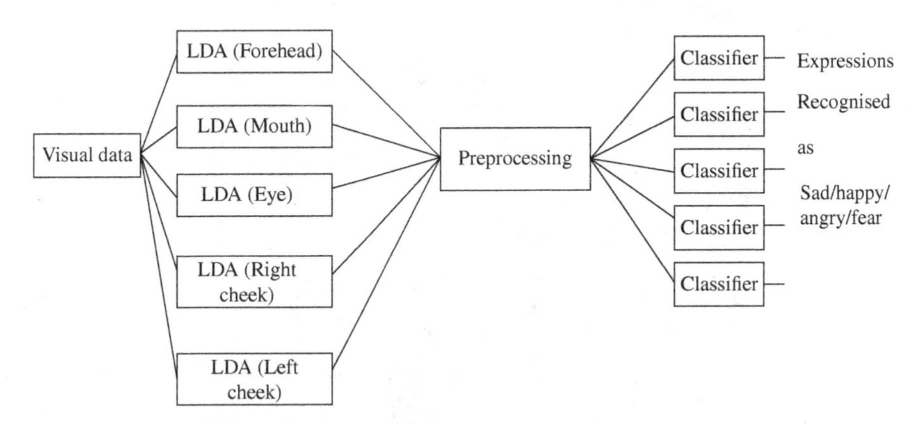

FIGURE 17.7 Facial expression feature quantification.

We compared both marker-based features and shape transformation/facial deformation features for facial expression quantification capability. The local coordinate centre of each frame was obtained by translating all the markers with reference to nose marker. Then the reference frame was picked by choosing the neutral frame with mouth closed. Each data frame is divided into five blocks: forehead, eye, lower mouth, right cheek, and left cheek area (see Figure 17.3b). For each block, the *x*-*y* coordinates of markers in that block were concatenated together to form a data vector. Then, linear discriminant analysis (LDA) method is used to reduce the number of features per frame into a 10D vector for each area. A 10D feature vector is obtained for each region. The next section describes results obtained from different experiments.

17.5 EXPERIMENTAL RESULTS AND DISCUSSION

17.5.1 Facial Expression Quantification

In this section, we compare the performance of the proposed shape transformation/facial deformation approach with the marker-based approach for the facial expression classification for high level expression data set of the DaFEx database. Figure 17.8 shows the SVD maps for the expressions happiness, sadness, and anger. For happy face (first row), the eyes contract, the mouth expands and the cheeks contract indicating a sideways expansion of the mouth. The lower lids expand depicting a cheek raise. The forehead shows no change and neither does the region between the eyes and the eyebrows. The contraction is indicated by a decrease in SVD values from the template to the subject images. The expansion of the regions is indicated by the increase in SVD values from the template to the subjects. The second row

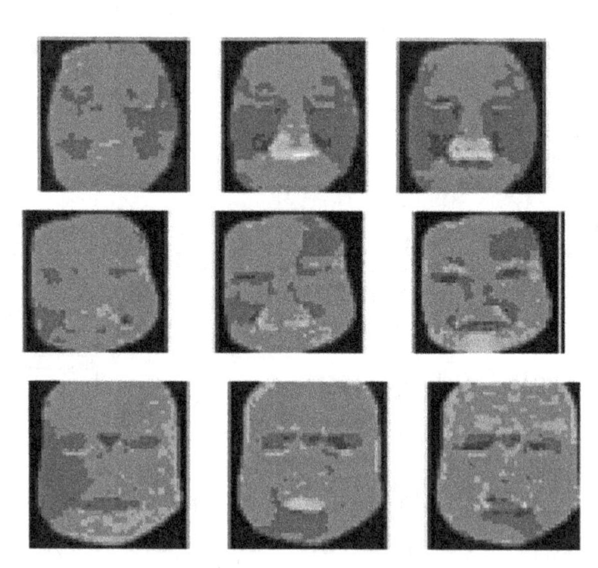

FIGURE 17.8 Quantification of expression for a subject as low, medium and high intensity (First row: happy; Second row: sadness, Third row: anger). The color version of the above figure is available at www.amitkonar.com/wileycolorimages.

TABLE 17.1 Comparing recognition performance for marker versus SVDF/VDF features

Expression	Marker	SVDF/VDF
Anger	86.24%	91.35%
Sadness	80.39%	81.66%
Happiness	88.38%	93.45%
Neutral	81.27%	81.66%

in Figure 17.6 shows sadness in varying intensities. The eyes contract, the region between the eyes and the eyebrows expand and the forehead contracts, indicating an eyebrow raise, the upper lip expands, the mouth expands and the cheeks contract. The chin expands and the whole face contracts, indicating a sideways expansion of the lips and the chin. The third row in Figure 17.8 shows the same analysis carried out for the expression of anger. The SVD values of pixels in the corresponding regions in the faces for anger indicate that the eyes contract, the forehead expands and the regions between the eyes contract, indicating a brow lowering. The cheeks contract and the lower lids expand, indicating a cheek raise.

17.5.2 Facial Expression Classification Using SVDF and VDF Features

In this set of experiments, the facial quantification features (SVDF and VDF features) were used for expression recognition. We used only one database (DaFEx database) for this study. Table 17.1 shows the recognition performance for the marker based approach as compared to SVDF/VDF features approach. As can be observed, from Table 17.1, anger, happiness, and neutral state are recognized with more than 90% of accuracy for the SVDF/VDF features as compared to the marker-based approach. For the marker-based approach, anger is classified with 86.24%, sadness with 80.39%, happiness with 88.38%, and neutral with 81.27% accuracy, whereas for SVDF/VDF features, the performance achieved is 91.35% for anger, 82.66% for sadness, 93.45% for happiness, and 81.66% for neutral expression. For neutral expressions, both marker-based and SVDF/VFD features perform similarly. The recognition rate of sadness was in general lower, 80.39% for the marker-based approach and 81.66% for the shape transformation approach. This emotion is confused with the neutral state, because the markers used for five regions on the face may not be accurately capturing the classes separately. The happiness recognition performs best for both the marker-based approach (88.28%) and the shape transformation (SVDF/VDF) features (93.45%). The shape transformation features perform significantly better for all expressions, but the best for happiness. These results suggest the shape transformation approach, (SVD+VDF) features, can explicitly quantify emotions and allow better recognition of difficult facial characteristics such as subtle or micro expressions.

17.5.3 Audiovisual Fusion Experiments

In this section, we examine whether multiple information sources can aid in recognizing emotions. For this, recognition performance with the unimodal acoustic and visual expression features is reported first, followed by experimental results for the

TABLE 17.2 Confusion matrix for emotion recognition based on acoustic features for dafex corpus

	Anger	Sadness	Happiness	Neutral
Anger	0.68	0.05	0.21	0.05
Sadness	0.07	0.64	0.06	0.22
Happiness	0.19	0.04	0.70	0.08
Neutral	0.04	0.14	0.01	0.81

fusion of audio and visual features based on feature-level, score-level and multilevel fusion. Further, we used marker-based facial features as they are easier to obtain as compared to SVDF/VDF features, and also due to poor performance, can benefit from the additional information channel (audio) to recognize emotions correctly.

17.5.3.1 Emotion Recognition Using Acoustic Features Table 17.2 shows the confusion matrix for recognizing four emotions based on acoustic information. For the DaFEx corpus, the overall recognition performance was 70.9%. The diagonal components of Table 17.1 reveal that all the emotions can be recognized with more than 64% accuracy, by using only the features of the speech. However, Table 17.2 shows that some pairs of emotions are usually confused more. Sadness is misclassified as neutral state (22%) and vice versa (14%). The same trend appears between happiness and anger, which are mutually confused (19% and 21% respectively). These results agree with the human evaluations done by De Silva *et al.* [7], and can be explained by similarity patterns observed in acoustic parameters of these emotions [26]. For example, speech associated with anger and happiness is characterized by longer utterance duration, shorter inter-word silence, higher pitch and energy values with wider ranges. On the other hand, in neutral and sad sentences, the energy and the pitch are usually maintained at the same level. Therefore, these emotions were difficult to be classified. For the eNTERFACE corpus, the recognition performance was around 10–15% lower than for the DaFEx corpus, though the performance trend for individual emotion states was similar to the DaFEx corpus.

17.5.3.2 Emotion Recognition Using Facial Expression Features Table 17.3 shows the performance of the emotion recognition for the DaFEx corpus based on

TABLE 17.3 Performance of the facial expression classifiers for DaFEx corpus

Area	Overall	Anger	Sadness	Happiness	Neutral
Forehead	0.73	0.82	0.66	1.00	0.46
Eyebrow	0.68	0.55	0.67	1.00	0.49
Low eye	0.81	0.82	0.78	1.00	0.65
Right cheek	0.85	0.87	0.76	1.00	0.79
Left cheek	0.80	0.84	0.67	1.00	0.67
Combined	0.85	0.79	0.81	1.00	0.81

TABLE 17.4 Confusion matrix of combined facial expression classifier for DaFEx corpus

	Anger	Sadness	Happiness	Neutral
Anger	0.79	0.18	0.00	0.03
Sadness	0.06	0.81	0.00	0.13
Happiness	0.00	0.00	1.00	0.00
Neutral	0.00	0.04	0.15	0.81

four facial expressions for each of the five facial regions shown in Figure 17.3b, and the combined facial expression classifier. This table reveals that the cheek areas give valuable information for emotion classification. It also shows that the eyebrows, which have been widely used in facial expression recognition, somehow give the poorest performance. Also, happiness seems to be classified without any mistake and much easier to recognize. Table 17.3 also reveals that the combined facial expression classifier has an accuracy of 85%, which is higher than most of the five facial region classifiers. For the eNTERFACE database, the performance for individual facial regions showed a similar trend as for DaFEx, though the performance of the combined classifier was 82%.

Table 17.4 shows the confusion matrix of the combined facial expression classifier to analyze in detail the limitation of this emotion recognition approach. The overall performance of this classifier was 85.1% for the DaFEx corpus and 82% for the eNTERFACE corpus. This table reveals that happiness is recognized with very high accuracy. The other three emotions are classified with 80% of accuracy, approximately. Table 17.4 also shows that in the facial expressions domain, anger is confused with sadness (18%) and neutral state is confused with happiness (15%). Notice that in the acoustic domain, sadness/anger and neutral/happiness can be separated with high accuracy, so it is expected that the bimodal fusion will give good performance for anger and neutral state. This table also shows that sadness is confused with neutral state (13%).

Unfortunately, these two emotions are also confused in the acoustic domain (22%), so it is expected that the recognition rate of sadness in the bimodal classifiers will be poor. Other discriminating information such as contextual cues are needed.

17.5.3.3 Emotion Recognition Using Audiovisual Fusion Table 17.5 displays the confusion matrix of the audiovisual fusion for the DaFEx corpus when the

TABLE 17.5 Confusion matrix of the feature-level fusion classifier for DaFEx corpus

	Anger	Sadness	Happiness	Neutral
Anger	0.95	0.00	0.03	0.03
Sadness	0.00	0.79	0.03	0.18
Happiness	0.02	0.91	1.00	0.08
Neutral	0.01	0.05	0.02	0.92

TABLE 17.6 Confusion matrix of the score-level fusion classifier for DaFEx corpus

	Anger	Sadness	Happiness	Neutral
Anger	0.84	0.08	0.00	0.08
Sadness	0.00	0.90	0.00	0.10
Happiness	0.00	0.00	0.98	0.02
Neutral	0.00	0.02	0.14	0.84

facial expressions and acoustic information were fused at feature level. The overall performance of this classifier was 89.1%. The overall performance of this classifier for the eNTERFACE corpus was 86.1%. It can be observed for both corpora, anger, happiness, and neutral state are recognized with more than 90% of accuracy. It supports our expectation that the recognition rate of anger and neutral state was higher than in unimodal systems. Sadness is the emotion with lower performance, which agrees with our previous analysis. This emotion is confused with neutral state (18%), because none of the modalities we considered can accurately separate these classes. Notice that the performance of happiness significantly decreased to 91%.

Table 17.6 shows the confusion matrix of the score-level bimodal classifier when the product-combining criterion was used. The overall performance of this classifier was 89.0% for the DaFEx corpus and 87.4% for the eNTERFACE corpus, which is very close to the overall performance achieved by the feature-level bimodal classifier (Table 17.5). However, the confusion matrices of both classifiers show important differences. Table 17.6 for DaFEx corpus shows that in this classifier, the recognition rate of anger (84%) and neutral states (84%) are slightly better than in the facial expression classifier (79% and 81%, Table 17.5), and significantly worse than in the feature-level bimodal classifier (95%, 92%, Table 17.5). However, happiness (98%) and sadness (90%) are recognized with high accuracy compared to the feature-level bimodal classifier (91% and 79%, Table 17.5). These results suggest that in the score-level fusion approach, the recognition rate of each emotion is increased, improving the performance of the audiovisual fusion.

Table 17.7 shows the confusion matrix of the multilevel classifier shown in Figure 17.4 comprising a fusion of score-level bimodal classifier (product-combining criterion) with acoustic features and facial expression classifiers. The overall performance of this classifier was 97.0% for the DaFEx corpus and 96.4% for the eNTERFACE corpus, which is better than the overall performance achieved by both the feature-level and score-level audiovisual fusion (Table 17.6). However, the

TABLE 17.7 Confusion matrix of the multilevel fusion classifier for DaFEx corpus

	Anger	Sadness	Happiness	Neutral
Anger	0.94	0.02	0.00	0.02
Sadness	0.00	0.96	0.00	0.08
Happiness	0.00	0.00	1.00	0.02
Neutral	0.00	0.00	0.02	0.94

confusion matrices of both classifiers show important differences. The happiness (100%) and sadness (96%) are recognized with high accuracy compared to other fusion approaches. These results suggest that for the multi-level fusion approach, the recognition performance of each emotion is increased, improving the overall performance of the system. Further, it was observed that the eNTERFACE data performed equally well as the DaFEx data in multilevel fusion mode, which means multilevel fusion makes the system more robust irrespective of the quality of the data and recording conditions.

Studies from neuroscience have indicated that humans use more than one modality to recognize emotions, so it can be expected that the performance of fusion of multiple modes at multiple levels will be higher than automatic unimodal systems. The results reported in this work confirm this hypothesis, since the multilevel fusion approach gave a significant improvement as compared to the performance of the acoustic or facial expression recognition schemes. The results show that pairs of emotions that were confused in one modality were easily classified in the other. For example, anger and happiness that were often misclassified in the acoustic domain were separated with greater accuracy in the facial expression emotion classifier. Therefore, when these two modalities were fused at feature-level, these emotions were classified with high precision. Unfortunately, sadness is confused with neutral state in both domains, so its performance was poor. Although the overall performance of the feature-level and decision-level bimodal classifiers was similar, the multilevel fusion approach resulted in better emotion recognition performance. Also, an analysis of the confusion matrices of both classifiers revealed that the recognition rate for each emotion type was totally different. For the multilevel fusion classifier, the recognition rate of each emotion increased compared to the facial expression classifier, which was the best unimodal recognition system. In the feature-level bimodal classifier, the recognition rate of anger and neutral state significantly increased. However, the recognition rate of happiness decreased by 9%. The preliminary results presented in this chapter suggest that the multilevel approach seems to be the best approach to fuse the modalities. Also, the results reveal that, even though the system based on audio information had poorer performance than the facial expression emotion classifier, its features have valuable information about emotions that cannot be extracted from the visual information. These results agree with the findings reported by Chen *et al.* [12], which showed that audio and facial expressions data present complementary information. On the other hand, it is reasonable to expect that some characteristic patterns of the emotions can be obtained by the use of either audio or visual features. This redundant information is very valuable to improve the performance of the emotion recognition system when the features of one of the modalities are inaccurately acquired as in the two different emotion databases examined here. For example, if a person has a beard, moustache or eyeglasses, the facial expressions will be extracted with a high level of error. In that case, audio features can be used to overcome the limitation of the visual information.

Although the use of facial markers is not suitable for real applications, and an automatic face detection and feature tracking approach is needed, the preliminary analysis presented in this chapter based on two emotion databases gives important clues about emotion discrimination contained in different blocks of the face. While the shapes and the movements of the eyebrows have been widely used for facial expression classification, the results presented in this chapter show that this facial area does not provide good emotion discrimination as compared to other facial areas such as the cheeks. However, the results reported here were for just three emotion states and the neutral state—though the experiments were performed for all six emotions in the corpora. For the other three emotions, eyebrows do play an important role, that is, for fear, disgust, and surprise. Also, the experiments were conducted by using two emotion databases, where the quality of the data and the recording conditions were not similar.

17.6 CONCLUSION

In this chapter, we have addressed two important issues in the facial expression and emotion recognition areas. Firstly, the need for quantification of expressions so as to detect subtle and micro expressions was addressed. And next, the need to utilize multiple channels of information for improvement in automatic recognition of emotions. We proposed a simple approach for quantifying the intensity of emotions. The applicability of our approach was demonstrated on various expressions at varying levels of intensity. Further, the technique is able to capture very subtle differences in facial expression change or capture micro expressions. The results show that the SVDF/VDF shape transformation features allow better quantification of facial expressions as compared to marker-based features. Further work involves a thorough testing of the proposed approach for all the subsets of rich expression data available in the DaFEx database.

Further, we reported the results of experiments on analysing the strengths and weaknesses of unimodal facial expression and acoustic emotion classification approaches, where some pairs of emotions are often misclassified. We proposed an audiovisual fusion approach at multiple levels to show that most of these confusions could be resolved. The further plans for this research will be to find better methods to fuse audiovisual information that can model the dynamics of facial expressions and speech. Segmental level acoustic information can be used to trace the emotions at a frame level. Also, it might be useful to find other kind of features that describe the relationship between both modalities with respect to temporal progression. For example, the correlation between the facial motions and the contour of the pitch and the energy might be useful to discriminate emotions.

REFERENCES

[1] A. Battocchi and F. Pianesi, "DaFEx: Un Database di Espressioni Facciali Dinamiche," in SLI-GSCP Workshop, Comunicazione Parlata e Manifestazione delle Emozioni, Padova, Italy, November–December 2004.

[2] N. Mana, P. Cosi, G. Tisato, F. Cavicchio, E. Magno, and F. Pianesi, "An Italian database of emotional speech and facial expressions," in 5th International Conference on Language, Resources and Evaluation (LREC2006), Genoa, Italy, May 24–26, 2006.

[3] O. Martin, J. Adell, A. Huerta, I. Kotsia, A. Savran, and R. Sebbe, "Multimodal caricatural mirror," in Proceedings of eNTERFACE 2005, Workshop, http://www.enterface.net/enterface05/docs/results/reports/project2.pdf

[4] C. Darwin, *The Expression of the Emotions in Man and Animals*, John Murray, London, 1872.

[5] C. G. Kohler, T. Turner, N. M. Stolar, W. B. Bilker, C. M. Bresinger, R. E. Gur, and R. C. Gur, "Differences in facial expression of four universal emotion," *Psychiatry Res.*, 128(3): 235–244, 2004.

[6] K. L. Schmidt and J. F. Cohn, "Human facial expressions as adaptations: evolutionary questions in facial expression research," *Yearb. Phys. Anthropol.*, 44: 3–24, 2001.

[7] G. Gainnoti, C. Caltagirone, and P. Zoccolotti, "Left/right and cortical/subcortical dichotomies in the neuropsychological study of human emotions," *Cogn. Emot.*, 7: 71–94, 1993.

[8] C. Davatzikos, "Measuring biological shape using geometry-based shape transformations," *Image Vis. Comput.*, 19: 63–74, 2001.

[9] C. Busso, D. Zhigang, Y. Serdar, B. Murtaza, C. M. Lee, A. Kazemzadeh, L. Sungbok, U. Neumann, and S. Narayanan, "Analysis of emotion recognition using facial expressions, speech and multimodal information," in Proceedings of ACM 6th International Conference on Mutlmodal Interfaces (ICMI 2004), State College, PA, October 2004.

[10] C. M. Lee, S. Narayana, and R. Pieraccini, "Classifying emotions in human–machine spoken dialogs," in International Conference on Multimedia and Expo (ICME), vol. 1, August 26–29, 2002, pp. 737–740.

[11] L. C. De Silva, T. Miyasato, and R. Nakatsu, "Facial emotion recognition using multimodal information," in IEEE International Conference on Information, Communications and Signal Processing (ICICS'97), Singapore, September 1997, pp. 397–401.

[12] L. S. Chen, T. S. Huang, T. Miyasato, and R. Nakatsu, "Multimodal human emotion expression recognition," in International Conference on Automatic Face and Gesture Recognition, Nara, Japan, IEEE Computer Society, April 1998.

[13] M. J. Black and Y. Yacoob, "Tracking and recognizing rigid and non-rigid facial motions using local parametric model of image motion," in International Conference on Computer Vision, IEEE Computer Society, Cambridge, MA, 1995, pp. 374–381.

[14] A. P. Essa Pentland, "Coding, analysis, interpretation, and recognition of facial expressions," *IEEE Transactions on Pattern Analysis and Machine Intelligence*, 19(7): 757–763, 1997.

[15] K. Mase, "Recognition of facial expression from optical flow," *IEICE Transactions on Information and Systems*, E74-D(10): 3474–3483, 1991.

[16] Y. L. Tian, T. Kanade, and J. Cohn, "Recognizing lower face action units for facial expression analysis," in 4th IEEE International Conference on Automatic Face and Gesture Recognition (FG'00), March 2000, pp. 484–490.

[17] Y. Yacoob and L. Davis, "Computing spatio-temporal representations of human faces," in IEEE Computer Society Conference on Computer Vision and Pattern Recognition, June 21–23, 1994, pp. 70–75.

[18] F. Dellaert, T. Polzin, and A. Waibel, "Recognizing emotion in speech. Spoken language," in Proceedings of the Fourth International Conference on ICSLP 96, vol. 3, October 3–6, 1996, pp. 1970–1973.

[19] T. L. Nwe, F. S. Wei, and L. C. De Silva, "Speech based emotion classification," in IEEE Region 10 International Conference on Electrical and Electronic Technology, vol. 1, August 19–22, 2001, pp. 297–301.

[20] C. M. Lee, S. Yildirim, M. Bulut, A. Kazemzadeh, C. Busso, Z. Deng, S. Lee, and S. Narayanan, "Emotion recognition based on phoneme classes," in Proceedings of ICSLP, 2004.

[21] T. Polzehl, S. Sundaram, H. Ketabdar, M. Wagner, and F. Metze, "Emotion classification in children's speech using fusion of acoustic and linguistic features," in Proceedings of the 10th Annual Conference of the International Speech Communication Association Interspeech-2009, Brighton, UK, September 2009.

[22] F. Metze, T. Polzehl, and M. Wagner, "Fusion of acoustic and linguistic speech features for emotion detection," in Proceedings of the 3rd IEEE International Conference on Semantic Computing (ICSC), Berkeley, CA, September 14–16, 2009.

[23] T. Polzehl, A. Schmitt, F. Metze, and M. Wagner, "Anger recognition in speech using acoustic and linguistic cues," *Speech Commun.*, 53(9–10): 1198–1209, 2011.

[24] L. C. De Silva and P. C. Ng, "Bimodal emotion recognition," in Proceedings of the Fourth IEEE International Conference on Automatic Face and Gesture Recognition, March 28–30, 2000, pp. 332–335.

[25] Y. Yoshitomi, S.-I. Kim, T. Kawano, and T. Kilazoe, "Effect of sensor fusion for recognition of emotional states using voice, face image and thermal image of face," in Proceedings of the 9th IEEE International Workshop on Robot and Human Interactive Communication, RO-MAN 2000, September 27–29, 2000, pp. 178–180.

[26] T. S. Huang, L. S. Chen, H. Tao, T. Miyasato, and R. Nakatsu, "Bimodal emotion recognition by man and machine," in ATR Workshop on Virtual Communication Environments, Kyoto, Japan, April 1998.

[27] L. S. Chen and T. S. Huang, "Emotional expressions in audiovisual human computer interaction," in IEEE International Conference on Multimedia and Expo, ICME 2000, vol. 1, July 30–August 2, 2000, pp. 423–426.

[28] R. Verma, C. Davatzikosa, J. Lougheadb, T. Indersmitten, H. Ranliang, C. Kohlerb, R. E. Gurb, and R. C. Gurb, "Quantification of facial expressions using high-dimensional shape transformations," *J. Neurosci. Methods*, 141(1): 61–73, 2005.

[29] W. E. Rinn, "The neuropsychology of facial expression: a review of the neurological and psychological mechanisms for producing facial expressions," *Psychol. Bull.*, 95(1): 52–77, 1984.

[30] M. R. Bartlett, J. C. Hager, P. Ekman, and T. J. Sejnowski, "Measuring facial expressions by computer image analysis," *Psychophysiology*, 36: 253–264, 1999.

[31] M. J. Black and Y. Yacoob, "Recognizing facial expressions in image sequences using local parameterized models of image motion," *Int. J. Comput. Vis.*, 25(1): 23–48, 1997.

[32] I. A. Essa and A. P. Pentland, "Coding, analysis, interpretation and recognition of facial expressions," *IEEE Transactions on Pattern Analysis and Machine Intelligence*, 19(7): 757–763, 1997.

[33] J. Lien, T. Kanade, J. Cohn, and C. Li, "Detection, tracking and classification of action units in facial expression," *J. Robot. Auton. Syst.*, 31: 131–146, 2000.

[34] C. Busso, Z. Deng, S. Yildrim, M. Bulut, C. M. Lee, A. Kazemzadeh, S. Lee, U. Neuamnn, and S. S. Narayanan, "Analysis of emotion recognition using facial expressions, speech and multimodal information," in Proceedings of the International Conference on Multimodal Interfaces (ICMI), State Park, PA, October 2004, pp. 205–211.

[35] P. Ekman and E. Rosenberg, *What the Face Reveals*, Oxford University Press, 1997.

[36] M. R. Bartlett, J. C. Hager, P. Ekman, and T. J. Sejnowski, "Measuring facial expressions by computer image analysis," *Psychophysiology*, 36: 253–264, 1996.

[37] M. J. Black and Y. Yacoob, "Recognizing facial expressions in image sequences using local parameterized models of image motion," *Int. J. Comput. Vis.*, 25(1): 23–48, 1997.

[38] I. A. Essa and A. P. Pentland, "Coding, analysis, interpretation and recognition of facial expressions," *IEEE Transactions on Pattern Analysis and Machine Intelligence*, 19(7): 757–763, 1997.

AUTHOR BIOGRAPHIES

Girija Chetty is an assistant professor and Head of Software Engineering in the Faculty of Information Sciences and Engineering, University of Canberra. She received her Bachelor's and Master's (Research) degrees in Electrical Engineering and Computer Science from India and her Doctorate in Information Sciences and Engineering from Australia. She has published more than 100 articles and papers in internationally peer-reviewed conferences and journals, and her research interests are in the area of image processing, information fusion, machine learning, pattern recognition, and software engineering.

Michael Wagner has a Diplomphysiker (M.Sc. in Physics) degree from the University of Munich (1973) and a Ph.D. in Computer Science from Australian National University (1979). Since 1979 he has held research and teaching positions at the Technical University of Munich, National University of Singapore, University of Wollongong, University of New South Wales/Australian Defence Force Academy, and Australian National University as well as commercial research and development positions at Siemens Nixdorf AG in Germany and ANUTech in Australia. Since 1996 he has been professor of computing at the University of Canberra, where he has also been Head of the School of Computing, Head of the Discipline of Software Engineering and Director of the National Centre for Biometric Studies. He is now the Director of the Human-Centred Computing Laboratory at UC. He has been a visiting professor at Deutsche Telekom Research Labs, Berlin and the Technical University of Berlin, University of Karlsruhe, Laboratoire Informatique of the University of Avignon (LIA), and Institute for Infocomm Research, and Nanyang Technological University, Singapore.

Roland Goecke is assistant professor of software engineering, Head of the Vision and Sensing Group, and Deputy Director of the Human-Centred Computing Laboratory at the Faculty of Information Sciences and Engineering, University of Canberra. He is also an adjunct fellow at the Information and Human-Centred Computing Group, Research School of Computer Science, Australian National University, Canberra. He received his Master's degree in Computer Science from the University of Rostock, Germany, in 1998, and his Ph.D. in Computer Science from Australian National University in 2004. His research interests are in affective computing, pattern recognition, computer vision, human–computer interaction, and multimodal signal processing.

18

FROM A DISCRETE PERSPECTIVE OF EMOTIONS TO CONTINUOUS, DYNAMIC, AND MULTIMODAL AFFECT SENSING

ISABELLE HUPONT AND SERGIO BALLANO

Aragon Institute of Technology, Zaragoza, Spain

EVA CEREZO AND SANDRA BALDASSARRI

Department of Computer Sciences and Systems Engineering, University of Zaragoza, Zaragoza, Spain

Natural human–human affective interaction is inherently continuous and multimodal: people experiment complex emotions over time and communicate them through multiple channels such as facial expressions, gestures, and dialogues. One of the main drawbacks of existing human affect analyzers is related to the poorness of the emotional information they provide as output. The great majority of studies confine themselves to give a single emotional label, such as "happy", "fear", or "sad" characterizing a whole user's video sequence. Moreover, this emotional information is often obtained from the study of a single channel rather than merging complementary information across channels. Human emotions are more complex and richer than simple emotional labels and can experiment strong variations over time. Therefore, those aspects of human affect should be captured and described by an ideal affect recognizer. In this chapter, a novel methodology to deal with multimodal fusion of emotional channels is defined. It first allows to pass from static to dynamic facial affect sensing, and then is expanded to fuse any other affective information coming from different channels over time. This is possible thanks to the use of a continuous evaluation–activation emotional space, which provides an algebra to fuse the different

Emotion Recognition: A Pattern Analysis Approach, First Edition. Edited by Amit Konar and Aruna Chakraborty.
© 2015 John Wiley & Sons, Inc. Published 2015 by John Wiley & Sons, Inc.

sources of affective information through mathematical formulation and obtain a 2D dynamic emotional path representing the user's emotional evolution as final output. To demonstrate its potential, the methodology is applied to different real human–computer interaction contexts.

18.1 INTRODUCTION

Affective computing is an emerging field aimed at developing intelligent systems able to provide a computer with the ability of recognizing, interpreting and processing human emotions. Since the introduction of the term Affective computing, the research community in this field has grown rapidly. Since late 1990s, an increasing number of efforts towards automatic human affect extraction have been reported in the literature. In particular, early efforts mainly focused on *facial affect recognition*, that is, the recognition of emotions by analyzing the user's facial expressions. The major impulse to investigate automatic facial expression analysis comes from the significant role of the face in our emotional and social lives. Facial expressions are the most powerful, natural, and direct way used by humans to communicate and to understand each other's affective state and intentions [1]. Thus, the interpretation of facial expressions is still today the most common method used for emotional detection and constitutes an indispensable part of current affective human–computer interfaces designs.

But developing a system that correctly interprets facial expressions is a difficult task. It initially involves extracting certain information (features) from the face that is subsequently used to feed a classification system that will judge the affectivity of a facial expression. Although the research in the field has seen a lot of progress in the past few years, several important issues must still be improved, related to:

- **The robustness of the classification mechanisms**. On the one hand, the implemented classification mechanisms should be robust enough to achieve accurate facial affect recognition, despite very frequent problems such as facial occlusions, changes in lighting conditions, large head rotations, and changes in the user's facial physiognomy (beard, hair, glasses, etc.). However, existing systems still make strong suppositions by assuming perfect laboratory settings where images are captured without noise inside *ad hoc* nonnaturalistic environments. On the other hand, the existing facial expression analyzers that obtain the best success rates for emotional classification make use of neural networks, rule-based expert systems, support vector machines or Bayesian nets-based classifiers. In Reference 2, an excellent state-of-the-art summary is given of the various methods recently used in facial expression emotional recognition. However, the majority of those studies confine themselves to selecting only one type of classifier for emotional detection, or at the most compare different classifiers and then use that which provides the best results [3]. The combination of the outputs of several classifiers (e.g., by averaging) may reduce the risk of an unfortunate selection of a poorly performing classifier, thus making the system more robust.

- **The validation and universality of the system**. A validation strategy must be correctly defined to evaluate the facial emotions recognizer, both in terms of classification performance and in terms of universality: it should be able to sense affective cues in any individual subject, of any ethnicity, age, gender, and physiognomy. Nevertheless, most current facial affect recognizers lack clear and complete validation mechanisms.
- **The type of emotional output**. One of the main drawbacks of existing facial expressions analyzers is related to the type of emotional information they provide as output. The great majority of studies confine themselves to giving a single emotional label or category characterizing a user's facial image or a whole facial video sequence. Actually, the most commonly used emotional categories in expression recognition systems are the six universal emotions proposed by Ekman [4] which include "joy", "sadness", "fear", "anger", "disgust", and "surprise". However, human emotions are more complex and richer than simple emotional labels and can experiment strong variations over time. Those aspects of human affect (complexity and dynamics of emotions) should be captured and described by an ideal facial affect recognizer.

Owing to the fact that efforts towards improving the aforementioned problems are to be made, facial affect sensing still forms an important current research trend in the field of Affective computing. However, another research focus is recently emerging in parallel: recognizing affect from multiple cues and modalities. Natural human–human affective interaction is inherently multimodal: people communicate emotions not only through facial expressions but also by means of other multiple channels such as body gestures, dialogues, texts and even physiological responses. Although several studies prove that multimodal fusion of affective information improves the robustness and accuracy of machine analysis of human emotion [5–7], most works still focus on increasing the success rates in sensing affect from a single channel rather than merging complementary information across channels [6].

The difficult task of *multimodal fusion of different affective channels* is still in its initial stage and really far from being solved to date [8]. The three aforementioned problems regarding facial analyzers (namely robustness, validation and type of emotional output) can be also extrapolated to multimodal systems. However, the main question to be answered is related to the accurate fusion of the information coming from different modalities over time. In most cases, those incoming inputs are unsynchronized, of very different natures, handling different time scales and metrics, etc. An ideal multimodal human affect analyzer should be able to fuse this information dynamically, by being adaptive to the eventual changes in the reliability of the quality of different inputs and being scalable enough to admit new channels without having to redefine and retrain the whole system. However, all those issues are still open in the current state of the art. Existing fusion strategies follow three main streams: feature-level fusion, decision-level fusion, and hybrid fusion.

Feature-level fusion combines the data (features) extracted from each channel in a joint vector before classification. Although several works have reported good performances when fusing different modalities at a feature level [7, 9, 10], this strategy becomes more challenging as the number of input features increases, and they are of very different natures (different timing, metrics, etc.). Adding new modalities implies a big effort to synchronize the different inputs and retrain the whole classification system. To overcome these difficulties, most researchers choose decision-level fusion, in which the inputs coming from each modality are modeled and classified independently, and these unimodal recognition results are integrated at the end of the process by the use of suitable criteria (expert rules, simple operators such as majority vote, sum, product, adaptation of weights, etc.).

Many studies have demonstrated the advantage of decision-level fusion over feature-level fusion, due to the uncorrelated errors from different classifiers [11] and the fact that time and feature dependence are abstracted. Various-mainly bimodal decision-level fusion methods have been proposed in the literature [12–14], but optimal fusion designs are still undefined. Most available multimodal recognizers have designed *ad hoc* solutions for fusing information coming from a set of given modalities but cannot accept new modalities without redefining and/or retraining the whole system. Moreover, in general, they are not adaptive to the input quality and therefore do not consider eventual changes in the reliability of the different information channels. Decision-level methods allow the integration of different algorithms without knowing their inner workings, which can be common when one or more of them are based on commercial software.

The hybrid methods try to combine the flexibility of the decision-level methods, by maintaining different classifiers for each modality, while using part of the information from every sensor in each modality, and taking advantage, when there is a statistical dependence between modalities, as in feature-level methods. For example, in Reference 15, a Multidimensional Dynamic Time Warping algorithm is used to improve speech recognition by fusing the audio channel with mouth gestures from a video channel. The common drawbacks of these methods with feature-level ones is the need to retrain the whole system when adding a new channel.

Finally, it is important to notice that the potential human–computer interaction (HCI) application areas of affective computing can range over very different fields. Some examples are natural interaction with virtual characters capable of understanding users' emotions and reacting accordingly [16], affect-based indexing of multimedia material [17], and educational video games [18]. However, since research is still focusing on the problem of robustly and automatically extracting human affect, real multimodal applications of affective computing are just beginning to appear in the literature and more interaction contexts should be exploited in order to show its potentialities.

Throughout this chapter, a methodology to deal with multimodal fusion of emotional channels is defined. It first allows to pass from static to dynamic facial affect sensing, and then is expanded to fuse any other dynamic affective information coming from different channels over time. First, a facial affect recognizer able to sense emotions from a user's captured static facial image is presented. Its inputs are a set of angles and distances between characteristic points of the face, chosen so that the face is modeled in a simple way without losing relevant facial expression information. It implements an emotional classification mechanism that combines in a novel and robust manner five classifiers, obtaining at the output an associated weight of the facial expression to each of the six Ekman's universal emotional categories plus "neutral". Secondly, this discrete output is enriched thanks to the use of a 2D continuous emotional space that provides both a way of describing a wider range of emotional states and an algebra to mathematically relate affective variations over time. Frame per frame of a video sequence, the exact 2D coordinates in the emotional space of the shown facial image are computed, and an "emotional path" characterizing the user's affective evolution is progressively built. A Kalman filtering technique is also exploited to control this path in real time and thus ensure temporal consistency and robustness of the system. This process is then expanded to deal with multimodality: a general and scalable methodology is defined to fuse any other affective information coming from different channels over time. The proposed methodolgy is applied to two real applications: it is used to enhance a tutoring system by making the distant teacher aware of the learner's emotional progress and to fuse dynamic affective information extracted from different channels of an instant messaging tool.

The structure of the chapter is the following. Section 18.2 describes the categorical facial classification method. In Section 18.3, the step from the discrete perspective of emotions to continuous and dynamic facial affect sensing is explained in detail. Section 18.4 proposes the novel multimodal fusion methodology, which is applied to different real HCI contexts in Section 18.5. Finally, Section 18.6 comprises the conclusions and future work.

18.2 A NOVEL METHOD FOR DISCRETE EMOTIONAL CLASSIFICATION OF FACIAL IMAGES

In this section, an effective method is presented for the automatic classification of facial expressions into discrete emotional categories. The method is able to classify the user's emotion in terms of the six Ekman's universal emotions (plus "neutral"), giving a confidence value to each emotional category. Section 18.2.1 explains the selection and extraction process of the features serving as inputs to the system. Section 18.2.2 describes the criteria taken into account when selecting the various classifiers and how they are combined. Finally, the obtained results are presented in Section 18.2.3.

18.2.1 Selection and Extraction of Facial Inputs

Facial Action Coding System (FACS) [19] was developed by Ekman and Friesen to code facial expressions in which the individual muscular movements in the face are described by action units (AUs). This work inspired many researchers to analyze facial expressions by means of image and video processing, where, by tracking of facial features and measuring a set of facial distances and angles, they attempt to classify different facial expressions. In particular, existing works demonstrate that a high emotional classification accuracy can be obtained by analyzing a small set of facial distances and angles [20–22].

Following that methodology, the initial inputs of our classifiers were established in a set of distances and angles obtained from characteristic facial points [23]. In fact, the inputs are the variations of these angles and distances with respect to the

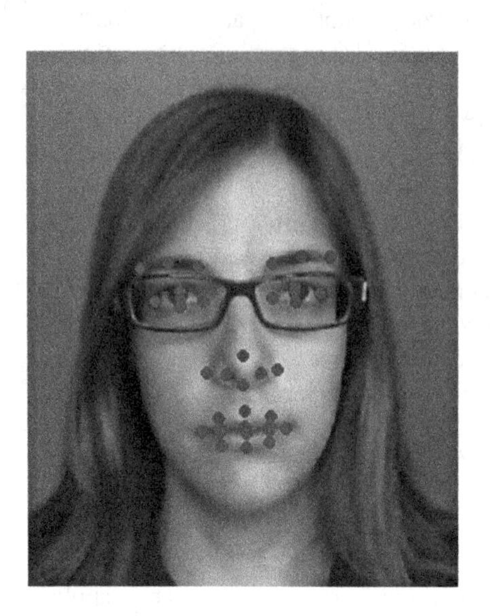

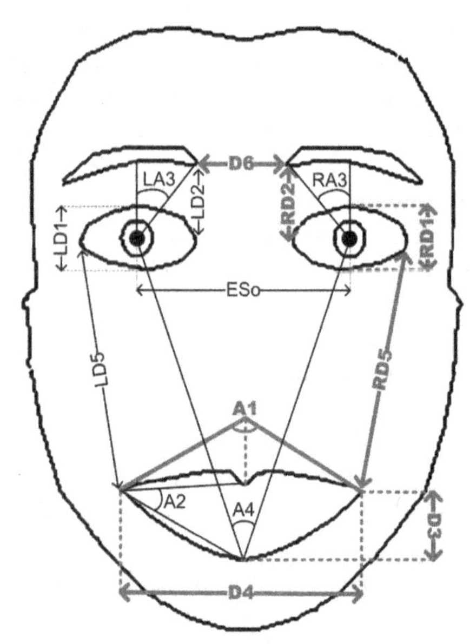

FIGURE 18.1 Facial characteristic points extracted thanks to FaceAPI (left) and facial parameters tested (right). In bold, the final selected parameters. Taken from FGNET.

"neutral" face. The chosen set of initial inputs compiles the distances and angles that have been proved to provide the best classification performance in existing works of the literature. The points are obtained thanks to faceAPI [24], a commercial real-time facial feature tracking program that provides Cartesian facial 3D coordinates. It is able to track up to +/− 90° of head rotation and is robust to occlusions, lighting conditions, presence of beard, glasses, etc. The initial set of parameters tested is shown in Figure 18.1. In order to make the distance values consistent (independently of the scale of the image, the distance to the camera, etc.) and independent of the expression, all the distances are normalized with respect to the distance between the eyes. The choice of angles provides a size invariant classification and saves the effort of normalization.

In order to determine the goodness and usefulness of the parameters, a study of the correlation between them was carried out using the data (distance and angle values) obtained from a set of training images. For this purpose, two different facial emotion databases were used: the FGNET database [25] that provides spontaneous (nonacted) video sequences of 19 different young Caucasian people, and the MMI Facial Expression Database [26] that holds 1280 acted videos of 43 different subjects from different races (Caucasian, Asian, South American and Arabic) and ages ranging from 19 to 62. Both databases show Ekman's six universal emotions plus the "neutral" one and provide expert annotations about the emotional apex frame of the video sequences. A new database has been built for this work with a total of 1500 static frames selected from the apex of the video sequences from the FGNET and MMI databases. It has been used as a training set in the correlation study and in the tuning of the classifiers.

A correlation-based feature selection technique [27] was carried out in order to identify the most influential parameters in the variable to predict (emotion) as well as to detect redundant and/or irrelevant features. Subsets of parameters that are highly correlated with the class while having low intercorrelation are preferred. In that way, from the initial set of parameters, only the most significant ones were selected to work with RD1, RD2, RD5, D3, D4, D6, and A1 (marked in bold in Figure 18.1). This reduces the number of irrelevant, redundant, and noisy inputs in the model and thus the computational time, without losing relevant facial information.

18.2.2 Classifiers Selection and Combination

In order to select the best classifiers, the Waikado Environment for Knowledge Analysis (WEKA) tool was used [28]. It provides a collection of machine learning algorithms for data mining tasks. From this collection, five classifiers were selected after tuning and benchmarking: RIPPER, Multilayer Perceptron, SVM, Naive Bayes and C4.5. The selection was based on their widespread use as well as on the individual performance of their WEKA implementation.

A 10-fold cross-validation test over the 1500 training images has been performed for each selected classifier. The success rates obtained for each classifier and each emotion are shown in the first five rows of Table 18.1. As can be observed, each classifier is very reliable for detecting certain specific emotions but not so much for others. For example, the C4.5 is excellent at identifying "joy" (92.90% correct) but is only able to correctly detect "fear" on 59.30% of occasions, whereas Naive Bayes is way above the other classifiers for "fear" (85.20%), but is below the others in detecting "joy" (85.70%) or "surprise" (71.10%). Therefore, an intelligent combination of the five classifiers is to be taken in such a way that the strong and weak points of each are considered for developing a method with a high success rate.

The classifier combination chosen follows a weighted majority voting strategy. The voted weights are assigned depending on the performance of each classifier for each emotion. From each classifier, a confusion matrix formed by elements $P_{jk}(E_i)$,

TABLE 18.1 Success rates obtained with a 10-fold cross-validation test over the 1500 training images for each individual classifier and each emotion (first five rows), and when combining the five classifiers (sixth row)

	Disgust	Joy	Anger	Fear	Sadness	Neutral	Surprise
RIPPER	50.00%	85.70%	66.70%	48.10%	26.70%	80.00%	80.00%
SVM	76.50%	92.90%	55.60%	59.30%	40.00%	84.00%	82.20%
C4.5	58.80%	92.90%	66.70%	59.30%	30.00%	70.00%	73.30%
Naive Bayes	76.50%	85.70%	63.00%	85.20%	33.00%	86.00%	71.10%
Multilayer Perceptron	64.70%	92.90%	70.40%	63.00%	43.30%	86.00%	77.80%
Combination of classifiers	**94.12%**	**97.62%**	**81.48%**	**85.19%**	**66.67%**	**94.00%**	**95.56%**

corresponding to the probability of having emotion i knowing that classifier j has detected emotion k, is obtained. The probability assigned to each emotion $P(E_i)$ is calculated as:

$$P(E_i) = \frac{P_{1k'}(E_i) + P_{2k''}(E_i) + \cdots + P_{5k^v}(E_i)}{5} \tag{18.1}$$

where: $k', k'' \ldots k^v$ are the emotions detected by classifiers 1, 2 … 5, *respectively.*

The assignment of the final output confidence value corresponding to each basic emotion is done following two steps:

1. First, the confidence value $CV(E_i)$ is obtained by normalizing each $P(E_i)$ to a 0 through 1 scale:

$$CV(E_i) = \frac{P(E_i) - min\{P(E_i)\}}{max\{P(E_i)\} - min\{P(E_i)\}} \tag{18.2}$$

   where:
   - $min\{P(E_i)\}$ is the greatest $P(E_i)$ that can be obtained by combining the different $P_{jk}(E_i)$ verifying that $k \neq i$ for every classifier j. In other words, it is the highest probability that a given emotion can reach without ever being selected by any classifier.
   - $max\{P(E_i)\}$ is that obtained when combining the $P_{jk}(E_i)$ verifying that $k = i$ for every classifier j. In other words, it is the probability that obtains a given emotion when selected by all the classifiers unanimously.
2. Second, a rule is established over the obtained affective weights in order to detect and eliminate emotional incompatibilities. The rule is based on the work of Plutchik [29], who assigned "emotional orientation" values to a series of affect words. For example, two similar terms (like "joyful" and "cheerful") have very close emotional orientation values while two antonymous words (like "joyful" and "sad") have very distant values, in which case Plutchik speaks of "emotional incompatibility." The rule to apply is the following: if emotional incompatibility is detected, that is, two nonnull incompatible emotions exist simultaneously, the one to be chosen would have closer emotional orientation to the rest of the nonnull detected emotions. For example, if "joy", "sadness", and "disgust" coexist, "joy" is assigned zero since "disgust" and "sadness" are emotionally closer according to Plutchik.

18.2.3 Results

The results obtained when applying the strategy explained in the previous section to combine the scores of the five classifiers with a 10-fold cross-validation test are shown in the sixth row of Table 18.1. As can be observed, the success rates for the "neutral", "joy", "disgust", "fear", and "surprise" emotions are very high (81.48%–97.62%). The lowest result of our classification is for "sadness": it is confused with

TABLE 18.2 Confusion matrix obtained combining the five classifiers

Emotion ⟶ is classified as	Disgust	Joy	Anger	Fear	Sadness	Neutral	Surprise
Disgust	**94.12%**	0.00%	2.94%	2.94%	0.00%	0.00%	0.00%
Joy	2.38%	**97.62%**	0.00%	0.00%	0.00%	0.00%	0.00%
Anger	7.41%	0.00%	**81.48%**	0.00%	7.41%	3.70%	0.00%
Fear	3.70%	0.00%	0.00%	**85.19%**	3.70%	0.00%	7.41%
Sadness	6.67%	0.00%	6.67%	0.00%	**66.67%**	20.00%	0.00%
Neutral	0.00%	0.00%	2.00%	2.00%	2.00%	**94.00%**	0.00
Surprise	0.00%	0.00%	0.00%	2.22%	0.00%	2.22%	**95.56%**

the "neutral" emotion on 20% of occasions, due to the similarity of their facial expressions. Nevertheless, the results can be considered positive as emotions with distant "emotional orientation" values (such as "disgust" and "joy" or "neutral" and "surprise") are confused on less than 2.5% of occasions and incompatible emotions (such as "sadness" and "joy" or "fear" and "anger") are never confused. Table 18.2 shows the confusion matrix obtained after the combination of the five classifiers.

18.3 A 2D EMOTIONAL SPACE FOR CONTINUOUS AND DYNAMIC FACIAL AFFECT SENSING

As discussed in the introduction, the use of a discrete set of emotions (labels) for emotional classification has important limitations. This section explains how the use of the 2D emotional space proposed by Whissell [30] allows to enrich the affective output of the system. It first serves to map the confidence values associated to each of Ekman's emotions obtained by the discrete emotional classification method to a 2D point in the space characterizing the analyzed facial expression (Section 18.3.1. By using a Kalman filtering technique, it then allows to build a dynamic "emotional" path in that space representing the user's affective progress over time (Section 18.3.2).

18.3.1 Facial Expressions Mapping to the Whissell Affective Space

Some researchers, such as Whissell [30] and Plutchik [31], prefer to view affective states not independent of one another but rather related to one another in a systematic manner. They consider emotions as a continuous 2D space whose dimensions are evaluation and activation. The evaluation dimension measures how a human feels, from positive to negative. The activation dimension measures whether humans are more or less likely to take some action under the emotional state, from active to passive. Besides categorical approach, dimensional approach is attractive because it provides a way of describing a wide range of emotional states and measuring the intensity of emotion. It is much more able to deal with nondiscrete emotions and variations in emotional states over time, since in such cases, changing from one universal emotion label to another would not make much sense in real-life scenarios.

However, in comparison with the category-based description of affect, very few works have chosen a dimensional description level, and the few that do are more related to the design of synthetic faces [32], data processing [33] or psychological studies [34] than to emotion recognition. Moreover, in existing affective recognition works, the problem is simplified to a two-class (positive vs. negative and active vs. passive) [35] or a four-class (quadrants of 2D space) classification [36], thereby losing the descriptive potential of 2D space.

To enrich the emotional output information from the system in terms of intermediate emotions, one of the most influential evaluation–activation 2D models has been used: that proposed by Whissell. In her study, Whissell assigns a pair of values (evaluation, activation) to each of the approximately 9000 selected affective words that make up her "Dictionary of Affect in Language" [30]. Figure 18.2 shows the position of some of these words in the evaluation–activation space. The next step is to build an emotional mapping so that an expressional face image can be represented as a point on this plane whose coordinates (x, y) characterize the emotion property of that face.

It can be seen that the words corresponding to each of Ekman's six emotions have a specific location $(x_i; y_i)$ in the Whissell space (in bold in Figure 18.2). Thanks to this, the output of the classifiers (confidence value of the facial expression to

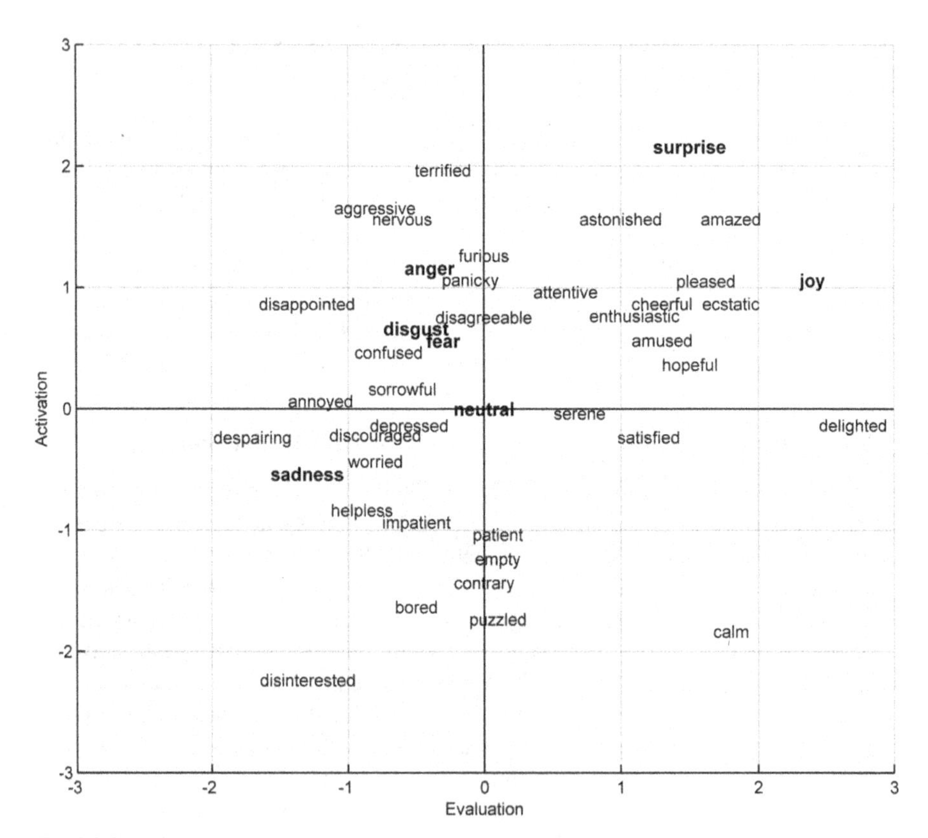

FIGURE 18.2 Several affective words positioned in Whissell's evaluation–activation space.

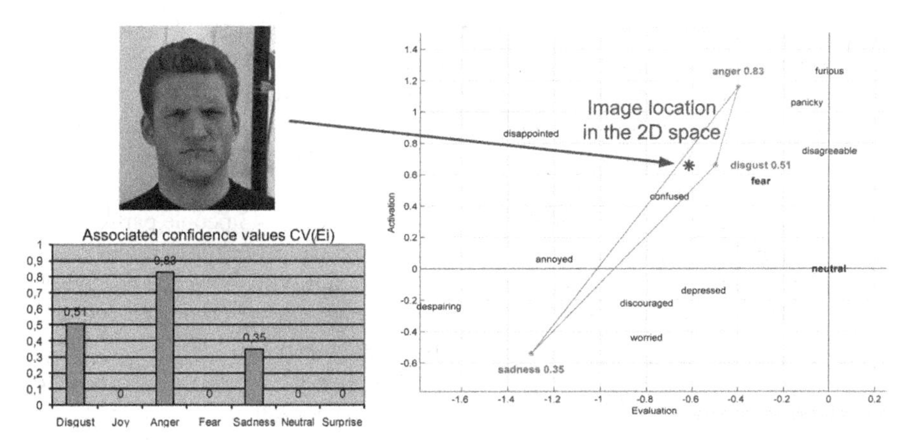

FIGURE 18.3 Example of emotional mapping. The facial images discrete classification method assigns to the facial expression shown a confidence value of 0.83 to "anger", 0.51 to "disgust," and 0.35 to "sadness" (the rest of Ekman's emotions are assigned zero). Its location in the 2D space is calculated as the center of mass of those weighted points (the asterisk inside the triangle). Taken from FGNET. The color version of the above figure is available at www.amitkonar.com/wileycolorimages.

each emotional category) can be mapped onto the space. This emotional mapping is carried out considering each of Ekman's six basic emotions plus "neutral" as weighted points in the evaluation–activation space. The weights are assigned depending on the confidence value $CV(E_i)$ obtained for each emotion. The final coordinates (x, y) of a given image are calculated as the centre of mass of the seven weighted points (see Figure 18.3). In this way, the output of the system is enriched with a larger number of intermediate emotional states.

The emotional mapping has been put into practice with the outputs of the classification system when applied to the database facial expressions images. Figure 18.4

FIGURE 18.4 Example of images from the database with their nearest label in the Whissell space after applying the 2D emotional mapping. Taken from FGNET.

shows several images of the database with their nearest label in the Whissell space after applying the proposed emotional mapping.

The database used in this work provides images labeled with one of the six Ekman universal emotions plus "neutral", but there is no a priori known information about their location in the Whissell 2D space. In order to evaluate the system results, there is a need to establish the region in the Whissell space where each image can be considered to be correctly located. For this purpose, a total of 43 people participated in one or more evaluation sessions (50 images per session). In the sessions, they were told to locate a set of images of the database in the Whissell space (as shown in Figure 18.2, with some reference labels). As a result, each one of the frames was located in terms of evaluation–activation by 16 different people.

The collected evaluation data have been used to define an ellipsoidal region where each image is considered to be correctly located. The algorithm used to compute the shape of the region is based on minimum volume ellipsoids (MVE). MVE looks for the ellipsoid with the smallest volume that covers a set of data points. Although there are several ways to compute the shape of a set of data points (e.g. using a convex hull, rectangle, etc.), we chose the MVE because of the fact that real-world data often exhibit a mixture of Gaussian distributions, which have equidensity contours in the shape of ellipsoids. First, the collected data are filtered in order to remove outliers: a point is considered an outlier if its coordinate values (in both dimensions) are greater than the mean plus three times the standard deviation. Then, the MVE is calculated following the algorithm described by Kumar and Yildrim [37]. The MVEs obtained are used for evaluating results at four different levels:

1. **Ellipse criteria.** If the 2D point coordinates detected by the system is inside the defined ellipse, it is considered a success; otherwise it is a failure.
2. **Quadrant criteria.** The output is considered to be correctly located if it is in the same quadrant of the Whissell space as the ellipse centre.
3. **Evaluation axis criteria.** The system output is a success if situated in the same semiaxis (positive or negative) of the evaluation axis as the ellipse centre. This information is especially useful for extracting the positive or negative polarity of the facial expression shown.
4. **Activation axis criteria.** The same criteria as used in evaluation axis above but projected to the activation axis. This information is relevant for measuring whether the user is more or less likely to take an action under the emotional state.

The results obtained following the different evaluation strategies are presented in Table 18.3. As can be seen, the success rate is 73.73% in the most restrictive case,

TABLE 18.3 Results obtained according to different evaluation criteria

	Ellipse criteria	Quadrant criteria	Evaluation axis criteria	Activation axis criteria
Sucess rate	73.73%	87.45%	94.12%	92.94%

that is, when the output of the system is considered to be correctly located when inside the ellipse. It rises to 94.12% when considering the evaluation axis criteria. Objectively speaking, these results are very good, especially when, according to Bassili [38], a trained observer can correctly classify facial emotions with an average of 87%. However, they are difficult to compare with other emotional classification studies that can be found in literature, given that either such studies do not recognize emotions in evaluation–activation terms, or they have not been tested under common experimental conditions (e.g., different databases or evaluation strategies are used).

18.3.2 From Still Images to Video Sequences through 2D Emotional Kinematics Modeling

As pointed out in the introduction, humans inherently display facial emotions following a continuous temporal pattern. With this starting postulate and thanks to the use of the 2D description of affect, which supports continuous emotional input, an emotional facial video sequence can be viewed as a point (corresponding to the location of a particular affective state in time t) moving through this space over time. In that way, the different positions taken by the point (one per frame) and its velocity over time can be related mathematically and modeled, finally obtaining an "emotional path" in the 2D space that reflects intuitively the emotional progress of the user throughout the video. In this section, a Kalman filtering technique is proposed to model the "emotional kinematics" of that point when moving along the Whissell space and thus enabling to both smooth its trajectory and improve the robustness of the method by predicting its future locations (e.g., in cases of temporal facial occlusions or inaccurate tracking).

Kalman filters are widely used in the literature for estimation problems ranging from target tracking to function approximation. Their purpose is to estimate a system's state by combining an inexact (noisy) forecast with an inexact measurement of that state, so that the most weight is given to the value with the least uncertainty at each time t. Analogously to classical mechanics, the "emotional kinematics" of the point in the Whissell space (x-position, y-position, x-velocity and y-velocity) are modeled as the system's state in the Kalman framework at time t_k. The output of the 2D classification system described in Section 18.3.1 is modeled as the measurement of the system's state. In this way, the Kalman iterative estimation process—that follows the well-known recursive equations detailed in Kalman's work [39]—can be applied to the recorded user's emotional video sequence, so that each iteration corresponds to a new video frame (i.e., to a new sample of the computed emotional path). For the algorithm initialization at t_0, the predicted initial condition is set equal to the measured initial state and the 2D point is assumed to have null velocity.

One of the main advantages of using Kalman filter for the 2D point emotional trajectory modeling is that it can be used to tolerate small occlusions or inaccurate tracking. As pointed out in Section 18.2.1, the input facial feature points of the classification method are obtained thanks to the commercial facial tracker faceAPI [24]. In general, existing facial trackers do not perform the detection with high accuracy: most of them are limited in terms of occlusions, fast movements, large head rotations, lighting, beards, glasses, etc. Although faceAPI deals with these problems quite robustly, on some occasions, its performance is poor, especially when working in real time. For that reason, its measurements include a confidence weighting, from 0 to 1, allowing the acceptability of the tracking quality to be determined. Thanks to it, when a low level of confidence is detected (lower than 0.5), the measurement will not be used and only the filter prediction will be taken as the 2D point position.

In order to demonstrate the potential of the proposed "emotional kinematics" model, it has been tested with a set of emotionally complex video sequences, recorded in a natural (unsupervised) setting. These videos are complex owing to three main factors.

1. An average user's home setup is used. A VGA resolution webcam placed above the screen is used with no special illumination, causing shadows to appear in some cases. In addition, the user placement, not covering the entire scene, reduces the actual resolution of the facial image.
2. Different emotions are displayed contiguously, instead of the usual "neutral" → "emotional-apex" → "neutral" affective pattern exhibited in the databases, so emotions such as "surprise" and "joy" can be expressed without neutral periods between them.
3. Some facial occlusions occur due to the user covering his/her face or looking away during a short period of time, causing the tracking program to lose the facial features. In these cases, only the prediction from the Kalman filter is used, demonstrating the potential of the "emotional kinematics" filtering technique.

15 videos from three different users were tested (Figure 18.5), ranging from 20 to 70 seconds from which a total of 127 key frames were selected by the user who recorded the video, looking for each of the "emotional apex" and "neutral" points. These key points were annotated in the Whissell space thanks to 18 volunteers. The collected evaluation data have been used to define a region where each image is considered to be correctly located, es explained in previous subsection. The results obtained following the different evaluation strategies are presented in Table 18.4. As can be seen, the success rate is 61.90% in the most restrictive case, that is, with ellipse criteria. It rises to 84.92% when considering the activation axis criteria.

18.4 EXPANSION TO MULTIMODAL AFFECT SENSING

In this section, the use of the bidimensional Whissell affective space is expanded to go beyond unimodal facial affect sensing, in order to define a general methodology for fusing the responses of multiple emotional recognition modules. The modules to be fused can be of very different natures, exploring different modalities, time scales, metric levels, etc. The proposed methodology is able to fuse the different sources of affective information over time and to obtain as final output a global 2D dynamic emotional path in the activation–evaluation space representing the user's affective progress. Moreover, it is scalable enough to add new modules coming from new channels without having to retrain the whole system.

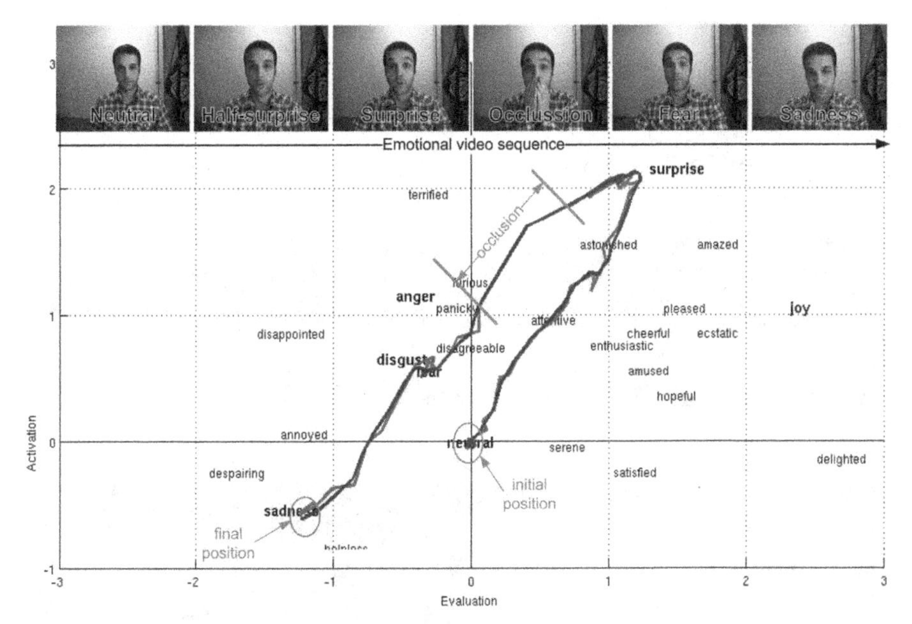

FIGURE 18.5 "Emotional kinematics" model response during the different affective phases of the video and the occlusion period. In red, emotional trajectory without Kalman filtering; In blue, reconstructed emotional trajectory using Kalman filter (visible clearly in the e-copy of the book).

Similar to the facial emotions recognition module presented in Section 18.2, every module i to be fused is assumed to output a list of one or more discrete emotional labels characterizing the affective stimulus recognized at a given time $t_0 i$. The possible output labels can be different for each module i. In this way, the modules' performances are maximized since, unimodal databases annotated in categorical terms are, to date, more complete and reliable than dimensional and/or multimodal ones, allowing the individual modules to be better trained and validated. Some interesting dimensional databases are publicly available [40,41] but, in comparison with categorical ones, they are limited in terms of number of modalities (in general, they explore audio and/or video channels exclusively), annotators, subjects, samples, etc. Moreover, manual dimensional annotation of ground truth is very time-consuming and unreliable, since a large labeling variation between different human raters is reported when working with the dimensional approach [35].

Figure 18.6 shows the general fusion scheme that will be explained step by step in next sections. Since the proposed methodology can combine any number of modules

TABLE 18.4 Results obtained in an uncontrolled environment according to different evaluation criteria

	Ellipse criteria	Quadrant criteria	Evaluation axis criteria	Activation axis criteria
Success rate	61.90%	74.60%	79.37%	84.92%

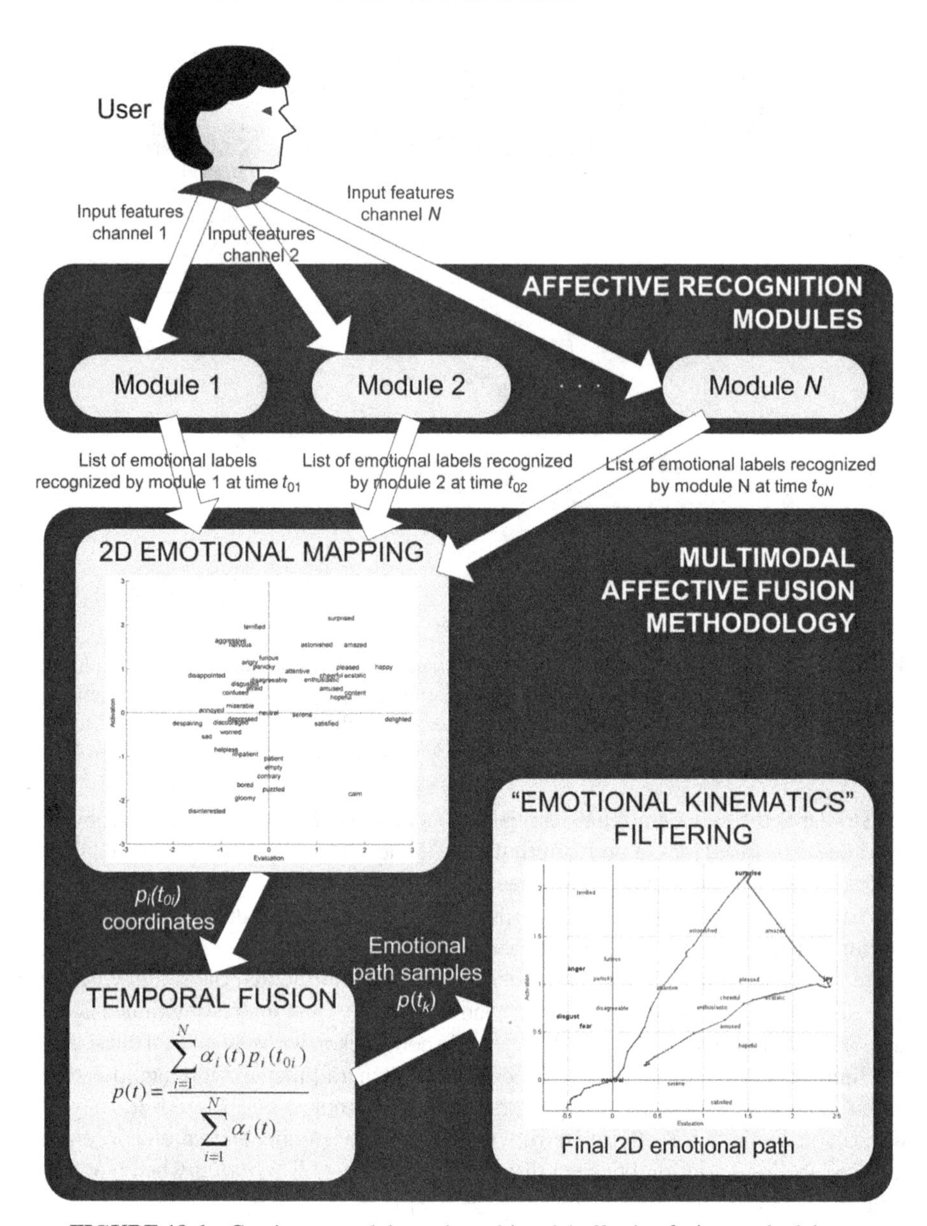

FIGURE 18.6 Continuous and dynamic multimodal affective fusion methodology.

covering different modalities and its overall performance is highly dependent on the accuracy of the individual modules to fuse, in this section, the methodology is presented from a theoretical point of view exclusively. However, in order to show its potential and usefulness, it will be applied in Section 18.5.2 to a real instant messaging interaction context by fusing the information coming from different affective channels.

18.4.1 Step 1: 2D Emotional Mapping to the Whissell Space

The first step of the proposed methodology expands the idea of mapping the output of the facial classification method to the evaluation–activation space to any categorical module i. The idea is to build an emotional mapping to the Whissell space so that the output of each module i at a given time t_{0i} can be represented as a 2D coordinates point $p_i(t_{0i}) = (x_i(t_{0i}); y_i(t_{0i}))$ that characterizes the affective properties extracted from that module at time t_{0i}. The majority of the categorical modules described in the literature provide as output a list of emotional labels with some associated weights at the time t_{0i} corresponding to the detection of the affective stimulus. Since the Whissell dictionary [30] is composed of more than 9000 affective words, whatever the labels used, each one has an associated 2D point in the Whissell space. Following the method explained in Section 18.3.1, the components $(x_i(t_{0i}); y_i(t_{0i}))$ of $p_i(t_{0i})$ are calculated.

18.4.2 Step 2: Temporal Fusion of Individual Modalities to Obtain a Continuous 2D Emotional Path

The second step of the methodology aims to compute a 2D emotional path characterizing the user's affective progress by fusing the different $p_i(t_{0i})$ coordinates obtained from each modality over time. The main difficulty to achieve multimodal fusion is related to the fact that t_{0i} affective stimulus arrival times may be known a priori or not, and may be very different for each module. To overcome this problem, the following equation is proposed to calculate the overall affective response $p(t) = (x(t); y(t))$ at any arbitrary time t:

$$p(t) = \frac{\sum_{i=1}^{N} \alpha_i(t) p_i(t_{oi})}{\sum_{i=1}^{N} \alpha_i(t)} \tag{18.3}$$

where N is the number of fused modalities, t_{0i} is the arrival time of the last affective stimulus detected by module i and $\alpha_i(t)$ is a [0;1] coefficient that will be further detailed. In this way, the overall fused affective response is the sum of each modality's contribution $p_i(t_{0i})$ modulated by the $\alpha_i(t)$ coefficients over time. Therefore, the definition of $\alpha_i(t)$ is especially important since it governs the temporal behavior of the fusion. As suggested by Picard [42], human affective responses are analogous to systems with additive responses with decay where, in the absence of input, the response decays back to a baseline. Following this analogy, the $\alpha_i(t)$ weights are defined as

$$\alpha_i(t) = \begin{cases} b_i \cdot c_i(t_{oi}) \cdot e^{-d_i(t-t_{oi})} & \text{if greather than } \epsilon \\ 0 & \text{elsewhere} \end{cases} \tag{18.4}$$

where:

- b_i is the general confidence that can be given to module i (e.g., the general recognition success rate of the module).

- $c_i(t_{0i})$ is the temporal confidence that can be assigned to the last output of module i due to external factors (i.e., not classification issues themselves). For instance, due to sensor errors, if dealing with physiological signals, or due to facial tracking problems if studying facial expressions (such as occlusions and changes in lighting conditions).
- d_i is the rate of decay (in s^{-1}) that indicates how quickly an emotional stimulus decreases over time for modality i.
- ϵ is the threshold below which the contribution of a module is assumed to disappear. Since exponential functions tend to zero at infinity but never completely disappear, ϵ indicates the $\alpha_i(t)$ value below which the contribution of a module is small enough to be considered nonexistent.

By defining the aforementioned parameters for each module i and applying Equations 18.3 and 18.4, the emotional path that characterizes the user's affective progress over time can be computed by calculating successive $p(t)$ values with any desired time between samples Δt. In other words, the emotional path is progressively built by adding $p(t_k)$ samples to its trajectory, where $t_k = k\Delta t$ (with k integer).

18.4.3 Step 3: "Emotional Kinematics" Path Filtering

The emotional path obtained after applying the second step of the methodology represents in an efficient, visual, and novel way the affective state of the user over time. However, two main problems threaten its calculation process:

1. If the contribution of every fused module is null at a given sample time, that is, every $\alpha_i(t)$ is null at that time, the denominator in Equation 18.3 is zero and the emotional path sample cannot be computed. Examples of cases in which the contribution of a module is null could be the failure of the connection of a sensor of physiological signals, the appearance of an occlusion in a facial/postural tracking system, or simply when the module is not reactivated before its response decays completely.
2. Large "emotional jumps" in the Whissell space can appear if emotional conflicts arise, such as if the distance between two close coordinates vectors $p_i(t_{0i})$ is long.

To solve both problems, the third and last step of the methodology applies the "emotional kinematics" filtering technique to the computed emotional path (Section 18.3.2). In this case, the successive emotional path samples $p(t_k)$ are modeled as the measurement of the system's state in the Kalman framework, that is, each iteration corresponds to a new sample of the computed emotional path. In this way, on the one hand, the "emotional kinematics" filter serves to smooth the emotional path's trajectory and thus prevent large "emotional jumps" in the Whissell space. On the other hand, situations in which the sum of $\alpha_i(t)$ is null are prevented by letting the filter prediction output be taken as the 2D point position for those samples.

18.5 BUILDING TOOLS THAT CARE

Affective computing has a wide range of potential application areas. In this section, the techniques proposed throughout this chapter are applied to two different human–computer interaction contexts: in Section 18.5.1, a t-learning tutoring tool has been provided with the capability of analyzing the students' facial expressions and extracting a 2D affective path from them; Section 18.5.2 presents an instant messaging application that allows to fuse the information coming from three different affective channels.

18.5.1 T-EDUCO: A T-learning Tutoring Tool

The growing success of interactive digital TV (IDTV) has given rise to new services that have not traditionally been associated with this medium. In particular, the term t-learning has been adopted [43] to denote TV-based interactive learning. Although the World Wide Web (WWW)-based distance learning methods seem to still be the current dominating trend, the utility of television itself as a learning tool is also well recognized [45]. This is especially true for the socially disadvantaged communities where television has far more penetration than WWW interaction.

T-learning is not just an adaptation for IDTV of the e-learning techniques used on the Internet [44]. It has its own distinctive characteristics, mostly related to the usability and technological constraints (e.g., low computer power) imposed by the television set and set-top boxes. For that reason, most t-learning applications have been more edutainment than formal learning [46]. More engaged and intelligent t-learning interactive applications are needed to achieve a more complete and efficient learning.

The main difference between an expert human teacher and a distance learning tutoring tool is that the former recognizes and addresses the emotional state of learners to, based upon that observation, take some action that positively impacts learning (e.g., by giving support to a learner who is likely to otherwise quit). Providing distance tutoring systems with these kinds of perceptual abilities would considerably benefit the learning process. However, to date, there is no t-learning tutoring system in the literature with the ability to intelligently recognize affective cues from the student.

In this section, T-EDUCO, the first t-learning affective aware tutoring tool, is proposed. T-EDUCO defies IDTV technical limitations by capturing through a camera the facial expressions of the student while performing the evaluations (tests) proposed by a broadcast interactive t-learning application at home. It integrates the method for continuous facial affect recognition from videos presented in Section 18.3.2 to intelligently extract emotional information from the captured video and present it to the distance tutor in a simple and efficient way so that he/she can be aware of the difficulties encountered by the student during the learning process.

18.5.1.1 T-EDUCO General Architecture The T-EDUCO tutoring tool has two main actors: the student and the tutor. The system's architecture can be explained from the points of view of both actors' environments (Figure 18.7).

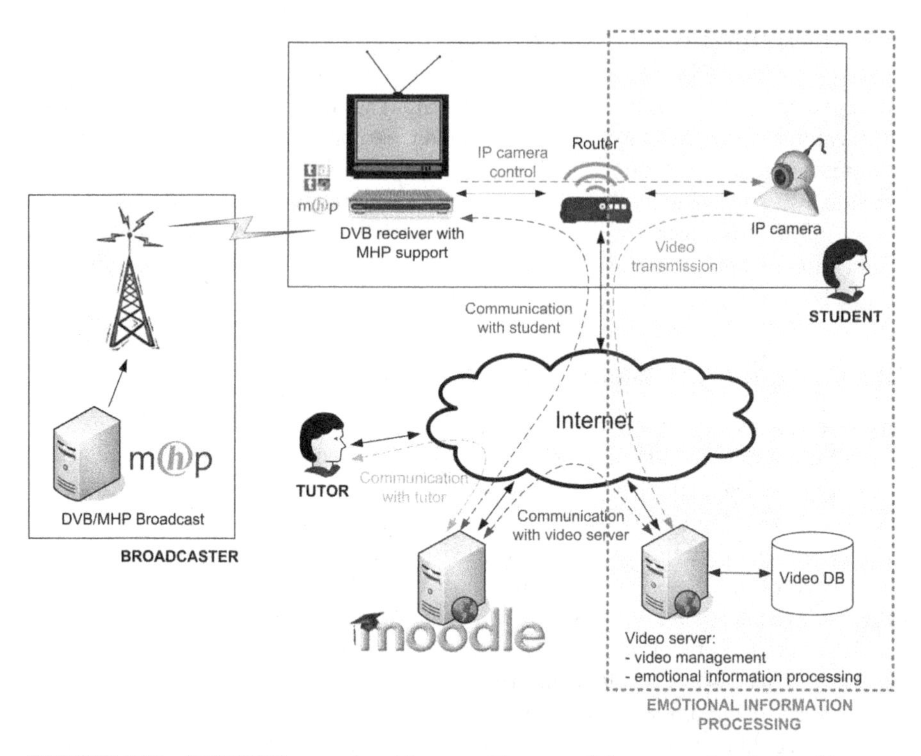

FIGURE 18.7 T-EDUCO general architecture. The part of the system that deals with emotional information processing is highlighted by the region surrounded by dotted line segments. The color version of the above figure is available at www.amitkonar.com/wileycolorimages.

- **Student environment.** The student is located at home. He/she accesses a broadcast t-learning interactive application through a set-top box (TELESystem TS7900HD), which has also IP communication capabilities, by logging in with his/her personal student ID and password. Besides showing the learning contents, the t-learning application is also able to control the management of an IP camera (LINKSYS WVC54GCA) via HTTP commands by accessing the set-top box middleware. This capability is exploited to record videos of the student that are stored in an external video server and further processed for automatically extracting emotional information about his/her affective state. The set-top box middleware manages the whole recording process as well as recorded video upload to the external video server.
- **Tutor environment.** The tutor interacts with T-EDUCO by accessing Moodle (Modular Object Oriented Developmental Learning Environment) [47], a free open source Learning Management System that helps educators to create effective online learning communities. Moodle allows to create, store, organize, integrate, and present educational contents. After performing some required adaptations in Moodle's source code, it has also been provided with the capability of both sending communications to the students' set-top boxes and receiving

information from the interactive t-learning application and the external video server via IP. In this way, T-EDUCO keeps the t-learning application, the external video server, and Moodle synchronized. This opens the door to the tutor to access each student's information—in particular, his/her current state in the course (number of accomplished lessons, answers, etc.) and the extracted emotional information which is available in the video server—and send him/her personal messages and contents.

This architecture provides the convergence between broadcast and broadband technologies. The interactive t-learning course is broadcast to every user with the same contents. Depending on the student's evolution throughout the course (both academic and emotional), the tutor can send additional personalized contents and exercises by means of broadband communications. In that way, both a global delivery of the learning contents and personalized communications to every user are assured. The broadcast t-learning interactive application is based on DVB-MHP (Multimedia Home Platform, version 1.1.3) [48] digital TV interactive standard.

18.5.1.2 Emotional Information Processing and Presentation to the Tutor Tutors are often in charge of a large number of students across different courses. Therefore, keeping close contact with each learner is difficult. For that reason, it turns out interesting to automatically extract emotional information from the student and present it to the tutor in a simple and efficient way so that problems during the learning process can be easily detected. This section focuses on detailing the part of the T-EDUCO system that deals with students' emotional information processing (the region surrounded by dotted line segments in Figure 18.7). The t-learning interactive course broadcast by T-EDUCO consists of a set of modules, and each module is composed of several lessons and a final evaluation test to be performed within a limited time period. Before starting the final test of each module, the application proposes the student to be recorded while answering the evaluation questions. This recorded video, which is stored in the video server, carries useful affective information since it captures the student's facial expressions and can considerably help the tutor to adopt an appropriate pedagogical strategy (e.g., by offering help or extra contents to a student that has shown frustration during the test). It is important to emphasize the need of getting affective information from the students in the "heat of the moment" while they are feeling emotions and they clearly look so [7].

Each student's emotional record is automatically processed in the video server by applying the method for continuous and dynamic facial affect recognition from videos presented in Section 18.3.2. In doing so, a continuous "emotional path" log characterizing the student's affective progress during the module assessment is obtained at the output for each analyzed video sequence (Figure 18.8). Timestamps and information about the beginning and the end of each exercise of the test are also included in the log. Finally, the tutor can easily access every emotional log, sorted by student, course, and module, through the Moodle platform.

The affective log provided to the tutor presents in a novel, efficient, and visual manner the emotional progress of the pupil over time. Its timestamps and the 2D

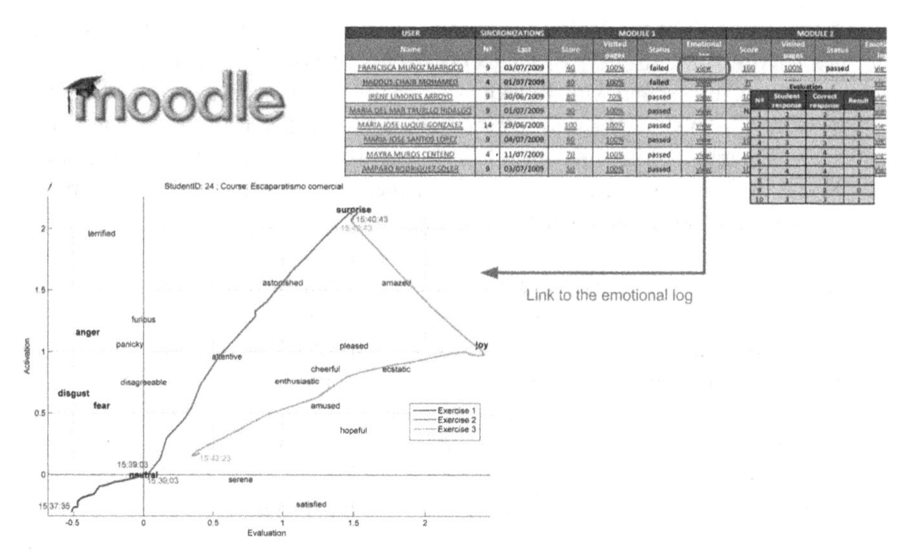

FIGURE 18.8 Example of emotional log that can be consulted by the tutor from the Moodle platform. Timestamps showing the beginning and the end of each exercise are included.

affective path allow the tutor to easily detect which exercise has caused problems to the student, in which task he/she has felt comfortable, etc. This kind of information is usually not considered in most distance learning systems, however, it is particularly important to enhance tutoring.

18.5.2 Multimodal Fusion Application to Instant Messaging

Instant Messaging (IM) is a widely used form of real-time text-based communication between people using computers or other devices. Advanced IM software clients also include enhanced modes of communication, such as live voice or video calling. As users typically experience problems in accurately expressing their emotions in IM text conversations (e.g., statements intended to be ironic may be taken seriously, or humorous remarks may not be interpreted exactly as intended), popular IM programs have resorted to providing mechanisms referred to as "smileys" or "emoticons" seeking to overcome the IM systems' lack of expressiveness [49]. This section aims to explore the potential of the multimodal affective fusion methodology presented in Section 18.4 through the use of an IM tool that combines different communication modalities: text, video, and "emoticons," each one with very different time scales.

18.5.2.1 Instant Messaging Tool and Fusion Modalities Although any publicly available IM tool could be used, a simple *ad hoc* IM tool has been designed. It allows two persons to communicate via text, live video, and "emoticons". Figure 18.9

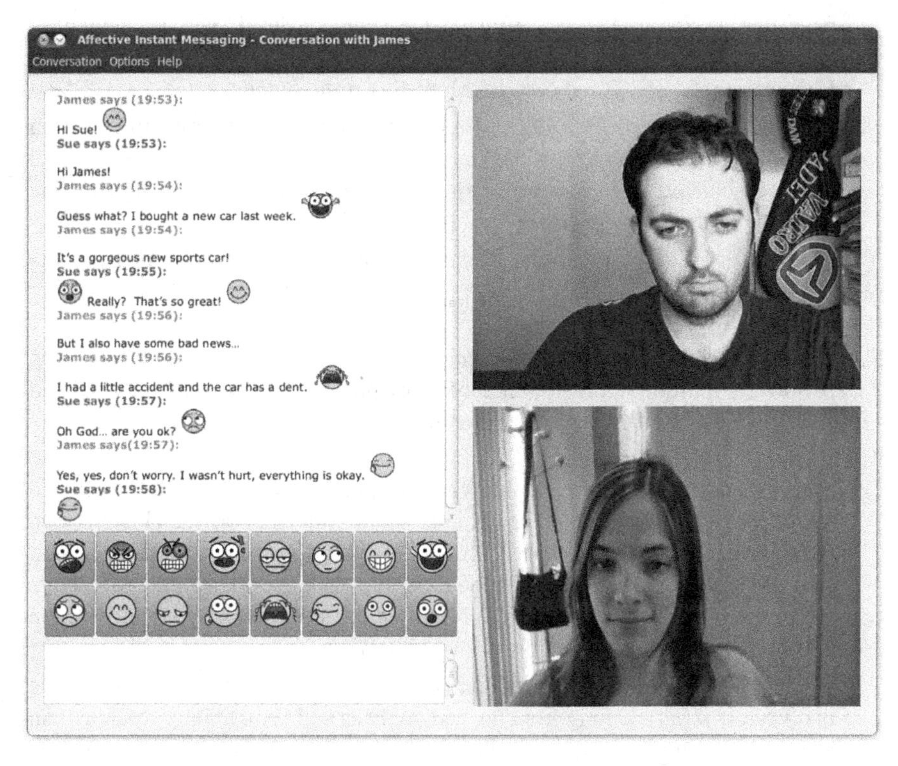

FIGURE 18.9 Snapshot of the IM tool during a conversation.

shows a snapshot of the tool during a conversation. The tool enables access in real time to:

1. The introduced text contents, when the user presses the "enter" key, that is when the user sends the text contents to his/her interlocutor.
2. The inserted "emoticons", when the user presses the "enter" key.
3. Each recorded remote user video frame (with a video rate $f = 25$ fps).

Three affective modules are used to extract emotional information from the IM tool. Each one explores a different modality and makes use of a different set of output emotional categories:

1. **Module 1**: Text analysis module. To extract affective cues from user's typed-in text, the "sentic computing" sentiment analysis paradigm presented in Reference 50 is exploited. By using artificial intelligence and Semantic Web techniques, the "sentic computing" module is able to process natural language texts to extract a "sentic vector" containing a list of up to 24 weighted emotional labels. "Sentic computing" enables the analysis of documents not only on the

page or paragraph-level but even on the sentence level (i.e., IM dialogues level), obtaining a very high precision (73%) and significantly good recall and F-measure rates (65% and 68% respectively) at the output.

2. **Module 2**: "Emoticon" module. "Emoticons" are direct affective information from the user. For this reason, this module simply outputs the list of emotional labels associated to the inserted "emoticons" (a total of 16 "emoticons" are available in the IM tool).
3. **Module 3**: Facial expressions analysis module. The one described in Section 18.2 is used.

18.5.2.2 Multimodal Fusion Methodology Tuning This section describes, step by step, how the multimodal fusion methodology presented in Section 18.4 is tuned to fuse the three different affect recognition modules in an optimal way.

1. **Step 1**: Emotional mapping to the Whissell space. Every output label extracted by the text analysis module, the "emoticon" module, and the facial expression analyzer has a specific location in the Whissell space. Thanks to this, the first step of the fusion methodology (Section 18.4.1) can be applied and vectors $p_i(t_{0i})$ can be obtained each time a given module i outputs affective information at time t_{0i} (with i comprised between 1 and 3).
2. **Step 2**: Temporal fusion of individual modalities. It is interesting to notice that vectors $p_i(t_{0i})$ coming from the text analysis and "emoticons" modules can arrive at any time t_{0i}, unknown a priori. However, the facial expression module outputs its $p_3(t_{03})$ vectors with a known frequency, determined by the video frame rate f. For this reason, and given that the facial expression module is the fastest acquisition module, the emotional path's time between samples is assigned to $\Delta t = 1/f$. The next step towards achieving the temporal fusion of the different modules (Section 18.4.2) is assigning a value to the parameters that define the $\alpha_i(t)$ weights, namely b_i, $c_i(t_{0i})$, d_i and ϵ. Table 18.5 summarizes the values assigned to each parameter for each modality and the reasons for their choice. It should be noted that it is especially difficult to determine the value of the different d_i given that there are no works in the literature providing data for this parameter. Therefore, it has been decided to establish the values empirically. Once the parameters are assigned, the emotional path calculation process can be started following Equations 18.3 and 18.4.
3. **Step 3**: "Emotional kinematics" path filtering. Finally, the "emotional kinematics" Kalman filtering technique (Section 18.4.3) is iteratively applied in real time each time a new sample is added to the computed emotional path. As in most of the works that make use of Kalman filtering, parameters σ and λ are established empirically. An optimal response has been achieved for $\sigma = 0.5 units.s^{-2}$ and $\lambda = 0.5 units^2$.

18.5.2.3 Experimental Results The fusion methodology has been applied to the James' side instant messaging conversation shown in Figure 18.9. This conversation

TABLE 18.5 Temporal fusion parameters

Module Modality	1 Text	2 "Emoticons"	3 Video
Total number of possible output labels	24 weighted emotional labels	16 emotional labels	6 Ekman's universal labels (plus "neutral") + confidence value to each output label
General confidence b_i	$b_1 = 0.65$ The general confidence is assigned the value of the module's recall rate	$b_2 = 1$ The maximum general confidence value is assigned since "emoticons" are the direct expression of user's affective state	$b_3 = 0.87$ The general confidence is assigned the value of the module's general success rate
Temporal confidence $c_i(t_{0i})$	$c_1(t_{01}) = c_2(t_{02}) = 1$ The temporal confidence is assigned constant value 1 since the modules do not depend on external factors		$c_3(t_{03})$ is assigned to the tracking quality confidence weighting, from 0 to 1, provided by the facial features tracking program for each analyzed video frame
Decay value d_i **Threshold value** ϵ	$d_1 = d_2 = 0.035s^{-1}$ Value established empirically $\epsilon = 0.1$ Value established empirically		Irrelevant since the emotional path sample rate is equal to the video frame rate

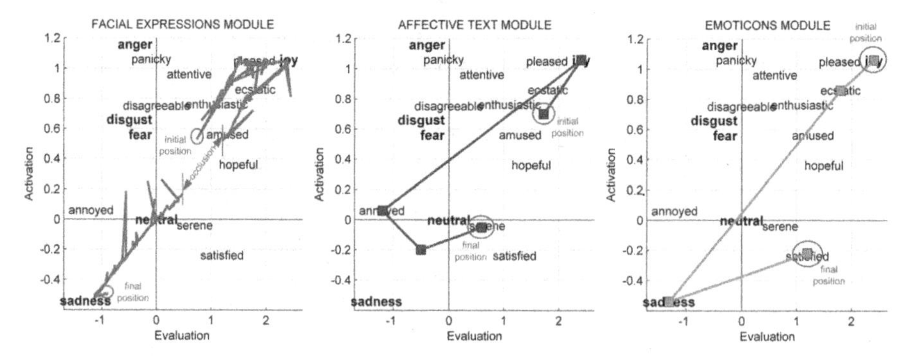

FIGURE 18.10 Emotional paths obtained when applying the methodology to each individual module separately without "emotional kinematics" filtering. Square markers indicate the arriving time of an emotional stimulus (not shown for facial expression module for figure clarity reasons).

is emotionally complex owing to the fact that contrasting emotions are displayed contiguously: at first, James is excited and happy about having bought a wonderful new car and shortly afterwards becomes sad when telling Sue he has dented it.

Figure 18.10 shows the emotional paths obtained when applying the methodology to each individual module separately (i.e., the modules are not fused, only the contribution of one module is considered) without using "emotional kinematics" filtering. At first sight, the timing differences between modalities are striking: the facial expressions module's input stimuli are much more numerous than those of the text and "emoticons", making the latter's emotional paths look more linear. Another noteworthy aspect is that the facial expression module's emotional path calculation is interrupted during several seconds (14 s approximately) due to the appearance of a short facial occlusion during the user's emotional display, causing the tracking program to temporarily lose the facial features.

Figure 18.11 presents the continuous emotional path obtained when applying the methodology to fuse the 3 modules, both without (left) and with (right) the "emotional kinematics" filtering step. As can be seen, the complexity of the user's affective progress is shown in a simple and efficient way. Different modalities complement each other to obtain a more reliable result. Although the interruption period of the emotional path calculation is considerably reduced with respect to the facial expressions module's individual case (from 14 s to 6 s approximately), it still exists since both the text and "emoticons" modules' decay process reaches the threshold ϵ before the end of the facial occlusion, causing the $\alpha_1(t)$ and $\alpha_2(t)$ weights to be null. Thanks to the use of the "emotional kinematics" filtering technique, the path is smoothed and the aforementioned temporal input information absence is solved by letting the filter prediction output be taken as the 2D point position for those samples.

18.6 CONCLUDING REMARKS AND FUTURE WORK

This paper first presents a facial affect recognizer able to sense emotions from a user's captured static facial image. It provides as output an associated weight of the facial

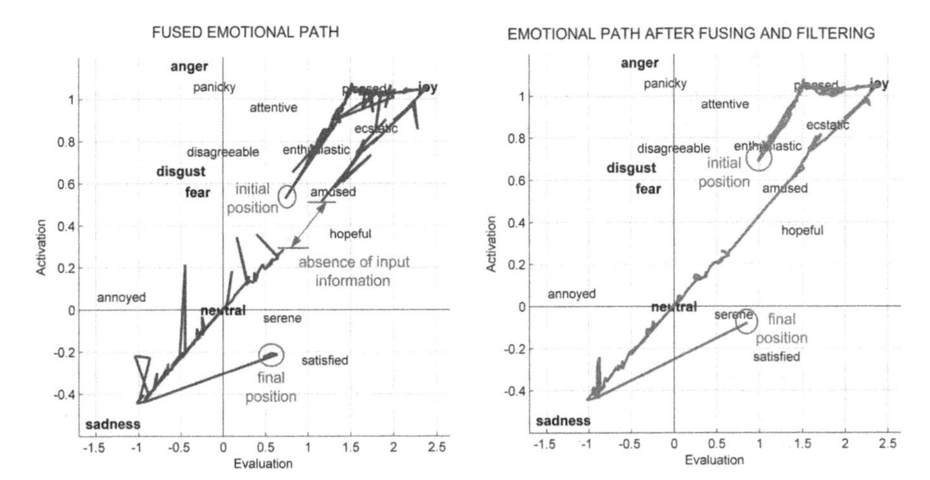

FIGURE 18.11 Continuous emotional path obtained when applying the multimodal fusion methodology to James' instant messaging conversation shown in Figure 18.9, without using "emotional kinematics" filtering (left), and using "emotional kinematics" filtering (right).

expression to each of the six Ekman's universal emotional categories plus "neutral". It has been exhaustively validated by means of statistical evaluation strategies and tested with an extensive database of 1500 images showing individuals of different races and gender, giving universal results with very promising levels of correctness.

The expansion to dynamic and multimodal affective computing is achieved thanks to the use of a 2D description of affect that provides the system with mathematical capabilities to face temporal and multisensory emotional issues. A novel methodology is presented to fuse any number of categorical modules, with very different time-scales and output labels. The proposed methodology outputs a 2D emotional path that represents the user's detected affective progress over time. A Kalman filtering technique controls this path in real-time to ensure temporal consistency and robustness to the system. Moreover, the methodology is adaptive to eventual temporal changes in the reliability of the different inputs' quality.

The potential of the methodology has been demonstrated by applying it to different real human–computer interaction contexts. Extracted affective information has helped to develop a tutoring tool where the distant teacher can be aware of the learner's emotional progress and to fuse dynamic affective information extracted from different channels of an IM tool.

This work brings a new perspective and invites further discussion on the still open issue of multimodal affective fusion. In general, evaluation issues are largely solved for categorical affect recognition approaches. The evaluation of the performance of dimensional approaches is, however, an open and difficult issue to be solved. In the future, our work is expected to focus in depth on evaluation issues applicable to dimensional approaches and multimodality. The proposed fusion methodology will be explored in different application contexts, with different numbers and natures of modalities to be fused.

ACKNOWLEDGMENTS

The authors wish to thank Dr. Cynthia Whissell for her explanations and kindness, Dr. Hussain and E. Cambria for the text analyzing tool, D. Abada for his participation in T-EDUCO and all the participants in the evaluation sessions, especially D. Carmona and T. Corts. This work has been partly financed by the University of Zaragoza through the AVIM project.

REFERENCES

[1] D. Keltner and P. Ekman, "Facial expression of emotion," *Handb. Emot.*, 2: 236–249, 2000.

[2] Z. Zeng, M. Pantic, G. I. Roisman, and T. S. Huang, "A survey of affect recognition methods: audio, visual, and spontaneous expressions," *IEEE Transactions on Pattern Analysis and Machine Intelligence*, 31: 39–58, 2009.

[3] G. Littlewort, M. S. Bartlett, I. Fasel, J. Susskind, and J. Movellan, "Dynamics of facial expression extracted automatically from video," *Image Vis. Comput.*, 24(6): 615–625, 2006.

[4] P. Ekman, T. Dalgleish, and M. Power, *Handbook of Cognition and Emotion*, Wiley, 1999.

[5] Z. Zeng, M. Pantic, and T. S. Huang, "Emotion recognition based on multimodal information," in *Affective Information Processing* (eds J. Tao and T. Tan), Springer, London, 2009, pp. 241–265.

[6] S. W. Gilroy, M. Cavazza, M. M. Niranen, E. E. André, T. T. Vogt, J. J. Urbain, M. M. Benayoun, H. H. Seichter, and M. M. Billinghurst, "PAD-based multimodal affective fusion," in Proceedings of the Conference on Affective Computing and Intelligent Interaction, 2009, pp. 1–8.

[7] A. Kapoor, W. Burleson, and R. W. Picard, "Automatic prediction of frustration," *Int. J. Hum.-Comput. Stud.*, 65(8): 724–736, 2007.

[8] H. Gunes, M. Piccardi, and M. Pantic, "From the lab to the real world: affect recognition using multiple cues and modalities," in *Affective Computing: Focus on Emotion Expression, Synthesis, and Recognition*, 2008, pp. 185–218.

[9] S. Caifeng, G. Shaogang, and P. W. McOwan, "Robust facial expression recognition using local binary patterns," in Proceedings of the IEEE International Conference on Image Processing, vol. 2, 2005, pp. 370–373.

[10] T. Pun, T. I. Alecu, G. Chanel, J. Kronegg, and S. Voloshynovskiy, "Brain–computer interaction research at the Computer Vision and Multimedia Laboratory, University of Geneva," *IEEE Transactions on Neural Systems and Rehabilitation Engineering*, 14(2): 210–213, 2006.

[11] L. I. Kuncheva, *Combining Pattern Classifiers: Methods and Algorithms*, Wiley-Interscience, 2004.

[12] Z. Zeng, J. Tu, M. Liu, T. S. Huang, B. Pianfetti, D. Roth, and S. Levinson, "Audio-visual affect recognition," *IEEE Transactions on Multimedia*, 9(2): 424–428, 2007.

[13] H. Gunes and M. Piccardi, "Bi-modal emotion recognition from expressive face and body gestures," *J. Netw. Comput. Appl.*, 30(4): 1334–1345, 2007.

[14] P. Pal, A. N. Iyer, and R. E. Yantorno, "Emotion detection from infant facial expressions and cries," in Proceedings of the IEEE International Conference on Acoustics, Speech and Signal Processing, vol. 2, 2006, pp. 721–724.

[15] M. Wöllmer, M. Al-Hames, F. Eyben, B. Schuller, and G. Rigoll, "A multidimensional Dynamic Time Warping algorithm for efficient multimodal fusion of asynchronous data streams," *Neurocomputing*, 73(3): 366–380, 2009.

[16] S. Baldassarri, E. Cerezo, and F. J. Seron, "Maxine: a platform for embodied animated agents," *Comput. Graph.*, 32(4): 430–437, 2008.

[17] R. Del-Hoyo, I. Hupont, F. J. Lacueva, and D. Abada, "Hybrid text affect sensing system for emotional language analysis," in Proceedings of the International Workshop on Affective-Aware Virtual Agents and Social Robots, ACM, 2009. Article no. 3.

[18] H. Boukricha, C. Becker, and I. Wachsmuth, "Simulating empathy for the virtual human Max," in Proceedings of the Second International Workshop on Emotion and Computing in Conjunction with the German Conference on Artificial Intelligence, 2007, pp. 22–27.

[19] P. Ekman, W. V. Friesen, and J. C. Hager, "Facial action coding system: investigator's guide," *Res. Nexus*, 2002.

[20] H. Soyel and H. Demirel, "Facial expression recognition using 3D facial feature distances," in *Image Analysis and Recognition*, Volume 4633 of Lecture Notes in Computer Science, 2007, pp. 831–838.

[21] Z. Hammal, L. Couvreur, A. Caplier, and M. Rombaut, "Facial expression classification: an approach based on the fusion of facial deformations using the transferable belief model," *Int. J. Approx. Reason.*, 46(3): 542–567, 2007.

[22] C. Y. Chang, J. S. Tsai, C. J. Wang, and P. C. Chung, "Emotion recognition with consideration of facial expression and physiological signals," in Proceedings of the Sixth Annual IEEE Conference on Computational Intelligence in Bioinformatics and Computational Biology, 2009, pp. 278–283.

[23] I. Hupont, S. Baldassarri, R. Del-Hoyo, and E. Cerezo, "Effective emotional classification combining facial classifiers and user assessment," *Articul. Motion Deform. Objects*, 5098: 431–440, 2008.

[24] SeeingMachines. Face API technical specifications brochure, 2008, http://www.seeingmachines.com/pdfs/brochures/faceAPI-Brochure.pdf

[25] F. Wallhoff, Facial expressions and emotion database, 2006, http://www.mmk.ei.tum.de/waf/fgnet/feedtum.html

[26] M. Pantic, M. Valstar, R. Rademaker, and L. Maat, "Web-based database for facial expression analysis," in Proceedings of the IEEE International Conference on Multimedia and Expo, 2005, pp. 317–321.

[27] M. A. Hall, *Correlation-Based Feature Selection for Machine Learning*, PhD thesis, Hamilton, New Zealand, 1998.

[28] I. Witten and E. Frank, *Data Mining: Practical Machine Learning Tools and Techniques*, 2nd edn, Morgan Kaufmann, San Francisco, CA, 2005.

[29] U. Hess, R. B. Adams, and R. E. Kleck, "Facial appearance, gender and emotion expression," *Emotion*, 4: 378–388, 2004.

[30] C. M. Whissell, *The Dictionary of Affect in Language, Emotion: Theory, Research and Experience*, vol. 4, Academic Press, 1989.

[31] R. Plutchik, *Emotion: A Psychoevolutionary Synthesis*, Harper & Row, 1980.

[32] N. Stoiber, R. Seguier, and G. Breton, "Automatic design of a control interface for a synthetic face," in Proceedings of the 13th International Conference on Intelligent User Interfaces, 2009, pp. 207–216.

[33] Y. Du, W. Bi, T. Wang, Y. Zhang, and H. Ai, "Distributing expressional faces in 2-d emotional space," in Proceedings of the 6th ACM International Conference on Image and Video Retrieval, 2007, pp. 395–400.

[34] F. Gosselin and P. G. Schyns, "Bubbles: a technique to reveal the use of information in recognition tasks," *Vis. Res.*, 41: 2261–2271, 2001.

[35] N. Fragopanagos and J. G. Taylor, "Emotion recognition in human–computer interaction," *Neural Netw.*, 18(4): 389–405, 2005.

[36] G. Caridakis, L. Malatesta, L. Kessous, N. Amir, A. Paouzaiou, and K. Karpouzis, "Modeling naturalistic affective states via facial and vocal expression recognition," in Proceedings of the International Conference on Multimodal Interfaces, 2006, pp. 146–154.

[37] P. Kumar and E. A. Yildirim, "Minimum-volume enclosing ellipsoids and core sets," *J. Optim. Theory Appl.*, 126: 1–21, 2005.

[38] J. N. Bassili, "Emotion recognition: the role of facial movement and the relative importance of upper and lower areas of the face," *J. Personal. Soc. Psychol.*, 37: 2049–2058, 1979.

[39] R. E. Kalman, "A new approach to linear filtering and prediction problems," *J. Basic Eng.*, 82(1): 35–45, 1960.

[40] E. Douglas-Cowie, R. Cowie, I. Sneddon, C. Cox, O. Lowry, M. McRorie, J. C. Martin, L. Devillers, S. Abrilian, A. Batliner, *et al.*, "The HUMAINE database: addressing the collection and annotation of naturalistic and induced emotional data," in Proceedings of the 2nd International Conference on Affective Computing and Intelligent Interaction, 2007, pp. 488–500.

[41] M. Grimm, K. Kroschel, and S. Narayanan, "The Vera am Mittag German audio-visual emotional speech database," in Proceedings of the IEEE International Conference on Multimedia and Expo, 2008, pp. 865–868.

[42] R. W. Picard, *Affective Computing*, The MIT Press, 1997.

[43] A. Dosi and B. Prario, "New frontiers of t-Learning: the emergence of interactive digital broadcasting learning services in Europe," in Proceedings of the World Conference on Educational Multimedia, Hypermedia & Telecommunications, vol. 2004, ED-Media, 2004, pp. 4831–4836.

[44] M. R. López, A. F. Vilas, R. P. D. Redondo, J. J. P. Arias, and J. B. Munoz, "Adaptive learning objects for T-learning," *IEEE Latin America Transactions*, 401–408, 2007.

[45] M. Lytras, C. Lougos, P. Chozos, and A. Pouloudi, "Interactive television and e-learning convergence: examining the potential of t-learning," in Proceedings of the European Conference on E-Learning, 2002.

[46] M. J. Damasio and C. Quico, "T-learning and interactive television edutainment: the Portuguese case study," in Proceedings of the Second European Conference on Interactive Television, 2004, pp. 4511–4518.

[47] Moodle—a free, open source course management system for online learning, http://www.moodle.org (last accessed September 2010).

[48] MHP standard draft TS 102 812 v1.3.1 - MHP 1.1.3, 2007.

[49] J. A. Sánchez, N. P. Hernández, J. C. Penagos, and Y. Ostrovskaya, "Conveying mood and emotion in instant messaging by using a two-dimensional model for affective states,"

in Proceedings of the Seventh Brazilian Symposium on Human Factors in Computing Systems, ACM, 2006, pp. 72–79.

[50] E. Cambria, A. Hussain, C. Havasi, and C. Eckl, "Sentic computing: exploitation of common sense for the development of emotion-sensitive systems," in *Development of Multimodal Interfaces: Active Listening and Synchrony*, vol. 5967, 2010, pp. 153–161.

AUTHOR BIOGRAPHIES

Isabelle Hupont received a B.Sc. degree in Telecommunications Engineering in 2006, an M.Sc. degree in Computer Science in 2008 and a Ph.D. degree in Computer Science in 2010 from the University of Zaragoza, Spain. She is currently working as an R&D+i project manager at the Aragon Institute of Technology, Spain. Her research interests include multimodal affective computing, multimodal human-computer interaction, artificial intelligence, and computer vision.

Sergio Ballano received a B.Sc. degree in Industrial Engineering in 2008. He is currently an M.Sc. student at the University of Zaragoza. Since 2007 he has worked as a researcher at the Aragon Institute of Technology, Spain. His research interests include multimodal affective computing, computer vision, and robotics.

Eva Cerezo obtained a B.S. degree in Physics in 1990 and an M.Sc. degree in Nuclear Physics in 1992 from the University of Zaragoza, Spain. She received a Ph.D. degree in Computer Science in 2002. She is currently a professor at the Computer Sciences and Systems Engineering Department at the University of Zaragoza and a member of the Advanced Computer Graphics Group. Her research fields are in computer animation, virtual humans, and affective multimodal human–computer interaction. Dr. Cerezo is a member of the Executive Board of the Spanish chapter of the Eurographics Association.

Sandra Baldassarri obtained a B.Sc. in Computer Science from the University of Buenos Aires in 1992 and a Ph.D. in Computer Science Engineering in 2004. Since 1996 she has been an associate professor at the Computer Science Department of the University of Zaragoza, Spain. She is a member of the Advanced Computer Graphics Group (GIGA) at the University of Zaragoza and the I3A (Institute of Research and Engineering of Aragon). Her research interests include realistic body modeling, animation and simulation of humans and characters, real-time animation, emotional interfaces, and multimodal interaction. Within the IEEE (Institute of Electrical and Electronic Engineers) she was chair of the Spanish chapter of Women in Engineering (WIE) from 1999 to 2004.

19

AUDIOVISUAL EMOTION RECOGNITION USING SEMI-COUPLED HIDDEN MARKOV MODEL WITH STATE-BASED ALIGNMENT STRATEGY

CHUNG-HSIEN WU, JEN-CHUN LIN, AND WEN-LI WEI

Department of Computer Science and Information Engineering, National Cheng Kung University, Tainan, Taiwan

With the increasing attention of audiovisual interfaces in human computer interaction applications, automatically recognizing human emotional states from audiovisual bimodal signals becomes more and more important. This chapter firstly introduces the current data fusion strategies among audiovisual signals for bimodal emotion recognition. Toward effective emotion recognition, a novel state-based alignment strategy, employed in a proposed semi-coupled hidden Markov model (SC-HMM), is then presented to align the temporal relation of states between audio and visual streams. Based on this alignment strategy, the SC-HMM can alleviate the problem of data sparseness and achieve better statistical dependency between states of audio and visual HMMs in most real-world scenarios. For performance evaluation, audiovisual signals with four emotional states (happy, neutral, angry, and sad) were collected. Each of the seven invited subjects was asked to utter 30 types of sentences twice to generate emotional speech and facial expression for each emotion. Subject-independent experiments were conducted and the experimental results show that the proposed SC-HMM outperforms other fusion-based bimodal emotion recognition methods.

Emotion Recognition: A Pattern Analysis Approach, First Edition. Edited by Amit Konar and Aruna Chakraborty.
© 2015 John Wiley & Sons, Inc. Published 2015 by John Wiley & Sons, Inc.

19.1 INTRODUCTION

Emotions are important in human intelligence, rational decision making, social interaction, perception, memory, and more [1]. Understanding human emotional states is indispensable for day-to-day functioning. In human–computer interaction systems, emotion recognition technology could provide harmonious interaction or heart-to-heart communication between computers and human beings. Hence, creating an emotion perception and recognition system to our daily lives are expected. The challenging research field, "affective computing," introduced by Picard [1] aims at enabling computer to recognize, express, and have emotions. In previous work, various studies in emotion recognition were focused only on considering single facial or vocal modality [2–7]. Based on psychological analysis [8, 9] human emotional states were transmitted by multiple channel processes such as face, voice, and speech content for communication. For this reason, exploring data fusion strategies between various channels will meet a token of emotional expression and therefore can improve recognition performance [10–13].

Many data fusion approaches have been developed in recent years. The fusion operations reported in the literature can be classified into three major categories, including feature-level, decision-level, and model-level fusion, for audiovisual emotion recognition [13]. In feature-level fusion [14–16], facial and vocal features were concatenated to construct the joint feature vectors and then modeled by a single classifier for emotion recognition. Although fusion at feature level can obtain the advantage of combining visual and audio cues, it will increase the dimensionality and may suffer from the problem of data sparseness. Hence, the advantage of combining the visual and audio modalities will be limited. In terms of decision-level fusion [14, 17–20], multiple signals can be modeled by its corresponding classifier first and then the recognition results are fused together to obtain the final result. The fusion-based method at the decision level, without increasing the dimensionality, can combine the various modalities by exploring the contributions of different emotion expressions. Since facial and vocal features are complementary to each other for emotion expression, the assumption of conditional independence among multiple modalities at decision level is inappropriate. Hence the mutual correlation between visual and audio modalities should be considered. Contrary to decision-level fusion, model-level fusion [11,21,22] emphasizes the information of mutual correlation among multiple modalities and explores the temporal relation between the visual and audio streams.

Recently, more and more studies have paid attention to model-level fusion in data fusion approaches. Coupled hidden Markov model (C-HMM) as one of the popular model-level fusion approaches has been proposed and was successfully used in different fields such as hand gesture recognition [23], 3D surface inspection [24], speech prosody recognition [25], audiovisual speech recognition [26], and speech animation [27]. In C-HMM, two component HMMs were linked through cross-time and cross-chain conditional probabilities. A state variable at time t is dependent on its preceding nodes of two component HMMs at time "$t-1$", which describes the temporal coupling relationship between two streams. This structure models the asynchrony of multiple modalities and preserves their natural correlations over time.

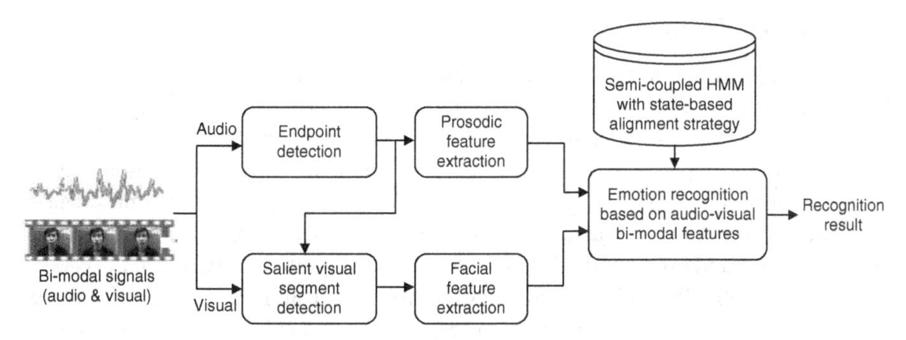

FIGURE 19.1 System diagram of the proposed audiovisual emotion recognition.

In bimodal audiovisual processing, however, the statistical dependency between audio and visual features were often not strong enough to be captured. For most real-world scenarios, the statistical dependency between two hidden states of the observation sequences is difficult to obtain from insufficient training samples for C-HMM training. Hence, it may suffer from the problem of data sparseness. To deal with this problem, an SC-HMM with simplified state-based alignment strategy is developed in this chapter.

Based on the previous discussion, a model which has the following characteristics is desirable.

1. It can model better statistical dependency between states of audio and visual HMMs for most real-world scenarios.
2. It has low computational complexity than C-HMM.
3. It can alleviate the problem of data sparseness.

Figure 19.1 illustrates an overview of the proposed emotion recognition system. First, the input bimodal signals are separated into audio and visual parts. Since endpoint detection based on speech is easier and more robust than mouth movement, the start and end points of a salient speech segment were determined first and then used to obtain the time-aligned salient visual segment. The extracted salient speech and visual segments are used to extract prosodic and facial features, respectively. Finally, using the prosodic and facial features, the proposed SC-HMM is employed for emotion recognition.

The rest of the chapter is organized as follows: Section 19.2 briefly outlines the method for feature extraction. Section 19.3 provides a detailed description of the proposed SC-HMM. Section 19.4 provides experiments and discussion. Section 19.5 gives the conclusion.

19.2 FEATURE EXTRACTION

In this section, the techniques of extracting the facial and prosodic features are introduced.

19.2.1 Facial Feature Extraction

Facial expression is one of the directly related cues to human emotions [8, 9, 13]. The facial action coding system (FACS) introduced by Ekman and Friesen aims at enabling the system to better understand and measure the human facial behaviors [28]. FACS decomposes all possible facial expressions into action units (AUs) through several facial muscle movements, that is, facial expression can be formed from various variations of facial regions such as eyebrow, eye, mouth, and so on. The MPEG-4 standard extends FACS to derive facial animation parameters (FAPs) [21, 29–31]. The FAPs have been proved to be useful to differentiate cases of varying activation of the same character emotion (e.g., happy and exhilaration) [30]. Hence, FAPs have been widely and successfully used in facial animation and expression recognition [21, 30–33]. Since FAPs are defined to measure the deformation of the facial feature points (FPs) [33] and FPs are also easily localized from each facial region, the extracted facial FPs are further transformed into FAPs for facial expression recognition.

In this chapter, face detection is performed based on the adaboost cascade face detector proposed in Reference 34 and can be used to provide initial facial position and reduce the time for error convergence in feature extraction. The active appearance model (AAM) [35] combines the two different characteristic features to effectively localize human facial features on a 2D visual image. The major structure of AAM can be divided into two statistical models: a shape model [36] and an appearance model [37]. It has achieved a great success on human face alignment even though the human faces may have nonrigid deformations. Hence, AAM has been widely used to extract facial feature points for facial expression recognition [38–40]. In this chapter, AAM was employed to extract the 68 labeled facial FPs from 5 facial regions including eyebrow, eye, nose, mouth, and facial contours for later FAPs calculation. In addition, due to head pose variations in the real 3D world, AAM may fail to converge in some cases such as a severe head's tilt or rotation causing the nonnear-frontal view [41]. However, AAM is useful to align the human faces on frontal or near-frontal view in most situations. Accordingly, only frontal or near-frontal view was considered in this chapter. From the results shown in Figure 19.2, AAM alignment achieved satisfactory results in our database.

In terms of FAPs [33], a 30-dimensional distance vector is first created containing vertical and horizontal distances from 24 of the 68 extracted FPs, as shown in Figure 19.3. For the purpose of normalization among different people, the scale-invariant MPEG-4 facial animation parameter units (FAPU) are used [30, 31, 33]. Measurement of FAPs requires the basis frame where the subject's expression is found to be neutral. It is manually selected from video sequences in the beginning. Thus, the FAPs from each frame are calculated by comparing FPs displacement with neutral frame [33]. For example as shown in Figure 19.1, the inner raised eyebrow FAPs are calculated as the distances, by projecting vertically from inner eyebrow FPs to the inner eye corners' FPs, that is, points 25 and 19 to points 30 and 35, which are further compared to their corresponding distances in the neutral frame.

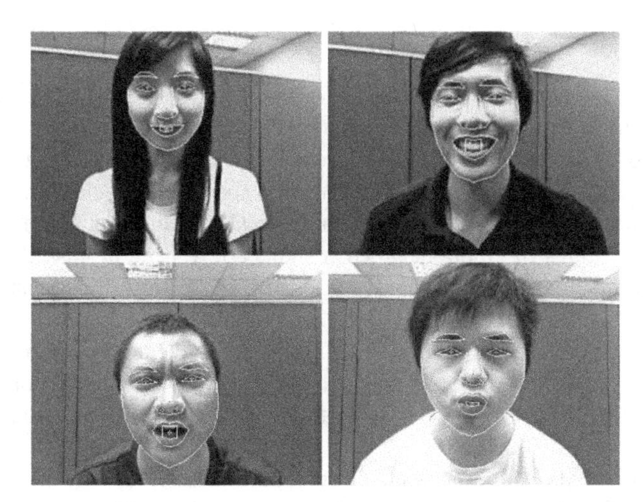

FIGURE 19.2 Examples of the extracted facial feature points using AAM. Extracted with permission from MHMC database.

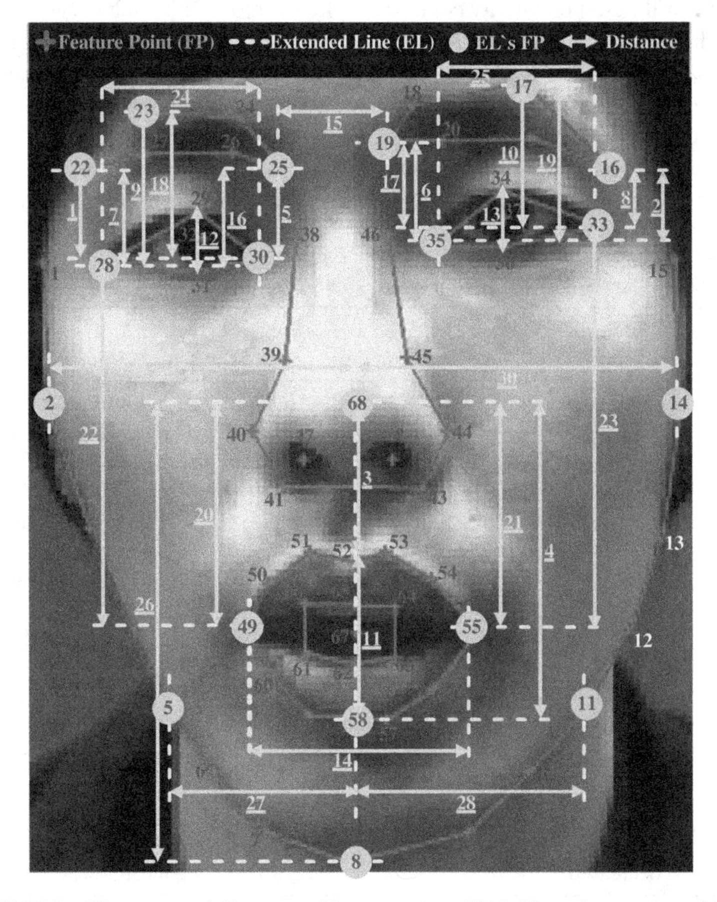

FIGURE 19.3 The extracted distances of feature points (FPs). The color version of the above figure is available at www.amitkonar.com/wileycolorimages.

19.2.2 Prosodic Feature Extraction

Emotions have some mechanical effects on physiology, which in turn have effects on the intonation of the speech. This is why it is possible in principle to predict some emotional information from the prosody of a sentence [42]. An important problem for all emotion recognition systems from speech is the selection of the set of relevant features to be used. Hence, several promising feature sets such as prosodic and acoustic features of emotional speech signals have been discussed over years. Among these features, prosody has been proven to be the primary and classical indicator of a speaker's emotional state [42, 43] and it was widely and successfully used for speech emotion recognition [7, 44–48]. Pitch and energy have been reported to contribute the most to speech emotion recognition [13, 49]. In addition, Morrison *et al.* [48] further summarized the correlations between prosodic features and the emotions as shown in Table 19.1.

In order to verify this argument, the prosodic features of a Chinese sentence with four emotional states are observed as shown in Figure 19.4. According to our observation, the emotional states happy and angry have higher intensity compared to sad and neutral based on energy contour. In pitch contour, the pitch ranges and pitch levels of happy and angry are much wider and higher than those of sad. Figure 19.4 highlights the difference of prosodic features between the four emotional states. This observation verified the analyses of the literature mentioned above.

In addition, formants and speaking rate among the prosodic features are also discussed frequently [48,50]. Although speaking rate is an important prosodic feature for emotion recognition, speech recognition error resulting in incorrect detection of

TABLE 19.1 Correlations among prosodic features and emotions [48]

	Pitch mean	Pitch range	Energy	Speaking rate	Formants
Anger	Increased	Wider	Increased	High	F1 mean increased; F2 mean higher or lower; F3 mean higher
Happiness	Increased	Wider	Increased	High	F1 mean decreased and bandwidth increased
Sadness	Decreased	Narrower	Decreased	Low	F1 mean increased and bandwidth increased; F2 mean lower
Surprise	Normal or increased	Wider	–	Normal	–
Disgust	Decreased	Wider or narrower	Decreased or normal	Higher	F1 mean increased and bandwidth decreased; F2 mean lower
Fear	Increased or decreased	Wider or narrower	Normal	Higher or low	F1 mean increased and bandwidth decreased; F2 mean lower

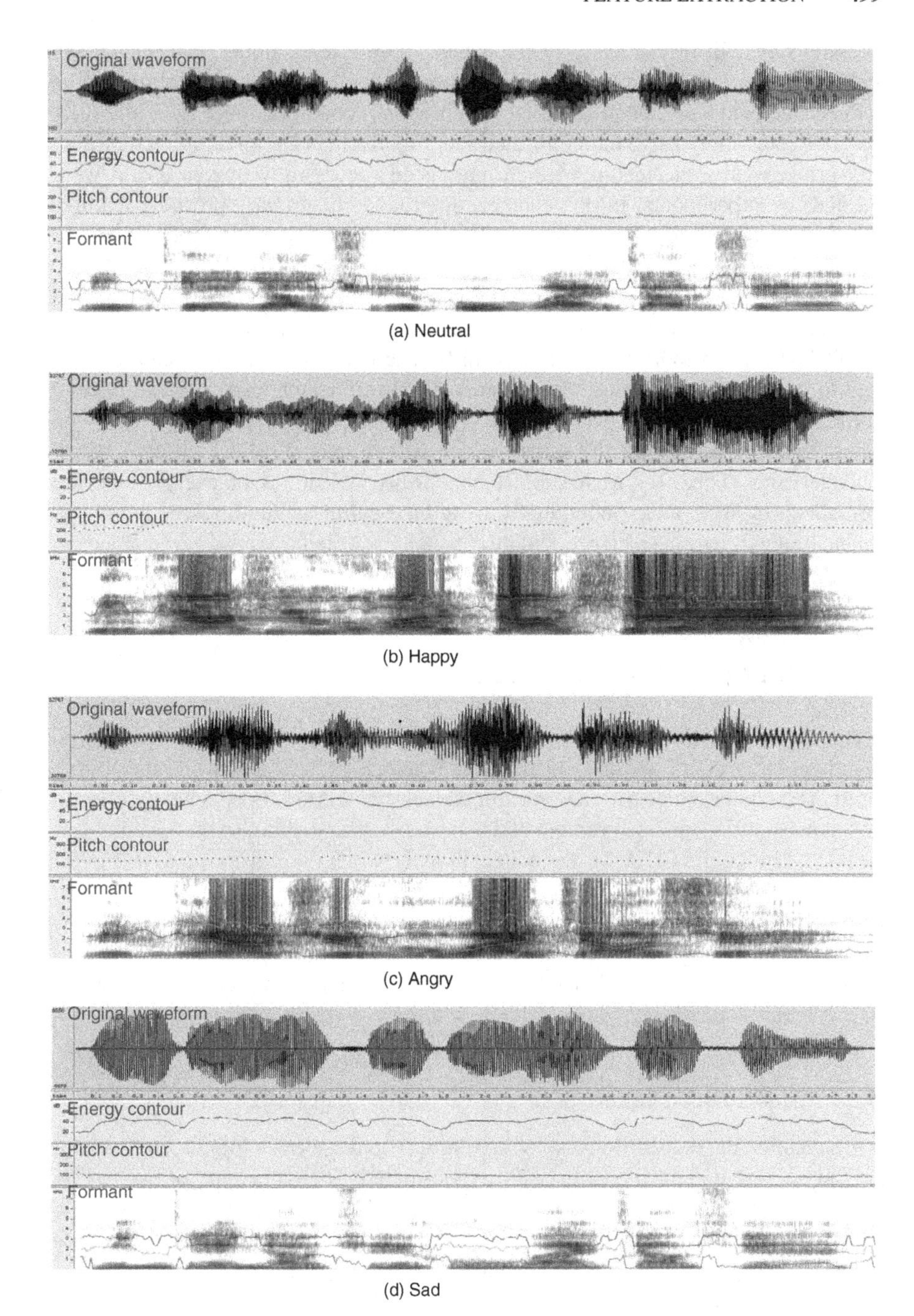

FIGURE 19.4 Examples of the prosodic features of a sentence with (a) neutral, (b) happy, (c) angry, and (d) sad emotions.

speaking rate will dramatically degrade the emotion recognition performance. In contrast to the popular prosodic features such as pitch and energy, recently, some studies have paid attention to use jitter for emotion recognition [45, 51]. While jitter has not been widely used and confirmed the effectiveness for emotion recognition, only primary prosodic features which have been successfully proven useful were selected for experiments. In this chapter, three kinds of primary prosodic features are adopted, including pitch, energy, and formants F1–F5 in each speech frame for emotion recognition.

Endpoint detection of speech signals based on energy and zero-crossing rate is first used to extract the salient segment of speech signals. For prosodic feature extraction, the pitch detection tool "Praat" is used [52]. The extracted prosodic features also need to be normalized because they often varies from person-to-person and depend on the recording condition. For example, in general, females have higher pitch than males. Even in the case of the same gender, people may still have different vocal characteristics. Thus, for each subject, the prosodic features of every frame are normalized by their mean individually from the neutral expression sequence [18]. The neutral expression sequence will be manually selected from video sequences in the beginning.

19.3 SEMI-COUPLED HIDDEN MARKOV MODEL

In this chapter, SC-HMM is proposed for emotion recognition based on state-based alignment strategy for audiovisual bimodal features. In C-HMM, while the frame-based cross-chain temporal relationship between two HMMs is completely considered, it may suffer from the problem of data sparseness and diminish the advantage of C-HMM. Unlike the C-HMM, the connections between two component HMMs for audio and visual streams in SC-HMM are chosen based on a simplified state-based alignment strategy which emphasizes on state-level alignment rather than the frame level. Hence, the proposed SC-HMM can alleviate the problem of data sparseness and obtain better statistical dependency between states of audio and visual HMMs for most real-world scenarios.

19.3.1 Model Formulation

In this chapter, the recognition task with four emotional states, happy, neutral, angry, and sad, represented by $\{e_h, e_n, e_a, e_s\}$ are considered. The probability of two tightly coupled audiovisual observation sequences O^a and O^v given the parameter set $\Lambda = \{\lambda^a_{e_k}, \lambda^v_{e_k}\}$ is obtained using (19.1). e^*_k represents the emotional state with maximum probability of the audiovisual observation sequences O^a and O^v given the parameter set Λ of the SC-HMM.

$$e^*_k = \underset{e_k \in \{e_h, e_n, e_a, e_s\}}{\operatorname{argmax}} P\left(O^a, O^v \mid \lambda^a_{e_k}, \lambda^v_{e_k}\right) \tag{19.1}$$

For SC-HMM with state sequence S^a and S^v for audio and visual HMMs respectively, the probability $P(O^a, O^v \mid \lambda^a_{e_k}, \lambda^v_{e_k})$ can be further inferred by (19.2).

$$P(O^a, O^v \mid \lambda^a_{e_k}, \lambda^v_{e_k}) = \sum_{S^a, S^v} P(O^a, O^v, S^a, S^v \mid \lambda^a_{e_k}, \lambda^v_{e_k}) \tag{19.2}$$

where $P(O^a, O^v \mid \lambda^a_{e_k}, \lambda^v_{e_k})$ can be approximated by selecting the optimal state sequence which maximizes the probability $P(O^a, O^v, S^a, S^v \mid \lambda^a_{e_k}, \lambda^v_{e_k})$ as follows.

$$P(O^a, O^v \mid \lambda^a_{e_k}, \lambda^v_{e_k}) \approx \max_{S^a, S^v} P(O^a, O^v, S^a, S^v \mid \lambda^a_{e_k}, \lambda^v_{e_k}) \tag{19.3}$$

Given O^a and O^v and the corresponding audio and visual HMMs, the problem addressed in this chapter is how to construct a useful structure linking the two component HMMs together. For SC-HMM, two component HMMs are modeled separately and their state sequences are related to each other for the same emotion state. The probability $P(O^a, O^v, S^a, S^v \mid \lambda^a_{e_k}, \lambda^v_{e_k})$ can be further approximated by (19.4).

$$P(O^a, O^v, S^a, S^v \mid \lambda^a_{e_k}, \lambda^v_{e_k}) \approx P(O^a, S^a, S^v \mid \lambda^a_{e_k}) P(O^v, S^v, S^a \mid \lambda^v_{e_k}) \tag{19.4}$$

where $P(O^a, S^a, S^v \mid \lambda^a_{e_k})$ can be divided into two parts using Bayes rule which consists of audio HMM output $P(O^a, S^a \mid \lambda^a_{e_k})$ and state alignment probability from audio to visual HMM $P(S^v \mid O^a, S^a, \lambda^a_{e_k})$. Conversely, $P(O^v, S^v, S^a \mid \lambda^v_{e_k})$ also can be divided into two parts which consists of visual HMM output $P(O^v, S^v \mid \lambda^v_{e_k})$ and the state alignment probability from visual to audio HMM $P(S^a \mid O^v, S^v, \lambda^v_{e_k})$. Therefore, (19.4) can be rewritten as (19.5).

$$\begin{aligned} P(O^a, O^v, S^a, S^v \mid \lambda^a_{e_k}, \lambda^v_{e_k}) &\approx P(O^a, S^a \mid \lambda^a_{e_k}) \\ &\times P(S^v \mid O^a, S^a, \lambda^a_{e_k}) P(S^a \mid O^v, S^v, \lambda^v_{e_k}) P(O^v, S^v \mid \lambda^v_{e_k}) \end{aligned} \tag{19.5}$$

where the state alignment probabilities $P(S^v \mid O^a, S^a, \lambda^a_{e_k})$ and $P(S^a \mid O^v, S^v, \lambda^v_{e_k})$ are difficult to obtain, since they need a sufficient number of training data. Accordingly, these two alignment probabilities are approximated as $P(S^v \mid S^a, \lambda^a_{e_k})$ and $P(S^a \mid S^v, \lambda^v_{e_k})$ in (19.6), respectively, by simplifying the state alignment conditions.

$$\begin{aligned} &P(O^a, O^v, S^a, S^v \mid \lambda^a_{e_k}, \lambda^v_{e_k}) \\ &\approx P(O^a, S^a \mid \lambda^a_{e_k}) P(S^v \mid S^a, \lambda^a_{e_k}) P(S^a \mid S^v, \lambda^v_{e_k}) P(O^v, S^v \mid \lambda^v_{e_k}) \end{aligned} \tag{19.6}$$

where $P(S^v \mid S^a, \lambda^a_{e_k})$ denotes the probability of visual state sequence S^v given that audio state sequence S^a has been decided by audio HMM $\lambda^a_{e_k}$ for the predicted emotional state. Conversely, $P(S^a \mid S^v, \lambda^v_{e_k})$ denotes the probability of audio state sequence S^a given that visual state sequence S^v has been decided by visual HMM $\lambda^v_{e_k}$ for the predicted emotional state.

Finally, we can combine (19.6) and (19.3) into (19.1) to obtain (19.7).

$$e_k^* \approx \underset{e_k \in \{e_h, e_n, e_a, e_s\}}{\operatorname{argmax}} \left(\max_{S^a, S^v} P(O^a, S^a \mid \lambda^a_{e_k}) P(S^v \mid S^a, \lambda^a_{e_k}) P(S^a \mid S^v, \lambda^v_{e_k}) P(O^v, S^v \mid \lambda^v_{e_k}) \right) \tag{19.7}$$

Clearly, the proposed method not only models the time series of audio and visual signals by individual HMM but also considers the state correlation between two component HMMs simultaneously to obtain an optimal recognition result.

In terms of training procedure (i.e., learning algorithm), the proposed SC-HMM consists of three parts.

1. Audio and visual HMMs were trained separately using the expectation maximization (EM) algorithm.
2. The best state sequences of the audio and visual HMMs are obtained using the Viterbi algorithm, respectively.
3. The audio and visual state sequences are aligned based on the alignment strategy which will be described in Subsection 19.3.2. Hence, the state alignment probabilities between audio HMM and visual HMM can be estimated over all training data.

An immediate advantage of this learning algorithm in SC-HMM over that for the C-HMM is computational efficiency and can alleviate the problem of insufficient data, while C-HMM attempts to optimize all the parameters "globally" by iteratively refining the parameters of the component HMMs and their coupling parameters.

In the test phase, the n-best output probabilities of state sequences of audio and visual HMMs in SC-HMM are first estimated, respectively. The state alignment probabilities are then considered to obtain the recognition result for each predicted emotional state. Finally, the maximum probability of the predicted emotional state was selected as the final result (19.7).

19.3.2 State-Based Bimodal Alignment Strategy

Figure 19.5 illustrates the alignment strategy between the states of the audio and visual HMMs in the proposed SC-HMM. In Figure 19.5, the upper and lower portions depict the input audio and visual observation sequences, respectively. The middle portion depicts the state alignment between two component HMMs. For the example shown in Figure 19.5, the first state of audio HMM covers two audio frames with state boundary ASB_1. Assuming the audio and image data are synchronized, based on the state boundary ASB_1, the two frames in the first audio state can be aligned to the first two frames in the first visual state (i.e., $Count_{a1 \rightarrow v1} = 2$). Conversely, the first state

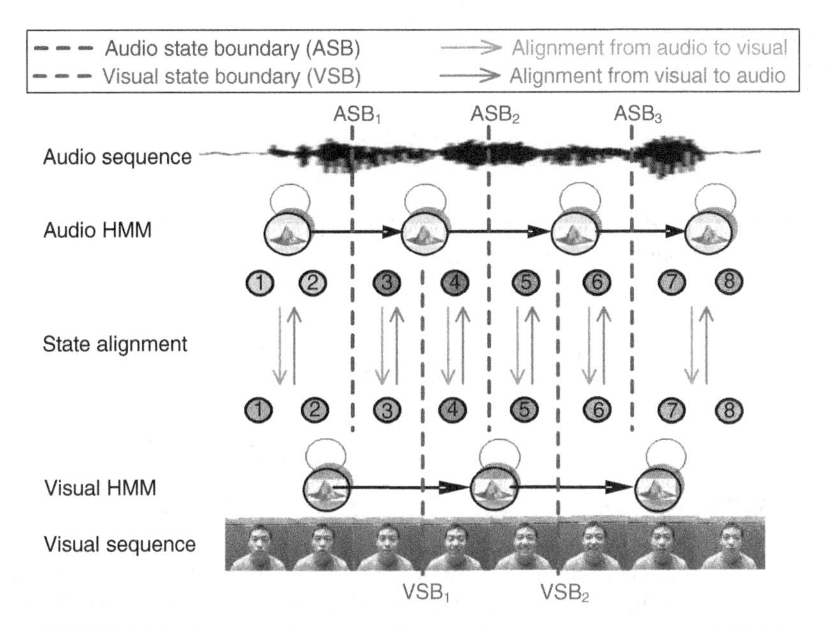

FIGURE 19.5 State-based bimodal alignment between audio and visual HMMs.

of visual HMM covers three image frames based on state boundary VSB_1; the first two frames are aligned to first audio state (i.e., $Count_{v1 \to a1} = 2$), and the third frame is aligned to second audio state (i.e., $Count_{v1 \to a2} = 1$).

Based on the above description of the asymmetric alignment between audio and visual states, the alignment probability from the ith audio state to the jth visual state can be estimated by (19.8) given that the ith audio state sequence has been decided by audio HMM $\lambda^a_{e_k}$.

$$P\left(S^v_j \mid S^a_i, \lambda^a_{e_k}\right) = \frac{1}{N} \sum_{n=1}^{N} \frac{Count^n_{ai \to vj}}{N^n_{ai}} \tag{19.8}$$

where N^n_{ai} represents the number of frames of the ith audio state of the nth training data. $Count^n_{ai \to vj}$ denotes the number of frames of the ith audio state aligned to the jth visual state for the nth training data. N denotes the total number of the training data. Conversely, the alignment probability from the jth visual state to the ith audio state $P(S^a_i \mid S^v_j, \lambda^v_{e_k})$ can also be estimated by (19.9). Hence, the alignment probability between audio and visual states can be constructed in the training phase.

$$P\left(S^a_i \mid S^v_j, \lambda^v_{e_k}\right) = \frac{1}{N} \sum_{n=1}^{N} \frac{Count^n_{vj \to ai}}{N^n_{vj}} \tag{19.9}$$

In the test phase, based on the audio and visual HMMs, the boundaries of audio and visual states can be estimated, respectively. The constructed alignment probability can be further applied for linking the two component HMMs. The alignment probability from audio state sequence to visual state sequence of (19.7) can be rewritten as (19.10).

$$P(S^v \mid S^a, \lambda_{e_k}^a) = \prod_{t=1}^{T} P(S_{j(t)}^v \mid S_{i(t)}^a, \lambda_{e_k}^a) \qquad (19.10)$$

where $P(S_{j(t)}^v \mid S_{i(t)}^a, \lambda_{e_k}^a)$ denotes the alignment probability of the ith audio state aligned to the jth visual state at time t. Conversely, the alignment probability from visual state sequence to audio state sequence of (19.7) can also be rewritten as (19.11).

$$P(S^a \mid S^v, \lambda_{e_k}^v) = \prod_{t=1}^{T} P(S_{i(t)}^a \mid S_{j(t)}^v, \lambda_{e_k}^v) \qquad (19.11)$$

19.4 EXPERIMENTS

In this section, the recording procedure of the MHMC database is first introduced. For performance evaluation, the current unimodal approaches and the bimodal fusion approaches are compared with the proposed SC-HMM using the MHMC database.

19.4.1 Data Collection

Models of discrete emotions have been proved useful in empirical studies [53]. These studies have focused on the use of facial expression [54] and the static picture [55–57] to show differences in some physiological signals between different discrete emotions. The discrete emotions including happiness, fear, disgust, sadness, and anger are further demonstrated to be universal across cultures and ages [58]. These findings clearly support the notion of categorical model of human emotion. Hence, for the field of emotion recognition, most of the current studies focused on the recognition emotions in some discrete categories [12,19,20,42,59,60]. However, the collections of audio or visual emotional signals in some categories are not easy such as fear or disgust. Accordingly, many studies [19,20,42,60] focused on recognition of more familiar emotion categories including happy, sad, angry, and neutral which is comparatively easy to collect. Based on literature mentioned above, in this chapter, the recognition work focused on the four discrete emotional categories.

Recently, although many efforts have been done toward the audiovisual emotion recognition, most of the audiovisual databases were not accessible [13]. Hence, for performance evaluation, the data was collected first in our Multimedia Human–Machine Communication (MHMC) Lab to construct the MHMC database which was recorded from seven actors of both genders. Totally, the MHMC database contains approximately 5 hours of emotional audiovisual data. Figure 19.6 shows some example images for four emotional states in the MHMC database.

For MHMC database, each emotional state contains 30 specific sentences which were designed and collected for speech and facial expression. During data collection,

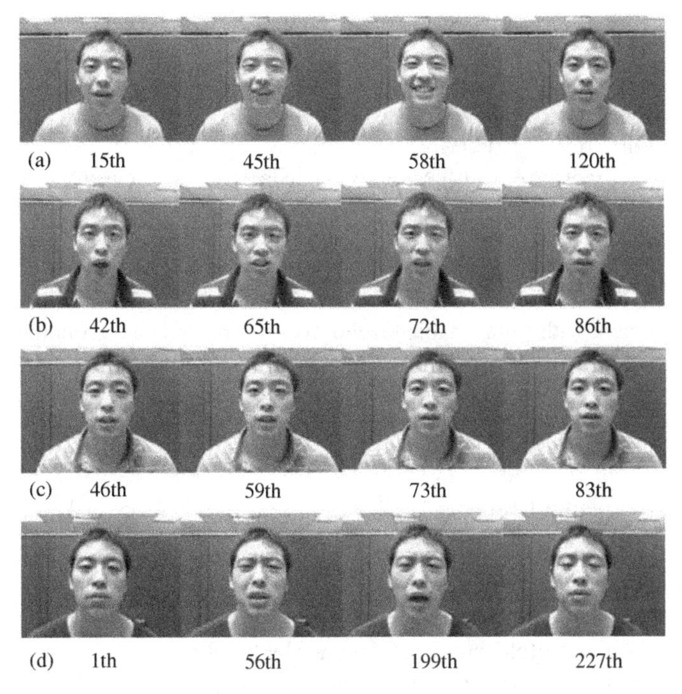

FIGURE 19.6 Examples for the posed expression with four emotional states: (a) happy, (b) angry, (c) neutral, and (d) sad. Extracted with permission from MHMC database.

each actor was asked to perform speech and facial expression while speaking each specific sentence. Each specific sentence is pronounced twice for recording in different sessions. Thus totally 1680 sentences were collected from 7 actors in the MHMC database. In order to avoid fatigue effect, for each actor, the recording procedure was divided into 6 sessions during 1 week (i.e., 40 sentences for each session). Hence, the actors had sufficient time to complete data recording. In terms of learning effect, before recording, each actor was asked to practice for 1 or 2 days to learn how to act properly. For data recording, each actor must ensure that the particular emotion is properly vocalized and expressed. The recorded data were also checked to ensure the consistency among the recorded data of the same emotion. The subjective tests were performed to set the ground truth for the recorded data. Four volunteers were recruited from our lab; each of them was asked to give the opinion of emotion label of the recorded data. During the labeling process, the subjects were asked to check the recorded samples more than once to ensure that the labels can truly reflect their feelings. Finally, the ground truth is set by averaging the opinions of all subjects. On average, if each data is labeled by less than three subjects, the actor will be asked to re-record the data.

For the data format of the MHMC database, the input images were captured with a Logitech QuickCam camera at a resolution of 320×240 pixels. The frame rate of recorded videos was 30 frames per second. Audio signals were sampled at 16 KHz and

TABLE 19.2 The properties of the MHMC database

Database	Actors	Size	Emotion description	Labeling	Video bandwidth	Audio bandwidth
MHMC	Seven	1680 videos	Category: happy, angry, neutral, sad	Subjects' judgment	320 × 240 pixels, 30 frames per second	16 KHz

processed at a frame length of 33 milliseconds. The recording was conducted in an office environment with a reasonable level of foreground and background noise, such as slight variation of actors' position, orientation, lighting conditions, white noise, and speech babble. Table 19.2 summarized the properties of the recorded MHMC database.

19.4.2 Experimental Results

For the MHMC database, four emotional states including neutrality (NEU), happiness (HAP), anger (ANG), and sadness (SAD) were considered. The database was first partitioned into seven subsets, one for each subject. The experiments are performed based on sevenfold cross-validation. Hence, for each fold, 360 sentences (from 6 subjects) and 60 sentences (from the remaining subject) were employed for training and testing, respectively, for each emotional state. Finally, the recognition performance was calculated by averaging the recognition results of seven subjects. In the experiments, the left-to-right HMM with eight hidden states was employed. Table 19.3 lists and describes all the model abbreviations used in the experiments.

The recognition rate for unimodal as well as bimodal fusion approaches is shown in Table 19.4. For the facial HMM (FA), the emotional states happy and angry achieve better recognition accuracy compared to neutral and sad, since the high active emotions are usually accompanied with high discrimination of facial features (e.g., happy). In addition, the recognition results of the emotional states neutral and sad in the prosodic HMM (PR) are better than those of happy and angry. From our observation, in general, the high active emotions usually have similar prosodic features such as energy and pitch and hence it is difficult to differentiate between the emotional states happy and angry. Based on the unimodal analyses, the result demonstrated that the facial and prosodic features are complementary to each other

TABLE 19.3 List and description of abbreviations

Abbreviations	Descriptions
FA	Facial HMM (unimodal)
PR	Prosodic HMM (unimodal)
FP	Facial–prosodic HMM (fusion at feature level)
C-HMM	Coupled HMM (fusion at model level)
SC-HMM	Semi-coupled HMM (fusion at model level)

TABLE 19.4 Recognition rates of four emotional states for different approaches

	NEU	HAP	ANG	SAD
FA	61.43	87.14	70.00	66.90
PR	71.43	65.71	70.24	76.67
FP	71.67	82.14	72.38	76.90
C-HMM	80.71	87.38	82.62	82.14
SC-HMM	82.38	88.57	84.29	85.71

for emotion recognition, and thus the bimodal fusion strategy is expected to improve the recognition accuracy.

For the average emotion recognition rate as shown in Figure 19.7, the facial HMM (FA) and prosodic HMM (PR) perform comparably to the feature-level fusion approach (FP). A reasonable explanation is that fusion at feature level will increase the dimensionality and may suffer from the problem of data sparseness. However, the performance of the FP is still better than unimodal approaches. It is therefore the advantage of combining audio and visual cues. Compared to FP, C-HMM improves approximately 8% recognition accuracy. The findings are in accordance with the descriptions mentioned above, that is, the information of mutual correlation among multiple signal streams is of great help for emotion recognition.

In terms of model-level fusion, the results shown in Figure 19.7 confirm that the proposed SC-HMM outperformed C-HMM. A reasonable explanation rests on the problem of data sparseness. In C-HMM, a state variable at frame t is dependent on its two predecessors (from two modalities) in previous frame "$t-1$", which describes the temporal coupling relationship between two streams. The dependency between the hidden states is often too unreliable to capture the statistical dependency of the observation sequences. Probably the major reason for unreliable statistical dependency is due to the problem of data sparseness. To further support our contention, training data analyses were conducted and also shown in Table 19.5. In the experiments, the training data were divided into two subsets: the first subset contains 240 training samples (from 4 subjects) per emotional state and the second subset contains 120 training samples (from 2 subjects) per emotional state. The test data were fixed

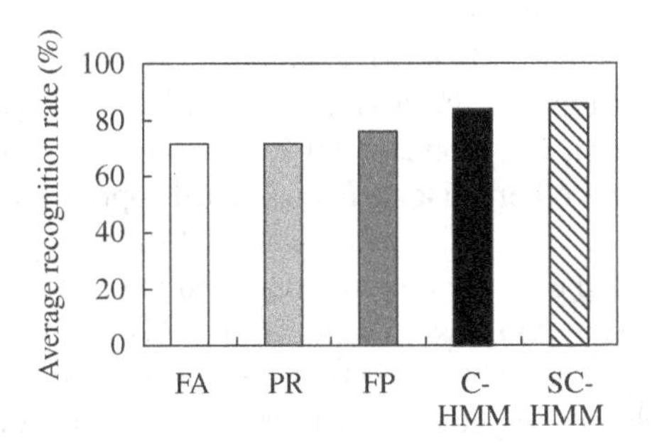

FIGURE 19.7 Average recognition rates for different approaches.

TABLE 19.5 Average recognition rates for insufficient training data condition

	Training subset I	Training ubset II
C-HMM	71.25	52.78
SC-HMM	78.75	66.39

at the same 180 samples (from 3 subjects) per emotional state. The experiments were performed on subject-independent mode.

The results in Table 19.5 indicate that the SC-HMM can achieve better performance than C-HMM in insufficient training data condition. These findings show that the proposed SC-HMM with state-based bimodal alignment strategy can obtain stronger statistical dependency between two HMMs in data sparseness condition. The data sparseness problem has a significant effect on the recognition results of C-HMM.

In terms of computational complexity, the C-HMM using "N-head" modified dynamic programming algorithm [61] can reduce the computational complexity from $O(TN^{2C})$ of the standard EM and Viterbi algorithms to $O(T(CN)^2)$ for C chains of N states apiece observing T data points. Compared to C-HMM with computational complexity of $O(T(CN)^2)$, the proposed SC-HMM can achieve $O(TCN^2)$. The lower computational complexity can significantly improve the efficiency of the system.

For the system performance, the system has been implemented on a personal computer with Pentium IV CPU and 1.25 GB memory. In the test mode, the time complexity of the overall system based on the proposed SC-HMM as shown in Figure 19.1 approximated 18 frames per second. According to observation, it should be noted that most of the time complexity is consumed in the preprocessing procedure (i.e., endpoint detection and feature extraction). Hence, the proposed bimodal fusion approach can be practically implemented on the application of human–computer interaction.

19.5 CONCLUSION

This chapter presented a new bimodal fusion approach called SC-HMM to align the temporal relation of the states of two component HMMs based on the simplified state-based alignment strategy. Five findings are summarized. First, the experimental results on unimodal methods motivate us to explore the different data fusion approaches toward effective emotion recognition. Second, joint audio and visual cues are significant to improve the performance of emotion recognition based on model-level fusion. Third, the proposed bimodal fusion approach outperforms the unimodal approaches and other existing fusion-based approaches. Fourth, compared to C-HMM, the proposed SC-HMM with simplified state-based alignment strategy can effectively alleviate the sparse data problem and obtain better statistical dependency between states for linking two component HMMs in insufficient training data condition. Finally, the proposed SC-HMM can also reach a good balance between performance and model complexity and hence is useful in different bimodal applications for most real-world scenarios.

REFERENCES

[1] R. W. Picard, *Affective Computing*, MIT Press, Cambridge, MA, 1997.

[2] Y. I. Tian, T. Kanade, and J. F. Cohn, "Recognizing action units for facial expression analysis," *IEEE Transactions on Pattern Analysis and Machine Intelligence*, 23(2), 2001.

[3] I. Cohen, N. Sebe, L. Chen, A. Garg, and T. S. Huang, "Facial expression recognition from video sequences: temporal and static modeling," *Comput. Vis. Image Underst.*, 91(1–2): 160–187, 2003.

[4] Y. Chang, C. Hu, R. Feris, and M. Turk, "Manifold based analysis of facial expression," *Image Vis. Comput.*, 24(6): 605–614, 2006.

[5] R. Cowie and R. Cornelius, "Describing the emotional states that are expressed in speech," *Speech Commun.*, 40(1–2): 5–32, 2003.

[6] B. Schuller, G. Rigoll, and M. Lang, "Hidden Markov model-based speech emotion recognition," in IEEE International Conference on Acoustics, Speech, and Signal Processing (ICASSP), Hong Kong, 2003.

[7] Y. L. Lin and G. Wei, "Speech emotion recognition based on HMM and SVM," in Fourth International Conference on Machine Learning and Cybernetics, Guangzhou, August 2005, pp. 18–21.

[8] A. Mehrabian, "Communication without words," *Psychol. Today*, 1968.

[9] N. Ambady and R. Rosenthal, "Thin slices of expressive behavior as predictors of interpersonal consequences: a meta-analysis," *Psychol. Bull.*, 1992.

[10] Z. J. Chuang and C. H. Wu, "Multi-modal emotion recognition from speech and text," *Int. J. Comput. Linguist. Chin. Lang. Process.*, 9(2): 45–62, 2004.

[11] N. Sebe, I. Cohen, T. Gevers, and T. S. Huang, "Emotion recognition based on joint visual and audio cues," in International Conference on Pattern Recognition (ICPR), Hong Kong, 2006.

[12] Y. Wang and L. Guan, "Recognizing human emotional state from audiovisual signals," *IEEE Transactions on Multimedia*, 10(4), 2008.

[13] Z. Zeng, M. Pantic, G. I. Roisman, and T. S. Huang, "A survey of affect recognition methods: audio, visual, and spontaneous expressions," *IEEE Transactions on Pattern Analysis and Machine Intelligence*, 31(1), 2009.

[14] C. Busso, Z. Deng, S. Yildirim, M. Bulut, C. M. Lee, A. Kazemzadeh, S. Lee, U. Neumann, and S. Narayanan, "Analysis of emotion recognition using facial expression, speech and multimodal information," in International Conference on Multimodal Interfaces (ICMI), State College, PA, October 13–15, 2004.

[15] B. Schuller, R. Muller, B. Hornler, A. Hothker, H. Konosu, and G. Rigoll, "Audiovisual recognition of spontaneous interest within conversations," in International Conference on Multimodal Interfaces (ICMI), Nagoya, Japan, November 12–15, 2007.

[16] Z. Zeng, Z. Zhang, B. Pianfetti, J. Tu, and T. S. Huang, "Audio-visual affect recognition in activation-evaluation space," in IEEE International Conference on Multimedia & Expo (ICME), Amsterdam, The Netherlands, July 6–8, 2005.

[17] S. Hoch, F. Althoff, G. McGlaun, and G. Rigoll, "Bimodal fusion of emotional data in an automotive environment," in IEEE International Conference on Acoustics, Speech, and Signal Processing (ICASSP), Philadelphia, PA, March 18–23, 2005.

[18] Z. Zeng, J. Tu, M. Liu, T. S. Huang, B. Pianfetti, D. Roth, and S. Levinson, "Audio-visual affect recognition," *IEEE Transactions on Multimedia*, 9(2), 2007.

[19] A. Metallinou, S. Lee, and S. Narayanan, "Audio-visual emotion recognition using gaussian mixture models for face and voice," in IEEE International Symposium on Multimedia (ISM), California, December 15–17, 2008.

[20] A. Metallinou, S. Lee, and S. Narayanan, "Decision level combination of multiple modalities for recognition and analysis of emotional expression," in IEEE International Conference on Acoustics, Speech, and Signal Processing (ICASSP), Texas, March 14–19, 2010.

[21] M. Song, M. You, N. Li, and C. Chen, "A robust multimodal approach for emotion recognition," *Neurocomputing*, 71(10–12): 1913–1920, 2008.

[22] N. Fragopanagos and J. G. Taylor, "Emotion recognition in human–computer interaction," *Neural Netw.*, 18: 389–405, 2005.

[23] M. Brand, N. Oliver, and A. Pentland, "Coupled hidden Markov models for complex action recognition," in Conference on Computer Vision and Pattern Recognition (CVPR '97), San Juan, Puerto Rico, June 17–19, 1997.

[24] F. Pernkopf, "3D surface inspection using coupled HMMs," in International Conference on Pattern Recognition (ICPR), Cambridge, UK, 2004.

[25] S. Ananthakrishnan and S. S. Narayanan, "An automatic prosody recognizer using a coupled multi-stream acoustic model and a syntactic-prosodic language model," in IEEE International Conference on Acoustics, Speech, and Signal Processing (ICASSP), Philadelphia, PA, March 18–23, 2005.

[26] A. V. Nefian, L. Liang, X. Pi, L. Xiaoxiang, C. Mao, and K. Murphy, "A coupled HMM for audio-visual speech recognition," in IEEE International Conference on Acoustics, Speech, and Signal Processing (ICASSP), Orlando, FL, May 13–17, 2002.

[27] L. Xie and Z. Q. Liu, "A coupled HMM approach to video-realistic speech animation," *Pattern Recognit.*, 40(8): 2325–2340, 2007.

[28] P. Ekman and W. Friesen, *The Facial Action Coding System: A Technique for the Measurement of Facial Movement*, Consulting Psychologists Press, Palo Alto, CA, 1978.

[29] MPEG Video: Facial Animation Parameters, FDIS 14496-2 Visual, ISO/IEC JTC1/SC29/WG11/N2502, October 1998.

[30] A. Raouzaiou, N. Tsapatsoulis, K. Karpouzis, and S. Kollias, "Parameterized facial expression synthesis based on MPEG-4," *EURASIP J. Appl. Signal Process.*, 2002(1): 1021–1038, 2002.

[31] A. M. Tekalp and J. Ostermann, "Face and 2-D mesh animation in MPEG-4," *Signal Process: Image Commun.*, 15(4–5): 387–421, 2000.

[32] S. V. Ioannou, A. T. Raouzaiou, V. A. Tzouvaras, T. P. Mailis, K. C. Karpouzis, and S. D. Kollias, "Emotion recognition through facial expression analysis based on a neurofuzzy network," *Neural Netw.*, 18(4): 423–435, 2005.

[33] K. Karpouzis, G. Caridakis, L. Kessous, N. Amir, A. Raouzaiou, L. Malatesta, and S. Kollias, "Modeling naturalistic affective states via facial, vocal, and bodily expressions recognition," in *Artificial Intelligence for Human Computing* (eds T. S. Huang *et al.*), LNAI 4451, Springer-Verlag, Heidelberg, 2007, pp. 91–112.

[34] P. Viola and M. Jones, "Rapid object detection using a boosted cascade of simple features," in Computer Vision and Pattern Recognition (CVPR), Hawaii, December 8–14, 2001.

[35] T. F. Cootes, G. J. Edwards, and C. J. Taylor, "Active appearance models," *IEEE Transactions on Pattern Analysis and Machine Intelligence*, 23(6): 681–685, 2001.

[36] T. F. Cootes, D. H. Cooper, C. J. Taylor, and J. Graham, "Trainable method of parametric shape description," *Image Vis. Comput.*, 10(5): 289–294, 1992.

[37] A. Lanitis, C. J. Taylor, and T. F. Cootes, "Automatic tracking, coding and reconstruction of human faces using flexible appearance models," *IEEE Electron. Lett.*, 30(19), 1994.

[38] F. Tang and B. Deng, "Facial expression recognition using AAM and local facial features," in International Conference on Natural Computation (ICNC), Haikou, August 24–27, 2007.

[39] H. C. Choi and S. Y. Oh, "Real-time recognition of facial expression using active appearance model with second order minimization and neural network," in IEEE International Conference on Systems, Man, and Cybernetics (ICSMC), Taipei, Taiwan, October 8–11, 2006.

[40] S. Lucey, A. B. Ashraf, and J. F. Cohn, "Investigating spontaneous facial action recognition through AAM representations of the face," in *Face Recognition* (eds K. Delac and M. Grgic), I-Tech Education and Publishing, Vienna, Austria, June 2007.

[41] C. W. Chen and C. C. Wang, "3D active appearance model for aligning faces in 2D images," in International Conference on Intelligent Robots and Systems (IROS), Nice, France, September 22–26, 2008.

[42] C. H. Wu, J. F. Yeh, and Z. J. Chuang, "Emotion perception and recognition from speech," in *Affective Information Processing* (eds J. H. Tao and T. N. Tan), Springer Science, 2009.

[43] G. Bailly, C. Benoit, and T. R. Sawallis, *Talking Machines: Theories, Models, and Designs*, Elsevier Science, Amsterdam, 1992.

[44] D. Ververidis, C. Kotropoulos, and I. Pitas, "Automatic emotional speech classification," in IEEE International Conference on Acoustics, Speech, and Signal Processing (ICASSP), Canada, May 17–21, 2004.

[45] I. Luengo, E. Navas, I. Hernez, and J. Snchez, "Automatic emotion recognition using prosodic parameters," in INTERSPEECH, Lisbon, Portugal, September 4–8, 2005.

[46] S. G. Kooladugi, N. Kumar, and K. S. Rao, "Speech emotion recognition using segmental level prosodic analysis," in International Conference on Devices and Communications (ICDECOM), Mesra, Ranchi, India, February 24–25, 2011.

[47] S. Kim, P. G. Georgiou, S. Lee, and S. Narayanan, "Real-time emotion detection system using speech: multi-modal fusion of different timescale features," in International Workshop on Multimedia Signal Processing (MMSP), Greece, October 1–3, 2007.

[48] D. Morrison, R. Wang, and L. C. De Silva, "Ensemble methods for spoken emotion recognition in call-centres," *Speech Commun.*, 49(2): 98–112, 2007.

[49] O. W. Kwon, K. Chan, J. Hao, and T. W. Lee, "Emotion recognition by speech signals," in 8th European Conference on Speech Communication and Technology, September 1–4, 2003, Geneva, Switzerland.

[50] C. M. Lee and S. Narayanan, "Toward detecting emotions in spoken dialogs," *IEEE Transactions on Speech and Audio Processing*, 13(2), 2005.

[51] X. Li, J. Tao, M. T. Johnson, J. Soltis, A. Savage, K. M. Leong, and J. D. Newman, "Stress and emotion classification using jitter and shimmer features," in IEEE International Conference on Acoustics, Speech, and Signal Processing (ICASSP), Hawaii, April 15–20, 2007.

[52] P. Boersma and D. Weenink, "Praat: doing phonetics by computer," http://www.praat.org/.2007

[53] R. A. Stevenson, J. A. Mikels, and T. W. James, "Characterization of the affective norms for English words by discrete emotional categories,"*Behav. Res. Methods*, 39(4): 1020–1024, 2007.

[54] P. Ekman, "Facial expression and emotion," *Am. Psychol.*, 48: 384–392, 1993.

[55] M. M. Bradley, M. Codispoti, B. N. Cuthbert, and T. J. Lang, "Emotion and motivation I: defensive and appetitive reactions in picture processing," *Emotion*, 1(3): 276–298, 2001.

[56] P. J. Lang, M. K. Greenwald, M. M. Bradley, and A. O. Hamm, "Looking at pictures: affective, facial, visceral, and behavioral reactions," *Psychophysiology*, 30: 261–273, 1993.

[57] J. A. Mikels, B. L. Fredrickson, G. R. Larkin, C. M. Lindberg, S. J. Maglio, and P. A. Reuter-Lorenz, "Emotional category data on images from the international affective picture system," *Behav. Res. Methods*, 37(4): 626–630, 2005.

[58] R. W. Levenson, "Autonomic specificity and emotion," in *Handbook of Affective Sciences* (eds R. J. Davidson, K. R. Scherer, and H. H. Goldsmith), Oxford University Press, New York, 2003.

[59] M. Yeasin, B. Bullot, and R. Sharma, "Recognition of facial expressions and measurement of levels of interest from video," *IEEE Transactions on Multimedia*, 8(3): 500–508, 2006.

[60] C. H. Wu and W. B. Liang, "Emotion recognition of affective speech based on multiple classifiers using acoustic-prosodic information and semantic labels," *IEEE Transactions on Affective Computing*, 2(1): 10–21, 2011.

[61] M. Brand, "Coupled hidden Markov models for modeling interacting processes," Perceptual Computing/Learning and Common Sense, Technical Report 405, MIT Media Lab, 1997.

AUTHOR BIOGRAPHIES

Chung-Hsien Wu received a Ph.D. degree in Electrical Engineering from National Cheng Kung University, Tainan, Taiwan, in 1991. Since August 1991, he has been with the Department of Computer Science and Information Engineering, National Cheng Kung University. He became a professor and a distinguished professor in August 1997 and August 2004, respectively. From 1999 to 2002, he served as chairman of the department. Currently, he is the Deputy Dean of the College of Electrical Engineering and Computer Science, National Cheng Kung University. He also worked at the Massachusetts Institute of Technology Computer Science and Artificial Intelligence Laboratory, Cambridge, MA, in Summer 2003 as a visiting scientist. He received the 2010 Outstanding Research Award of National Science Council (NSC), Taiwan. He is currently an associate editor of *IEEE Transactions on Audio, Speech, and Language Processing*, *IEEE Transactions on Affective Computing*, and *ACM Transactions on Asian Language Information Processing*. His research interests include speech recognition, text-to-speech, and spoken language processing. Dr. Wu is a senior member of IEEE and a member of International Speech Communication Association (ISCA). He has been the President of the Association for Computational Linguistics and Chinese Language Processing (ACLCLP) since September 2009.

Jen-Chun Lin received a B.S. degree in Electrical and Computer Engineering from National Formosa University, Yunlin, Taiwan, in 2005, and an M.S. degree in Information Engineering from I-Shou University, Kaohsiung, Taiwan, in 2007. He is currently working toward the Ph.D. degree in the Institute of Computer Science and Information Engineering, National Cheng Kung University, Tainan, Taiwan.

His research interests include multimedia signal processing, motion estimation and compensation, gesture recognition, emotion recognition, and multimodal fusion.

Wen-Li Wei received B.S. and M.S. degrees in Information Engineering from I-Shou University, Kaohsiung, Taiwan, in 2006 and 2008, respectively. She is currently working toward a Ph.D. degree in the Institute of Computer Science and Information Engineering, National Cheng Kung University, Tainan, Taiwan.

Her research interests include multimedia signal processing, motion estimation and compensation, gesture recognition, emotion recognition, and multimodal fusion.

20

EMOTION RECOGNITION IN CAR INDUSTRY

CHRISTOS D. KATSIS

Deptartment of Applications of Information Technology in Administration and Economy, Technological Educational Institute of Ionian Islands, Lefkada, Greece

GEORGE RIGAS

Unit of Medical Technology and Intelligent Information Systems, Deptartment of Materials Science and Engineering, University of Ioannina, Ioannina, Greece

YORGOS GOLETSIS AND DIMITRIOS I. FOTIADIS

Deptartment of Economics,University of Ioannina, Ioannina, Greece

Emotion recognition has gained large attention by the automotive field, both industry and research community. The inability to manage one's emotions while driving is identified as one of the major causes of accidents. When drivers are overwhelmed by anger or stress, their thinking, perceptions, and judgments are impaired, leading to misinterpretation of events. Driving in real traffic conditions is a complex task, since fast decisions need to be taken given limited information. The driving task poses differing demands on the driver. According to the American Highway Traffic Safety Administration, high stress influences adversely drivers' reaction in critical conditions, thus, it is one of the most important reasons for car accidents along with fatigue, intoxication, and aggressive driving [1]. In addition, driving events can alter drivers' emotions, which in turn can change drivers' performance. In this chapter we present the most recent advances in emotion recognition, focusing on emotions elicited from the driving task, or those influencing the driver's performance. We describe the methods presented in the literature categorizing them, firstly based on

Emotion Recognition: A Pattern Analysis Approach, First Edition. Edited by Amit Konar and Aruna Chakraborty.
© 2015 John Wiley & Sons, Inc. Published 2015 by John Wiley & Sons, Inc.

the information used in order to recognize emotion, and secondly based on emotions that the method is recognizing. Finally two exemplar systems are described, which have incorporated emotion recognition as core functionality in order to provide a better driving experience.

20.1 INTRODUCTION

Emotions influence various cognitive processes in humans, including perception, reasoning, and intuition. Driving presents a context in which a user's emotional state plays a significant role. Driving behavior adapts based on the complexity of traffic situations and driver risk tolerance, which is influenced by driver motives and emotional state. Driver's emotions often influence driving performance; the latter could be improved if the car actively responds/assesses the emotional state of the driver [2]. Research suggests that drivers do emote and their driving performance is affected by their emotions [3]. Car driving is in general considered an instrumental activity. Once the destination is set, the driving task is carried out rationally until the goal of reaching the destination is accomplished. This rational behavior can sometimes be interrupted by intense emotions either generated by other road users, or traffic situations. Thus, we may be angered by another driver or become nervous when faced with a complicated intersection. When being emotional, our judgment may be biased, and we may change our priorities to act [4]. Eyben *et al.* [5], indicate that essential driving abilities and attributes can be affected by emotion: perception and organization of memory, categorization and preference, goal generation evaluation, decision making, strategic planning, focus and attention, motivation and performance, intention, communication, and learning.

According to Lazarus [6], emotions prepare the person for adaptive action and promote adaptation to the environment. In order to drive safer, a person needs to be better aware of his/hers emotions and possess the ability to manage them effectively [7]. Although the causes of car accidents can be diverse, most researchers agree that the human factor accounts for most of them. Even a small disturbance can have severe consequences in the performance of a complex cognitive task as car driving. Emotion is a key factor that can be expected to affect cognitive functioning and therefore to increase task demand. However, it is still not clear whether emotions indeed constitute a serious problem for road safety. Since emotions are not registered as accident causes on standard accident registration forms, a direct link between emotions and road accidents cannot be made based on accident statistics. Therefore, accident data do not provide enough information. On the other hand, numerous studies have proved that emotional states extremely influence driving behavior and increase the risk of causing an accident [8–14].

The number of traffic fatalities worldwide is estimated to be around 1.2 million per year, and the number of people injured in traffic accidents about 50 million [15]. To improve overall safety, numerous technologies have been developed and deployed over the last years. Nowadays, it is possible to buy a vehicle with infrared night vision systems [16]—to increase seeing distance beyond headlamp range, adaptive cruise control [17]—which maintains a safe distance from the vehicle in front, lane departure warning systems [18]—to alert the driver of an unintended departure from the intended lane of travel, automatic braking systems [19]—to sense an imminent collision with another vehicle, person, or obstacle, and a variety of systems aiming to minimize the occurrence and consequences of automobile accidents. A major drawback of available vehicular systems, however, is that they do not include the driver in the loop of the decision-making processes. For example, even if the driver is heavily cognitively loaded or distracted, the decision threshold of safety systems and the human–computer interface's information exchange protocol remain the same [20] although it should have been increased. The aim of emotion recognition in the car industry is therefore to provide a kind of "state variable" which serves as input for subsequent processing in emotion-sensitive accessories, aiming to improve not only driving comfort but also safety [21]. Thereby safe driving can be supported by either attempting to improve the affective state of the driver (making the driver "happy") or directing him into a neutral emotional state or even adapting the car safety systems taking into account the emotion of the driver [22].

Recognizing emotions in a vehicle is not an easy task due to specificities of the car: noise, movement, need for minimum distraction of the driver, etc. As a result many researchers have attempted to provide drivers' emotion recognition based on different modalities such as video, speech, or even biosignals. The goal of this chapter is therefore to present the state of the art related to emotion recognition in car industry, including both methodological approaches/algorithms and systems. In addition, two innovative applications in the car industry are presented, one related to car racing drivers and one related to everyday driving. In the following paragraphs, we first present an overview of state-of-the-art approaches divided into two parts, one per modality and one per emotion. We then present the two exemplar cases. Finally, open issues, needs, and future trends are discussed.

20.2 AN OVERVIEW OF APPLICATION FOR THE CAR INDUSTRY

Literature review reveals that pattern recognition techniques have been applied to detect emotions. Different modalities have been examined either separated or in a multimodal scheme adapting to the different characteristics of each type of emotion. In the next paragraphs we provide an overview of state of the art per modality and per emotion.

20.3 MODALITY-BASED CATEGORIZATION

In this section we categorize the methods presented in the literature based on the information used, in order to recognize emotions. The most common information sources, which are presented in the following paragraphs, are video, speech, biosignals, and their combination (multimodal approach).

20.3.1 Video-Image-Based Emotion Recognition

Advanced image-video processing techniques are often used to extract driver's facial characteristics. Visual information processing, usually acquired though in-car cameras, is considered as a good approach to assess fatigue, arousal/drowsiness, and distraction. Li [23] presented a method for drivers' fatigue assessment from facial expressions using hidden Markov models (HMM). Dixit [24] proposed a warning system for vehicle driver fatigue detection combining image-processing techniques with fuzzy logic inference. Li [25] suggested a method which is based on combined Adaboost algorithm, particle filter, and prior knowledge to detect and track eyes' features in order to assess drivers' fatigue. In another work, Li *et al.* [26] used haar-like features to detect driver's mouth geometric characteristics correlated to yawning which is an important character to assess driver's fatigue. Azim *et al.* presented a nonintrusive fatigue detection system based on the video analysis of drivers using the Viola-Jones face detection algorithm in combination with fuzzy c-means (s-FCM) clustering to determine subjects' yawning state [27]. Yin *et al.* [28] used Gabor filters to get a multiscale representation for drivers' image sequences. Then, they extracted local binary patterns from each multiscale image and applied a statistical learning algorithm to extract most discriminative features from the multiscale dynamic features to construct a classifier for fatigue detection. Hachisuka *et al.* [29] used a combination of an active appearance model (for measuring the 3D coordinates of the feature points on the drivers facial images) and k-nearest-neighbor algorithm to classify drowsiness into six levels. Other techniques are based on the analysis of eye-blink characteristics formed by image sequences to realize the real-time assessment of driver's arousal state [30, 31]. Kutila *et al.* [32] proposed a module able to detect drivers' distraction workload by fusing stereo vision and lane tracking data, running both rule-based and support-vector machine (SVM) classification methods. Finally, to estimate driver attention, Kaminski [33] introduced a system to compute both head orientation and gaze detection from single images.

20.3.2 Speech Based Emotion Recognition

Generally speaking, speech, although easy to be acquired through a microphone, requires driver's participation, which is not always present especially for a sole driver. Speech based emotion recognition is useful for adapting driver–car speech-based interaction, for example, by changing the way directions are pronounced from a GPS. Attention theory suggests that speech-based interactions are less distracting to the driver than interactions with a visual display [34]. Speech-based emotion recognition is particularly useful for in-car board systems where information of the mental state of the driver may be provided to the system to initiate his/her safety [35]. Speech analysis has been applied for the assessment of driver's affective state: Malta *et al.* used Bayesian networks to estimate driver's frustration in real conditions [20], Boril *et al.* [36] used a fusion of Gaussian mixture models and SVM classifiers to analyze and detect cognitive load and frustration on drivers' speech. Grimm *et al.*

[37] extracted acoustic features from the drivers' speech signal and mapped them to an emotion state in a multidimensional, continuous-valued emotion space. Jones and Jonsson [2] proposed a system for driver's emotional assessment, which uses pitch, volume, rate of speech and other spectral coefficients. Their system maps these features to emotions using statistical and neutral network classifiers.

20.3.3 Biosignal-Based Emotion Recognition

Previous studies [38] have demonstrated that emotional arousal and valence stimulate different brain regions and in turn affect peripheral systems of the body. Significant physiological responses to emotions have been studied, showing, for example, measurable changes in heart rate, phalange temperature, blood volume pulse and respiration in fearful, angry, and joyful states [39,40]. Numerous methods have been used to infer affective states, but only few of them focus on drivers. Helander [41], used electrocardiogram (ECG), skin conductivity (EDR), and two electromyogram (EMG) sensors in order to monitor drivers in rural roads. Time sequence analyses of the drivers' physiological responses and motor activity showed that electrodermal responses (EDR) are induced by the mental effort of the driving task rather the physical effort necessary to maneuver the vehicle. Healey and Picard [42], proposed a system which uses heart rate, electroencephalographic, and respiration data in order to assess drivers' stress level in real time. Yamakoshi *et al.* [43] examined physiological variables, including cardiovascular parameters to measure drivers' awareness level. Nasoz *et al.* [44] proposed a multimodal driver affective interface where galvanic skin response, heartbeat, and temperature were used to assess certain emotions (neutral, anger, fear, sadness, and frustration). Katsis *et al.* proposed that specific metrics of surface EMGs such as mean frequency (MNF), median frequency (MDF), and the signal RMS amplitude can be used as valid fatigue indicators [45]. Finally, Murugappan *et al.* [46] applied a combination of surface Laplacian (SL) filtering, time–frequency analysis of wavelet transform (WT), and linear classifiers to electroencephalogram (EEG) signal to classify discrete emotions (happy, surprise, fear, disgust, and neutral).

20.3.4 Multimodal Based Emotion Recognition

Recent reports propose the fusion of different modalities in order to estimate the driver's affective state. Most approaches usually employ a combination of data acquired from the driver (e.g., video, speech, biosignals). Katsis *et al.* [47] proposed a methodology for the assessment of the emotional states: high stress, low stress, disappointment, and euphoria, using as classifiers SVM and an adaptive neuro-fuzzy inference system (ANFIS) fed with features extracted from facial electromyograms (fEMGs), ECG, respiration, and EDR. Rothkrantz [48] presented a multimodal system to assess the affective state of car drivers based on analysis of facial expressions and speech signals. Latest research combines the above-mentioned modalities with environmental conditions (e.g., temperature, humidity) and driver's interaction with the car features (i.e., steering wheel grip force) in order to increase the emotions' classification accuracy. Clarion [49] proposed an integrated device which exploits the use of specific physiological data, video recordings of the driver and environment, parameters from the vehicle, and contextual data to evaluate a driver's functional state. Leng *et al.* [50] designed and carried out experiments to find the mapping relation among heart rate, skin conductance, griping force, and skin temperature on two kinds of emotions: fear and amusement. Rigas *et al.* [51] exploited a set of features obtained from physiological signals from the driver (ECG, electrodermal activity-EDA, and respiration), video recordings from the driver's face, and environmental information in order to detect drivers' stress and fatigue.

20.4 EMOTION-BASED CATEGORIZATION

In this section we categorize the driver emotion recognition literature based on the emotions recognized. Those emotions are mainly those affecting driving performance and include stress, fatigue, confusion, nervousness. We also present methods recognizing distraction, which can be related to the emotional state of the driver and has obvious implications on the driving performance.

20.4.1 Stress

According to Wickens and Hollands [52] stress is a physiological reaction to both the internal and external influences of a task. Matthews [53] expanded this definition to include the interactions that can exist between the external demands of and the internal perceptions about the driving task. Driver stress can be either short term, including in this way anger or irritation, or long term referring to a general inconvenience with the car use [54]. Numerous studies link highly aroused stress states with impaired decision-making capabilities [55], decreased situational awareness [56], and degraded performance which could impair driving ability [57].

As stress can bring about changes in physiological measurements such as skin conductivity, respiration muscle activity, and heart rate, several works in the literature focus on driver stress recognition based on biosignal processing. ECG, EMG), respiration, skin conductivity, blood pressure, and body temperature are the most common signals collected from the driver in order to estimate the workload and the levels of stress he/she experiences. Healey *et al.* [58] presented a real-time method for data collection and analysis in real driving conditions to detect the driver stress status. According to them, there is a strong correlation between driver status and selected physiological signal (EMG, ECG, skin conductivity, and respiration effort). In another study, Healey *et al.* [59] specified an experimental protocol for data collection. Four stress level categories were created according to the results of the subjects self-report questionnaires. A linear discriminant function was used to rank each feature individually based on the recognition performance and a sequential forward floating selection (SFFS) algorithm was used to find an optimal set of features to recognize driver stress.

Healey *et al.* [42], proposed a slightly different protocol, while the results showed that for most drivers the skin conductivity and the heart rate are most closely correlated to driver stress level. Zhai *et al.* [60] developed a system for stress detection using blood volume pressure, skin temperature variation, electrodermal activity (EDA), and pupil diameter[1].

Rani *et al.* [61] presented a real-time method for driver's stress detection based on the heart rate variability (HRV) using Fourier and wavelet analyses. Liao *et al.* [62] presented a probabilistic model for driver's stress detection based on probabilistic inference using features extracted from multiple sensors.

20.4.2 Fatigue

The estimation of fatigue is well studied in the literature [42,47,58–72]. The majority of relative works is based on in-lab experiments, mainly focusing on face monitoring and blink detection to calculate eye activation [70], while the vehicular experiments serve for indirect fatigue recognition through its impact on driving issues (speed maintenance, steering control). These methods, however, are suitable for the recognition of rather late stages of the fatigue (drowsiness) when the effects on driver's face are quite noticeable and performance change has already become critical.

In the road environment, even earlier fatigue stages can affect driving performance. This is because even lower fatigue levels still cause declines in physiological vigilance/arousal, slow sensorimotor functions (i.e., slower perception and reaction times), and information processing impairments, which in turn diminish driver's ability to respond to unexpected and emergency situations [71]. Therefore, the impact of fatigue on the driver's performance should not be estimated using only driving measures, but additional parameters, associated with the driving performance, are needed [67].

According to Crawford [64], physiological measures are the most appropriate indicators of driver fatigue. This has been confirmed by numerous studies, which followed similar approaches for driver fatigue estimation, making use of biosignals obtained from the driver [63, 68, 72].

Bittner *et al.* [63] presented an approach for the detection of fatigue based on biosignals acquired from the driver EEG, ECG, electrooculogram (EOG) and video monitoring. They examined different features that might be correlated with fatigue, such as the spectrum of the EEG, the percentage of eye closure (PERCLOS), and the fractal properties of HRV. They concluded that the first two are more correlated with instant fatigue levels of the driver, while the last is most suitable for the detection of the permanent state of the driver. Li [72] addressed the estimation of driver's mental fatigue using HRV spectrum analysis using a simulator for data collection. The features obtained from HRV indicated high correlation with the mental fatigue of the driver. Yang *et al.* [68] used heterogeneous information sources to detect driver's fatigue. The information sources included fitness, sleep deprivation, environmental

[1] The measure through the center of the adjustable opening in the iris of the eye, terminated at both ends by its circumference.

information (traffic, road condition, etc.), physiological signals (ECG, EEG), and video monitoring parameters (head movement, blink rate, and facial expressions). In order to combine all the above-mentioned information they used the Dempster–Shafer theory and rules for determining whether the driver is in fatigue state or not. Ji *et al.* [66] proposed a probabilistic framework for modeling and real-time inferencing of human fatigue by integrating data from various sources and certain relevant contextual information. They used a dynamic Bayesian network which encapsulates the time-dependent development of fatigue symptoms. The estimation is based on visual cues and behavioral variables.

As research in the field progresses, a variety of physiological signals has been used for fatigue detection. The most informative measures in terms of fatigue recognition are those extracted from the EEG signal, which have been used for the quantification of task specific performance changes [73–80].

20.4.3 Confusion and Nervousness

Another driver state often responsible for traffic violations and even road accidents is confusion or irritation, as it is related to loss of self-control and therefore loss of vehicle control. Confusion can be provoked by nonintuitive user interfaces or defective navigation systems as well as by complex traffic conditions, mistakable signs, and complicated routing. Moreover, the amount of information that needs to be processed simultaneously during driving is a source of confusion especially for older people [81], who have slower perception and reaction times.

Just like stress, confusion or irritation leads to impairment of driving capabilities including driver's perception, attention, decision making, and strategic planning.

Nervousness corresponds to a level of arousal above the "normal" one, which best suits to the driving task. It is an affective state with negative impact both on decision-making process and strategic planning. Nervousness can be induced by a variety of reasons either directly related to the driving task like novice drivers or by other factors like personal/ physical conditions. In Reference 82, the nervousness induced by the use of drugs is examined with respect to effects on driving. According to these findings, nervousness mainly affects driver's concentration resulting in impaired driving performance.

Li *et al.* [21] presented an approach to detect fatigue, nervousness, and confusion based on a probabilistic framework using DBNs. Their method was based on the analysis of heterogeneous data such as context, user information as well as vision information including facial expressions, gaze, and PERCLOS parameters. This method was evaluated on synthetic and real data for fatigue detection only, whereas it has not become apparent whether it works well for nervousness and confusion estimation.

20.4.4 Distraction

Driver inattention, although it cannot be consider as an emotion, is one of the major human factors in traffic accidents. The US National Highway Traffic Safety Administration estimates that in 25% of all crashes some form of inattention is involved [1].

Distraction (besides drowsiness) as one form of driver inattention may be characterized as: "any activity that takes a driver's attention away from the task of driving" [83].

Eyben *et al.* [84] presented a distraction detection method using driving behavior information and Long Short-Term Memory (LSTM) recurrent neural networks [85]. The driving behavior information included vehicle context (steering wheel angle, throttle position, speed, heading angle, and lateral deviation) and head position and rotation measurements. The information with the highest discriminative power was the head rotation, followed by steering wheel angle, heading angle, throttle position, speed, and lateral deviation.

Liang *et al.* [86] also presented a method for detection of driver distraction, based on eye movements and simple measures of driving performance. Using SVMs they were able to detect driver distraction with high accuracy.

The above-mentioned studies demonstrate that video monitoring of driver's face and simple measures of driving performance can be used to detect driver distraction.

20.5 TWO EXEMPLAR CASES

Putting machine intelligence and pattern recognition techniques into practice and providing real-time integrated systems to the car industry is the missing link for closing the loop of car–driver interaction. In the following paragraphs we demonstrate the work done in two exemplar cases. The first one is oriented toward car racing, while the second one is toward everyday driving. In the first system, emotion recognition can contribute to the holistic view of the car–driver system, which is currently only partially provided by existing telemetry systems. Although recent advances in car racing technology have made it feasible to collect (in real time) a great number of parameters regarding the optimal car operation (e.g., shock-absorber travel, steering wheel position, brake pressure, etc.), there is still a lack of systems which provide information about the driver's emotional status. Recognizing emotions and associating them with driving behavior and car status can be of great value for car racing teams trying to optimize the car/driver/environment setting. Such tools, apart from the demanding car racing environment, can also assist toward the evolution of production of cars.

In the second case, emotion recognition is used in advanced car safety systems (active or passive), which on the one hand can identify hazardous emotional driver states and on the other can provide tailored (according to each state and the associated hazard) suggestions and warnings to the driver.

20.5.1 AUBADE

20.5.1.1 System Overview The main goal of the system is to automatically monitor and classify the psychological condition of drivers, operating under extreme stress conditions (for example, car racing drivers) by applying pattern recognition techniques.

The system, named AUBADE (wearable EMG AUgmentation system for roBust behAvioral understanding) estimates the emotional state using features extracted from fEMGs, Respiration (RESP), EDA, and ECG. In addition, a real-time generic 3D facial mesh that represents driver's expressions is provided to the car racing team. The usual way to assess human emotion is by employing advanced image-processing techniques in order to extract the facial characteristics. In our case, it is very difficult to apply image-processing techniques, since for safety reasons the users are wearing a mask and above it a casque. The proposed system realizes an alternative method in order to monitor the facial expressions of the subject. Instead of using image-processing techniques AUBADE utilizes the processing of surface EMG sensors, placed on the fireproof mask that the drivers are currently wearing.

20.5.1.2 System's Design The AUBADE system consists of: (a) the multisensorial AUBADE wearable, (b) the data acquisition module, (c) the communication module, (d) the facial representation module which is used for the projection of the obtained data through user specific 3D face meshes, (e) the AUBADE repositories: where the acquired signals, the facial animation videos, and personal medical information are stored and (f) the AUBADE intelligence module which is used for the real-time assessment of the subjects' emotional state based on predefined emotional classes. The architecture of AUBADE system is presented in Figure 20.1. A more detailed description of the AUBADE system's functionalities and modules follows.

1. The multisensorial AUBADE wearable: It is a noninvasive, ergonomic, comfortable, and easy to use wearable that includes a number of sophisticated biosensors gathering the following physiological data: fEMG[2] signals, respiration signal (RESP), EDR, and ECG. It is composed of three pieces: (i) the balaclava containing 16 EMG textile fireproof sensors, (ii) the ECG and RESP sensors on the thorax of the driver and (iii) the EDR textile and fireproof sensor placed inside the driver's glove.
2. The data acquisition module: The signal acquisition unit consists of both hardware (data acquisition card) and software components that appropriately collect and preprocess all the biosignals obtained from the sensors of the wearable. Since body movement may affect readings from sensors, a preprocessing stage is employed to compensate for any artifacts generated. The preprocessing consisted of low-pass filters at 100 and 500 Hz for the ECG and the fEMGs, respectively, and smoothing (moving average) filters for the RESP and EDA signals. The data acquisition card is small, highly integrated, and scalable in order to fit into small places inside the car. The filtered signals are send wirelessly to the centralized system for further processing and decision making.
3. The communication module: It is activated by the system end user and is responsible for the secure transfer of the biosignals collected and processed by the data acquisition module. The user measurements are transferred through a wireless LAN to the centralized system for further analysis. Furthermore, to

[2]EMG signals are acquired from the following facial muscles: frontalis, corrugators, nasali s-alaeque nasi, zygomaticus major, masseter, orbicularis oris, depressor labii, and mentalis.

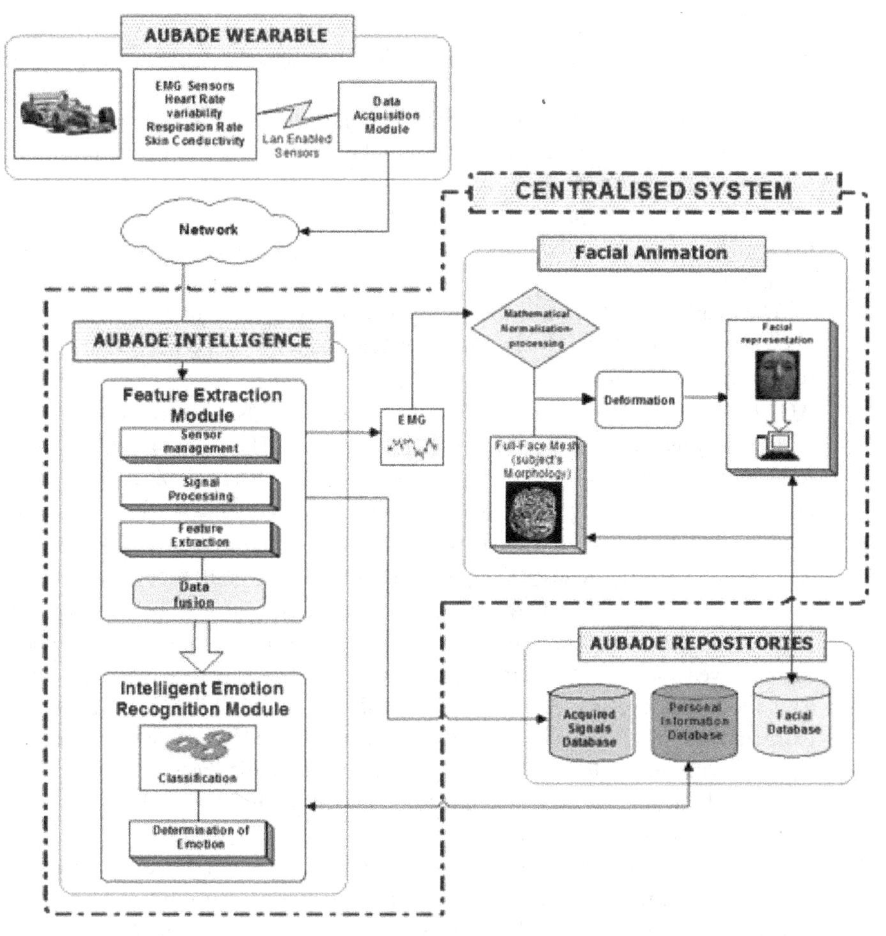

FIGURE 20.1 The AUBADE system architecture.

ensure protection of personal data over wireless networks, appropriate security and privacy mechanisms have been implemented.

4. The facial representation module: The facial animation module models the deformation of skin tissue according to a three-layer model, consisting of skull, muscle, and skin layers. Each layer consists of a number of nodes, which are connected with neighboring nodes of the same layer and nodes in the layers above/below. Each node represents a mass and each link between nodes is modeled as a spring. A more detailed description of the facial expression modeling procedure can be found in References 87 and 88. The module flow goes through several processing stages before producing the 3D reconstruction.

The features of the EMG signals, as extracted by the feature extraction module, are used to estimate the contraction of the subject's monitored muscles. The outcome of this procedure is the quantification of muscle contraction for the eight muscles being monitored, in the range 0.0 ··· 1.0 (where value 0 means that the muscle is not contracted and value 1 means that the muscle is fully contracted).

The contraction level drives the muscle model, to calculate the new position of muscle nodes. The muscle model is simulating linear and sphincter muscles, which are the kinds of muscles involved in AUBADE. Numerical methods, through the attachment of muscle nodes in the face's geometry, solve the mathematical model of the mass–spring network, given the new position of the muscle nodes.

The displacement of each node of the skin mesh is then applied to the face's geometry, as calculated by the mathematical model in the previous step. The resulting mesh is then presented on the user's screen.

5. AUBADE repositories: The system's databases store the acquired raw signals which are ranked per user, per date, per event, etc. They can be recalled any time from this database and can be analyzed by specialists and researchers who are able to draw statistical and other information. The databases also store the medical history of the subjects as well as their facial animation videos.
6. The AUBADE intelligence module: It is the core module of the AUBADE system. It is divided in two major modules namely the feature extraction and the intelligent emotion recognition modules. The feature extraction module receives the data from the sensors through the signal acquisition unit. This module converts the selected biosignals (fEMG, RESP, EDA, ECG) into extracted features (shown in Table 20.1) that can be used by the intelligent emotion recognition module in order to determine subject's basic emotions. The extracted biosignals are also intended to compensate for day-to-day variations and differences between individuals and are described next.

   The mean value and the root mean square (RMS) are computed for a considered time window of each fEMG signal. The mean of the absolute values of the first differences (Mean_abs_fd). For a signal $X_N = (x_1, x_2, x_N)$ the Mean_abs_fd is defined as:

$$Mean\_abs\_fd = \sum_{1}^{N} (|x_2 - x_1| + |x_3 - x_2| + +|x_N - x_(N-1)|)/N \qquad (20.1)$$

where N is the number of samples. This feature is the average magnitude of the change in measures across considered time window. The mean amplitude

TABLE 20.1 Features extracted for each of the acquired biosignals

Facial EMGs	ECG	Respiration	EDR
Mean value	Mean_abs_fd	Mean_abs_fd	Mean_abs_fd
RMS	Mean amp	Mean amp	Mean amp
	rate	rate	rate

(Mean amp) for the ECG, RESP, and EDR signals are defined as the average value of the computed amplitudes in a time window.

Rate: It is the heart rate, respiration rate, and number of skin conductance responses (SCRs), calculated in a specific time window. The whole feature vector:

$$u = \{feature1, feature2, \cdots, featureN\} \tag{20.2}$$

is composed of 41 features (N = 41) produced from 10 second biosignal recordings, which is fed to the intelligent emotion recognition module. Moreover, every 1/20 second the RMS of each fEMG signal is computed and a vector is produced and is passed to the facial animation module.

7. The intelligent emotion recognition module: The intelligent emotion recognition module performs a real-time classification of the subject's emotional state based on predefined emotional classes. Four emotional classes have been investigated: stress level (decomposed into high stress and low stress), and valence level (decomposed into euphoria and dysphoria). Stress and valence level states are chosen since they affect drivers' behavior [89]. In psychology, valence is used in order to examine the intrinsic attractiveness (positive valence) or aversiveness (negative valence) of an event or situation. On the other hand, stress level refers to any event or situation that makes heightened demands on a person's mental or emotional resources.

   In order to assess the emotional state into one of the pre-defined categories a two-stage classifier is employed. First, a C4.5 regression tree classifier [90] is used to classify the driver's state as high stress, low stress, and valence level. Then, a tree augmented naive Bayesian classifier—TAN—[91] categorizes the driver's valence level as euphoria or dysphoria. The C4.5 and TAN Bayesian classifiers have been selected, since they are fast to train, they demand small training sets, offer the ability for decision interpretation, and have been successfully applied in many fields [92]. The two-stage classification procedure is described below.

   (a) First stage: high stress, low stress/valence classification. The construction of the decision tree is implemented using the C4.5 inductive algorithm. The essence of the algorithm is to construct a decision tree from the training data. Each internal node of the tree corresponds to a principal component, while each outgoing branch corresponds to a possible range of that component. The leaf nodes represent the class to be assigned to a sample. The C4.5 algorithm is applied to a set of data and generates a decision tree, which minimizes the expected value of the number of tests for the classification of the data.

   (b) Second stage: euphoria/dysphoria emotional state classification. Tree augmented Bayesian classifier belongs to the class of Bayesian network classifiers [91]. A Bayesian network is a directed acyclic graph that encodes a joint probability distribution over the variables X. The probability distribution of each variable can be determined by considering the distribution of its parents. Using these assumptions, the joint probability of the variable set X using the chain rule can be written as the following product:

$$p(X) = \prod_{i=1}^{n} p(x_i|parents(x_i)) \tag{20.3}$$

Bayesian networks can be used for classification tasks. A well-known Bayesian network classifier is naive Bayes classifier [93] where all the attributes are independent. Tree augmented naive Bayesian classifier approximates the interactions between attributes by using a tree structure imposed on the naive Bayesian structure. The main advantages of TAN are: (i) it drops the independence assumption between attributes which does not hold in many cases and (ii) the complexity of learning a TAN classifier is only polynomial. Experiments with TAN classifier indicated very good performance in a variety of classification problems [94].

20.5.1.3 Results The selected biosignals are obtained in a controlled environment. This is implemented through a professional 3D projection system and specialized visualization glasses, providing a realistic racing environment simulation. During the experiments subjects interact with force feedback devices reproducing the mechanical responses of the primary controls (gear shift, steering wheel, pedals). Data come from 10 healthy males which have average or high level driving skills. Four subjects, aged 2–35, had above average driving skills and six persons, aged 29–34, professional driving skills. Each driver participated in eight driving rounds with average duration 3–4 minutes. Driving routes are randomly selected for each subject and comprise different difficulty levels and weather conditions (rain, snow, etc.), thus resulting in different emotional states and driving behaviors. Three cameras are used to record each driver's round. The first captures the driver's face, the second the driver's round event, and the third a general view of the whole experiment. Three experienced psychologists from the Neurological Clinic of Modena, Italy, annotated the person's emotional state every ten seconds. Since only in 57% of the cases the opinions of the three experts coincided, annotation conflicts were resolved using the captured videos for re-evaluation of the experts' decision. During the experimental procedure the multisensorial balaclava is not used (only the sensors are attached in subjects' face), thus, the experts can assess the subjects' emotional state according to their facial expression.

The raw signals (fEMGs, ECG, respiration, and EDR), obtained previously, are initially segmented in windows of 10 seconds. Then, for each 10 second period, the feature extraction module extracts from the signals the features described in Section 20.5.1.1. Then, a vector is constructed containing the extracted features along with the experts' annotation (for the specific 10 second period). These vectors, for each subject and driving round (1301 vectors in total: high stress (HS)—327, low stress

TABLE 20.2 The average obtained classification results for the decision tree in terms of Se, PPA and Acc

Affective state	Se (%)	PPA (%)	Acc (%)
HS	64.8	80.3	
LS	77.0	70.64	**80.90**
VAL	88.8	83.88	

TABLE 20.3 Decision tree confusion matrix

Emotional states	Classified as HS	Classified as LS	Classified as VAL
HS	212	6	46
LS	23	154	41
VAL	92	40	687

(LS)—200, dysphoria (Dys)—365, euphoria (Eup)—409, constitute our dataset used to validate our classification scheme.

Initially, the decision tree (C4.5) is used for the drivers' affective state classification into high stress (HS), low stress (LS), and valence (euphoria, dysphoria) (VAL). The classifying performance is measured in terms of sensitivity (Se) and positive predictive accuracy (PPA). Tables 20.2 and 20.3 present the obtained (average) results and the confusion matrix for the decision tree, respectively. The obtained average overall classification accuracy (Acc) is 80.9%

The classification results of TAN classifier, into euphoria and dysphoria are given next. Table 20.4 provides the average Se, specificity (Sp), and Acc and Table 20.5 the confusion matrix, respectively. These results were obtained without taking into consideration the false classification (false positives and false negatives) of the C4.5 classifier. The obtained average overall accuracy was 82.4%.

The average classifications results for the two-stage classifier are presented in Table 20.6, while the confusion matrix is shown in Table 20.7. The overall accuracy of the two-stage classifier is 71.9%. It should be noted that in order to minimize the bias associated with the random sampling of the training and testing data samples, 10-fold cross-validation was used in all cases. Finally, it should be also mentioned that different experiments were conducted, before employing the two-stage classifier. Several classifiers (naive Bayes classifier, TAN, K-NN, C4.5, support vector machines (SVM), adaptive-network-based fuzzy inference System (ANfiS)) have been tested for the 4-class problem (LS, HS, Eup, Dys) but reported lower overall accuracy.

TABLE 20.4 The average obtained classification results for the TAN classifier in terms of Se, Sp, and Acc. The DT classification errors are not taken into account

Affective state	Se (%)	Sp (%)	Acc (%)
Eup	72.1	88.6	**82.4**
Dys	91.7	78.6	

TABLE 20.5 TAN confusion matrix

Emotional states	Classified as Eup	Classified as Dys
Eup	263	102
Dys	34	375

TABLE 20.6 The overall classifications results for the two-stage classifier

Affective state	Se (%)	PPA (%)	Acc (%)
HS	64.5	79.9	**71.9**
LS	76.5	68.9	
Dys	50.4	79.3	
Eup	94.6	66.4	

20.5.1.4 Future work At present, the AUBADE system classifies the driver's affective state as high stress, low stress, dysphoria, and euphoria. As reported by the psychologists evaluating the experiment, emotions do not always correlate with specific driving events and subjects may experience different emotions during the same driving event, depending on their personalities. Furthermore, a subject driving in real driving conditions usually expresses fear instead of dysphoria. This indicates that in real-life conditions, the emotional states might be altered and thus, new classes may be added. Moreover, it must be noted that AUBADE currently can assess only the driver's dominant emotional state. On-going work of ours focuses on the simultaneous identification of other co-occurring emotional states. In addition, future work will target an autonomic system that will provide the driver's emotional state using information arising both from the generated facial model (driver's facial expression) and the selected biosignals. In any case, real-condition experiments with a large number of subjects are needed for confirming our results.

20.5.2 I-Way

20.5.2.1 System Overview The I-Way Project was a multipurposed automotive project, funded by the EU. The main goal of the project was to develop an intelligent

TABLE 20.7 The confusion matrix for the two-stage classifier

	Classified as			
Affective state	HS	LS	Dys	Eup
HS	211	25	21	70
LS	7	153	16	24
Dys	40	39	184	102
Eup	6	5	11	387

environment-aware driver warning and information system. To achieve this goal several novel technologies enlarging the driver's perception were adopted. Those technologies include V2V, V2I, and I2V communication, proceeding vehicles and object detection using vision and radar technologies, as well as an alerting mechanism taking into account the environment information (acquired using the above-mentioned technologies) and the physical state of the driver, using on the wheel sensor and video monitoring driver's face. The focus of this chapter is on the alerting mechanism and the driver state recognition (DSR.) which includes driver's stress and fatigue.

The I-Way alerting mechanisms takes into account driver's stress and fatigue in order to provide alerts that would leverage the vigilance of the driver when fatigued and produce less mental effort when stressed.

In the first section we will shortly describe the architecture of the I-WAY DSR system. Then we present with details the methods used and the corresponding results [51]. We conclude with the outcome of this effort and further directions.

20.5.2.2 System Design

1. I-Way communication module. A very important feature of the I-Way system is the communication between vehicles and vehicles with the road infrastructure. This allowed vehicles to exchange information about perceived events and early notification of drivers.
2. Preprocessing for decision support (PDS) module. The preprocessing for decision support (PDS) module is responsible for gathering data from the near and medium range of the I-WAY vehicle and preparing them appropriately for the I-WAY situation assessment module (SAM). The scope of the PDS module is to facilitate the decision-making process and enable fast reactions to issue alerts on time. The information arriving at PDS come from a front radar sensor, a camera monitoring the road and the vehicle itself, using the CAN-BUS. Moreover, the combining and preprocessing of the above gathered information within a separate module gives the overall system both modularity and reduced complexity.
3. Situation assessment module (SAM). SAM is the central part of the I-WAY system that encapsulates all the intelligence required from an automated driver assistance system. The functionality of SAM is twofold.
    - Proactive risk assessment: preview an event/situation and inform the driver well in advance with supportive recommendations when necessary. The decision of what type of support to give to the driver is based on the overall situation assessment and varies from informative messages to warnings and roaring alerts depending on how critical the situation is and driver's emotion (stress and fatigue).
    - Scouting: full perception of the surrounding environment with the scope to update other vehicles and the road infrastructure about conditions/events prevailing at specific highway locations.

A very important submodule of SAM is the driver status assessment module, responsible for the classification of driver state in fatigue and stress levels. This module we will be further described in the remaining of this section.

20.5.2.3 Driver Status Assessment Module As described in Section 20.3 there are several approaches on affine state recognition based on sensory information used. A multimodal approach was adopted in the I-WAY, with a further goal to investigate the potential use of wheel mounted physiological sensors for driver status assessment.

In order to set up a real-time system for driver stress and fatigue monitoring in real driving conditions, the sensors for the physiological signal acquisition should be minimally obtrusive. The following signals were incorporated: (i) ECG through a g.ECG sensor which is placed on the subject's chest, (ii) EDA through two Ag/AgCl EDA sensors attached on the subject's middle and index fingers of the right hand, and (iii) the respiration rate using a g.RESP piezoelectric respiration sensor. In the in-vehicle system an ECG and grip force sensors were mounted on the wheel.

The grip force was considered as a suitable replacement of the EDA signal.[3] The system architecture is presented in Figure 20.2.

The methodology followed consisted of three main steps (depicted in Figure 20.2).

1. *Preprocessing and feature extraction* which included: (a) signal acquisition, (b) preprocessing and feature extraction, (c) video acquisition processing and feature extraction and (d) environment information extraction. The features are extracted in time windows of 5 minutes, that is a reasonable compromise between the need of sufficient sample size in order to have reliable statistic properties and the need of small window to capture the changes in the psychophysiology of the driver [65].
2. *Feature selection*
3. *Classification*

Signal acquisition/preprocessing and feature extraction: During the experimental phase, the Biopac MP-100 system was used for signal acquisition. The ECG signal is acquired at sampling frequency 300 Hz, while the EDA and the respiration signal at 50 Hz. The resolution is set to 12-bit for all signals.

ECG signal. Initially, a low-pass Butterworth signal filtering, Butterworth filter is applied to the ECG signal to remove the baseline wonder. Then the R peaks are detected, using the procedure described in Reference 95. The R–R intervals constitute the RR variability signal (RRV). The features extracted from the RRV signal where interpolated and resampled to 1 Hz. The features extracted from the RRV signal are summarized in Table 20.8.

EDA signal. The EDA signal is downsampled to 1 Hz. A smoothing filter is applied, since in many cases noise is evident in the signal; then the low frequency 0.01 Hz of the signal is removed which is considered as the skin conductance level

[3]EDA signal was not suitable for in-vehicle use due to security restrictions.

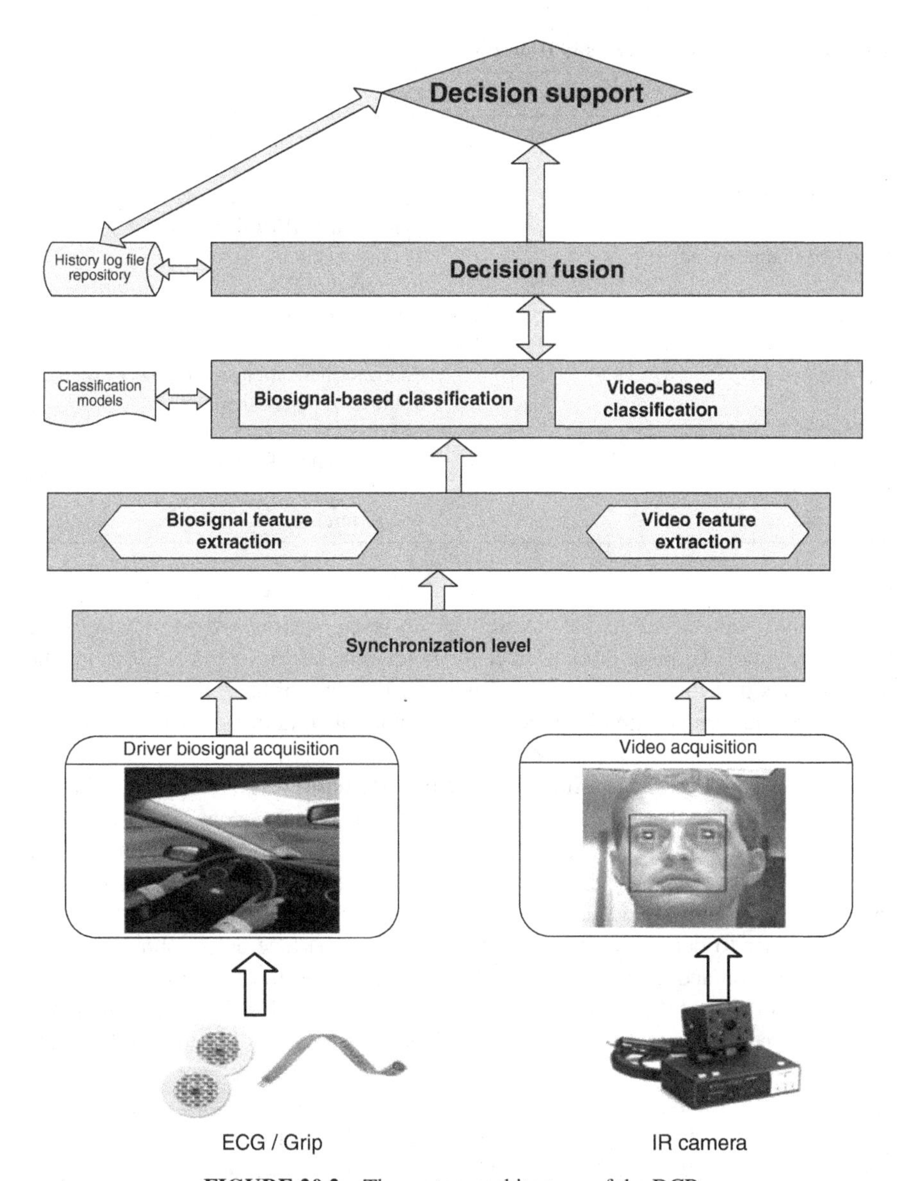

FIGURE 20.2 The system architecture of the DCR.

(SCL). The first absolute difference (FAD) of the remaining signal is calculated, giving a measure of the SCR.

Respiration signal. The signal was downsampled to 10 Hz and the wonder is removed. The power spectrum of the signal, using FFT transform, is extracted. A smoothing of the power spectrum follows, and the maximum energy frequency between 0.1 Hz and 1.5 Hz is selected as the dominant respiration frequency (DRF).

TABLE 20.8 Features extracted from the RRV signal

Feature	Description
Very low frequency (VLF)	Energy on the 0.01–0.05 Hz interval of RRV power spectrum
Low frequency (LF)	Energy on the 0.05–0.2 Hz interval of RRV power spectrum
High low frequency (HF)	Energy on the 0.2–0.4 Hz interval of RRV power spectrum
LF/HF	The ratio of the LF to the HF component
Spectrum entropy	After normalization of the power spectrum the entropy is calculated
Detrended fluctuation analysis (DFA)	See References 96–98
Lyapunov mean and max exponents	See References 99

Furthermore, we extract another feature which is the ratio of the heart rate to the respiration rate. As respiration is a main modulator of the cardiac function, the hypothesis is that for normal/relaxed conditions the ratio of heart to respiration rate is constant and changes are observed only in abnormal conditions, such as stress and fatigue.

Video acquisition processing and feature extraction: The subcomponent that is responsible to actively monitor the driver's state using video captured frames considers the eyes activity (blinking rate) as the most reliable measurement for detecting the fatigue level. This is in accordance to state of the art techniques presented in scientific literature. In a truly automatic system there are three distinct stages that occur in order to detect eye blinking; these include: (i) face detection/acquisition, (ii) eye detection/acquisition, and (iii) blink detection/classification. In the last stage the detected states of the eyes are categorized into open or closed. Depending on the period of time that the subject's eyes remain closed we can deduce the blinking rate of that particular subject.

The video of the face of the driver is processed following the approach described in References 100 and 101. The first step is the detection of the face and the second is the detection of eyes. The information of interest is: (i) the movement of the head, which could be an indicator for both stress and fatigue and (ii) the mean level of eye opening as an indicator of fatigue. An estimation of PERCLOS is also estimated, considering eye closure when the confidence of eye presence is less than zero. As a measure of head movement, the standard deviation of the face position in the video frame, and as a measure of eye opening the confidence of eye detection (provided in Reference 101), are used. If the eyes are wide open this confidence is high, while for near close eyes it is quite low.

Environment information extraction: As discussed earlier, in the I-WAY project there are numerous information sources (V2V, V2I, I2V, and sensors). Thus a rich representation of the environment is feasible. Since driver state is closely related to the environment such information could be very useful for the overall driver state assessment.

Step I: Data acquisition: From the forward looking camera, road monitoring is employed. From road monitoring video, useful information about driving environment conditions during each session are extracted. This information concerns weather, road visibility, and traffic conditions. Bad weather and low visibility are reported as important stress factors [102]. Another important stress factor is traffic density [52, 103]. All environmental variables are categorized in two states (*good/bad* weather, *low/good* visibility, and *low/high* traffic density).

Step II: Feature selection: Feature selection is a very common practice before classification. Such an approach is a prerequisite in cases where the ratio of data to features is low. Furthermore, introducing redundant features or features highly correlated can deteriorate the classification performance. Therefore, to build a robust classifier for stress and fatigue detection, features should be evaluated on their contribution as indicators of the examined states. Each feature contribution is evaluated based on the DAUC measure, which is the difference in the area under curve (AUC) of a classifier based on the specific feature and a random classifier. The DAUC is used as a metric of discrimination power of a feature. The DAUC of the optimal classifier is 0.5, thus features with DAUC near 0.5 are considered to be optimal[4]. In order to select the optimal feature set for more than one classification problem, the average DAUC of each feature is calculated. Then features are sorted according to their average DAUC, obtaining a feature ranking.

Step III: Classification: The third step of the DSR is classification. The performance of four different classifiers was examined: (i) SVM, (ii) decision trees using the C4.5 inductive algorithm [90], (iii) naive Bayes classifier, and (iv) general Bayesian classifier.

20.5.3 Results

During the experimental phase data were collected from a subject participating in a large number of sessions (over 35 sessions of 1 hour) on real and various driving conditions. The first step was the evaluation of the extracted features on both fatigue and stress classification, as described in Section 20.5.2.3. The physiological features with higher average DAUC were *mean RR*, *std of RR*, *LF/HF ratio*, *mean EDA level*, *FAD of EDA*, *mean respiration rate,* and *HR (bpm)/Resp. Rate (bpm)*. More complex RRV features (DFA, approximate entropy, and Lyapunov exponents) lack discrimination power. However, those features are more used in medical applications, extracted from long recordings, and are related with problematic heart function [104, 105]. For within-individual variations, simple RRV characteristics have proved to be rather informative [59]. From the video features, *std of eye activation* and PERCLOS were better indicators for fatigue, whereas *std of head positions* was a better indicator of stress. Finally from environmental conditions, the weather conditions seem to be the most important.

[4]The area of the optimal classifier is 1 and the area of a random classifier 0.5, thus the difference of an optimal from a random classifier is also 0.5.

TABLE 20.9 Results for the *fatigue* classification problem using selected features. For each classifier, the sensitivity (Sens.) and the specificity (Spec.) per class as well as the total accuracy (Acc.) are given

	Normal		Low fatigue		High fatigue		
	Sens.	Spec.	Sens.	Spec.	Sens.	Spec.	Acc.
SVM	0.89	0.87	0.79	0.84	0.96	0.92	0.88
Decision trees	0.74	0.74	0.71	0.70	0.94	0.95	0.80
Naive Bayes	0.70	0.86	0.65	0.64	0.92	0.79	0.76
Bayes classifier	0.76	0.83	0.73	0.71	0.93	0.88	0.81

Next we describe the obtained results using the classifiers described in section 20.5.2.3. The dataset is unbalanced and to address this problem, the following procedure was used; 50 balanced datasets from the original one were extracted and for each dataset a stratified 10-crossvalidation was performed.

Tables 20.9 and 20.10 present the results for fatigue and stress classification using three sets of features: (i) only physiological features, (ii) physiological and video features, and (iii) physiological, video, and environmental features. In these tables the sensitivity and specificity per class, as well as the total accuracy for all classifiers and feature sets are given. For the two-class stress problem the information provided is sufficient to evaluate the performance of the classification. However, for the three-class fatigue problem a better insight is given through the confusion matrix of the classification. From Tables 20.9 and 20.10 we observe that SVM had the best performance in all feature sets for classification of both states. In Table 20.9 we observe that the highest accuracy 88%, for fatigue classification, was obtained also using SVM.

20.6 OPEN ISSUES AND FUTURE STEPS

Before driving cars that can efficiently recognize our emotions and adapt accordingly, researchers still have to deal with certain issues: First of all, the set of most

TABLE 20.10 Results for the *stress* classification problem using the selected features. For each classifier, the sensitivity (Sens.) and the specificity (Spec.) per class as well as the total accuracy (Acc.) are given

	Normal		Stress		
	Sens.	Spec.	Sens.	Spec.	Acc.
SVM	0.88	0.85	0.84	0.88	0.86
Decision trees	0.82	0.81	0.80	0.81	0.81
Naive Bayes	0.85	0.76	0.74	0.83	0.79
Bayes classifier	0.86	0.76	0.73	0.83	0.79

important emotions related to the driving performance should be automatically recognized. Linguists have defined inventories of the emotional states most encountered in our lives.

Of course not all of them are related to driving performance. Apart from determining the most important ones the simultaneous existence of multiple emotions is an often neglected fact by the researchers, however it is quite common for a subject (driver) to simultaneously experience more than one emotion. An automatic recognition system should be able to identify multiple emotional states. Moreover, mapping of multiple emotional states to the driving performance is still at infant stage. Most of the reported works are performed during simulated conditions. Therefore, more experiments are needed in real driving conditions. Furthermore, tests should be randomized or generalized by employing different drivers, cars, and environmental conditions. At technical level, sensors and efficient techniques of capturing and processing data without even been noticed by the driver are needed. Cost can also be a decisive factor here, but as the systems become more applicable to car industry it is expected to fall rapidly. The most critical issue however is how the recognized emotions can be exploited and how they are optimally incorporated into human machine interfaces or active safety systems. The collaboration of experts from many different fields is required here.

20.7 CONCLUSION

Recent advances in artificial intelligence have provided the means for efficient drivers' state assessment. Numerous researchers have reported promising results. Putting these results into practice by developing real-life systems deserves further attention as it is expected to bring about significant benefits in road safety and driving comfort.

REFERENCES

[1] J. Wang, R. Knipling, and M. Goodman, "The role of driver inattention in crashes; new statistics from the 1995 Crashworthiness Data System (CDS)," in Annual Conference of the Association for the Advancement of Automotive Medicine, Des Plaines, IL, 1996.

[2] C. M. Jones and I. M. Jonsson, "Detecting emotions in conversations between driver and in-car information systems," in *Affective Computing and Intelligent Interaction*, Volume 3784: Lecture Notes in Computer Science, Springer, 2005, pp. 780–787.

[3] R. G. Burns and M. A. Katovich, "Examining road rage/aggressive driving—media depiction and prevention suggestions," *Environ. Behav.*, 35(5): 621–636, 2003.

[4] J. Mesken, M. P. Hagenzieker, T. Rothengatter, and D. de Waard, "Frequency, determinants, and consequences of different drivers' emotions: an on-the-road study using self-reports, (observed) behaviour, and physiology," *Transp. Res. F*, 10(6): 458–475, 2007.

[5] F. Eyben, M. Wöllmer, T. Poitschke, B. Schuller, C. Blaschke, B. Färber, and N. Nguyen-Thien, "Emotion on the road: necessity, acceptance, and feasibility of affective computing in the car," *Adv. Hum.-Comput. Int.*, 2010: 1–17, 2010.

[6] R. S. Lazarus, *Relational Meaning and Discrete Emotions*, Oxford University Press, New York, 2001.

[7] L. James, *Road Rage and Aggressive Driving*, Prometheus Books, Amherst, NY, 2000.

[8] S. Wright, G. Underwood, P. Chapman, and D. Crundall, "Anger while driving," *Transp. Res. F*, 2(13): 55–68, 1999.

[9] D. Jovanovic, K. Lipovac, P. Stanojevic, and D. Stanojevic, "The effects of personality traits on driving-related anger and aggressive behaviour in traffic among Serbian drivers," *Transp. Res. F*, 14(1): 43–53, 2011.

[10] T. Lajunen and D. Parker, "Are aggressive people aggressive drivers? A study of the relationship between self-reported general aggressiveness, driver anger and aggressive driving," *Accid. Anal. Prev.*, 33(2): 243–255, 2001.

[11] S. S. McLinton and M. F. Dollard, "Work stress and driving anger in japan," *Accid. Anal. Prev.*, 42(1): 174–181, 2010.

[12] S. M. Nesbit and J. C. Conger, "Evaluation of cognitive responses to anger-provoking driving situations using the articulated thoughts during simulated situations procedure," *Transp. Res. F*, 14(1): 54–65, 2011.

[13] S. M. Nesbit, J. C. Conger, and A. J. Conger, "A quantitative review of the relationship between anger and aggressive driving," *Aggress. Violent Behav.*, 12(2): 156–176, 2007.

[14] A. Villieux and P. Delhomme, "Driving anger and its expressions: further evidence of validity and reliability for the driving anger expression inventory French adaptation," *J. Saf. Res.*, 41(5): 417–422, 2010.

[15] WHO, "Road casualties worldwide—WHO report," 2004.

[16] R. O'Malley, E. Jones, and M. Glavin, "Detection of pedestrians in far-infrared automotive night vision using region-growing and clothing distortion compensation," *Infrared Phys. Technol.*, 53(6): 439–449, 2010.

[17] L. Y. Xiao and F. Gao, "A comprehensive review of the development of adaptive cruise control systems," *Veh. Syst. Dyn.*, 48(10): 1167–1192, 2010.

[18] W. Kwon and S. Lee, "Performance evaluation of decision making strategies for an embedded lane departure warning system," *J. Robot. Syst.*, 19(10): 499–509, 2002.

[19] C. Ramos, "Ambient intelligence—a state of the art from artificial intelligence perspective," in *Progress in Artificial Intelligence*, Volume 4874 of Lecture Notes in Computer Science (eds J. Neves, M. F. Santos, and J. M. Machado), Springer, Berlin/Heidelberg, 2007, pp. 285–295.

[20] L. Malta, C. Miyajima, N. Kitaoka, and K. Takeda, "Analysis of real-world driver's frustration," *IEEE Transactions on Intelligent Transportation Systems*, 12(1): 109–118, 2011.

[21] X. Li and Q. Ji, "Active affective state detection and user assistance with dynamic bayesian networks," *IEEE Transactions on Systems, Man and Cybernetics, Part A (Systems and Humans)*, 35(1): 93–105, 2005.

[22] I. M. Jonsson, H. Harris, and C. Nass, "Improving automotive safety by pairing driver emotion and car voice emotion," in Conference on Human Factors in Computing Systems (CHI '05), Portland, OR, 2005, pp. 1973–1976.

[23] H. Li, "Computer recognition of human emotions," in Proceedings of 2001 International Symposium on Intelligent Multimedia, Video and Speech Processing (ISIMP), Hong Kong, 2001, pp. 490–493.

[24] V. V. Dixit, A. V. Deshpande, and D. Ganage, "Face detection for drivers' drowsiness using computer vision," in International Federation for Medical and Biological Engineering (IFMBE), vol. 35, 2011, pp. 308–311.

[25] L. Li, Y. Chen, and L. Xin, "Driver's eyes detection and tracking system using two cameras," in Proceeding of the 10th Int. Conf. of Chinese Transportation Professionals, vol. 382, 2010, pp. 2098–2108.

[26] L. Li, Y. Chen, and Z. Li, "Yawning detection for monitoring driver fatigue based on two cameras," in Proceedings of the 12th International IEEE Conference on Intelligent Transportation Systems, 2009, pp. 12–17.

[27] T. Azim, M. A. Jaffar, M. Ramzan, and A. M. Mirza, "Automatic fatigue detection of drivers through yawning analysis," in Signal Processing, Image Processing and Pattern Recognition, Springer, 2009, pp. 125–132.

[28] B. C. Yin, X. Fan, and Y. F. Sun, "Multiscale dynamic features based driver fatigue detection," *Int. J. Pattern Recognit.*, 23(3): 575–589, 2009.

[29] S. Hachisuka, K. Ishida, T. Enya, and M. Kamijo, "Facial expression measurement for detecting driver drowsiness," in *Engineering Psychology and Cognitive Ergonomics*, Volume 6781 of Lecture Notes in Computer Science (ed. D. Harris), Springer, Berlin/Heidelberg, 2011, pp. 135–144.

[30] Y. Noguchi, R. Nopsuwanchai, M. Ohsuga, and Y. Kamakura, "Classification of blink waveforms towards the assessment of driver's arousal level—an approach for hmm based classification from blinking video sequence," in *Engineering Psychology and Cognitive Ergonomics*, Volume 4562 of Lecture Notes in Computer Science (ed. D. Harris), Springer Berlin/Heidelberg, 2007, pp. 779–786.

[31] Y. Noguchi, K. Shimada, M. Ohsuga, Y. Kamakura, and Y. Inoue, "The assessment of driver's arousal states from the classification of eye-blink patterns," in *Engineering Psychology and Cognitive Ergonomics*, Volume 5639 of Lecture Notes in Computer Science (ed. D. Harris), Springer, Berlin/Heidelberg, 2009, pp. 414–423.

[32] M. Kutila, M. Jokela, G. Markkula, and M. R. Ru, "Driver distraction detection with a camera vision system," in IEEE International Conference on Image Processing (ICIP 2007), vol. 6, San Antonio, TX, 2007, pp. VI201–VI204.

[33] J. Y. Kaminski, D. Knaan, and A. Shavit, "Single image face orientation and gaze detection," *Mach. Vis. Appl.*, 21(1): 85–98, 2009.

[34] H. Lunenfeld, "Human factor considerations of motorist navigation and information systems," in Vehicle Navigation and Information Systems Conference, Toronto, 1989.

[35] B. Schuller, G. Rigoll, and M. Lang, "Speech emotion recognition combining acoustic features and linguistic information in a hybrid support vector machine-belief network architecture," in IEEE Conf. on Acoustics, Speech and Signal Processing, vol. 1, 2004, pp. I-577–I-580.

[36] H. Boril, S. O. Sadjadi, T. Kleinschmidt, and J. H. Hansen, "Analysis and detection of cognitive load and frustration in drivers' speech," in Proceedings of INTERSPEECH, Japan, 2010, pp. 502–505.

[37] M. Grimm, K. Kroschel, H. Harris, C. Nass, B. Schuller, G. Rigoll, and T. Moosmayr, "On the necessity and feasibility of detecting a driver's emotional state while driving," in *Affective Computing and Intelligent Interaction* (eds A. C. R. Pavia, R. Prada, and R. W. Picard), Springer, Berlin, Heidelberg, 2007, pp. 126–138.

[38] E. Z. Izard and A. J. Flidlund, "Electromyographic studies of facial expressions of emotions and patterns of emotions," in *Social Psychophysiology: A Sourcebook* (eds J. T. Cacioppo and R. E. Petty), Guilford Press, New York, 1983, pp. 287–306.

[39] M. Benovoy, J. R. Cooperstock, and J. Deitcher, "Biosignals analysis and its application in a performance setting: towards the development of an emotional-imaging generator," in Proceedings of the First International Conference on Biomedical Electronics and Devices, BIOSIGNALS 2008, vol. 1, Funchal, Madeira, Portugal, January 28–31, 2008.

[40] P. Ekman, R. W. Levenson, and W. V. Friesen, "Autonomic nervous-system activity distinguishes among emotions," *Science*, 221(4616): 1208–1210, 1983.

[41] M. Helander, "Applicability of drivers electrodermal response to design of traffic environment," *J. Appl. Psychol.*, 63(4): 481–488, 1978.

[42] J. A. Healey and R. W. Picard, "Detecting stress during real-world driving tasks using physiological sensors," *IEEE Transactions on Intelligent Transportation Systems*, 6: 156–166, 2005.

[43] P. Rolfe, Y. Yamakoshi, H. Hirose, and T. Yamakoshi, "A novel physiological index for driver's activation state derived from simulated monotonous driving studies," *Transp. Res. Part C: Emerg. Technol.*, 17(1): 69–80, 2009.

[44] F. Nasoz, O. Ozyer, C. L. Lisetti, and N. Finkelstein, "Multimodal affective driver interfaces for future cars," in Proceedings of the Tenth ACM International Conference on Multimedia, 2002, pp. 319–322.

[45] C. D. Katsis, N. E. Ntouvas, C. G. Bafas, and D. I. Fotiadis, "Assessment of muscle fatigue during driving using surface EMG," in Proc. of the IASTED Int. Conf. on Biomedical Engg., vol. 262, 2004.

[46] M. Murugappan, R. Nagarajan, and S. Yaacob, "Combining spatial filtering and wavelet transform for classifying human emotions using EEG signals," *J. Med. Biol. Eng.*, 31(1): 45–51, 2011.

[47] C. D. Katsis, G. Ganiatsas, and D. I. Fotiadis, "An integrated telemedicine platform for the assessment of affective physiological states," *Diagn. Pathol.*, 1: 1–16, 2006.

[48] L. J. M. Rothkrantz, D. Datcu, and N. Absil, "Multimodal affect detection of car drivers," *Neural Netw. World*, 19(3): 293–305, 2009.

[49] A. Clarion, C. Ramon, C. Petit, A. Dittmar, J. P. Bourgeay, A. Guillot, C. Gehin, E. McAdams, and C. Collet, "An integrated device to evaluate a driver's functional state," *Behav. Res. Methods*, 41(3): 882–888, 2009.

[50] H. Leng, Y. Lin, and L. Zanzi, "An experimental study on physiological parameters toward driver emotion recognition," in *Ergonomics and Health Aspects of Work with Computers*, Volume 4566 of Lecture Notes in Computer Science (ed. M. Dainoff), Springer, Berlin/Heidelberg, 2007, pp. 237–246.

[51] G. Rigas, Y. Goletsis, P. Bougia, and D. I. Fotiadis, "Towards driver's state recognition on real driving conditions," *Int. J. Veh. Technol.*, 2011.

[52] C. M. Wickens and D. L. Wiesenthal, "State driver stress as a function of occupational stress, traffic congestion, and trait stress susceptibility," *J. Appl. Behav. Res.*, 10(2): 83–97, 2005.

[53] G. Matthews, "Towards a transactional ergonomics for driver stress and fatigue," *Theor. Issues Ergon.*, 3(2): 195–211, 2002.

[54] S. Kumakura and M. Kataniwa, "A method for analyzing car driver's sensitivity to stress and comfort," Technical Report 19, 2007.

[55] A. D. Baddeley, "Selective attention and performance in dangerous environments," *Br. J. Psychol.*, 63(4): 537–546, 1972.

[56] M. A. Vidulich, M. Stratton, M. Crabtree, and G. Wilson, "Performance-based and physiological measures of situational awareness," *Aviat. Space Environ. Med.*, 65(5): A7–A12, 1994.

[57] T. Chidster, H. Foushee, S. Gregorich, and R. Helmreich, "How effective is cockpit resource management training? Issues in evaluating the impact of programs to enhance crew coordination," *Flight Saf. Dig.*, 19(5): 1–17, 1990.

[58] J. Healey, "Wearable and automotive systems for affect recognition from physiology," PhD thesis, 2000.

[59] J. A. Healey and R. Picard, "SmartCar: detecting driver stress," in 15th International Conference on Pattern Recognition, 2000, pp. 218–221.

[60] J. Zhai and A. Barreto, "Stress detection in computer users based on digital signal processing of noninvasive physiological variables," in EMBS '06, 28th Annual International Conference of the IEEE, 2006, pp. 1355–1358.

[61] P. Rani, J. Sims, R. Brackin, and M. Sarkar, "Online stress detection using psychophysiological signals for implicit human–robot cooperation," *Robotica*, 20(6): 673–685, 2002.

[62] W. Liao, W. Zhang, Z. Zhu, and Q. Ji, "A decision theoretic model for stress recognition and user assistance," in Proceedings of Association for the Advancement of Artificial Intelligence (AAAI), 2005, pp. 529–534.

[63] R. Bittner, P. Smrcka, M. Pavelka, P. Vysok, and L. Pousek, "Fatigue indicators of drowsy drivers based on analysis of physiological signals," in ISMDA '01: Proceedings of the Second International Symposium on Medical Data Analysis, 2001, pp. 62–68.

[64] A. Crawford, "Fatigue and driving," *Ergonomics*, 4: 143–54, 1961.

[65] M. E. Dawson, A. M. Schell, and D. L. filion, *Handbook of Psychophysiology*, Cambridge University Press, New York, 2000.

[66] J. Qiang, P. Lan, and C. Looney, "A probabilistic framework for modeling and real-time monitoring human fatigue," *IEEE Transactions on Systems, Man and Cybernetics*, 36(5): 862–875, 2006.

[67] A. M. Williamson, A. M. Feyer, and R. Friswell, "The impact of work practices on fatigue in long distance truck drivers," *Accid. Anal. Prev.*, 28(6): 709–719, 1996.

[68] G. Yang, Y. Lin, and P. Bhattacharya, "A driver fatigue recognition model using fusion of multiple features," in 2005 IEEE International Conference on Systems, Man and Cybernetics, vol. 2, 2005, pp. 1777–1784.

[69] H. Storm, K. Myre, M. Rostrup, O. Stokland, M. D. Lien, and J. C. Raeder, "Skin conductance correlates with perioperative stress," *Acta Anaesthesiol. Scand.*, 46(7): 887–895, 2002.

[70] H. Tan and Y. J. Zhang, "Detecting eye blink states by tracking iris and eyelids," *Pattern Recognit. Lett.*, 27(6): 667–675, 2006.

[71] D. J. Mascord and R. A. Heath, "Behavioral and physiological indices of fatigue in a visual tracking task," *J. Saf. Res.*, 23(1): 19–25, 1992.

[72] Z. Li, "Spectral analysis of heart rate variability as a quantitative indicator of driver mental fatigue, Technical Report, SAE International, 2002.

[73] H. J. Eoh, M. K. Chung, and S.-H. Kim, "Electroencephalographic study of drowsiness in simulated driving with sleep deprivation," *Int. J. Ind. Ergon.*, 35(4): 307–320, 2005.

[74] M. Horvth, E. Frantik, K. Kopriva, and J. Meissner, "EEG theta activity increase coinciding with performance decrement in a monotonous task," *Activ. Nerv. Super.*, 18(3): 207–210, 1976.

[75] B. T. Jap, S. Lal, P. Fischer, and E. Bekiaris, "Using EEG spectral components to assess algorithms for detecting fatigue," *Expert Syst. Appl.*, 36(2), Part 1: 2352–2359, 2009.

[76] S. K. Lal and A. Craig, "Driver fatigue: electroencephalography and psychological assessment," *Psychophysiology*, 39(3): 313–321, 2002.

[77] S. K. Lal and A. Craig, "Reproducibility of the spectral components of the electroencephalogram during driver fatigue," *Int. J. Psychophysiol.*, 55(2): 137–143, 2005.

[78] S. K. Lal, A. Craig, P. Boord, L. Kirkup, and H. Nguyen, "Development of an algorithm for an EEG-based driver fatigue countermeasure," *J. Saf. Res.*, 34(3): 321–328, 2003.

[79] J. C. Miller, "Batch processing of 10000 h of truck driver EEG data," *Biol. Psychol.*, 40(1–2): 209–222, 1995.

[80] A. Subasi, "Automatic recognition of alertness level from EEG by using neural network and wavelet coefficients," *Expert Syst. Appl.*, 28(4): 701–711, 2005.

[81] K. Ball and G. Rebok, "Evaluating the driving ability of older adults," *J. Appl. Gerontol.*, 13(1): 20–38, 1994.

[82] S. MacDonald, R. Mann, M. Chipman, B. Pakula, P. Erickson, A. Hathaway, and P. MacIntyre, "Driving behavior under the influence of cannabis or cocaine," *Traffic Injury Prevent.*, 9(3): 190–194, 2008.

[83] T. Ranney, W. R. Garrott, and M. J. Goodman, "NHTSA driver distraction research: past, present, and future," 2001.

[84] M. Wöllmer, F. Eyben, S. Reiter, B. Schuller, C. Cox, E. Douglas–Cowie, and R. Cowie, "Abandoning emotion classes—towards continuous emotion recognition with modelling of long range dependencies," in INTERSPEECH, 2008, pp. 597–600.

[85] S. Hochreiter, J. Schmidhuber, "Long short-term memory," *Neural Comput.*, 9(8): 1735–1780, 1997.

[86] Y. Liang, M. L. Reyes, and J. D. Lee, "Real-time detection of driver cognitive distraction using support vector machines," *IEEE Transactions on Intelligent Transportation Systems*, 8(2): 340–350, 2007.

[87] K. Watersy and D. Terzopoulos, "Physically-based facial modeling, analysis and animation," *J. Vis. Comput. Animat.*, 1(2): 8, 1990.

[88] E. C. Prakash, E. Sung, and Y. Zhang, "Anatomically accurate individual face modeling," *Stud. Health Technol. Inform.*, 94: 3, 2003.

[89] A. H. Taylor and L. Dorn, "Stress, fatigue, health, and risk of road traffic accidents among professional drivers: the contribution of physical inactivity," *Annu. Rev. Public Health*, 27: 371–391, 2006.

[90] J. R. Quinlan, *C4.5: Programs for Machine Learning (Morgan Kaufmann Series in Machine Learning)*, Morgan Kaufmann, 1993.

[91] N. Friedman, D. Geiger, and M. Goldszmidt, "Bayesian network classifiers," *Mach. Learn.*, 29(2–3): 131–163, 1997.

[92] X. H. Yu and W. W. Recker, "Stochastic adaptive control model for traffic signal systems," *Transp. Res. Part C: Emerg. Technol.*, 14(4): 263–282, 2006.

[93] S. K. Andersen, "Probabilistic reasoning in intelligent systems: networks of plausible inference: Judea Pearl," *J. Artif. Intell.*, 48(1): 117–124, 1991.

[94] D. A. Morales, E. Bengoetxea, P. Larranaga, M. Garcia, Y. Franco, M. Fresnada, and M. Merino, "Bayesian classification for the selection of in vitro human embryos using morpholagical and clinical data," *Comput. Methods Progr. Biomed.*, 90(2): 104–116, 2008.

[95] R. M. Rangayyan, *Biomedical Signal Analysis: A Case-Study Approach*, IEEE Press, NJ, 2001.

[96] C. Heneghan and G. McDarby, "Establishing the relation between detrended fluctuation analysis and power spectral density analysis for stochastic processes," *Phys. Rev. E*, 62(5): 6103–6110, 2000.

[97] E. R. Bojorges-Valdez *et al.*, "Scaling patterns of heart rate variability data," *Physiol. Meas.*, 28(6): 721, 2007.

[98] J. C. Echeverria, M. S. Woolfson, J. A. Crowe, B. R. Hayes-Gill, G. D. Croaker, and H. Vyas, "Interpretation of heart rate variability via detrended fluctuation analysis and alphabeta filter," *Chaos*, 13(2): 467–75, 2003.

[99] J. H. A. Qader, L. M. Khadra, and H. Dickhaus, "Nonlinear dynamics in HRV signals after heart transplantations," in Proceedings of the Fifth International Symposium on Signal Processing and Its Applications, ISSPA '99, vol. 1, 1999, pp. 231–234.

[100] I. Fasel, B. Fortenberry, and J. Movellan, "A generative framework for real time object detection and classification," *Comput. Vis. Image Underst.*, 98(1): 182–210, 2005.

[101] P. Viola and M. Jones, "Robust real-time object detection," *Int. J. Comput. Vis.*, 2001.

[102] M. Chipman and Y. L. Jin, "Drowsy drivers: the effect of light and circadian rhythm on crash occurrence," *Saf. Sci.*

[103] J. D. Hill and L. N. Boyle, "Driver stress as influenced by driving maneuvers and roadway conditions," *Transp. Res. Part F: Traffic Psychol. Behav.*, 10(3): 177–186, 2007.

[104] M. Meyer, A. Rahmel, C. Marconi, B. Grassi, P. Cerretelli, and J. E. Skinner, "Stability of heartbeat interval distributions in chronic high altitude hypoxia," *Integr. Physiol. Behav. Sci.*, 33(4): 344–362, 1998.

[105] P. K. Stein, Q. Le, and P. P. Domitrovich, "Development of more erratic heart rate patterns is associated with mortality post-myocardial infarction," *J. Electrocardiol.*, 41(2): 110–115, 2008.

[106] R. Banuls and L. Montoro, "Motivational and emotional aspects involved in driving," in *Traffic Psychology Today* (ed. P.-E. Barjone), Springer, Berlin, Germany, 2001, pp. 138–318.

[107] G. Rusconi, M. C. Brugnoli, P. Dosso, K. Kretzschmar, P. Bougia, D. I. Fotiadis, L. Salgado, F. Jaureguizar, and M. De Feo, "I-way, intelligent co-operative system for road safety," in Intelligent Vehicles Symposium, 2007 IEEE, 2007, pp. 1056 –1061.

AUTHOR BIOGRAPHIES

Christos D. Katsis received a Diploma in Physics in 1998 and a Ph.D. degree in Medical Physics from the University of Ioannina, Ioannina, Greece, in 2008. He is currently a visiting assistant professor in the Department. of Applications of Information Technology in Administration and Economy of the Technological Educational Institute of Ionian Islands. He has many years of research experience in biomedical engineering, emotion recognition, computer modeling, computational medicine, automated diagnosis, and telemedicine.

George Rigas received a Ph.D. in Computer Science from the Department of Computer Science, University of Ioannina, Greece, in 2009. His thesis was on the assessment of driver physiological state using physiological signals. Since 2010 he has been a postdoc researcher in the Unit of Medical Technology and Intelligent Information Systems in the University of Ioannina. His research interests involve signal processing, machine learning, and pattern recognition.

Yorgos Goletsis holds a Diploma in Electrical Engineering and a Ph.D. degree in Operations Research, both from the National Technical University of Athens, Athens, Greece. He is a lecturer in the Department of Economics, University of Ioannina. His research interests include operations research, decision-support systems, multi-criteria analysis, quantitative analysis, data mining, artificial intelligence, project evaluation.

Dimitrios I. Fotiadis (M '01-SM '07) received a Diploma in Chemical Engineering from the National Technical University of Athens, Athens, Greece, in 1985, and a Ph.D. degree in Chemical Engineering from the University of Minnesota, Minneapolis, in 1990. He is currently a professor in the Department of Materials Science and Technology, University of Ioannina, Ioannina, Greece. His current research interests include biomedical technology, biomechanics, scientific computing, and intelligent information systems.

INDEX

Emotion Recognition: A Pattern Analysis Approach, First Edition. Edited by Amit Konar and Aruna Chakraborty.
© 2015 John Wiley & Sons, Inc. Published 2015 by John Wiley & Sons, Inc.